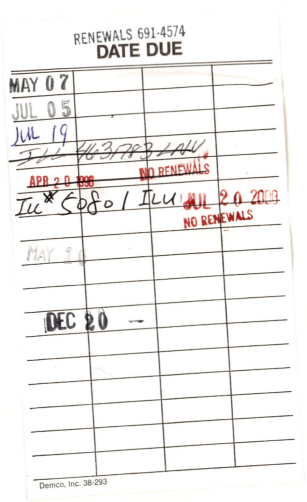

# Boundary Element Methods in Engineering Science

# Boundary Element Methods in Engineering Science

P. K. Banerjee

*Professor of Civil Engineering,
State University of New York at Buffalo*

and

R. Butterfield

*Professor and Head of the Department of
Civil Engineering, University of Southampton*

McGRAW-HILL Book Company (UK) Limited

London · New York · St Louis · San Francisco · Auckland · Bogotá · Guatemala
Hamburg · Johannesburg · Lisbon · Madrid · Mexico · Montreal · New Delhi
Panama · Paris · San Juan · São Paulo · Singapore · Sydney · Tokyo · Toronto

Published by McGraw-Hill Book Company (UK) Limited

MAIDENHEAD · BERKSHIRE · ENGLAND

British Library Cataloguing in Publication Data

Banerjee, P. K.
    Boundary element methods in engineering science.
    1. Boundary value problems
    2. Engineering mathematics
    I. Title    II. Butterfield, R.
    620′.001′515353    TA347.B69    80-40983

    ISBN 0-07-084120-9

Library of Congress Cataloging in Publication Data

Banerjee, P. K.
    Boundary element methods in engineering science.
    Bibliography: p.
    Includes index.
    1. Boundary value problems
    2. Engineering mathematics
    I. Butterfield, Roy, joint author.    II. Title
    TA347.B69B36    620′.001′51535    80-40983
    ISBN 0-07-084120-9    AACR1

Typeset in Great Britain by John Wright & Sons Ltd, Bristol
Printed and bound in the United States of America

# CONTENTS

## CHAPTER 10 TRANSIENT PROBLEMS IN ELASTICITY

## CHAPTER 11 PLATE-BENDING PROBLEMS

## CHAPTER 15 COMPUTER IMPLEMENTATION OF BOUNDARY ELEMENT METHODS

# PREFACE

One feature common to almost all quantitative investigations of realistic problems in engineering and applied science is that the boundary geometry of the region of interest is far too irregular for analytical solutions to be feasible and some form of numerical solution becomes necessary. The most commonly used numerical techniques involve extensive subdivision of the region, either by grids of lines to define nodal points, as in finite difference methods, or by sectioning it into large numbers of discrete simplified components, as in finite element methods.

The latter technique has now reached such a stage of development and popularity that one might well doubt whether there is any other approach which can offer comparable power and simplicity. This book is about such an alternative, equally versatile, method based on a study of the equations governing field problems in the form of boundary integral equations rather than the more usual differential ones. The most striking feature of boundary integral equation methods is that, in principle, only the boundaries of the region being investigated have to be discretized, which therefore leads to many fewer discrete elements than any scheme requiring internal subdivision of the whole body. Consequently, these methods generate much smaller systems of algebraic equations for final solution than do other methods.

Whereas the finite element method has been successively refined, following its first introduction as a physically reasonable approach, boundary integral equation methods have been very much more the province of the mathematician and the literature, although extensive, is mostly in a form not immediately attractive to the engineering analyst.

This book is an attempt to redress the balance. Its title, *Boundary Element Methods in Engineering Science* (BEM), has been chosen to suggest, correctly, that the main process involved is one of boundary subdivision of systems into convenient elements. All the concepts involved are introduced initially through physical and intuitive arguments followed by more rigorous formulation in the hope that the reader will be impressed by their underlying simplicity. Indeed, anyone acquainted with either influence lines, or matrix methods of structural analysis, or the superposition of unit solutions to problems (Green's functions, etc.) will find that the main ideas underlying BEM are already familiar.

The book is meant to be self-contained and studied sequentially apart from a few mathematical operations not covered in all undergraduate courses in engineering and applied science, which are relegated to Appendices. Comprehensive reference lists are provided since, although boundary integral equation methods have already been applied to a wide range of problems, it is only recently that much of the work has been seen to have a common basis, practical computer programme suites compiled, and interdisciplinary interest aroused.

The analyses discussed in the book range widely over the linear, non-linear, steady state, and transient problems of solid and fluid mechanics together with hybrid solutions using BEM in conjunction with other numerical methods. Numerous examples of solved problems are also presented in order to demonstrate that BEM have indeed already been applied throughout virtually the whole of engineering science.

We would also mention that one form of a boundary element method algorithm for a homogeneous region generates what is equivalent to a single 'finite element' enveloping the complete region. Such 'superelements' can be incorporated into any conventional finite element scheme using standard assembly procedures to produce a hybrid formulation. One immediate attraction of the hybrid approach is that infinitely distant boundaries can then be accurately and simply modelled, a facility unique to BEM.

P. K. Banerjee
R. Butterfield

# ACKNOWLEDGEMENTS

Many of the examples used in the book and some of the computational ideas were provided by colleagues of ours among whom were Drs G. R. Tomlin, J. O. Watson, R. M. Driscoll, T. G. Davies, D. N. Cathie, and G. Mustoe, and it has been our privilege to have worked with them. We are also indebted to many other leading workers who have generously allowed us to include numerous examples of their work, and to Drs T. A. Cruse, R. P. Shaw, J. O. Watson, and R. B. Wilson for valuable discussions and criticisms of the manuscript.

Permission to reproduce material from their publications has been granted by the American Institute of Physics, Applied Science Publishers, Cambridge University Press, Douglas Aircraft Company, Institution of Electrical Engineers, Pergamon Press, Plenum Publishing Company, Prentice-Hall Inc., the Society of Naval Architects and Marine Engineers, State University of New York, Thomas Telford Ltd., and John Wiley & Sons, and we gratefully acknowledge this assistance. Whilst every effort has been made to identify the true copyright holder of certain diagrams and tables, it has not always been possible to do this with absolute certainty and we offer our apologies for any errors in this respect.

Finally, we wish also to acknowledge the computational facilities provided by the Civil Engineering Departments of the University of Southampton and University College, Cardiff, and financial support from the Military Vehicles Experimental Establishment, Christchurch; the Science Research Council and the Department of the Environment, London; and the Department of Ocean Engineering, Lloyd's Register of Shipping—a list which also accounts for the undisguised civil engineering emphasis of some of the examples presented in the book. In view of our own special interest in geotechnical engineering we have also selected a number of examples which are particularly relevant to this field.

# CHAPTER ONE

# AN INTRODUCTION TO BOUNDARY ELEMENT METHODS

## 1-1 BACKGROUND

When an engineer or scientist constructs a quantitative mathematical model of almost any kind of system he usually starts by establishing the behaviour of an infinitesimally small, differential element of it based on assumed relationships between the major variables involved. This leads to a description of the system in the form of a set of differential equations. Once the basic model has been constructed and the properties of the particular differential equation understood, subsequent efforts are then directed towards obtaining a solution of the equations within a particular region which is often of a very complicated shape and composed of zones of different materials each with complex properties. Various conditions will have been specified on the boundaries of the region and these may be either constant or variable with time, etc. It is not at all surprising therefore that the solution of such differential equations has been a major concern of analysts for over two centuries.

The irregular boundaries of the majority of practical problems preclude any analytical solution of the governing equations and numerical methods become the only feasible means of obtaining adequately precise and detailed results.

The numerical methods most widely used at present tackle the differential equations directly in the form in which they were derived, without any further mathematical manipulation, in one of two ways: *either* by approximating the differential operators in the equations by simpler, localized algebraic ones valid at a series of nodes within the region *or* by representing the region itself by non-infinitesimal (i.e., finite) elements of material which are assembled to provide an approximation to the real system.

The finite difference method[1] is the progenitor of the former approach and was, until about fifteen years ago when it began to be superseded by methods of the second kind, much the most commonly used technique. Finite difference methods have attractions in that they can, in principle, be applied to any system of differential equations, but, unfortunately, incorporation of the problem boundary conditions is very often an unwieldly and not conveniently computerized operation. The precision of the numerical solution obtained is entirely dependent on the fineness of the mesh used to define the nodal points and consequently large systems of simultaneous algebraic equations are always generated as part of the solution procedure.

By far the most popular approach at present is the alternative one of reverting to physical subdivision of the body into elements of finite size—the larger the better from the point of view of minimizing the number of simultaneous equations generated for solution. Each element reproduces approximately the behaviour of the small region of the body which it represents, but complete continuity between the elements is only enforced in an overall sense (at nodes usually) rather than throughout the full extent of adjacent interfaces (i.e., such methods essentially approximate the body and its articulation). The finite element method[2] epitomizes this approach and, in recent years, has reached such a stage of development that many workers would doubt that any equivalent, let alone superior, technique might ever appear. The range and power of finite element methods, together with the relative ease with which realistic boundary conditions can be incorporated, does indeed present a formidable challenge to any other contending system. Its weakest aspects are that it is conceptually a whole body discretization scheme which inevitably leads to very large numbers of finite elements, especially in three-dimensional problems with distant boundaries, within each of which the solution variables do not all vary continuously and frequently show unrealistic jumps in value between adjoining elements.[2]

## 1-2 AN ALTERNATIVE APPROACH

An obvious alternative approach to the set of differential equations would be to attempt to integrate them analytically in some way before either proceeding to any discretization scheme or introducing any approximations. We are, of course, attempting to integrate the differential equations to find a solution whatever method we use, but the essence of boundary integral equation techniques is the transformation of the differential equations into equivalent sets of integral ones as the first step in their solution. Intuitively one would expect from such an operation (if it were successful) a set of equations which would involve only values of the variables at the extremes of the range of integration (i.e., on the boundaries of the region). This, in turn, would imply that any discretization scheme needed subsequently would only involve subdivisions of the bounding surface of a body. This is exactly what happens with the consequence that any homogeneous region requires only surface, rather than whole-body, discretization (hence the name, the boundary element method) and therefore becomes merely one large, sophisticated 'element' in the finite element sense. The solution variables will then vary continuously throughout the region and all approximations of geometry, etc., will only occur on its outer boundaries.

Another intuitive expectation would be that the development of the boundary integral equations and their solution might well involve more mathematical complexity than the other methods mentioned above. Fortunately this is only partially true, although it does account for the fact that boundary integral equation methods have been developed mainly by mathematicians in the past. The published literature, although extensive, tends to have an obvious mathematical bias without the final, compensatory 'carrot' that at the end of it all a comprehensive technique will emerge which the analyst can use in a general way. However, the situation has improved, from the utilitarian point of view, over the last few years, and boundary element methods (BEM) of analysis, developed essentially from integral equation ideas, are now available which are generally applicable without recourse to proofs of existence and uniqueness for each individual solution. As a result the method is now gaining very considerable popularity and being incorporated into high-speed digital computer algorithms immediately useful to the practising analyst.

## 1-3 THE HISTORICAL DEVELOPMENT OF BOUNDARY ELEMENT METHODS

Whereas the major properties of differential equations were well established by the nineteenth century the first rigorous investigation of the classical kinds of integral equation was published by Fredholm as late as 1905. Since then they have been studied intensively, particularly in connection with field theory, and there are many texts dealing with these developments,[3,4] although we shall not need to refer to them very much.

A major contribution to the formal understanding of integral equations generally has been made more recently by Mikhlin[5-7] who discusses such equations with both scalar and vector (multidimensional) integrands and in particular those with singularities and discontinuities within the range of integration. All of this is presented within a rigorous mathematical framework, most of which is rather unfamiliar to the majority of applied scientists. Despite the great advances that have been made in the classification and analysis of the properties of integral equations, none of the major authors appear to have considered the possibility that a general numerical algorithm for solving a wide range of practical problems might be based on them. The impetus for such a development has been provided by the high-speed digital computer and one result has been the emergence of the boundary element method.

Although all BEM have a common origin they divide naturally into three different but closely related categories.

1. *The direct formulation of BEM.* In this formulation the unknown functions appearing in the integral equations are the actual physical variables of the problem. Thus, for example, in an elasticity problem such an integral equation solution would yield all the tractions and displacements on the system boundary directly and those within the body can be derived from the boundary values by numerical integration. Some of the recently developed algorithms based on this approach have been described by Cruse, Lachat, Rizzo, Shaw, Watson, and others,[8-23] which they refer to as the boundary integral methods.

2. *Semi-direct formulations of BEM.* Alternatively, the integral equations can be formulated in terms of unknown functions analogous to stress functions in elasticity or stream functions in potential flow. When the solution has been obtained in these terms simple differentiation will yield, for example, the internal stress distribution. This approach, known as the semi-direct method, has been developed by Henry, Jaswon, Ponter, Rim, and Symm.[24-28]

3. *Indirect formulations of BEM.* In the indirect formulation the integral equations are expressed entirely in terms of a unit singular solution of the original differential equations distributed at a specific density over the boundaries of the region of interest. (The unit singular solution may be, for example, the 'free-space Green function' of the differential equation which implies, correctly, that BEM and what are often called Green's function methods are also closely related.) The density functions themselves have no specific physical significance but once they have been obtained from a numerical solution of the integral equations the values of the solution parameters anywhere within the body can be calculated from them by simple integration processes. Recently developed algorithms based on such an approach are described by Banerjee, Butterfield, Hess, Jaswon, Massonnet, Oliviera, Symm, Tomlin, Watson, and others,[29-43] who have used them to solve a wide range of engineering problems.

This book is entirely concerned with a comprehensive demonstration of the power and simplicity of these methods together with an adequate, but not formally rigorous, exposition of the mathematical background. We shall deal predominantly with the direct and indirect formulations of BEM because, in our opinion, these are much more generally useful than the semi-direct approach. The indirect BEM provides a particularly clear and simple physical illustration of the basic solution procedure, and therefore in the subsequent chapters we shall describe this method of solution first.

The references given above relate only to recent publications using BEM from which very general problem-solving algorithms have been developed although many other workers have solved specific problems using closely similar methods which are listed in the Bibliography at the end of this chapter. Problems which have been solved successfully include the majority of those governed by the classical partial differential equations of continuum mechanics in one, two, and three space dimensions including anisotropy, inhomogeneity, and, more recently, non-linearity.

## 1-4 THEIR RANGE OF APPLICATION

In principle these methods can be applied to any problem for which the governing differential equation is either linear or incrementally linear.[44-49] In problems involving elliptic differential equations the solutions are direct, whereas for parabolic and hyperbolic systems of equations marching processes in time have to be introduced. Thus a very wide range of physical problems is encompassed; e.g., those of steady state and transient potential flow, elastostatics, elastodynamics, elastoplasticity, acoustics, etc., can all be solved by either the direct or the indirect formulations of BEM.[8-49] BEM can also be used in conjunction with other numerical techniques,[44] such as finite element or finite difference methods, in a hybrid formulation. Such composite solutions extend the range of application almost indefinitely since the BEM have very distinct advantages for problems of large physical dimensions whereas finite element methods are an attractive means of incorporating finite size bodies into such systems or fine detail in regions with rapidly varying properties. A more comprehensive comparison of these attributes will now be provided as a conclusion to this introductory chapter.

## 1-5 A COMPARISON OF THE ATTRIBUTES OF FINITE ELEMENT AND BOUNDARY ELEMENT METHODS

### 1-5-1 Applicability

All boundary integral equation methods utilize the principle of superposition and are therefore only applicable to either completely linear systems or those which are, or can be approximated as, incrementally linear systems. This latter category therefore extends their compass to a great many problems of interest in engineering science. There appear to be very few problems solvable by finite element methods which cannot be solved at least as efficiently by BEM. These are problems in which either the properties of almost every individual material element are different or those in which the general geometry of the problem is such that one or more spatial dimensions are disproportionately small in relation to others but not sufficiently so to genuinely reduce its effective dimensionality (e.g., moderately thick plates and shells, narrow thin strips, etc.).

## 1-5-2  Problem Dimensionality

BEM reduce the dimensionality of the basic process by one; i.e., for two-dimensional problems the analysis generates a one-dimensional boundary integral equation and for three-dimensional problems only two-dimensional surface integral equations arise.

Each distinct bounded zone in a BEM analysis has to be treated as homogeneous and therefore, for problems in which the inhomogeneity is so great that very large numbers of small homogeneous zones are needed to model it adequately, the BEM zonal boundary scheme degenerates into one of essentially whole-body subdivision. In this case the BEM and finite element schemes become virtually indistinguishable from each other.

If, in a homogeneous region problem, either distributed body forces are to be included or the governing differential equations are only quasi-linear (as in the case of elastoplasticity, for example), then the boundary integrals have to be augmented by a volume integral involving arbitrary subdivisions of the interior of the body. However, in these cases the internal subdivisions do not result in any increase in the order of the final system of algebraic equations to be solved and the advantage remains with BEM. The reader should distinguish carefully between the latter situation in which the interior subdivisions arise from known distributions of body forces† (or pseudo-incremental body forces in plasticity) in otherwise homogeneous zones and the former which reflects the fundamental initial inhomogeneity of the problem.

Thus for the great majority of practical cases the simple boundary discretization necessarily leads to a very much smaller system of simultaneous equations than any scheme of whole-body discretization. On the other hand, the system matrices generated by BEM are fully populated for a homogeneous region and block banded when more than one is involved, whereas the much larger matrices which arise in finite element solutions are relatively sparsely populated.

The evaluation of each component of the matrices in a BEM solution does, however, involve much more arithmetic calculation than its finite element counterpart, which offsets some of the computer time saved by the much reduced matrix reduction requirements. Nevertheless, this also means that, as bigger and bigger problems are tackled, the overall computer costs increase very much less dramatically with problem size with BEM than for finite element schemes. From studies undertaken by various workers[13, 29] it can be concluded that comparable solution times between finite element and boundary element methods on three-dimensional problems solved with similar precision generally show a time advantage of from four to ten to one in favour of the latter. This difference could be very much greater in certain classes of problem which are particularly amenable to BEM, for example:

1. Systems with some boundaries at infinity. Since the BEM solution procedure automatically satisfies admissible boundary conditions at infinity, no subdivi-

---

† Volume integrals of continuously distributed conservative body forces can very often be transformed into equivalent boundary integrals using the Gauss theorem (Chapter 6).

sions of these boundaries arise whereas with the finite element method infinite boundaries have to be approximated by an appreciable number of distant elements.
2. Those involving semi-infinite regions with portions of the free surface 'unloaded'. Again by choosing the appropriate singular solution to use with BEM the 'unloaded' areas, which are usually the greater part of the free surface, do not need to be discretized at all.[32]

### 1-5-3 Continuous Interior Modelling

BEM involve modelling only the boundary geometry of the system. Once the necessary boundary information has been derived values of the solution variables can then be calculated at any subsequently selected interior points. Furthermore, the solution is fully continuous throughout the interior of the body. Both of these features appear to be unique to BEM among the possible alternatives. As a result of the latter facility the analyst can obtain values of variables at any specific interior point he may care to choose subsequent to the main analysis and with very high resolution—near, for example, stress concentrations in elastic or elastoplastic bodies.

### 1-5-4 Accuracy and Error Distribution

The boundary integral equation itself is a statement of the exact solution to the problem posed and errors due to discretization and numerical approximations arise only on, and adjacent to, the boundaries due to our inability to carry out the required integrations in closed form. If the numerical integration procedure is made sufficiently sophisticated (by using, for example, curved boundary elements and continuously varying distributions of functions over the boundary), then the errors so introduced can be very small indeed. Numerical integration is, of course, always a much more stable and precise process than numerical differentiation and neither the direct nor the indirect BEM require any differentiation of numerical quantities whatsoever.

It should by now be quite clear that, in the absence of body forces, the analyst need only specify the boundary geometry data of a region (in addition to the boundary conditions, material properties, etc., common to all solution methods). The effort devoted to data preparation is therefore substantially less than that required by any method involving internal geometrical modelling. Thus for the great majority of practical problems BEM offer very substantial advantages over finite element methods of analysis.

## 1-6 CONCLUDING REMARKS

In this chapter we have described the historical development of BEM as a practical problem-solving tool and discussed their usefulness in comparison to other currently popular alternatives. From these comparisons we conclude that BEM have great potential advantages over other methods, some of which have now been realized. It is our hope that the subsequent chapters of this book will help to accelerate this process both by illustrating their power through the solution of a wide range of practical problems and by emphasizing the very simple physical ideas, already familiar to most engineers and applied scientists, on which they are based.

## 1-7 REFERENCES

1. Southwell, R. V. (1946) *Relaxation Methods in Theoretical Physics*, Oxford University Press.
2. Zienkiewicz, O. C. (1971) *The Finite Element Method in Engineering Science*, McGraw-Hill, London.
3. Kellog, O. D. (1929) *Foundations of Potential Theory*, Springer, Berlin; also published by Dover, New York, in 1953.
4. Kupradze, V. D. (1963) *Potential Methods in the Theory of Elasticity*, translated from Russian by Israel Program for Scientific Translation, Jerusalem.
5. Mikhlin, S. G. (1957) *Integral Equations*, Pergamon Press, Oxford.
6. Mikhlin, S. G. (1965) *Multidimensional Singular Integrals and Integral Equations*, Pergamon Press, Oxford.
7. Mikhlin, S. G. (1965) *Approximate Solutions of Differential and Integral Equations*, Pergamon Press, Oxford.
8. Cruse, T. A. (1969) 'Numerical solutions in three-dimensional elastostatics', *Int. J. Solids and Structs*, **5**, 1259–1274
9. Cruse, T. A. (1972) 'Application of the boundary integral equation method in solid mechanics', in H. Tottenham and C. Brebbia (eds), *Proc. Int. Conf. Southampton Univ.*, Vol. 2.
10. Cruse, T. A. (1974) 'An improved boundary integral equation method for three-dimensional stress analysis', *Computers and Structs*, **4**, 741–757.
11. Cruse, T. A., and Rizzo, F. J. (1968) 'A direct formulation and numerical solution of the general transient elasto-dynamic problem', *J. Math. Anal. Appl.*, **22**, 244–259.
12. Cruse, T. A., and Rizzo, F. J. (eds) (1975) 'Boundary integral equation methods—computational applications in applied mechanics', *Proc. ASME Conf. on Boundary Integral Methods*, ASME, New York.
13. Lachat, J. C. (1975) 'Further developments of the boundary integral techniques for elasto-statics', Ph.D. thesis, Southampton University.
14. Lachat, J. C., and Watson, J. O. (1975) 'A second generation boundary integral equation program for three-dimensional elastic analysis', in T. A. Cruse and F. J. Rizzo (eds), *Proc. ASME Conf. on Boundary Integral Equation Methods*, ASME, New York.
15. Shaw, R. P., and Friedman, M. B. (1962) 'Diffraction of a plane shock wave by a free cylindrical obstacle at a free surface', *Fourth U.S. Nav. Congr. of Appl. Mech.*, pp. 371–379.
16. Friedman, M. B., and Shaw, R. P. (1962) 'Diffraction of a plane shock wave by an arbitrary rigid cylindrical obstacle', *J. Appl. Mech.*, **29**(1), 40–46.
17. Banaugh, R. P., and Goldsmith, W. (1963) 'Diffraction of steady acoustic waves by surfaces of arbitrary shape', *J. Acoust. Soc. Am.*, **35**(10), 1590–1601.
18. Mitzner, K. M. (1967) 'Numerical solution for transient scattering from a hard surface of arbitrary shape—retarded potential technique', *J. Acoust. Soc. Am.*, **42**(2), 391–397.
19. Shaw, R. P. (1966) 'Diffraction of acoustic pulses by obstacles of arbitrary shape with a robin boundary condition—Part A', *J. Acoust. Soc. Am.*, **41**(4), 855–859.
20. Shaw, R. P. (1969) 'Diffraction of pulses by obstacles of arbitrary shape with an impedance boundary condition', *J. Acoust. Soc. Am.*, **44**(4), 1962–1968.

21. Rizzo, F. J., and Shippy, D. J. (1977) 'An advanced boundary integral equation method for three-dimensional thermo-elasticity', *Int. J. Num. Meth. in Engng*, **11**, 1753.
22. Rizzo, F. J., and Shippy, D. J. (1979) 'Recent advances of the boundary element method in thermoelasticity', in P. K. Banerjee and R. Butterfield (eds), *Developments in Boundary Element Methods*, Vol. I, Chap. VI, Applied Science Publishers, London.
23. Lachat, J. C., and Watson, J. O. (1976) 'Effective numerical treatment of boundary integral equations: a formulation for three-dimensional elasto-statics', *Int. J. Num. Meth. in Engng*, **10**, 991–1005.
24. Jaswon, M. A. (1963) 'Integral equation methods in potential theory—I', *Proc. Roy. Soc.*, **273**(A), 23–32.
25. Jaswon, M. A., and Ponter, A. R. (1963) 'An integral equation method for a torsion problem', *Proc. Roy. Soc.*, **273**, 237–246.
26. Rim, K., and Henry, A. S. (1967) 'An integral equation method in plane elasticity', NASA Report No. CR–779–1967.
27. Symm, G. T. (1963) 'Integral equation methods in potential theory', *Proc. Roy. Soc.*, **275**(A), 33–46.
28. Symm, G. T. (1964) 'Integral equation methods in elasticity and potential theory', Ph.D. thesis, London University.
29. Banerjee, P. K. (1976) 'Integral equation methods for analysis of piece-wise non-homogeneous three-dimensional elastic solids of arbitrary shape', *Int. J. Mech. Sci.*, **18**, 293–303.
30. Banerjee, P. K., and Butterfield, R. (1976) 'Boundary element methods in geomechanics', in G. Gudehus (ed.), *Finite Elements in Geomechanics*, Chap. 16, Wiley, London.
31. Tomlin, G. R., and Butterfield, R. (1974) 'Elastic analysis of zoned orthotropic continua', *Proc. ASCE, Engng Mech. Div.*, **EM3**, 511–529.
32. Butterfield, R., and Banerjee, P. K. (1971) 'The problem of pile-cap pile-group interaction', *Géotechnique*, **21**(2), 135–142.
33. Massonnet, C. E. (1965) 'Numerical use of integral procedures', in O. C. Zienkiewicz and G. S. Holister (eds), *Stress Analysis*, Chap. 10, Wiley, London.
34. Oliviera, E. R. A. (1968) 'Plane stress analysis by a general integral method', *J. ASCE, Engng Mech. Div.*, **February**, 79–85.
35. Watson, J. O. (1973) 'Analysis of thick shells with holes by using integral equation method', Ph.D. thesis, Southampton University.
36. Banerjee, P. K., and Driscoll, R. M. C. (1976) 'Three-dimensional analysis of raked pile groups', *Proc. Inst. Civ. Eng., Res. and Theory*, **91**(2), 653–671.
37. Chen, L. H., and Schweikert, J. (1963) 'Sound radiation from an arbitrary body', *J. Acoust. Soc. Am.*, **35**, 1626–1632.
38. Banerjee, P. K. (1971) 'Foundations within a finite elastic layer—application of the integral equation method', *Civ. Engng.* **November**, 1197–1202.
39. Hess, J. L., and Smith, A. M. O. (1964) 'Calculations of nonlifting potential flow about arbitrary three-dimensional bodies', *J. Ship Res.*, **8**(2), 22–44.
40. Jaswon, M. A., and Symm, G. T. (1977) *Integral Equation Methods in Potential Theory and Elasto-statics*, Academic Press, London.
41. Hess, J. L., and Smith, A. M. O. (1966) 'Calculations of potential flow about arbitrary bodies', in *Progress in Aeronautical Sciences*, Vol. 8, pp. 1–138, Pergamon Press, New York.
42. Hess, J. L. (1974) 'The problem of three-dimensional lifting potential flow and its solution by means of surface singularity distributions', *Computer Meth. in Appl. Mech. Engng*, **4**, 283–319.
43. Hess, J. L. (1975) 'Improved solution for potential flow about arbitrary axi-symmetric bodies by the use of a higher order surface source method', *Computer Meth. in Appl. Mech. Engng*, **5**, 297–308.
44. Zienkiewicz, O. C. (1978) *The Finite Element Method*, 3rd ed., McGraw-Hill, London.
45. Banerjee, P. K., and Davies, T. G. (1979) 'Analysis of some case histories of laterally loaded pile groups', *Proc. Int. Conf. on Num. Meth. in Offshore Piling*, Institute of Civil Engineers, London.
46. Davies, T. G. (1979) 'Linear and nonlinear analyses of pile groups', Ph.D. thesis, University of Wales, University College, Cardiff.

47. Swedlow, J. L., and Cruse, T. A. (1971) 'Formulation of boundary integral equations for three-dimensional elasto-plastic flow', *Int. J. Solids and Structs*, **7**, 144–151.
48. Marjaria, M., and Mukherjee, S. (1980) 'Improved boundary integral equation method for time-dependent inelastic deformation in metals', *Int. J. Num. Meth. in Engng*, **15**(1), 97–112.
49. Chaudonneret, M. (1977) 'Boundary integral equation method for visco-plasticity analysis' (in French), *J. de Méch. Appliq.*, **1**(2), 113–131.

## 1-8 BIBLIOGRAPHY

Chicurel, R., and Suppiger, E. W. (1964) 'The reflection method in elasticity and bending of plates', *ZAMP*, **15**, 629–638.

Gruters, H. (1971) 'Berechnung des Spannungszustandes in homogenen anisotropen Scheiben mit Hilfe einer Integralgleichungsmethode', Thesis, Aachen.

Heise, U. (1969) 'Eine Integralgleichungsmethode zur Lösung des Scheibenproblems mit gemischten Randbedingungen', Thesis, Aachen.

——— (1975) 'The calculations of Cauchy principal values in integral equations for boundary value problems of the plane and three-dimensional theory of elasticity', *J. Elasticity*, **5**, 99–110.

——— (1976) 'Non-integral terms in integral equations in the plane and three-dimensional theory of elasticity', *Mech. Res. Comm.*, **3**, 119–124.

——— (1978) 'The spectra of some integral operators for plane elastostatical boundary value problems', *J. Elasticity*, **8**, 47–49.

——— (1978) 'Numerical properties of integral equations in which the given boundary values and the sought solutions are defined on different curves', *Computers and Structs*, **8**, 199–205.

Herrera, I. (1978) 'Theory of connectivity: a systematic formulation of boundary element methods', *Proc. Int. Conf. on Boundary Element Methods*, Southampton University, Pentech Press.

——— and Sabina, F. J. (1978) 'Connectivity as an alternative to boundary integral equations', *Proc. Natn. Acad. Sci., U.S.A.*, **75**, 5.

Kompis, V. (1970) 'Integralgleichungsverfahren zur Lösung der ersten Randwertaufgabe der ebenen Elastizitätstheorie', Thesis, Aachen.

Massonnet, Ch. (1949) 'Résolution graphomécanique des problèmes généraux de l'élasticité plane', *Bull. CERES Liège*, **4**, 3–183.

——— (1956) 'Solution générale du problème aux tensions de l'élasticité tridemensionnelle', *Proc. Ninth Congr. Appl. Mech.*, Brussels, pp. 168–180.

Miche, R. (1926) 'Le calcul pratique de problèmes élastiques à deux dimensions par la méthode des équations intégrales', *Proc. Second Int. Congr. Tech. Mech.*, Zurich, pp. 126–130.

Rieder, G. (1962) 'Iterationsverfahren und Operatorgleichungen in der Elastizitätstheorie', *Abh. Braunschweig. Wiss. Ges.*, **14**, 109–343.

——— (1968) 'Mechanische Deutung und Klassifizierung einiger Integralverfahren der ebenen Elastizitätstheorie I, II', *Bull. Acad. Pol. Sci., Sér. Sci. Technol.*, **16**, 101–114.

——— (1969) 'Eine Variante zur Integralgleichung von Windisch für das Torsionproblem', *ZAMM*, **49**, 351–358.

——— (1972) 'Über Eingrenzungsverfahren und Integralgleichungsmethoden für elastische Scheiben, Platten und verwandte Probleme', *Wiss. Z. der Hochsch. für Archit. und Bauwesen Weimar*, **19/2**, 217–222.

——— (1974) *On Kupradze's Generalised Stress—Its Applications to Certain Integral Operators of Plane Elasticity*, Anniversary volume. H. Parkus, Vienna.

——— (1975) 'Adjoint integral equations in elasticity', in *Schiffstechnisches Symposium, Experimentelle und mathematische Methoden der Grundlagenforschung in der Schiffstechnik*, Rostock, pp. 141–151.

Weinel, E. (1931) 'Die Integralgleichung des ebenen Spannungszustandes und der Platten theorie', *ZAMM*, **11**, 349–360.

# TWO

## SOME ONE-DIMENSIONAL PROBLEMS

### 2-1 INTRODUCTION

In order to introduce the ideas underlying the use of BEM, and the character of the unit solutions of the governing differential equations which arise, in the simplest possible way the following sections present, in some detail, the solutions to a series of one-dimensional problems. At this stage formal mathematical statements of the various procedures are dispensed with, the solutions are developed using essentially intuitive reasoning, and emphasis is placed upon the physical significance of the operations, particularly in the case of the indirect BEM solutions. We would emphasize that we are not in any way recommending BEM as a preferred way of solving such elementary problems. Indeed, for one-dimensional systems generally, BEM are not efficient problem-solving tools at all. Nevertheless, the endpoint reached via the one-dimensional examples is a series of systematic steps, a solution algorithm, which illustrates all the essential features of procedures that can be used almost unchanged to solve vastly more complicated two- and three-dimensional problems. Because of this the reader is advised to work step by step through each of the initial simple examples and to study the notation most carefully, which, as far as possible, corresponds to that adopted elsewhere throughout the book.

### 2-2 THE METHOD OF INFLUENCE FUNCTIONS

In order to introduce the key ideas of superposition of solutions we shall first explore the similarities and differences between BEM and the already well-established method of influence functions (or Green's function method) and consider the problem of potential flow in one dimension.

## 2-2-1 Potential Flow in One Dimension

Figure 2-1 represents a one-dimensional, homogeneous field of length $L$ and unit cross section. The boundaries of the system are merely the two endpoints $P, Q$ with only one 'boundary element' at each point. These are maintained at zero potential, $p(Q) = p(P) = 0$. A point source, intensity $\psi$, is applied at any position $B$, a 'load point', and located by the coordinate $\xi$. A general point $P'$, a 'field point', within the body is located by coordinate $x$.

We could be considering here either the flow of electricity or heat along a uniform conductor or non-viscous incompressible fluid flow along a uniform conduit. In all these cases the potential $p(x)$, representing voltage, temperature, or total head respectively, will be governed by the Laplace equation at all points within $PQ$ other than $B$. Thus in one dimension

$$\frac{d^2p}{dx^2} = 0 \tag{2-1}$$

If the conductivity of the medium is $k$ then the current intensity, heat flux, or fluid flow velocity $v(x)$ will be given by

$$v = -k\frac{dp}{dx} \tag{2-2}$$

Again, Eqs (2-1) and (2-2) could equally well describe the deflection $p(x)$ and the slope $v(x)$ of a tightly stretched horizontal weightless string under a high tension $k$ and a small vertical load $\psi$ as shown in Fig. 2-2. It is a very elementary exercise[1] to solve Eqs (2-1) and (2-2) under the given boundary conditions at the ends $P, Q$ to obtain

$$p(x) = \psi\frac{(L-\xi)\,x}{L\ k} \qquad v(x) = -\psi\frac{(L-\xi)}{L} \qquad \text{for } 0 \leqslant x < \xi \tag{2-3a}$$

and

$$p(x) = \psi\frac{(L-x)\,\xi}{L\ k} \qquad v(x) = \psi\frac{\xi}{L} \qquad \text{for } \xi < x \leqslant L \tag{2-3b}$$

At $x = \xi$, $p(x)$ is uniquely defined by either (2-3a) or (2-3b) but there is a step change in $v(x)$ equal to $\psi$ as $P'$ moves from one side to the other side of $B$ [i.e., as $x$ increases from $(\xi - \varepsilon)$ to $(\xi + \varepsilon)$ with $\varepsilon \to 0$]. Such a step change in one of the dependent variables as the 'load point' and 'field point' coincide are key features of all BEM and of integral equation methods generally. It is therefore of the utmost importance that the subsequent mathematical manipulation of such quantities is supported by a clear understanding of the physical process, which in this case is the bifurcation of the influx from $\psi$ partly towards $P$ and partly towards $Q$.

If $\psi = 1$ is substituted in Eqs (2-3) they then define 'the influence functions' for $p(x)$ and $v(x)$ in the systems shown in Figs 2-1 and 2-2 under the prescribed boundary conditions. Since these functions are linear in the source strength $\psi$ we can use them, together with the principle of superposition, to solve a problem

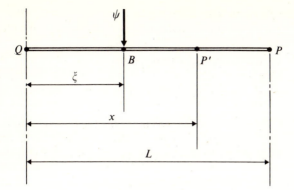

**Figure 2-1**

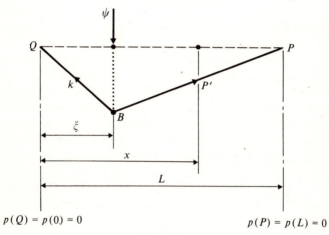

$p(Q) = p(0) = 0$                                        $p(P) = p(L) = 0$

**Figure 2-2**

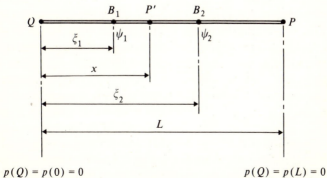

$p(Q) = p(0) = 0$                                        $p(Q) = p(L) = 0$

**Figure 2-3**

such as that in Fig. 2-3. This represents the same system, under the same boundary conditions, with specified multiple sources ($\psi_1, \psi_2, ...$) acting at points ($\xi_1, \xi_2, ...$) as shown.

Superposition of the appropriate multiples of Eq. (2-3) using ($\psi_1 ..., \xi_1 ...$) pairs will clearly provide the required $p(x)$ and $v(x)$ throughout $PQ$. This is the basic principle underlying the use of influence functions, or Green's functions, of all kinds.

If we are now presented with a problem which is identical to those illustrated in Figs 2-1 and 2-2, apart from a simple change of boundary conditions, to $p(Q) = p^*$ and $v(P) = v^*$, say, then a slightly more involved solution[1] of Eqs (2-1) and (2-2) will yield

$$p(x) = \left(p^* - \frac{v^* x}{k}\right) + \frac{\psi x}{k} \qquad v(x) = (v^* - \psi) \qquad \text{for } 0 \leqslant x < \xi \quad (2\text{-}4a)$$

and

$$p(x) = \left(p^* - \frac{v^* x}{k}\right) + \frac{\psi \xi}{k} \qquad v(x) = v^* \qquad \text{for } \xi < x \leqslant L \quad (2\text{-}4b)$$

Unfortunately simple superposition of these equations will no longer yield a solution to a problem such as the one defined in Fig. 2-3 under the altered boundary conditions of $p(0) = p^*$ and $v(L) = v^*$. By using appropriate combinations of Eqs (2-3) a solution can, of course, be obtained even to 'mixed' boundary-value problems of the above kind.

In a more general situation, such as that depicted in Fig. 2-4, which represents the analogous problem for an irregularly shaped region, even the basic influence functions are not readily found and it is not usually convenient to obtain a solution by the technique described above. An alternative approach has to be developed if a similar concept is to be used to solve the generalized problem efficiently.

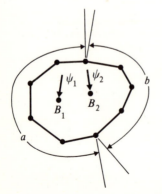

a = boundary elements representing point $Q$

b = boundary elements representing point $P$

**Figure 2-4**

## 2-3 APPLICATIONS OF THE INDIRECT BOUNDARY ELEMENT METHOD

The starting point for one technique which achieves this (the boundary element method) is the realization that we do have available, for virtually all classical field equations, the solutions for unit excitation applied within a homogeneous region of infinite extent (variously called unit solutions, fundamental singular solutions, or Green's functions for infinite regions, free-space Green's functions, etc.) The BEM incorporates such solutions very effectively, via the principle of superposition, into a highly efficient numerical scheme of great power and versatility.

### 2-3-1 Potential Flow in One Dimension

If we return to our one-dimensional potential flow problem and consider a field of infinite length, as shown in Fig. 2-5, we can write down immediately the relevant solution of Eqs (2-1) and (2-2) for this case as

$$p(r) = \frac{\phi}{2k}(\ell - |r|) \tag{2-5a}$$

$$V(r) = -k\frac{dp}{dx} = \frac{\phi}{2}(\text{sgn } r) \tag{2-5b}$$

The coordinate $r$ in this simple infinite region solution is the local distance between the load point $R$ and the field point $P'$. For a global origin of coordinates at $O$, as shown in Fig. 2-5, $r = (x - \xi)$. In the infinite system, $p(r)$ can only be defined relative to a zero value $p(r) = 0$ at some arbitrary point which we have taken to be at $r = \pm l$. For systems of infinite extent, which play a major role in all BEM, we shall use the symbol $\phi$ for general sources as in Eqs (2-5) and reserve $\psi$ for sources of known strength applied at specific points within the system.

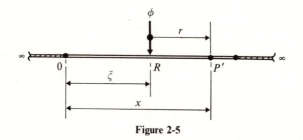

**Figure 2-5**

We now rewrite Eqs (2-5a, b) to introduce two forms of a general notation which will appear frequently in the formal development of solutions to more complex problems:

$$p(P') = \phi(R)\,G(P', R) \quad \text{or} \quad p(x) = \phi(\xi)\,G(x, \xi) \tag{2-5c}$$

$$V(P') = \phi(R)\,F(P', R) \quad \text{or} \quad V(x) = \phi(\xi)\,F(x, \xi) \tag{2-5d}$$

The first form in (2-5c) expresses the functional relationship between quantity $\phi$ at $R$ and another entity $G$ which depends upon an ordered pair of points $(P', R)$ in such a way that their product $\phi(R) G(P', R)$ yields the value of some parameter at $P'$, in this case $p(r) \equiv p(P')$. The second version of these operations is an exactly equivalent statement expressed in terms of the global coordinates of $R$ (that is, $x$) and $P'$ (that is, $\xi$).

In the same way $F(P', R) \equiv F(x, \xi)$ is the function which when multiplied by $\phi(R) \equiv \phi(\xi)$ yields the flow velocity at $P'$, $V(r) \equiv V(P') \equiv V(x)$.

The function $(\mathrm{sgn}\, r)$ used in Eq. (2-5b) is defined to have the following properties:

$$(\mathrm{sgn}\, r) = 1 \text{ for positive } r, \; x > \xi$$

$$(\mathrm{sgn}\, r) = -1 \text{ for negative } r, \; x < \xi$$

$$(\mathrm{sgn}\, r) \text{ is undefined at } r = 0, \; x = \xi$$

although

$$(r\, \mathrm{sgn}\, r) = 0 \text{ at } r = 0$$

Use of this function ensures not only that $V(r)$ will change sign with $r$, as it should, but also that a distinction has to be made between $r = \pm \varepsilon$, say, where $\varepsilon \to 0$, in order to overcome the problem of the step change in $V(r)$ as $r \to 0$ [i.e., to deal with the problem of $V(0)$ being multivalued]. Conversely, the use of $|r|$ in Eq. (2-5a) ensures that $p(r)$ does not change sign with $r$ [i.e., it preserves the single valued nature of $p(0)$].

The following steps illustrate a solution method based on Eqs (2-5), which is, in fact, an application of the indirect BEM. The result is an algorithm applicable without modification to any one-dimensional steady state potential flow problem. For clarity, we shall demonstrate the solution of the mixed boundary-value problem shown in Fig. 2-6. A key conceptual step is to embed the 'real' system of Fig. 2-6 within the infinite field to produce the 'fictitious' system depicted in Fig. 2-7. The reason for the asterisks (*) attached to all symbols in the 'real' system of Fig. 2-6 is now clear, as identical symbols appear unadorned in the 'fictitious' system within which the actual solution procedure develops.

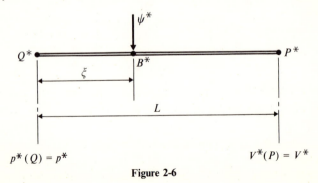

Figure 2-6

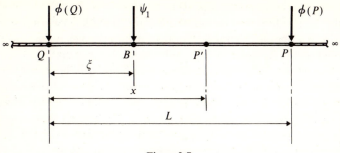

**Figure 2-7**

The boundaries of the one-dimensional field $(QP)$ are merely the two points $(P, Q)$ themselves and therefore only two 'boundary elements' are involved, one at $P$ and one at $Q$. We shall call sources in the fictitious system 'fictitious sources' and apply one to each boundary element $\phi(Q)$, $\phi(P)$ as a general procedure. These sources are initially of unknown strength; nevertheless their effect at any interior point $P'$ can be written down algebraically directly from Eqs (2-5). In particular, the reciprocal effects of $\phi(Q)$ and $\phi(P)$ at both $Q$ and $P$ together with the effect of $\psi(B)$ (located at $B$) on the same points will be of interest to us. We can select our arbitrary length $l = L$, for convenience, and use Eqs (2-5a, b) to write the following expressions for the potentials and velocities at any point $P'$ in the fictitious system. Using a mixed form of the formal notation we have

$$p(x) = G(x,0)\,\phi(Q) + G(x,L)\,\phi(P) + G(x,\xi)\,\psi(B) \qquad (2\text{-}6a)$$

$$V(x) = F(x,0)\,\phi(Q) + F(x,L)\,\phi(P) + F(x,\xi)\,\psi(B) \qquad (2\text{-}6b)$$

The distinction between our use of both the symbols $\psi$ and $\phi$ for sources should now be clear in that $\psi$ will be used for known, specified source strengths, as mentioned previously, whereas $\phi$ will be reserved solely for the 'fictitious sources' [here $\phi(P)$ and $\phi(Q)$] applied at the boundaries of the 'fictitious' system.

If we now take the field point $P'$ towards the boundary, to within a distance $\to 0$ of points $P$ and $Q$ respectively [i.e., to points located at $P(L-\varepsilon), Q(0+\varepsilon)$ with $\varepsilon \to 0$], we can write

At $Q$:
$$\frac{1}{2k}[L\phi(Q) + 0\phi(P) + (L-\xi)\,\psi(B)] = p(Q) \qquad (2\text{-}7a)$$

At $P$:
$$\frac{1}{2k}[0\phi(Q) + L\phi(P) + \xi\psi(B)] = p(P) \qquad (2\text{-}7b)$$

At $Q$:
$$\tfrac{1}{2}[\phi(Q) - \phi(P) - \psi(B)] = V(Q) \qquad (2\text{-}7c)$$

At $P$:
$$\tfrac{1}{2}[\phi(Q) - \phi(P) + \psi(B)] = V(P) \qquad (2\text{-}7d)$$

It is of crucial importance to appreciate that the field point has to approach the boundary elements, here the points $P$ and $Q$, from inside the region of interest whence $p(Q), p(P), V(P), V(Q)$ are evaluated within a small neighbourhood $\varepsilon$ of $Q$

and $P$, as $\varepsilon \to 0$. The negative sign attached to $\phi(P)$ in Eq. (2-7$d$) for $V(P)$ and conversely the positive sign on $\phi(Q)$ in Eq. (2-7$c$) should be studied carefully. Whereas the $p$ values are defined unambiguously, even when the load point ($R$) and the field point ($P'$) coincide, $V$ is multivalued in this situation, as mentioned previously, and the value of $V(0)$ is ambiguous in Eq. (2-5$b$). The ambiguity in $V(0)$ is resolved by our stipulation that our field point ($P'$) always lies within $PQ$ (i.e., we treat $PQ$ as a closed interval) and $P'$ approaches either $P$ or $Q$ from within $PQ$ when we are dealing with the interior region $PQ$. The correct sign on $\phi(P)$ in Eq. (2-6$d$) and on $\phi(Q)$ in (2-7$c$) then follows automatically since $r$ has the values $(-\varepsilon)$ and $(+\varepsilon)$ respectively in the two cases. Conversely, if our interest is in the 'exterior problem' (i.e., the region extending to infinity outside $PQ$), then $r$ will have values $(+\varepsilon)$ and $(-\varepsilon)$ as $P'$ approaches $P$ and $Q$ in turn.

We now specify that the boundary conditions at $P$ and $Q$ in the fictitious system are to be precisely those in the real problem, whence $p(Q) = p^*$ and $V(P) = V^*$ in Eqs (2-7$a$) and (2-7$d$) respectively. These equations then become

$$L\,\phi(Q)+0\,\phi(P)+(L-\xi)\,\psi(B) = 2kp^* \tag{2-8$a$}$$

$$\phi(Q)-\phi(P)+\psi(B) = 2V^* \tag{2-8$b$}$$

two equations from which, in principle, the two unknowns $\phi(Q)$ and $\phi(P)$ can be found. The values of $\phi$ can then be used in conjunction with Eqs (2-6) again to calculate $p(x)$ and $V(x)$ at any selected interior point $x$ within $PQ$.

Before describing this step in more detail it is worth noting that we have only used the two equations (2-7$a, b$) from the complete set of four equations (2-7)—those which relate to the two known boundary values $p^*$ and $V^*$. (In a properly posed problem governed by a linear second-order differential equation we shall always have one or other of the parameters $p$, $V$, or a linear relationship between them, specified at every point on the boundary.) We deliberately chose as our example the more complicated 'mixed boundary-value problem' with $p^*$ known at $Q$ and $V^*$ known at $P$. Had we selected a problem similar to that in Fig. 2- with both $p^*(Q)$ and $p^*(P)$ specified we could obtain equations similar to (2-8) by requiring equivalence of $p(Q)$ with $p^*(Q)$ and $p(P)$ with $p^*(P)$ in Eqs (2-7$a, b$). In fact, we shall find that even in the most complex problems we can always select sufficient equations from sets closely similar to Eqs (2-7) to build up a solvable set (2-8) from which the relevant $\phi$ values can be calculated.

Rather surprisingly it will be found that sets of equations such as (2-8) will yield the fictitious sources directly in all three-dimensional regions, whereas in most one- and two-dimensional problems they have to be slightly modified. This modification arises from the fact that Eq. (2-5$a$) deals with relative potential only, as already mentioned. In all situations where only relative values of variables can be calculated we shall find that a very minor modification of the equation becomes necessary. It can be seen from Eqs (2-5) that as $r$ approaches infinity so does $p(r)$ become infinite, but for $\phi(P)$ and $\phi(Q)$ and $\psi$ to yield a unique solution to our problem their combined effect has to produce zero nett flux across all infinitely remote boundaries. This physically reasonable requirement[2] will be discussed in more detail later. For the time being we note that the condition is equivalent to requiring the sum of all applied sources $\phi(P)$, $\phi(Q)$, and $\psi$ to be zero. Since the datum from which the potentials have been measured is quite arbitrary [as, for example, in the choice of $l$ in Eq. (2-5a)] we can adjust this uniformly throughout the whole infinite field by a constant $C$ (initially unknown) in the same way that the datum used for hydraulic potentials, or the 'earth' for electrical potentials, can be set to any level we care to choose.

The datum adjustment will merely modify all potentials throughout the system by $C$, and have no effect whatsoever on $V(x) = -k(dp/dx)$. If we therefore insert modified sources $\phi'$ in Eq. (2-5a) such that $\phi = (\phi' + 2kC)$ and require $C$ to have that unique value which will ensure zero sum of all the $\phi'$ and $\psi$ sources, Eq. (2-5a) becomes $p(r) = (\phi'/2k)(l-|r|)+C$ from which Eqs (2-8) can be rewritten as

$$L\phi'(Q)+0\phi'(P)+(L-\xi)\psi(B)+2kC = 2kp^* \qquad (2\text{-}9a)$$

$$\phi'(Q)-\phi'(P)+\psi(B) = 2V^* \qquad (2\text{-}9b)$$

$$\phi'(Q)+\phi'(P)+\psi(B) = 0 \qquad (2\text{-}9c)$$

or in matrix form as

$$\begin{bmatrix} L & 0 & 2k \\ 1 & -1 & 0 \\ 1 & 1 & 0 \end{bmatrix} \begin{Bmatrix} \phi'(Q) \\ \phi'(P) \\ C \end{Bmatrix} = \begin{Bmatrix} 2kp^* \\ 2V^* \\ 0 \end{Bmatrix} - \begin{Bmatrix} L-\xi \\ 1 \\ 1 \end{Bmatrix} \psi(B) \qquad (2\text{-}9d)$$

where the final equation is the zero sum condition and the set of three equations will provide the unknown $\phi'(P)$, $\phi'(Q)$, and $C$. Since $\psi$ and $p^*$ are numerically specified quantities they have to remain unaltered by this datum adjustment. Although the foregoing explanation of the adjustment of the potential datum $C$ is possibly rather verbose, the reader should note that the resultant changes in Eqs (2-8) to produce Eqs (2-9) are quite trivial ones and can be incorporated into them as they are being assembled.

The solution of Eqs (2-9) is easily found to be

$$\phi'(Q) = V^*-\psi \qquad \phi'(P) = -V^* \qquad 2kC = 2kp^*-LV^*+\xi\psi$$

By substituting these values into Eqs (2-6) once more we can calculate $p(x)$, $V(x)$ at any point $P'(x)$ within $PQ$. Thus, for $0 \leqslant x < \xi$,

$$p(x) = \frac{1}{2k}[(L-x)(V^*-\psi)+(L-\xi+x)\psi-V^*x]+C$$

Therefore

$$p(x) = p^* - \frac{x}{k}V^* + \frac{x}{k}\psi \qquad (2\text{-}10a)$$

and

$$V(x) = \tfrac{1}{2}(V^*-\psi+V^*-\psi) = V^*-\psi$$

and, similarly, for $\xi < x \leqslant L$,

$$p(x) = p^* - \frac{x}{k}V^* + \frac{\xi}{k}\psi$$

(2-10b)

and
$$V(x) = V^*$$

Equations (2-10) are seen to be identical to the previous solution of the same problem given in Eqs (2-4).

Before proceeding to further illustrative examples it may be helpful to summarize the few, very simple steps involved in the previous solution, since they inevitably become rather submerged in the explanations which accompany them. There are really only five such steps:

1. Statement of the infinite region unit solutions, Eqs (2-5)
2. Assembly of the required set of Eqs (2-8) directly from (2-5) to relate the unknown fictitious boundary element potentials $\phi$ and the specified internal source strengths $\psi$ to the given boundary information
3. Augmentation of these equations by the datum adjustment $C$ and the zero nett flux sum condition to produce Eqs (2-9)
4. Solution of (2-9) for all the known $\phi'$ values and $C$
5. Backsubstitution of the $\phi'$ and $C$ into (2-6) to obtain the values of $p(x)$ and $V(x)$ at any point $P'(x)$ in the field

It is worth remarking that the above algorithm can be applied identically to potential flow problems in two and three dimensions with the additional attraction that step 3 does not arise in three-dimensional problems.

Inspection of the solution will also reveal that any number of specified internal sources $(\psi_1, \psi_2, ..., \psi_q)$ can be dealt with equally easily and merely add to the second right-hand-side vector in Eq. (2-9d) without increasing the number of unknowns to be determined, which remain equal to the number of boundary elements plus one. The presence of a zero term in the diagonal of the left-hand-side matrix in Eq. (2-9d) does not imply that the matrix is singular and, with care, it can be inverted in any standard way.

## 2-3-2 The Simple Beam Problem

As an example of the indirect BEM applied to a system governed by a fourth-order differential equation we shall consider the problem of a simple uniform beam supporting a system of concentrated loads and moments as shown in Fig. 2-8. The beam is of length $L$, second moment of area $I$ in the plane of bending, and made from material with Young's modulus $= E$. One end of the beam $Q^*$ is encastré and therefore both the slope $\theta^*(Q^*) = (dw^*/dx)(Q^*)$ and the deflection $w^*(Q^*)$ are zero. The other end $P^*$ is simply supported, that is, $m(P^*) = -EI(d^2 w^*/dx^2)(P^*) = 0$, but held at a level $w^*(P^*) = w^*$ below $Q^*$.

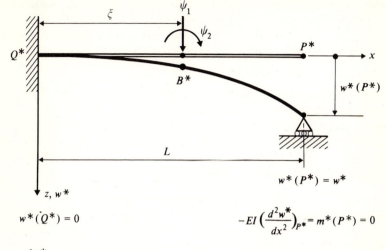

$$w^*(Q^*) = 0$$

$$\left(\frac{dw^*}{dx}\right)_{Q^*} = \theta^*(Q^*) = 0$$

$$w^*(P^*) = w^*$$

$$-EI\left(\frac{d^2w^*}{dx^2}\right)_{P^*} = m^*(P^*) = 0$$

**Figure 2-8**

The problem is therefore one of a statically indeterminate beam with mixed boundary conditions governed by the fourth-order differential equation

$$\frac{d^4 w}{dx^4} = 0 \tag{2-11}$$

everywhere other than at the loading points $B_1^*, B_2^*$ etc. There are still only two boundary elements, one at $Q^*$ and one at $P^*$, and, in a properly posed problem involving a linear fourth-order differential equation, two boundary conditions will always be specified at each of them. The applied loading functions can be either point loads, as $\psi_1$, or point moments, as $\psi_2$, in Fig. 2-8.

Following the solution procedure outlined in Sec. 2-3-1 we now embed our beam within a similar infinite one-dimensional field and consider the effect at any field point $P'$ of the two types of fictitious loads $(\phi_1, \phi_2)$ applied at a point $R$ (see Fig. 2-9). Again in the infinite system certain parameters such as displacements

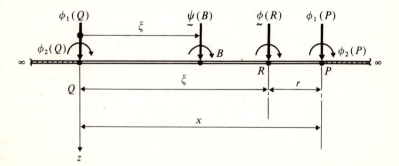

**Figure 2-9**

$w(r)$ and rotations $\theta(r)$ are calculable only as relative values, and we shall once more measure them relative to zero values at $|r| = l$. Writing for convenience $\lambda = 1/12EI$ and $\rho = r/l$, where $r = (x - \xi)$, our unit solutions of Eq. (2-11) for the infinite system under $\phi_1$ and $\phi_2$ are easily shown to be:[3]

Due to $\phi_1$ (a concentrated load):

$$w(x) = \phi_1 \lambda l^3 (2 + |\rho|^3 - 3|\rho|^2) = \phi_1(\xi) G(x, \xi) \qquad (2\text{-}12a)$$

$$\theta(x) = \frac{dw}{dx} = \phi_1 \cdot 3\lambda l^2 |\rho|(|\rho| - 2)(\operatorname{sgn}\rho) = \phi_1(\xi) F(x, \xi) \qquad (2\text{-}12b)$$

$$m(x) = -EI\frac{d^2 w}{dx^2} = \phi_1 \frac{l}{2}(1 - |\rho|) = \phi_1(\xi) E(x, \xi) \qquad (2\text{-}12c)$$

$$s(x) = -EI\frac{d^3 w}{dx^3} = -\phi_1 \frac{(\operatorname{sgn}\rho)}{2} = \phi_1(\xi) D(x, \xi) \qquad (2\text{-}12d)$$

Due to $\phi_2$ (a concentrated moment):

$$w(x) = -\phi_2 \lambda l^2 |\rho|(|\rho|^2 - 3|\rho| + 2)(\operatorname{sgn}\rho) = \phi_2(\xi) K(x, \xi) \qquad (2\text{-}13a)$$

$$\theta(x) = \frac{dw}{dx} = -\phi_2 \lambda l(3|\rho|^2 - 6|\rho| + 2) = \phi_2(\xi) \bar{L}(x, \xi) \qquad (2\text{-}13b)$$

$$m(x) = -EI\frac{d^2 w}{dx^2} = -\phi_2(1 - |\rho|)\frac{(\operatorname{sgn}\rho)}{2} = \phi_2(\xi) M(x, \xi) \qquad (2\text{-}13c)$$

$$s(x) = -EI\frac{d^3 w}{dx^3} = \phi_2\left(\frac{1}{2l}\right) = \phi_2(\xi) N(x, \xi) \qquad (2\text{-}13d)$$

As before, we have written out the complete set of solutions, although they will not all be required for any specific problem. Equations (2-12d) and (2-13c) involving $(\operatorname{sgn}\rho)$ terms are again those which are multivalued at $\rho = 0$ (i.e., at the coincidence of the load point $R$ and the field point $P'$) and hence need to be interpreted in the sense of $R$ approaching $P'$, usually on the boundary, from inside the region of interest (that is, $PQ$).

In keeping with having two specified boundary conditions at each end (boundary) element we shall apply two $\Phi$ components on each of them also which leads to two sets of two component $\Phi$ vectors, as shown in Fig. 2-9:

$$\Phi(Q) = \begin{Bmatrix} \phi_1(Q) \\ \phi_2(Q) \end{Bmatrix} \quad \text{and} \quad \Phi(P) = \begin{Bmatrix} \phi_1(P) \\ \phi_2(P) \end{Bmatrix}$$

Step 2 in the solution procedures is to assemble equations in terms of the fictitious $\Phi(P)$, $\Phi(Q)$ and all the specified loadings

$$\psi(B) = \begin{Bmatrix} \psi_1(B) \\ \psi_2(B) \end{Bmatrix}$$

etc., using equations from sets (2-12) and (2-13) which relate them to the known boundary conditions, in this case $w^*(Q), \theta^*(Q), w^*(P), m^*(P)$. Since $l$ is arbitrary we shall again choose $l = L$ for simplicity. Before doing this, for any interior point

$P'$ we can write the following equations in the formal notation:

$$w(x) = \phi_1(Q)\,G(x,0) + \phi_2(Q)\,K(x,0) + \phi_1(P)\,G(x,L) + \phi_2(P)\,K(x,L)$$
$$+ \psi_1(B)\,G(x,\xi_1) + \psi_2(B)\,K(x,\xi_1)$$

$$\theta(x) = \phi_1(Q)\,F(x,0) + \phi_2(Q)\,\bar{L}(x,0) + \phi_1(P)\,F(x,L) + \phi_2(P)\,\bar{L}(x,L)$$
$$+ \psi_1(B)\,F(x,\xi_1) + \psi_2(B)\,\bar{L}(x,\xi_1)$$

$$m(x) = \phi_1(Q)\,E(x,0) + \phi_2(Q)\,M(x,0) + \phi_1(P)\,E(x,L) + \phi_2(P)\,M(x,L)$$
$$+ \psi_1(B)\,E(x,\xi_1) + \psi_2(B)\,M(x,\xi_1)$$

$$s(x) = \phi_1(Q)\,D(x,0) + \phi_2(Q)\,N(x,0) + \phi_1(P)\,D(x,L) + \phi_2(P)\,N(x,L)$$
$$+ \psi_1(B)\,D(x,\xi_1) + \psi_2(B)\,N(x,\xi_1)$$

When set out at length in this way these equations look extremely unattractive, but a careful study of them will reveal the underlying simplicity of their form whereby the coefficients of the various $\phi$ and $\psi$ terms can be obtained, in a numerical solution, by simple substitution of coordinate values in the known 'kernel functions' $(D, E, F, G, K, \bar{L}, M, N)$ which are provided by the 'point load' and 'point moment' solutions for an infinite beam set out in Eqs (2-12) and (2-13).

Using matrix notion these equations can be written compactly as

$$
\begin{Bmatrix} w(x) \\ \theta(x) \\ m(x) \\ s(x) \end{Bmatrix}
=
\begin{bmatrix}
G(x,0) & K(x,0) & G(x,L) & K(x,L) \\
F(x,0) & \bar{L}(x,0) & F(x,L) & \bar{L}(x,L) \\
E(x,0) & M(x,0) & E(x,L) & M(x,L) \\
D(x,0) & N(x,0) & D(x,L) & N(x,L)
\end{bmatrix}
\begin{Bmatrix} \phi_1(Q) \\ \phi_2(Q) \\ \phi_1(P) \\ \phi_2(P) \end{Bmatrix}
$$

$$
+
\begin{bmatrix}
G(x,\xi_1) & K(x,\xi_1) \\
F(x,\xi_1) & \bar{L}(x,\xi_1) \\
E(x,\xi_1) & M(x,\xi_1) \\
D(x,\xi_1) & N(x,\xi_1)
\end{bmatrix}
\begin{Bmatrix} \psi_1(B) \\ \psi_2(B) \end{Bmatrix}
\qquad (2\text{-}14)
$$

By taking the field point $x$ successively to the boundary elements $Q$ and $P$ as before such that $x = 0 + \varepsilon$ and $x = L - \varepsilon$, $\varepsilon \to 0$, respectively and using only those equations for which the left-hand side is specified on the boundaries, we obtain

$$
\begin{Bmatrix} w(Q) \\ \theta(Q) \\ w(P) \\ m(P) \end{Bmatrix}
=
\begin{bmatrix}
2\lambda L^3 & 0 & 0 & 0 \\
0 & -2\lambda L & 3\lambda L^2 & \lambda L \\
0 & 0 & 2\lambda L^3 & 0 \\
0 & 0 & L/2 & \tfrac{1}{2}
\end{bmatrix}
\begin{Bmatrix} \phi_1(Q) \\ \phi_2(Q) \\ \phi_1(P) \\ \phi_2(P) \end{Bmatrix}
$$

$$
+
\begin{bmatrix}
\lambda L^3(2 + \mu_1^3 - 3\mu_1^2) & \lambda L^2\,\mu_2(\mu_2^2 - 3\mu_2 + 2) \\
-3\lambda L^2\,\mu_1(\mu_1 - 2) & -\lambda L(3\mu_2^2 - 6\mu_2 + 2) \\
\lambda L^3(2 + v_1^3 - 3v_1^2) & -\lambda L^2\,v_2(v_2^2 - 3v_2) \\
(L/2)(1 - v_1) & -(\tfrac{1}{2})(1 - v_2)
\end{bmatrix}
\begin{Bmatrix} \psi_1 \\ \psi_2 \end{Bmatrix}
\qquad (2\text{-}15)
$$

$$\text{where} \quad \begin{Bmatrix} w(Q) \\ \theta(Q) \\ w(P) \\ m(P) \end{Bmatrix} = \begin{Bmatrix} 0 \\ 0 \\ w^* \\ 0 \end{Bmatrix}$$

and $\mu_1 = \xi_1/L$, $\mu_2 = \xi_2/L$, $v_1 = 1 - \xi_1/L$, $v_2 = 1 - \xi_2/L$. The reader is advised to check through all of the terms in Eq. (2-15), paying particular attention to the sign of each of them.

The entire procedure is, of course, directed towards the construction of a completely general algorithm, which can be programmed to solve large, complicated problems. Almost inevitably, processes which are designed to operate primarily as numerical procedures are very cumbersome indeed when used to obtain algebraic answers to very simple problems such as the present one. However, we believe that a clear physical appreciation of the various operations is necessary before attempting sophisticated solutions, and we shall therefore follow through the algebraic solution of our propped cantilever problem under the simplified condition of zero external loading (that is, $\psi_1 = \psi_2 = 0$). Thus, in principle, the reduced set of Eqs (2-15) can be solved to determine the complete four-component ($\phi$) vector. Thereafter the individual $\phi$ values can be substituted into the relevant equations in the sets (2-12) and (2-13) or (2-14) to calculate, for instance, the bending moment at $Q$, $m(Q)$, or shear force $s(P)$ at $P$. For example, $m(Q)$ will be given from Eq. (2-14) by the limit, as $\varepsilon \to 0$, of

$$m(Q) = E(0+\varepsilon, 0)\,\phi_1(Q) + M(0+\varepsilon, 0)\,\phi_2(Q) + E(0, L)\,\phi_1(P) + M(0, L)\,\phi_2(P) \tag{2-16}$$

If the reduced set of Eqs (2-15) are solved for $\phi$ as they stand and the values of $\phi$ substituted into Eq. (2-16), the reader will find that $m(Q) = -w^*/8\lambda L^2$ is obtained, which is incorrect! The correct answer is easily shown to be $m(Q) = -w^*/4\lambda L^2$.

The error has arisen from the fact that the various parameters calculated from Eqs (2-12) and (2-13) do not approach constant values as $P$ approaches infinity but continue to increase. Thus the cumulative effect of all the $\phi$ and $\psi$ contributions will not always be unique on an infinitely distant boundary, as they must be if the problem is to be solved correctly. Fortunately, this can be achieved very easily by a device identical in principle to that used for the potential flow example. Once more we have freedom in our infinite system, this time to superimpose an arbitrary rigid-body displacement (in the direction of $z$), $C_1$ say, throughout the entire field and require $C_1$ to have the unique value which ensures that

$$\phi_1(Q) + \phi_1(P) + \psi_1(B_1) = (\Sigma\phi_1 + \Sigma\psi_1) = 0.$$

This latter equation is in effect an overall equilibrium equation in the $z$ direction which could otherwise only be achieved by reactions at the infinite boundary. We can also determine a further constant, $C_2$ say, which is a whole field rigid-body rotation so chosen that

$$\phi_2(Q) + \phi_2(P) + \psi_2(B_2) = (\Sigma\phi_2 + \Sigma\psi_2) = 0$$

which is a form of rotational equilibrium equation. By St Venant's principle the nett effect of both of these zero sum quantities will be zero at an infinitely remote boundary. Furthermore, since $C_1$ and $C_2$ are an independent rigid-body displacement and a rotation respectively, their spatial derivatives are zero and the augmented set of equations to replace Eq. (2-14) can be written down immediately as

$$
\begin{bmatrix}
\boxed{\phantom{xxx}} & \begin{matrix} 1 & 0 \\ 0 & 1 \\ 1 & 0 \\ 0 & 1 \end{matrix} \\
\begin{matrix} 1 & 0 & 1 & 0 & 0 & 0 \\ 0 & 1 & 0 & 1 & 0 & 0 \end{matrix}
\end{bmatrix}
\begin{Bmatrix} \phi(Q) \\ \phi(P) \\ C_1 \\ C_2 \end{Bmatrix}
+
\begin{bmatrix}
\boxed{\phantom{xxx}} \\
\begin{matrix} 1 & 0 \\ 0 & 1 \end{matrix}
\end{bmatrix}
\begin{Bmatrix} \psi_1 \\ \psi_2 \end{Bmatrix}
=
\begin{Bmatrix} 0 \\ 0 \\ w^* \\ 0 \\ 0 \\ 0 \end{Bmatrix}
\tag{2-17}
$$

where the hatched blocks are identical to those in Eq. (2-14). Solution of these equations will now yield the unique $\phi$, $C_1$, and $C_2$ values which solve our problem. For the case when $\psi_1 = \psi_2 = 0$ it is left as an exercise for the reader to show that

$$
-\phi_1(Q) = \phi_1(P) = \frac{w^*}{4\lambda L^3} \qquad \phi_2(Q) = -\phi_2(P) = \frac{w^*}{4\lambda L^2}
$$
$$
C_1 = \frac{w^*}{2} \qquad C_2 = 0
\tag{2-18}
$$

whence, from Eq. (2-16),

$$
m(Q) = \frac{L}{2} - \frac{w^*}{4\lambda L^3} + \left(-\frac{1}{2}\right)\left(\frac{w^*}{4\lambda L^2}\right) + 0 + 0 = -\frac{w^*}{4\lambda L^2}
$$

and
$$
s(P) = \frac{1}{2}\left(\frac{w^*}{4\lambda L^3} + \frac{w^*}{4\lambda L^3} - \frac{w^*}{4\lambda L^2}\frac{1}{L} + \frac{w^*}{4\lambda L^2}\frac{1}{L}\right) = \frac{w}{4\lambda L^3}
$$

which are the correct solutions for $m(Q)$ and $s(P)$.

It is worth noting that the $6 \times 6$ matrix involved in the solution of Eq. (2-17) is the same irrespective of either the magnitudes of the prescribed boundary conditions in the right-hand-side vector of the loading vector $\psi$. The latter can clearly contain any number of point loading components $\psi$ without intrinsically increasing the complexity of the solution. Furthermore, in two-dimensional problems where the number of boundary elements, and hence $\phi$ components, is greatly increased we shall find that one $C$ value only has to be introduced to deal with the case of potential flow and two parameters $(C_1, C_2)$ are sufficient for problems in plane elasticity. Consequently, the relatively much larger numbers of equations involved have to be rather insignificantly increased, by only one and two respectively, to meet the infinite boundary condition.

An alternative and common boundary condition, at $P$ say, is elastic spring support such that $s(P) = K(P)w(P)$, where $K(P)$ is the stiffness of the support at $P$. The reader is recommended to augment Eq. (2-17) by that expressing $s(P)$ in terms of $(\phi, \psi)$ and hence see that these equations together with the above $s(P) = K(P)w(P)$ relationship are sufficient to solve this problem also.

We shall now use the same simple problems to illustrate an alternative BEM approach.

## 2-4 APPLICATIONS OF THE DIRECT BOUNDARY ELEMENT METHOD

The direct formulation of BEM uses exactly the same unit solutions of the governing differential equations as the indirect method. These solutions are also superimposed somewhat similarly but the solution is carried through 'directly' in terms of the physical variables of the problem (i.e., fictitious distributions of potentials, forces, etc., are not introduced). A direct BEM solution therefore yields the unknown boundary values directly, which is an attraction, but the solution variables at interior points are then more difficult to determine than they are using the indirect method.

In the following sections we shall solve the previous one-dimensional problems again, this time using the direct method. In order to simplify the algebraic expressions as much as possible without loss of clarity the values of $k$ and $EI$ will be taken as unity throughout and $L$ retained as the beam length.

More satisfactorily, when solving problems numerically one can transform the governing equation to dimensionless variables by envisaging:

For potential flow $p \to \dfrac{p}{p_0}$: $\quad x \to \dfrac{x}{L}$ whence $\psi \to \dfrac{L^2 \psi}{kp_0}$, etc.

For the beam $w \to \dfrac{w}{L}$: $\quad x \to \dfrac{x}{L}$ whence $\psi \to \dfrac{L^3 \psi}{EI}$, etc.

where $p_0$ is any convenient arbitrary potential and $L$ a characteristic dimension of the system.

### 2-4-1 Potential Flow in One Dimension

The governing differential equation for the problem can be written as

$$\frac{d^2 p(x)}{dx^2} = -\psi(x) \tag{2-19}$$

where now $\psi$ is a specified distributed source intensity along $x$ and the velocity $V(x)$ is simply $-dp(x)/dx$. In order to investigate the possibility of actually integrating Eq. (2-19), over the range $0 \leqslant x \leqslant L$, we shall introduce a function $G(x, \xi)$ which is, as yet, undefined except that it is sufficiently continuous to be differentiable as often as required.

If we multiply both sides of (2-19) by $G$ and integrate by parts twice we obtain[4]

$$\int_0^L \frac{d^2 p(x)}{dx^2} G \, dx = -\int_0^L \psi(x) G \, dx$$

Therefore,

$$\left[ G \frac{dp(x)}{dx} \right]_0^L - \int_0^L \left[ \frac{dG}{dx} \frac{dp(x)}{dx} \right] dx = -\int_0^L \psi(x) G \, dx \tag{2-20}$$

and

$$\left[ G \frac{dp(x)}{dx} - p(x) \frac{dG}{dx} \right]_0^L + \int_0^L p(x) \frac{d^2 G}{dx^2} \, dx = -\int_0^L \psi(x) G \, dx$$

We now specify $G$ to be a solution of

$$\frac{d^2 G(x, \xi)}{dx^2} = -\delta(x, \xi) \tag{2-21}$$

where $\delta(x, \xi)$ is the Dirac 'delta function' (or impulse function) which is mathematically equivalent to the effect of a unit concentrated source applied at the point $\xi$. The key property of the delta function is that it is zero at all $x$ except in the neighbourhood of $x = \xi$ where it becomes infinitely large in such a way that[4, 5]

$$\int_{-\infty}^{\infty} \delta(x, \xi) \, dx = \int_0^L \delta(x, \xi) \, dx = 1(\xi)$$

The delta function is therefore an operator with a sifting property, a 'needle' (Fig. 2-10) which 'points' to a specific value, $p(\xi)$ say, of any function $p(x)$ which it operates on as in

$$\int_0^L p(x) \delta(x, \xi) \, dx = p(\xi)$$

If we substitute Eq. (2-21) into Eq. (2-20) we get

$$-\int_0^L \psi(x) G \, dx = \left[ G \frac{dp(x)}{dx} - p(x) \frac{dG}{dx} \right]_0^L - \int_0^L p(x) \delta(x, \xi) \, dx$$

which simplifies, using the above property of the $\delta$ operator, to

$$-\int_0^L \psi(x) G \, dx = \left[ G \frac{dp(x)}{dx} - p(x) \frac{dG}{dx} \right]_0^L - p(\xi) \tag{2-22}$$

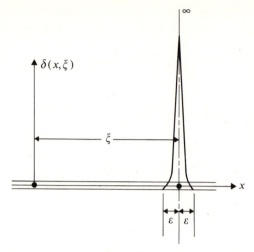

**Figure 2-10**

Our function $G$, which is the solution of Eq. (2-21), is clearly precisely that previously quoted in Eq. (2-5a) with $\phi = 1$ (i.e., the fundamental singular solution of the governing differential equation); thus

$$G(x, \xi) = 0.5(l - |r|) \qquad\qquad \kappa = 1 \qquad (2\text{-}23a)$$

and
$$-\frac{dG(x, \xi)}{dx} = F(x, \xi) = 0.5(\operatorname{sgn} r) \qquad \kappa = 1 \qquad (2\text{-}23b)$$

where, as before, $r$ is the distance $(x - \xi)$ and at $r = \pm l$ the function $G = 0$. If we choose the arbitrary length $l = L$ and recall that here the velocity $V(x) = -dp(x)/dx$, we can write Eq. (2-22) as

$$p(\xi) = -[G(x, \xi) V(x) - F(x, \xi) p(x)]_0^L + \int_0^L \psi(x) G(x, \xi)\, dx$$

or
$$p(\xi) = -[G(L, \xi) V(L) - G(0, \xi) V(0)] + [F(L, \xi) p(L) - F(0, \xi) p(0)]$$

$$+ \int_0^L \psi(x) G(x, \xi)\, dx \qquad (2\text{-}24)$$

Equation (2-24) is seen to provide the potential $p(\xi)$ at any interior field point $(\xi)$ consistent with a set of boundary potentials $(p(0), p(L))$, velocities $(V(0), V(L))$, and a specified internal distribution of sources $\psi(x)$. If there is merely a single point source $\psi(x_o)$ at a point $x_o$ within the region then the integral on the right-hand side of Eq. (2-24) becomes simply $\psi(x_o) G(x_o, \xi)$.

By taking the first derivative of $p(\xi)$ with respect to the field variable $\xi$ we have

$$-\frac{dp(\xi)}{d\xi} = v(\xi) = \left[\frac{dG}{d\xi}(L,\xi)\,v(L) - \frac{dG}{d\xi}(0,\xi)\,v(0)\right] - \left[p(L)\frac{dF}{d\xi}(L,\xi) - p(0)\frac{dF(0,\xi)}{d\xi}\right]$$

$$- \int_0^L \psi(x)\frac{dG(x,\xi)}{d\xi}\,dx$$

or, more formally, we can write

$$v(\xi) = [G'(L,\xi)\,v(L) - G'(0,\xi)\,v(0)] - [p(L)\,F'(L,\xi) - p(0)\,F'(0,\xi)]$$

$$- \int_0^L \psi(x)\,G'(x,\xi)\,dx \tag{2-25}$$

where $G'(L,\xi)$, $F'(L,\xi)$, etc., are the derivatives of $G(L,\xi)$, $F(L,\xi)$ with respect to the field variable $\xi$. Equation (2-25) thus provides the velocity $v(\xi)$ at any field point ($\xi$) consistent with the boundary potentials and velocities and known internal source distribution.

Now by taking the field point $\xi$ to the boundary points $P$ and $Q$ (see Fig. 2-3) such that $\xi = L - \varepsilon$ for $P$ and $\xi = 0 + \varepsilon$ for $Q$ we can use Eq. (2-24) to write

$$\left\{\begin{array}{c} p(L-\varepsilon) \\ p(0+\varepsilon) \end{array}\right\} = -\left[\begin{array}{cc} G(L,L-\varepsilon) & -G(0,L-\varepsilon) \\ G(L,0+\varepsilon) & -G(0,0+\varepsilon) \end{array}\right]\left\{\begin{array}{c} v(L) \\ v(0) \end{array}\right\}$$

$$+\left[\begin{array}{cc} F(L,L-\varepsilon) & -F(0,L-\varepsilon) \\ F(L,0+\varepsilon) & -F(0,0+\varepsilon) \end{array}\right]\left\{\begin{array}{c} p(L) \\ p(0) \end{array}\right\}$$

$$+\left\{\begin{array}{c} \int_0^L \psi(x)\,G(x,L-\varepsilon)\,dx \\ \int_0^L \psi(x)\,G(x,0+\varepsilon)\,dx \end{array}\right\} \tag{2-26a}$$

Substituting values for $G$ and $F$ from Eqs (2-23a, b) we obtain in the limit, as $\varepsilon \to 0$,

$$\left\{\begin{array}{c} p(L) \\ p(0) \end{array}\right\} = -\left[\begin{array}{cc} 0.5L & 0 \\ 0 & -0.5L \end{array}\right]\left\{\begin{array}{c} v(L) \\ v(0) \end{array}\right\} + \left[\begin{array}{cc} 0.5 & 0.5 \\ 0.5 & 0.5 \end{array}\right]\left\{\begin{array}{c} p(L) \\ p(0) \end{array}\right\}$$

$$+\left\{\begin{array}{c} \int_0^L \psi(x)\,G(x,L)\,dx \\ \int_0^L \psi(x)\,G(x,0)\,dx \end{array}\right\}$$

or
$$-\begin{bmatrix} 0.5L & 0 \\ 0 & -0.5L \end{bmatrix} \begin{Bmatrix} v(L) \\ v(0) \end{Bmatrix} - \begin{bmatrix} 0.5 & -0.5 \\ -0.5 & 0.5 \end{bmatrix} \begin{Bmatrix} p(L) \\ p(0) \end{Bmatrix}$$

$$+ \begin{Bmatrix} \int_0^L \psi(x)\,G(x,L)\,dx \\ \int_0^L \psi(x)\,G(x,0)\,dx \end{Bmatrix} = 0 \qquad\qquad (2\text{-}26b)$$

Equation (2-26b) can be used to calculate the initially unknown boundary-value data from the known values and $\psi(x)$, for our one-dimensional field. For example,

1. With $v(L)$, $v(0)$, and $\psi(x)$ specified, then solution of (2-26b) will yield the unknown values $p(L)$ and $p(0)$.
2. With $p(L)$, $p(0)$, and $\psi(x)$ specified, then solution will provide the unknowns $v(L)$ and $v(0)$.
3. With $v(L), p(0), \psi(x)$ or $v(0), p(L), \psi(x)$ specified, the solution process will provide respectively the values of the set $v(0), p(L)$ or $v(L), p(0)$.

If we return to our initial problem, shown in Fig. 2-1 [that is, $p(L) = p(0) = 0$ and a point source of intensity $\psi$ specified at a distance $\xi_1$ from the left-hand end], then Eqs (2-26) become

$$-\begin{bmatrix} 0.5L & 0 \\ 0 & -0.5L \end{bmatrix} \begin{Bmatrix} v(L) \\ v(0) \end{Bmatrix} = -\begin{Bmatrix} \psi(\xi_1)\,G(\xi_1,L) \\ \psi(\xi_1)\,G(\xi_1,0) \end{Bmatrix} = -0\cdot5 \begin{Bmatrix} \psi(\xi_1) \\ \psi(L-\xi_1) \end{Bmatrix}$$

or
$$v(L) = \psi\xi_1 \qquad v(0) = -\psi(L-\xi_1)$$

which agrees with Eqs (2-3). The potentials and velocities at selected interior points can then be obtained by substituting the values of $p(L)$, $p(0)$, $v(L)$, $v(0)$ in Eqs (2-24) and (2-25) respectively.

On the other hand, if we wish to solve the mixed boundary-value problem, shown in Fig. 2-6 [that is, $p(0) = p^*$, $v(L) = v^*$, and a point source $\psi$ applied at a distance $\xi_1$ from the left-hand end], we can write Eq. (2-26b) as

$$\begin{bmatrix} 0.5L & 0 \\ 0 & -0.5L \end{bmatrix} \begin{Bmatrix} v^* \\ v(0) \end{Bmatrix} + \begin{bmatrix} 0.5 & -0.5 \\ -0.5 & 0.5 \end{bmatrix} \begin{Bmatrix} p(L) \\ p^* \end{Bmatrix} = 0.5 \begin{Bmatrix} \psi\xi_1 \\ \psi(L-\xi_1) \end{Bmatrix}$$

the solution of which provides

$$p(L) = \psi\xi_1 - Lv^* + p^* \qquad v(0) = v^* - \psi$$

which agrees with the previous answer obtained using the indirect method. Once again Eqs (2-24) and (2-25) can be used to obtain the potentials and velocities at any internal point ($\xi$) by simple substitution of these boundary values.

## 2-4-2 The Simple Beam Problem

The governing differential equation for the problem is

$$\frac{d^4 w}{dw^4} = \psi_1(x) \qquad \text{for } EI = 1 \tag{2-27}$$

where $w(x)$ is the vertical deflection of the longitudinal axis of the beam and $\psi_1$ the distributed vertical loading intensity. As before, we can multiply both sides of Eq. (2-27) by a suitably continuous function $G$ and integrate by parts; thus

$$\int_0^L \frac{d^4 w(x)}{dx^4} G \, dx = \int_0^L \psi_1(x) G \, dx$$

Therefore,

$$\left[ G \frac{d^3 w(x)}{dx^3} \right]_0^L - \int_0^L \frac{dG}{dx} \frac{d^3 w(x)}{dx^3} dx = \int_0^L \psi_1(x) G \, dx$$

$$\text{and} \quad \left[ G \frac{d^3 w(x)}{dx^3} - \frac{dG}{dx} \frac{d^2 w(x)}{dx^2} \right]_0^L + \int_0^L \frac{d^2 G}{dx^2} \frac{d^2 w(x)}{dx^2} dx = \int_0^L \psi_1(x) G \, dx \tag{2-28}$$

Once more the direct BEM formulation requires that $G$ be a solution of the governing differential equation for an infinite field (this time an infinitely long beam) under unit 'loading' (i.e., a unit point load at $\xi$ on the beam). Thus $G(x, \xi)$ is a solution of

$$\frac{d^4 G}{dx^4} = \delta(x, \xi) \tag{2-29}$$

The form of $G$ is clearly exactly analogous to that of a displacement $w$. We shall designate $G$ to be identical to a (virtual) displacement $w^*$ and use the related quantities for slopes ($\theta^*$) and moments ($m^*$), etc., given by $\theta^* \equiv dG/dx$, $m^* \equiv -d^2 G/dx^2$, $s^* \equiv -d^3 G/dx^3$, $w^* \equiv G$. If we rewrite Eq. (2-28) using these symbols we obtain

$$\left[ -w^*(x, \xi) s(x) + \theta^*(x, \xi) m(x) \right]_0^L + \int_0^L m^*(x, \xi) m(x) \, dx = \int_0^L \psi_1(x) w^*(x, \xi) \, dx \tag{2-28a}$$

which is, in fact, simply a form of virtual work equation for the loaded beam system.

If we continue with the integration of Eq. (2-28) by parts a further twice this leads to

$$\left[ G \frac{d^3 w}{dx^3} - \frac{dG}{dx} \frac{d^2 w}{dx^2} + \frac{d^2 G}{dx^2} \frac{dw}{dx} - \frac{d^3 G}{dx^3} w \right]_0^L + \int_0^L \frac{d^4 G}{dx^4} w(x) \, dx = \int_0^L \psi_1(x) G \, dx \tag{2-30}$$

which can also be written, using the alternative notation based on $G(x, \xi) \equiv w^*(x, \xi)$, as

$$[-w^*(x)\,s(x) + \theta^*(x)\,m(x) - m^*(x)\,\theta(x) + s^*(x)\,w(x)]_0^L + \int_0^L \psi_1^*(x)\,w(x)\,dx$$

$$= \int_0^L \psi_1(x)\,w^*(x)\,dx$$

$$[-w^*(x)\,s(x) + \theta^*(x)\,m(x)]_0^L - \int_0^L \psi_1(x)\,w^*(x)\,dx$$

$$= [-w(x)\,s^*(x) + \theta(x)\,m^*(x)]_0^L - \int_0^L \psi_1^*(x)\,w(x)\,dx \qquad (2\text{-}31)$$

Equation (2-31) states that the work done by the forces of system 1 (the real system) on the displacements of system 2 (any other admissible system*) is equal to the work done by the forces of system 2 on the displacements of system 1 (i.e., the well-known 'reciprocal theorem' due to Betti, 1872).[8] We could have used this theorem, or the virtual work equation, as the starting point for our solution, but the more general treatment starting from equations such as (2-27) and (2-29) preserves a unity of approach throughout all the problems presented. By using the basic integration property of the delta operator equation (2-29) when substituted into (2-31) yields

$$-w(\xi) = \left[ G\frac{d^3 w}{dx^3} - \frac{dG}{dx}\frac{d^2 w}{dx^2} + \frac{d^2 G}{dx^2}\frac{dw}{dx} - \frac{d^3 G}{dx^3}w \right]_0^L - \int_0^L \psi_1(x)\,G\,dx \quad (2\text{-}32)$$

an equation analogous to (2-22).

We note that Eq. (2-12a) with $\Phi = 1$ is the solution of (2-29) which we require, i.e.,

$$G(x, \xi) = \lambda l^3 (2 + |\rho|^3 - 3|\rho|^2) \qquad (2\text{-}33a)$$

Differentiation of (2-33a) with respect to $x$ produces

$$\frac{dG(x, \xi)}{dx} = F(x, \xi) = 3\lambda l^3 |\rho| \{|\rho| - 2\} \qquad (2\text{-}33b)$$

$$-\frac{d^2 G(x, \xi)}{dx^2} = E(x, \xi) = 0.5l(1 - |\rho|) \qquad (2\text{-}33c)$$

$$-\frac{d^3 G(x, \xi)}{dx^3} = D(x, \xi) = -0.5(\text{sgn}\,\rho) \qquad (2\text{-}33d)$$

where $\rho = r/l$, $r = x - \xi$, and $l$ is an arbitrary distance at which the function $G(x, \xi) = 0$. For convenience we shall again use $l = L$. We shall also need the functions $G'$, $F'$, $E'$, and $D'$ which are defined as the derivatives of functions $G$, $F$,

$E$, and $D$ respectively with respect to $\xi$. Thus

$$G'(x, \xi) = \frac{dG}{d\xi} = -3\lambda l^2 \,|\rho|\,\{\,|\rho|-2\}\,(\text{sgn } \rho) \qquad (2\text{-}34a)$$

$$F'(x, \xi) = \frac{dF}{d\xi} = 6\lambda l(1 - |\rho|) = 0.5l(1 - |\rho|) \qquad (2\text{-}34b)$$

$$E'(x, \xi) = \frac{dE}{d\xi} = 0.5(\text{sgn } \rho) \qquad (2\text{-}34c)$$

$$D'(x, \xi) = \frac{dD}{d\xi} = 0 \qquad (2\text{-}34d)$$

By substituting Eqs (2-33$a$) to (2-33$d$) in Eq. (2-32) we obtain

$$-w(\xi) = [-G(x, \xi)\,s(x) + F(x, \xi)\,m(x) - E(x, \xi)\,\theta(x) + D(x, \xi)\,w(x)]_0^L$$

$$-\int_0^L G(x, \xi)\,\psi_1(x)\,dx$$

or
$$-w(\xi) = [-G(L, \xi)\,s(L) + G(0, \xi)\,s(0) + F(L, \xi)\,m(L) - F(0, \xi)\,m(0)]$$

$$+[-E(L, \xi)\,\theta(L) + E(0, \xi)\,\theta(0) + D(L, \xi)\,w(L) - D(0, \xi)\,w(0)]$$

$$-\int_0^L G(x, \xi)\,\psi_1(x)\,dx \qquad (2\text{-}35)$$

Equation (2-35) provides the transverse displacement at any interior point $\xi$ due to boundary values of shears, moments, rotations, and displacements and the specified internal distribution of transverse loading intensity $\psi_1(x)$. By differentiating Eq. (2-35) with respect to the field point $\xi$ and using Eq. (2-34$a$) to (2-34$d$) we can obtain

$$-\theta(\xi) = [-G'(L, \xi)\,s(L) + G'(0, \xi)\,s(0) + F'(L, \xi)\,m(L) - F'(0, \xi)\,m(0)]$$

$$+[-E'(L, \xi)\,\theta(L) + E'(0, \xi)\,\theta(0) + D'(L, \xi)\,w(L) - D'(0, \xi)\,w(0)]$$

$$-\int_0^L G'(x, \xi)\,\psi_1(x)\,dx \qquad (2\text{-}36)$$

where the functions $G'(L, \xi)$, $G'(0, \xi)$, etc., are the derivatives of $G(L, \xi)$, $G(0, \xi)$, etc., with respect to $\xi$.

If we wish to incorporate an applied point moment at $x_o$, say $\psi_2(x_o)$, then this can be done mostly simply by representing it as the limiting case of two opposed transverse point loadings as shown in Fig. 2-11. If, as before, we take the field point $\xi$ to the boundaries such that $\xi = 0 + \varepsilon$ and $\xi = L - \varepsilon$ we can use Eqs (2-35)

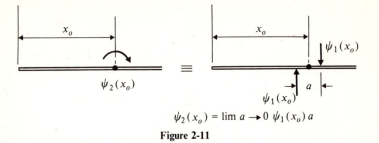

$$\psi_2(x_o) = \lim a \to 0 \; \psi_1(x_o) a$$

**Figure 2-11**

and (2-36) to interrelate all the boundary data and the applied loading via

$$
-\begin{Bmatrix} w(L-\varepsilon) \\ w(0+\varepsilon) \\ \theta(L-\varepsilon) \\ \theta(0+\varepsilon) \end{Bmatrix}
$$

$$
=\begin{bmatrix} -G(L,L-\varepsilon) & G(0,L-\varepsilon) & F(L,-\varepsilon) & -F(0,L-\varepsilon) \\ -G(L,0+\varepsilon) & G(0,0+\varepsilon) & F(L,0+\varepsilon) & -F(0,0+\varepsilon) \\ -G'(L,L-\varepsilon) & G'(0,L-\varepsilon) & F'(L,L-\varepsilon) & -F'(0,L-\varepsilon) \\ -G'(L,0+\varepsilon) & G'(0,0+\varepsilon) & F'(L,0+\varepsilon) & -F'(0,0+\varepsilon) \end{bmatrix}
\begin{Bmatrix} s(L) \\ s(0) \\ m(L) \\ m(0) \end{Bmatrix}
$$

$$
+\begin{bmatrix} D(L,L-\varepsilon) & -D(0,L-\varepsilon) & -E(L,L-\varepsilon) & E(0,L-\varepsilon) \\ D(L,0+\varepsilon) & -D(0,0+\varepsilon) & -E(L,0+\varepsilon) & E(0,0+\varepsilon) \\ D'(L,L-\varepsilon) & -D'(0,L-\varepsilon) & -E'(L,L-\varepsilon) & E'(0,L-\varepsilon) \\ D'(L,0+\varepsilon) & -D'(0,0+\varepsilon) & -E'(L,0+\varepsilon) & E'(0,0+\varepsilon) \end{bmatrix}
\begin{Bmatrix} w(L) \\ w(0) \\ \theta(L) \\ \theta(0) \end{Bmatrix}
$$

$$
-\begin{Bmatrix} \displaystyle\int_0^L G(x,L-\varepsilon)\,\psi_1(x)\,dx \\[2mm] \displaystyle\int_0^L G(x,0+\varepsilon)\,\psi_1(x)\,dx \\[2mm] \displaystyle\int_0^L G'(x,L-\varepsilon)\,\psi_1(x)\,dx \\[2mm] \displaystyle\int_0^L G'(x,0+\varepsilon)\,\psi_1(x)\,dx \end{Bmatrix}
\tag{2-37}
$$

Equation (2-37) summarizes four equations which interconnect the values of the eight boundary condition parameters $(w, \theta, m, s$ at $x = 0$ and $x = L)$ and the specified applied loading intensity $\psi_1(x)$. In a well-posed problem four of the boundary condition parameters will be specified and Eq. (2-37) can be used to determine the others.

The reader is once more advised to study the form of Eq. (2-37). Although the whole procedure is most unattractive for simple one-dimensional problems it is important to appreciate that equations such as (2-37) [and (2-26a)] can be assembled systematically and that the components of the matrices are merely numerical values calculated by simple coordinate substitutions in the various singular solutions $(E, F, ...; E', F', ...)$. One of the basic attractions of BEM is that, no matter how complicated the geometry of the problem, the boundary conditions on each boundary element can be interrelated by sets of equations of this kind, all assembled by a systematic procedure incorporating the fundamental unit solutions, so that, although the complexity of the problem may increase enormously, the complexity of the underlying solution algorithm does not.

If we now return to our problem of the propped cantilever for which the external loading is zero and $w(0) = \theta(0) = 0$, $w(L) = w^*$, $m(L) = 0$, we can write Eq. (2-37) as

$$
\begin{bmatrix}
-G(L, L-\varepsilon) & G(0, L-\varepsilon) & -F(0, L-\varepsilon) & -E(L, L-\varepsilon) \\
-G(L, 0+\varepsilon) & G(0, 0+\varepsilon) & -F(0, 0+\varepsilon) & -E(L, 0\rightarrow\varepsilon) \\
-G'(L, L-\varepsilon) & G'(0, L-\varepsilon) & -F'(0, L-\varepsilon) & 1-E'(L, L-\varepsilon) \\
-G'(L, 0+\varepsilon) & G'(0, 0+\varepsilon) & -F'(0, 0+\varepsilon) & -E'(L, 0+\varepsilon)
\end{bmatrix}
\begin{Bmatrix}
s(L) \\
s(0) \\
m(0) \\
\theta(L)
\end{Bmatrix}
$$

$$
+
\begin{bmatrix}
[1 + D(L, L-\varepsilon)] w^* \\
[D(L, 0+\varepsilon)] w^* \\
[D'(L, L-\varepsilon)] w^* \\
[D'(L, 0+\varepsilon)] w^*
\end{bmatrix}
= 0
$$

By substituting values from Eqs (2-33a) to (2-33d) and (2-34a) to (2-34d) we have

$$
\begin{bmatrix}
-2\lambda L^3 & 0 & -3\lambda L^2 & -0.5L \\
0 & -2\lambda L^3 & 0 & 0 \\
0 & -3\lambda L^2 & 0 & 0.5 \\
-3\lambda L^2 & 0 & -0.5L & -0.5
\end{bmatrix}
\begin{Bmatrix}
s(L) \\
s(0) \\
m(0) \\
\theta(L)
\end{Bmatrix}
= -
\begin{Bmatrix}
0.5w^* \\
-0.5w^* \\
0 \\
0
\end{Bmatrix}
$$

which can be rearranged to eliminate the zero diagonal component to produce

$$
\begin{bmatrix}
-2\lambda L^3 & 0 & -3\lambda L^2 & -0.5L \\
0 & -2\lambda L^3 & 0 & 0 \\
-3\lambda L^2 & 0 & -0.5L & -0.5 \\
0 & -3\lambda L^2 & 0 & 0.5
\end{bmatrix}
\begin{Bmatrix}
s(L) \\
s(0) \\
m(0) \\
\theta(L)
\end{Bmatrix}
= -
\begin{Bmatrix}
-0.5w^* \\
0.5w^* \\
0 \\
0
\end{Bmatrix}
$$

the solution of which is

$$
s(L) = \frac{w^*}{4\lambda L^3} \qquad s(0) = \frac{w^*}{4\lambda L^3} \qquad m(0) = -\frac{w^*}{4\lambda L^2} \qquad \text{and} \qquad \theta(L) = \frac{3w^*}{2L}
$$

which of course agrees with the solution obtained by the indirect method.

Now that the values of all eight boundary parameters have been calculated the displacements and rotations at any interior point $\xi$ can be evaluated by backsubstituting them into Eqs (2-35) and (2-36). The values of moment $m(\xi)$ and shear $s(\xi)$ at the same point $\xi$ can also be obtained by backsubstituting the boundary values into expressions similar to Eq. (2-36) derived by differentiating Eq. (2-36) further with respect to the field variable $\xi$ [that is, $m(\xi) = d\theta(\xi)/d\xi$ and $s(\xi) = dm(\xi)/d\xi$].

## 2-5 COMPARISONS BETWEEN DIRECT AND INDIRECT BOUNDARY ELEMENT METHODS

It is probably rather evident to the reader by now that although the basic solutions utilized in the direct and indirect formulations are the same (i.e., the solutions for unit excitation in an infinite region) there are some very significant differences between the two approaches. These are summarized below:

1. Whereas we were able to construct a solution algorithm for the indirect BEM using simple physical reasoning, together with the concept of a 'fictitious' system embedded in an infinite region, the direct method required a more sophisticated approach which proved to be closely related to the use of 'integral identities'[6] [e.g., Green's second identity, Eq. (2-20), and Betti's reciprocal theorem, Eq. (2-30)]. Nevertheless, both methods utilized the same infinite region unit solutions to generate the components of the matrix kernels in the final systems of equations. If, however, the developments are followed through closely the reader will find that the roles of $x$ and $\xi$ are reversed in the two analyses. Thus, in the indirect BEM, $p(x)$, $w(x)$, etc., are being evaluated whereas in the direct formulation operations such as

$$\int_0^L p(x)\,\delta(x,\xi)\,dx = p(\xi)$$

cause the subsequent solution to be developed in terms of $p(\xi)$, $w(\xi)$, etc. Since, in general, expressions such as $dG/dx$ and $dG/d\xi$, for example, will be quite different the two methods may diverge more than one might superficially expect, as will be seen later.

2. A further consequence of this is that rather more computational effort is usually required to assemble the final equations for the direct BEM. Although the solution does have the advantage of yielding the real, rather than 'fictitious', boundary parameter values this is partly offset by the increase in computation required to evaluate solutions at interior points.

3. The direct method does not require special consideration of cases where $G(x,\xi)$ does not satisfy the infinite boundary 'zero radiation condition', although this minor modification to the indirect method does not arise in three-dimensional problems—nor indeed in all one- and two-dimensional situations. For example, the fundamental solution in one-dimensional 'spring-supported

beam systems'[7] does satisfy equilibrium conditions at infinity and the augmented indirect method equations are not needed for such problems.

4. One might question at this stage whether or not both the indirect (IBEM) and direct (DBEM) are indeed equally rigorous methods of analysis since the former has usually been introduced in a rather intuitive fashion. It will be shown, more appropriately in Chapter 3, that they are indeed formally equivalent procedures.

## 2-6 CONCLUDING REMARKS

In this chapter we have demonstrated explicit solutions of two one-dimensional problems by means of both the indirect and direct BEM. Our objective was to show that the underlying ideas, closely related to influence functions and Green's function methods, are rather simple and well known. Each formulation involves a systematic series of steps which will be used again and again throughout the book, and the reader is strongly advised to master these before proceeding further.

It is clearly our intention to show in subsequent chapters how these basic ideas can be extended to solve increasingly complicated problems with surprisingly little modification. The final result is an elegant, powerful, and practical solution technique which has already been applied to problems in steady state and transient potential flow, elastostatics, elastodynamics, elastoplasticity and fluid mechanics, etc., in any number of space dimensions.

Although the problems solved in Chapter 2 were selected to illustrate the key features of the two alternative methods we are well aware that neither inhomogeneity nor anisotropy have yet been introduced. This will be rectified in the next chapter in connection with the solution of general two-dimensional steady state potential flow problems. It is our hope that this chapter has, at one and the same time, encouraged the reader by emphasizing the underlying simplicity of BEM and provided him with a good grasp of the physical basis of the solution procedures.

## 2-7 REFERENCES

1. Mikhlin, S. G. (1957) *Integral Equations*, Pergamon Press, Oxford.
2. Sommerfield, A. (1949) *Partial Differential Equations in Physics*, Academic Press, New York.
3. Hughes, W. F., and Gaylord, W. (1964) *Basic Equations of Engineering Science*, McGraw-Hill, New York.
4. Greenberg, M. D. (1971) *Application of Green's Function in Science and Engineering*, Prentice-Hall, Englewood Cliffs, N.J.
5. Wylie (Jr), C. R. (1960) *Advanced Engineering Mathematics*, 3rd ed., McGraw-Hill, New York.
6. Malvern, L. E. (1969) *Introduction to the Mechanics of Continuous Medium*, Prentice-Hall, Englewood Cliffs, N.J.
7. Hetenyi, M. (1946) *Beams on Elastic Foundation*, University of Michigan Press.
8. Betti, E. (1872) 'Teori dell elasticita', *Il Nuovo Ciemento*, 7–10.

# THREE
## TWO-DIMENSIONAL PROBLEMS OF STEADY STATE POTENTIAL FLOW

### 3-1 INTRODUCTION

The main objective of our book is to guide the reader through classes of important problems which can be solved efficiently by BEM and which become progressively more complicated, either by virtue of their dimensionality, their governing equations, or purely by the use of higher-order numerical discretization schemes.

This chapter is the first of these steps and in it linear two-dimensional steady state potential flow is to be analysed. That is, the flow is in two space dimensions, the governing equations are linear, and analysis of the system at any time instant produces identical solutions. The governing partial differential equations for such problems are elliptical (Laplace's equation of Poisson's equation) and relate to the simplest mathematical models of hydraulic flow and the flow of electricity, heat, etc. In each of these the differential equation has to be satisfied by a potential function $p$ (electrical or hydraulic potential or temperature), the spatial gradient of which is linearly related by a conductivity or permeability parameter to a flux or flow (electrical current density, fluid flow velocity, or heat flux respectively).

Although throughout this chapter, and subsequent ones, the singular solutions become progressively more cumbersome algebraically, the basic steps in the solution procedure remain identical to those explained in detail in Chapter 1.

The idea of formulating the solution to problems in potential theory in terms of integral equations appears in early texts,[1] but it is only very much more recently that this approach has been translated into a general numerical solution procedure. Jaswon[2] and Symm[3] published a semi-direct boundary element algorithm for two-dimensional potential flow and Jaswon and Ponter[4] an analogous formulation of the shaft torsion problem. This was also discussed more fully in relation to each of the direct, semi-direct, and indirect boundary element methods by Mendleson.[5] The indirect method was first extended to encompass zoned anisotropic media by Butterfield and Tomlin[6-8] while Niwa Kobayashi, and Fukui[9] published a direct BEM solution of free-surface seepage problems. At the same time separate but parallel developments were taking place in the field of heat conduction (e.g., Chang, Kang, and Chen[10]). All of these applications are encompassed in the indirect and direct BEM algorithms established in this chapter.

## 3-2 GOVERNING EQUATIONS

The general problem which we are to solve can be posed as follows (Fig. 3-1). A two-dimensional homogeneous region ($D^*$) with isotropic permeability ($\kappa$) is bounded by a surface ($S^*$), over one portion of which the boundary potential ($p^*$) is specified and the normal velocity (flux) component is ($u^*$) over the remainder of $S^*$. Sources, or sinks, of known intensity $\psi$ per unit area may also exist within $D^*$.

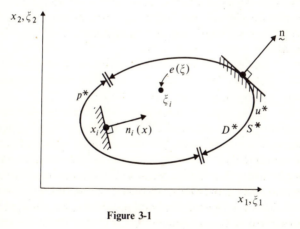

**Figure 3-1**

We are to determine the flow velocity and the potential existing at any specified point within $D^*$ or on $S^*$.

Referred to a rectangular cartesian coordinate system ($x_1, x_2$; or $x_i, i = 1, 2$) the continuity equation for the flow at all points, other than sources or sinks, is

Laplace's equation which has to be satisfied by the potential $p(x)$ at any point $(x_i)$:

$$\frac{\partial^2 p(x)}{\partial x_1^2} + \frac{\partial^2 p(x)}{\partial x_2^2} = \frac{\partial^2 p(x)}{\partial x_i \partial x_i} = 0 \qquad (3\text{-}1)$$

The corresponding flow 'velocity' vector components $v_i(x)$ are given variously by Darcy's law, Ohm's law, etc., as

$$v_i(x) = -\kappa \frac{\partial p(x)}{\partial x_i} \qquad (3\text{-}2)$$

We are to solve Eq. (3-1) within $D^*$ subject to the specified boundary conditions on $S^*$. Thus if all the boundary potentials were given around $S^*$ (the Dirichlet problem) we would have, say,

$$p^*(x_o) = g(x_o) \qquad x_o \text{ on } S^* \qquad (3\text{-}3a)$$

or if all the velocity components normal to the boundary were specified (the Neumann problem) we would have, say,

$$u^*(x_o) = h(x_o) \qquad x_o \text{ on } S^* \qquad (3\text{-}3b)$$

where

$$u^*(x_o) = v_1(x_o) n_1(x_o) + v_2(x_o) n_2(x_o) = v_i(x_o) n_i(x_o)$$

with $n_i(x_o)$ the components of the outward normal unit vector at $(x_o)$. The more general, mixed boundary-value problem would have either $p^*$ or $u^*$ specified on each portion of $S^*$.

Before considering our fundamental solution of Eq. (3-1) we should note that all the following analysis is directly applicable to homogeneous anisotropic regions for which the permeability will be a second-rank tensor quantity $\kappa_{ij}$ (see Appendix A). The generalization of the flux equation becomes

$$v_i(x) = -\kappa_{ij} \frac{\partial p(x)}{\partial x_j}$$

which leads to the continuity equation, via $\partial v_i(x)/\partial x_i = 0$,

$$\kappa_{ij} \frac{\partial^2 p(x)}{\partial x_i \partial x_j} = 0$$

for spatially constant $\kappa_{ij}$.

Here we are using Einstein's indicial summation convention on $i$ and $j$. In order to keep various expressions as concise as possible this convention will insinuate itself more and more as the book progresses. Anyone unfamiliar with it will find a brief explanation in Appendix A.

However, if we ensure that the axes of $x_i$ are directed along the principal axes of $\kappa_{ij}$ (that is, $\kappa_{ij}$ is diagonalized with principal permeabilities of $\kappa_1$ and $\kappa_2$, say) this equation simplifies to

$$\kappa_1 \frac{\partial p(x)}{\partial x_1^2} + \kappa_2 \frac{\partial p(x)}{\partial x_2^2} = 0$$

whence by substituting scaled coordinates $(\eta_i)$ such that $[\eta_1 = x_1,$ $\eta_2 = x_2 \sqrt{(\kappa_1/\kappa_2)}]$ Eq. (3-1) is recovered as $\partial^2 p(\eta)/\partial \eta_i \partial \eta_i = 0$ and the problem has been transformed into an isotropic one with $\kappa = \sqrt{(\kappa_1 \kappa_2)}$. If this prior operation is assumed then the following analyses cover problems involving both isotropic and anisotropic regions.

## 3-3 SINGULAR SOLUTIONS

The fundamental solution of Eq. (3-1), which is a basic component of all the subsequent analysis, relates the potential $p(x)$ generated at a 'field point' $(x_i)$ by a unit source $e(\xi)$, say, applied at a load point $(\xi_i)$. Although the origin for the coordinates $(\xi_i)$ is identical to that for $(x_i)$ it is absolutely essential that each set be reserved for a specific purpose. Thus the classical singular solution can be written as

$$p(x) = G(x, \xi) e(\xi) \tag{3-4}$$

where

$$G(x, \xi) = -\frac{1}{2\pi\kappa} \ln \left| \frac{r}{r_o} \right|$$

with $r_o$ a constant such that $G = 0$ at $r = r_o$ and

$$r^2 = (x_1 - \xi_1)^2 + (x_2 - \xi_2)^2 = (x - \xi)_i (x - \xi)_i = y_i y_i$$

if $y_i = (x - \xi)_i$. $G(x, \xi)$ is a 'two-point' function involving the coordinates $(x, \xi)$ of two points and since we shall sometimes need to differentiate, or integrate, $G$ with respect to $x$ and sometimes with respect to $\xi$, a distinction must be preserved between the two arguments.

Equation (3-4) only specifies the potential $p(x)$ *relative* to a zero datum at $r = r_o$, a consequence of which is that $p$ does not go to zero as $r \to \infty$. This 'logarithmic behaviour at infinity' means that in the indirect BEM the augmented equations will be required (see Sec. 2-3-1).

By differentiating (3-3) with respect to $x_i$ we obtain the flux (velocity) vector components $v_i(x)$:

$$v_i(x) = -\kappa \frac{\partial p(x)}{\partial x_i} = \frac{1}{2\pi} \frac{y_i}{r^2} e(\xi) \tag{3-5}$$

If $n_i(x)$ are the components of a unit vector at $x_i$ defining the outward normal direction to a line element through $(x_i)$ then the velocity $u(x)$ along $n$ is then

$$u(x) = v_i n_i = \frac{y_i n_i}{2\pi r^2} e(\xi) = \frac{n_1 y_1 + n_2 y_2}{2\pi r^2} e(\xi) \tag{3-6a}$$

or

$$u(x) = F(x, \xi) e(\xi) \qquad \text{say}$$

where

$$F(x, \xi) = \frac{(x - \xi)_i \, n_i(x)}{2\pi r^2} = \frac{y_i \, n_i(x)}{2\pi r^2} \tag{3-6b}$$

It is worth noting that whereas $G(x, \xi)$ is symmetrical in its arguments $(x, \xi)$, $F(x, \xi)$ is antisymmetrical and changes sign when $x$ and $\xi$ are interchanged. Also as $x_i \rightarrow \xi_i$ (i.e., as the field and load points coalesce) so both the singular expressions, Eqs (3-4) and (3-6), for $p$ and $u$ become infinite. The first expression, involving $\ln r$, is only 'weakly singular' when integrated along a line element and it will be found that the singularity is suppressed when such a function is integrated across it in the normal way. However, the second expression, $F(x, \xi)$, involving a singularity of order $1/r$ is 'strongly singular' when integrated along a line through the singular point and cannot be integrated successfully in the same way. Such singular functions play a major role in all BEM and, as will be seen later, ensure the diagonal dominance of the resulting matrix equations.

## 3-4 INDIRECT BOUNDARY ELEMENT FORMULATION FOR A HOMOGENEOUS REGION

In order to keep the conceptual model clear we show the 'fictitious' system related to the real system in Fig. 3-2 (that is, $A$ and $S$ are in every way identical to $A^*$ and $S^*$ except that they are embedded in an infinite two-dimensional region of our material). We shall designate field points on $S$ by $(x_o)_i$ and again require that the boundary potentials $p$ and velocities $u$ on $S$ be precisely those specified on $S^*$. Thus

$$p^*(x_o) \equiv p(x_o) \qquad \text{and} \qquad u^*(x_o) \equiv u(x_o)$$

To complete the problem specification known sources $\psi(z)$ per unit area of $A$ (and $A^*$) will also be included. For clarity a separate coordinate $(z_i)$ analogous to

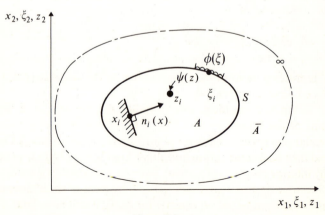

**Figure 3-2**

$\xi$ will be used to locate them with the origin for $z$ coinciding with that for $x$ and $\xi$. We now use $(\xi_i)$ to locate load (source) points on $S$ and introduce fictitious sources of initially unknown intensity $\phi(\xi)$ per unit length of $S$. The nett result of both the $\phi$ and $\psi$ source distributions at any field point $(x_i)$ (i.e., the values of the potential $p$ and any directed flux component $u$ generated there) can be deduced by integrating the unit solutions over $S$ and $A$ respectively. Thus to obtain $p(x)$ we substitute $\phi(\xi)$ for $e(\xi)$ in Eq. (3-4), integrate over $S$, and add to it the result of substituting $\psi(z)$ for $e(\xi)$, also in Eq. (3-4), integrating this over the domain $A$ to produce

$$p(x) = \int_S G(x, \xi)\, \phi(\xi)\, dS(\xi) + \int_A G(x, z)\, \psi(z)\, dA(z) + C \qquad (3\text{-}7)$$

where the constant $C$ arises from the second term in $G(x, \xi)$, Eq. (3-4). As in the one-dimensional example, in Sec. 2-2-1, $C$ will eventually take on the specific value which ensures that the nett 'radiation' across the infinite boundary will be zero, which is necessary to guarantee the uniqueness of our solution. This will be achieved when

$$\int_S \phi(\xi)\, dS(\xi) + \int_A \psi(z)\, dA(z) = 0$$

Precisely similar operations using $F(x, \xi)$, etc., will yield the normal boundary velocity components across $S$ via

$$u(x) = \int_S F(x, \xi)\, \phi(\xi)\, dS(\xi) + \int_A F(x, z)\, \psi(z)\, dA(z) \qquad (3\text{-}8)$$

In principle the only remaining formal step to arrive at a 'solution' to the problem is to bring the point $(x_i)$ onto the boundary $S$ (that is, $x \to x_o$), whence Eqs (3-7) and (3-8) become

$$p(x_o) = \int_S G(x_o, \xi)\, \phi(\xi)\, dS(\xi) + \int_A G(x, z)\, \psi(z)\, dA(z) + C \qquad (3\text{-}9)$$

$$u(x_o) = \oint_S F(x_o, \xi)\, \phi(\xi)\, dS(\xi) + \int_A F(x_o, z)\, \psi(z)\, dA(z) \qquad (3\text{-}10)$$

where $\oint$ represents an improper integral due to the singularity in $F$ as $\xi \to x_o$. This will be discussed more fully later.

In a 'well-posed' problem, one of either $p(x_o)$ or $u(x_o)$ will be known at every point of the boundary and Eqs (3-9) and (3-10) are two simultaneous integral equations which can be solved for the only unknown $\phi(\xi)$. Once $\phi(\xi)$ has been found then backsubstitution into either Eqs (3-7) or (3-8) will produce $p(x)$ or $u(x)$ at any interior point $(x_i)$ of interest. Before we proceed to show how Eqs (3-9) and (3-10) can be approximated numerically by a set of simultaneous algebraic equations [which can be solved directly for $\phi(\xi)$], it will be helpful if we look at Eqs (3-9) and (3-10) a little more closely.

First, they are scalar integral equations since all the kernel functions $(G, F)$, the sources $(\phi, \psi)$, and therefore also $(p, u)$ are scalar quantities. In contrast, if we wish to calculate the velocity vector components $v_i(x)$ we can write Eq. (3-5) as

$$v_i(x) = \frac{y_i}{2\pi r^2} e(\xi) = H_i(x, \xi) e(\xi)$$

and integrate over $S$ and $A$ again to obtain

$$v_i(x) = \int_S H_i(x, \xi) \phi(\xi) \, dS(\xi) + \int_A H_i(x, \xi) \psi(z) \, dA(z) \qquad (3\text{-}11)$$

Once $\phi(\xi)$ has been found then $v_i(x)$ can be calculated at any point by carrying out these integrations. Equation (3-11) necessarily has a vector kernel function $(H_i)$ to yield the vector components $(v_i)$ and is therefore a multidimensional integral equation of the first rank. In elasticity problems the integral equations which appear involve kernel functions of higher rank, e.g., second-rank kernels of the form $G_{ij}(x, \xi)$, etc.

Furthermore, the integral equations (3-9) and (3-10) are singular in that, within the ranges of integration involved (i.e., over $S$ and $A$), each of the kernel functions $(G, F)$ become infinite at some point. The terms involving integration over $A$, which have singularities when $z = x_o$, are not really troublesome because it will be found that although $1/r$ is a strong singularity when integrated over a line element $(S)$ it becomes 'weak', and therefore not a problem, when integrated over an area $(A)$ as in Eq. (3-10). On this basis we would anticipate that all the integrals (or numerical summations) required in Eqs (3-9) and (3-10) can be evaluated in the normal (Lebesgue) sense with the exception of the first, line integral in Eq. (3-10) which has a strong, $1/r$, singularity when $\xi = x_o$. Consequently, this term has to be treated as a Cauchy principal value integral with an added 'free term' from the singularity.

In essence this means that normal integration is carried out around $S$ with the exception of a small neighbourhood, $\pm \varepsilon$ along $S$, of $x_o$ and the limiting value of this process as $\varepsilon \to 0$ is defined as the Cauchy principal value of the integral. The value of the free term is, in fact, obvious in some cases on physical grounds. For example, in our problem when $\xi = x_o$ the point source $\phi(\xi)$ bifurcates half into the interior region $(A)$ and half into the exterior region $(\bar{A})$. Its contribution to $u(x)$ as $x$ approaches $x_o$ from inside $A$ is therefore clearly $-\frac{1}{2}\phi(x_o)$, provided that $x_o$ is not located on any corner of $S$ (that is, $x_o$ on $S$ must possess a unique tangent direction). The value of the 'free term' on corners is discussed later. We can now rewrite Eq (3-10) replacing the special integral ($\oint$) by a Cauchy integral over $S$ together with the $-\frac{1}{2}\phi(x_o)$ term as

$$u(x_o) = -\tfrac{1}{2}\phi(x)_o + \int_S F(x_o, \xi) \phi(\xi) \, dS(\xi) + \int_A F(x_o, z) \psi(z) \, dA(z) \qquad (3\text{-}12)$$

where the first integral has to be interpreted as a Cauchy integral and the second as a conventional one.

In order to carry out the integration and develop a numerical solution to our problem, which is principally the determination of $\phi(\xi)$ from Eqs (3-9) and (3-12), we shall need to discretize the surface $S$ of the region and, if there are internal sources $\psi$, also the internal area $A$.

### 3-4-1 Discretization of the Surface and Volume Integrals

If we were able to integrate Eqs (3-9) and (3-12) in closed form and solve them for $\phi(\xi)$ then our solution would be exact; this is virtually impossible in practical problems and approximations have to be introduced. Therefore in BEM inaccuracies arise from, and only from, numerical discretization and integration procedures which means that by refining these approximations any degree of precision is theoretically achievable. In practice there has to be a trade-off between computing time and effort and solution accuracy. The following algorithm is probably the simplest that will provide worthwhile practical results. In addition to producing accurate solutions to theoretical test cases it has also been used to solve quite ambitious problems in groundwater flow (Sec. 3-9).

This discretization scheme utilizes linear boundary elements, characterized by their midpoints, along any one of which, say the $q$th element, the fictitious source $\phi(\xi^q)$ is uniformly distributed. For simplicity we shall also consider the $\psi$ sources to be uniformly distributed over triangular elements (of any convenient size) and characterized by $\psi(z^l)$ on the $l$th of them. If we approximate $S$ and $A$ by $N$ linear boundary segments and $M$ triangular internal cells, as in Fig. 3-3, we can write discrete approximations to Eqs (3-9) and (3-12) for $p(x_o^p)$ and $u(x_o^p)$, the potential and normal velocity components on the $p$th boundary element, as

$$p(x_o^p) = \sum_{q=1}^{N} \phi(\xi^q) \int_{\Delta S} G(x_o^p, \xi^q)\, dS(\xi^q)$$

$$+ \sum_{l=1}^{M} \psi(z^l) \int_{\Delta A} G(x_o^p, z^l)\, dA(z^l) + C \qquad (3\text{-}13)$$

$$u(x_o^p) = -\tfrac{1}{2}\phi(x_o^p) + \sum_{q=1}^{N} \phi(\xi^q) \int_{\Delta S} F(x_o^p, \xi^q)\, dS(\xi^q)$$

$$+ \sum_{l=1}^{M} \psi(z^l) \int_{\Delta A} F(x_o^p, z^l)\, dA(z^l) \qquad (3\text{-}14)$$

where $(x_o^p)_i$ are the coordinates of the midpoint of the $p$th boundary element, $\Delta S$ the length of the $q$th boundary element, and $\Delta A$ the area of the $l$th internal cell.

### 3-4-2 Formation of the System Matrices

Equation (3-13) represents the potential generated at the midpoint (centroid) of boundary element $p$ by the totality of all sources $\phi$ and $\psi$. Equation (3-14) provides the corresponding outward velocity component normal to element $p$.

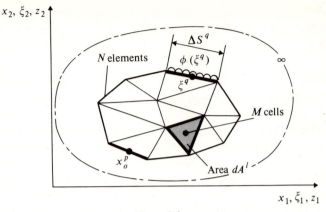

**Figure 3-3**

We have deliberately allowed the notation to become quite elaborate so that the roles of the various coordinates, the location of field and load points, etc., are defined unambiguously. Once this has been achieved then it becomes very much more convenient to rewrite Eqs (3-13) and (3-14) in matrix notation as

$$p^p = \left( \int_{\Delta S} G^{pq}\, dS \right)\boldsymbol{\phi}^q + \left( \int_{\Delta A} G^{pl}\, dA \right)\boldsymbol{\psi}^l + C \qquad (3\text{-}15)$$

$$u^p = -\tfrac{1}{2}\phi^p + \left( \int_{\Delta S} F^{pq}\, dS \right)\boldsymbol{\phi}^q + \left( \int_{\Delta A} F^{pl}\, dA \right)\boldsymbol{\psi}^l \qquad (3\text{-}16)$$

where $\boldsymbol{\phi}^q$ is an $N \times 1$ column vector, $\boldsymbol{\psi}^l$ an $M \times 1$ vector, and the terms in brackets are compatible row vectors. Each term in these row vectors is the result of integrating, for example, the $G$ kernel function over $\Delta S$, etc. Thus the first term in the first row in Eq. (3-15) would be

$$G^{p1} = \int_{\Delta S_1} G(x_o^p, \xi^1)\, dS(\xi^1).$$

The detailed evaluation of these intermediate integrals will be discussed later; for the moment we note that the final form of Eqs (3-15) and (3-16) will always be equivalent to

$$p^p = (G^{pq})\,\boldsymbol{\phi}^q + (G^{pl})\,\boldsymbol{\psi}^l + C \qquad (3\text{-}17)$$

$$u^p = (F^{pq})\,\boldsymbol{\phi}^q + (F^{pl})\,\boldsymbol{\psi}^l \qquad (3\text{-}18)$$

where the $-\tfrac{1}{2}\phi^p$ term in Eq. (3-16) has now been added into the $(p = q)$ term of $(F^{pq})$.

If precisely similar operations are carried out for all elements $p(p = 1, 2, \cdots, N)$ the total sets of all such equations for **p** and **u** can be assembled to yield simply

$$\mathbf{p} = \mathbf{G}^S \boldsymbol{\phi} + \mathbf{G}^A \boldsymbol{\psi} + \mathbf{I}C \tag{3-19}$$

$$\mathbf{u} = \mathbf{F}^S \boldsymbol{\phi} + \mathbf{F}^A \boldsymbol{\psi} \tag{3-20}$$

where clearly $\mathbf{p}$, $\mathbf{u}$, and $\boldsymbol{\phi}$ are $N \times 1$ vectors of surface quantities and $\boldsymbol{\psi}$ is an $M \times 1$ vector of sources within $A$.

$\mathbf{G}^S$, $\mathbf{F}^S$ are $N \times N$ matrices which, being derived from integrals of $\mathbf{G}^{pq}$, etc., over the boundary $S$, are distinguished by the superfix $S$ from the matrices $\mathbf{G}^A$, $\mathbf{F}^A$ which are $N \times M$ matrices with an $A$ superfix denoting that they originated from integrals of $G^{pl}$, etc., over area elements $\Delta A$. $\mathbf{I}$ is an $N \times 1$ unit column vector.

Before assembling Eqs (3-19) and (3-20) into a global set we need to return to our discussion of $C$. As mentioned previously, $C$ has arisen as an arbitrary parameter related to our freedom of choice of datum from which to measure the potentials. If we so chose $C$ that the algebraic sum of all sources, $\phi$ and $\psi$, applied within $A$ and over $S$ is zero then we shall have overcome the intrinsic problem that the $\ln(r)$ terms within the $G$ kernel functions do not go to zero as $r \to \infty$.

We therefore need an auxiliary equation to ensure this: viz.,

$$\sum_{q=1}^{N} \int_{\Delta S} \phi(\xi^q)\, dS(\xi^q) + \sum_{l=1}^{M} \int_{\Delta A} \psi(z^l)\, dA(z^l) = \sum_{q=1}^{N} (\phi^q \Delta S) + \sum_{l=1}^{M} (\psi^l \Delta A) = 0$$

If both $\phi$ and $\psi$ are uniformly distributed over boundary elements and internal cells then this equation simplifies to, say,

$$(\mathbf{b}_n)\boldsymbol{\phi} + (\mathbf{b}_m)\boldsymbol{\psi} = 0 \tag{3-21}$$

where $(\mathbf{b}_n)$ and $(\mathbf{b}_m)$ are therefore $N \times 1$ and $M \times 1$ row vectors respectively whose components are merely element lengths, in $(\mathbf{b}_n)$, and cell areas, in $(\mathbf{b}_m)$.

When we assemble a global set of equations from (3-19), (3-20), and (3-21) to solve a specific problem we shall not require every component equation. In general, boundary potentials will only be specified on some, $\mathbf{p}^S$ say, of the total set of $\mathbf{p}$ boundary potentials; similarly, the normal velocity will be specified on $\mathbf{u}^S$ of the total $\mathbf{u}$ set. However, the total number of components in $\mathbf{p}^S$ and $\mathbf{u}^S$ together will always be $N$. We therefore select from (3-19) and (3-20) just the $N$ equations for which boundary information has been provided and assemble them, together with (3-21), into a global set as

$$\begin{Bmatrix} \alpha \mathbf{p}^S \\ \mathbf{u}^S \\ 0 \end{Bmatrix} = \begin{bmatrix} \mathbf{I}_o & \alpha \mathbf{G}^{SS} \\ \mathbf{0} & \mathbf{F}^{SS} \\ \mathbf{0} & \mathbf{b}_n \end{bmatrix} \begin{Bmatrix} \alpha C \\ \boldsymbol{\phi} \end{Bmatrix} + \begin{bmatrix} \alpha \mathbf{G}^{SA} \\ \mathbf{F}^{SA} \\ \mathbf{b}_m \end{bmatrix} \{\boldsymbol{\psi}\} \tag{3-22}$$

$$(N+1) \qquad (N+1)\times(N+1) \quad (N+1) \quad (N+1)\times M \quad (M\times 1)$$

where $\mathbf{G}^{SS}$, $\mathbf{G}^{SA}$ are submatrices, selected from the total $\mathbf{G}^S$, $\mathbf{G}^A$ sets, corresponding to $\mathbf{p}^S$; $\mathbf{F}^{SS}$, $\mathbf{F}^{SA}$ are similarly selected from $\mathbf{F}^S$, $\mathbf{F}^A$ to correspond with $\mathbf{u}^S$; $\mathbf{I}_o$, $\mathbf{0}$ are unit and zero column vectors with numbers of components corresponding to $\mathbf{p}^S$ and $\mathbf{u}^S$ respectively; and $\alpha$ is a scaling factor which multiplies the components shown in order to ensure that all matrix elements are of the same order of magnitude ($\alpha \approx \kappa$).

### 3-4-3 Calculation of Interior Potentials and Velocities

The only unknown member of Eq. (3-22) is $\phi$, which can be found by simple matrix algebra involving the reduction of an $(N + 1)^2$ matrix. Back substitution of $\phi$ into other equations of the (3-19), (3-20) sets will determine the remaining boundary potentials and fluxes if required. If these are not needed as part of the solution then clearly only the $\mathbf{p}^S$, $\mathbf{u}^S$ component equations of (3-19) and (3-20) ever need to be assembled. Potentials or velocities at any internal node $(x_i)$ can be calculated using $\phi$ and the discretized counterpart of Eq. (3-17). This would be identical to Eq. (3-15) except that the $p$ element superfix would now identify the particular internal point of interest at $(x_i^p)$. The velocity component at the same point in any $n_i(x^p)$ direction can be found similarly by using Eq. (3-16).

It is important to note that:

1. The solution of Eq. (3-22) for $\phi$ involves the reduction of only an $(N \times 1)^2$ matrix (i.e., the dimensions of the matrix are governed entirely by $N$, the number of boundary elements, and are quite unrelated to the number of internal subdivisions, $M$, used in the interior of the body). This will be found to be true for all specified sources and sinks, and indeed for body forces generally, in any number of space dimensions. The fact that a series of solutions can be obtained for different values of $\psi$ without repeating the reduction process will play a major role in later chapters where non-linear effects such as occur, for example, in elastoplasticity will be incorporated as 'pseudo' body forces.
2. Essentially the same algorithm that has been used to compile all the $F$, $G$ terms in Eq. (3-22) is used again to evaluate $p(x_i)$, $u(x_i)$ at selected interior points once $\phi$ has been found.

The direct BEM solution algorithm will now be developed. This again involves integrations of the singular solutions around $S$ and over $A$ leading, finally, to matrix equations relating boundary conditions, although the subsequent evaluation of $u(x_i)$ is not so straightforward as in condition 2 above.

### 3-5 DIRECT BOUNDARY ELEMENT FORMULATION FOR A HOMOGENEOUS REGION

Once more we start from the governing differential equation equivalent to (2-19) in two space dimensions:

$$\kappa \frac{\partial^2 p(x)}{\partial x_i \, \partial x_i} = -\psi(x) \tag{3-23}$$

where $\Psi(x)$ is now a specified distribution of source strengths over the domain $(A)$ of our problem. The flux at any point will again be $v_i(x) = -\kappa\,\partial p/\partial x_i$ and we shall assume that an anisotropic region will have been transformed geometrically (Sec. 3-2) into an equivalent isotropic one of permeability $\kappa$.

The following analysis is a generalization of the one developed in Sec. 2-4-1 where now $G(x,\xi)$ is a two-point function specified to be the solution of the equation

$$\kappa\frac{\partial^2\,G(x,\xi)}{\partial x_i\,\partial x_i} = -\delta(x,\xi) \tag{3-24}$$

Here $\delta(x,\xi)$ is the two-dimensional Dirac delta function which is zero unless all corresponding components of $x_i$ and $\xi_i$ are identical. When $x_i \equiv \xi_i$, $\delta(x,\xi)$ has a sifting property such that, for example,

$$\int_A p(x)\,\delta(x,\xi)\,dA(x) = p(\xi) \qquad \xi_i \in A \tag{3-25}$$

i.e., when the left-hand-side product in this equation is integrated over $A$ the value of $p(\xi)$ at the specific point $x_i = \xi_i$ is sifted out as the only non-zero resultant. In Eq. (3-24) the function $G$ has to be the ubiquitous singular solution for the potential generated at $x_i$ by a unit point source applied at the point $\xi_i$ in an infinite two-dimensional body [Eq. (3-4), repeated below]:

$$G(x,\xi) = -\frac{1}{2\pi\kappa}\ln\left|\frac{r}{r_o}\right| \qquad r^2 = (x-\xi)_i(x-\xi)_i \tag{3-4a}$$

We shall also need the function $F$, Eq. (3-6), for the velocity component $u(x)$, due to $G$, in the direction of a unit vector with components $n_i(x)$ at $x_i$:

$$F(x,\xi) = \frac{n_i(x)(x-\xi)_i}{2\pi r^2} = -\kappa n_i(x)\frac{\partial G(x,\xi)}{\partial x_i}$$

We now multiply both sides of Eq. (3-23) by $G$ and integrate by parts twice over the region $A$ (cf. Sec. 2-4-1). This will again generate an equation expressing $p(\xi)$ in terms of the boundary information on $S$ and derivatives of $G$. Since operations of this kind occur throughout the basic development of all direct BEM solutions we shall write the various steps out fully for this particular case and consider the two terms on the left-hand side of our equation separately. We therefore start from

$$\int_A\left(G\frac{\partial^2 p}{\partial x_i\,\partial x_i}\right)dA(x) = \int_A G\left(\frac{\partial^2 p}{\partial x_1\,\partial x_1} + \frac{\partial^2 p}{\partial x_2\,\partial x_2}\right)dA(x) = -\int_A \frac{G\psi}{\kappa}\,dA(x) \tag{3-26}$$

where $dA(x) = dx_1\,dx_2$ is an infinitesimal element of $A$. Integrating solely the $G(\partial^2 p/\partial x_1\,\partial x_1)$ term by parts twice and using the limits defined in Fig. 3-4 leads to

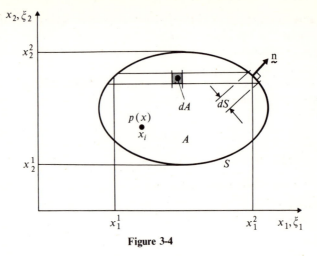

**Figure 3-4**

the following series of equations:

$$\int_A \left( G \frac{\partial^2 p}{\partial x_1 \partial x_1} \right) dA = \int_{x_2^1}^{x_2^2} \int_{x_1^1}^{x_1^2} \left( G \frac{\partial^2 p}{\partial x_1 \partial x_1} \right) dx_1 \, dx_2$$

$$= \int_{x_2^1}^{x_2^2} \left\{ \left[ G \frac{\partial p}{\partial x_1} \right]_{x_1^1}^{x_1^2} \right.$$

$$\left. - \int_{x_1^1}^{x_1^2} \left( \frac{\partial G}{\partial x_1} \frac{\partial p}{\partial x_1} \right) dx_1 \right\} dx_2$$

$$= \int_{x_2^1}^{x_2^2} \left\{ \left[ G \frac{\partial p}{\partial x_1} - \frac{\partial G}{\partial x_1} p \right]_{x_1^1}^{x_1^2} + \int_{x_1^1}^{x_1^2} \left( \frac{\partial^2 G}{\partial x_1 \partial x_1} p \right) dx_1 \right\} dx_2$$

$$(3\text{-}27)$$

The second integral in (3-27) is clearly identical to

$$\int_A p \left( \frac{\partial^2 G}{\partial x_1 \partial x_1} \right) dA$$

and the first one can be interpreted as a boundary integral over $S$ by noting that, at $(x_1^1, x_1^2)$ on $S$, $dx_2 = n_1 \, dS$ (Fig. 3-4). Equation (3-27) can therefore be written as

$$\int_A \left( G \frac{\partial^2 p}{\partial x_1 \partial x_1} \right) dA = \int_S \left( G \frac{\partial p}{\partial x_1} - p \frac{\partial G}{\partial x_1} \right) n_1 \, dS + \int_A p \left( \frac{\partial^2 G}{\partial x_1 \partial x_1} \right) dA$$

If we perform exactly similar operations on the $(\partial^2 G / \partial x_2 \, \partial x_2)$ term and add the results, we obviously obtain

$$\int_A \left( G \frac{\partial^2 p}{\partial x_i \partial x_i} \right) dA = \int_S \left( G \frac{\partial p}{\partial x_i} - p \frac{\partial G}{\partial x_i} \right) n_i \, dS + \int_A \left( p \frac{\partial^2 G}{\partial x_i \partial x_i} \right) dA \quad (3\text{-}28)$$

where from Eqs (3-24) and (3-25) the last term reduces to $-p(\xi)/\kappa$.

By equating (3-26) and (3-28) we obtain, when written out fully,

$$p(\xi) = \kappa \int_S \left[ G(x,\xi) \frac{\partial p(x)}{\partial x_i} - p(x) \frac{\partial G(x,\xi)}{\partial \xi} \right] n_i(x) \, dS(x) + \int_A G(x,\xi) \, \psi(x) \, dA(x)$$

$$(3\text{-}29)$$

If, as before, we define

$$F(x,\xi) = -\kappa [\partial G(x,\xi)/\partial x_i] \, n_i(x)$$

and recall that the flux in the $n_i$ direction is

$$u(x) = -\kappa [\partial p(x)/\partial x_i] \, n_i(x)$$

then Eq. (3-29) can be written more concisely as

$$p(\xi) = \int_S [p(x) F(x,\xi) - u(x) G(x,\xi)] \, dS(x) + \int_A \psi(x) G(x,\xi) \, dA(x) \quad (3\text{-}30)$$

an equation which enables us to calculate the potential at any point $\xi$ from a knowledge of both the potential and the flux at all points around the boundary $S$ and the specified internal source distribution.

Equations (3-29) and (3-30) are essentially the two-dimensional equivalent of Eq. (2-22) and the reader may therefore speculate, correctly, that three-dimensional problems will generate an equation identical to (3-30) with the suffix range extending over 1, 2, 3, with $S$ becoming a surface area and $A$ a volume $V$.

There are some features of (3-30) which need to be appreciated before we use it in its boundary form to solve the general two-dimensional problem. These arise from the fact that the argument in the $p(\xi)$ term is now $\xi$ rather than the $x$ we have become accustomed to in the indirect BEM and therefore we shall be calculating potentials at the point $\xi_i$ (that is, $x$ and $\xi$ have been interchanged). As a consequence of this, integration is now with respect to $x$ and therefore the unit normal $n(x)$ is to be taken at each point around the boundary and *not*, as in IBEM, at the specific field point. In fact, the load/field point idea is now not at all helpful and the DBEM equation is best considered as determining $p(\xi)$ at any point ($\xi$) by a summation of effects from other points ($x$) over $S$ and within $A$.

If we now imagine the point $\xi_i$ to approach the boundary $S$ from inside $A$, where, say, $\xi = \xi_o, \xi_o \in S$, then Eq. (3-31) becomes the limiting value, as $\xi \to \xi_o$, of

$$p(\xi_o) = \oint_S p(x) F(x,\xi_o) \, dS(x) - \int_S u(x) G(x,\xi_o) \, dS(x)$$

$$+ \int_A \psi(x) G(x,\xi_o) \, dA(x) \quad (3\text{-}31)$$

This equation is again a singular scalar integral equation of the type discussed in Sec. 3-4, which interrelates all the boundary potentials $p(x)$, boundary fluxes $u(x)$, and the specified internal distribution of sources $\psi(x)$. All the integrals have singularities whenever $x$ and $\xi_o$ coincide, although it will be found subsequently that those involving $G$, with a logarithmic singularity, can be evaluated without difficulty either analytically or numerically. However, the $F$ boundary integral has a strong singularity in two dimensions of order $1/r$ and has to be evaluated as

$$\oint_S p(x)\,F(x,\xi_o)\,dS(x) = \tfrac{1}{2}p(\xi_o) + \int_S p(x)\,F(x,\xi_o)\,dS(x) \qquad (3\text{-}32)$$

in which the final integral has to be interpreted as a Cauchy principal-value integral. The 'free term' in (3-32) will always be positive for both the interior and exterior problems if we ensure that $\xi \to \xi_o$ from inside the region of interest and associate $u(\xi_o)$ with the 'outward' normal to $S$.

The following simple evaluation of the 'free term'[11] $\tfrac{1}{2}p(\xi_o)$ in Eq. (3-32) may help the reader to understand why it is necessary to interpret the integral involving $F$ in this way. Figure 3-5a shows a smooth portion of the boundary $S$

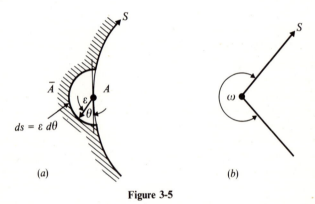

*(a)*            *(b)*

**Figure 3-5**

near $\xi_o$ which has been extended by a small semi-circular region, of radius $\varepsilon \to 0$, to envelope $\xi_o$ which is actually on the boundary. We are to calculate the singular portion of $\int_S p(\xi_o)\,F(x,\xi_o)\,dS(x)$ which arises when $\xi \to \xi_o$ on $S$ from inside $A$. If $\xi_o$ is the origin of a local coordinate system then, from Eq. (3-6b) around $S$,

$$F(x,\xi_o) = \frac{(x-\xi_o)_i}{2\pi r^2}\,n_i(x)$$

$$F(x,\xi_o) = \frac{x_i n_i}{2\pi\varepsilon^2} = \frac{1}{2\pi\varepsilon}$$

and $dS(x) = \varepsilon\,d\theta$ whence

$$\int_{\Delta S} p(\xi_o)\,F(x,\xi_o)\,dS(x) = p(\xi_o)\int_0^\pi \frac{1}{2\pi\varepsilon}\varepsilon\,d\theta = \tfrac{1}{2}p(\xi_o) \qquad (3\text{-}33)$$

where the sign is now positive when the normal $n$ is 'outward', from $A$ into $\bar{A}$. Equation (3-32) therefore accounts for this 'free term' separately and the integral around the remaining boundary of $S$, where $x \neq \xi_o$, becomes a Cauchy principal value one. It is worth noting here (Fig. 3-5$b$) that if $S$ does not have a unique tangent at $\xi_o$ then the limits of integration for the free term will not be 0 to $\pi$ and the coefficient $p(\xi_o)$ will not be $\frac{1}{2}$. This situation can arise at corners or junctions of linearly approximated boundary elements; e.g., from Fig. 3-5$b$ the limits of integration are now 0 to $\omega$ and the 'free term' is $(\omega/2\pi)p(\xi_o)$. It is perhaps appropriate to mention here that, for potential flow problems in three dimensions, the 'free term' can be evaluated in an almost identical fashion by considering a hemi-spherical region, of radius $\varepsilon \to 0$, enclosing $\xi_o$. We then have $G(x, \xi) = 1/4\pi kr$ (Chapter 5) and therefore $F(x, \xi) = y_i n_i(x)/4\pi r^3$ so that, as $\xi \to \xi_o$,

$$F(x, \xi_o) = \frac{x_i n_i}{4\pi\varepsilon^3} = \frac{1}{4\pi\varepsilon^2} \quad \text{and now} \quad dS = 2\pi\varepsilon^2 \cos\theta \, d\theta$$

whence

$$\int_{\Delta S} p(\xi_o) F(x, \xi_o) \, dS(x) = p(\xi_o) \int_0^{\pi/2} \frac{\cos\theta}{2} \, d\theta = \tfrac{1}{2}p(\xi_o) \qquad (3\text{-}34)$$

once more. For the case of a corner corresponding to Fig. 3-5$b$ the 'free term' $= (\omega/4\pi)p(\xi_o)$ where $\omega$ is now the solid angle at the corner.

Substitution of (3-32) into (3-31) produces the boundary integral equation in the form we require:

$$\tfrac{1}{2}p(\xi_o) = \int_S [p(x) F(x, \xi_o) - u(x) G(x, \xi_o)] \, dS(x) + \int_A \psi(x) G(x, \xi_o) \, dA(x) \quad (3\text{-}35)$$

In principle this identity enables us to use the specified boundary values and internal sources to calculate the remaining unspecified boundary data. Once all the boundary information is known (i.e., both $p$ and $u$ around $S$) it can be used in Eq. (3-30) to calculate $p(\xi)$ at any point inside $A$. We can also calculate $u(\xi)$ by recollecting that $u(\xi) = -\kappa[\partial p(\xi)/\partial \xi_i] n_i(\xi)$ and differentiating the $\xi$ dependent terms in Eq. (3-30) under the integral sign to produce

$$u(\xi) = \int_S [p(x) H(x, \xi) - u(x) F(x, \xi)] \, dS(x) + \int_A \psi(x) F(x, \xi) \, dA(x) \quad (3\text{-}36)$$

which now involves the differentiation of $F(x, \xi)$ at $\xi_i$ and a new expression

$$H(x, \xi) = -\kappa \frac{\partial F(x, \xi)}{\partial \xi_j} n_j(\xi) = \kappa^2 \frac{\partial^2 G(x, \xi_j)}{\partial x_i \, \partial \xi_j} n_i(x) n_j(\xi)$$

whence

$$H(x, \xi) = \frac{\kappa}{2\pi r^2} \left( \frac{2 y_i y_j}{r^2} - \delta_{ij} \right) n_i(x) n_j(\xi) \qquad (3\text{-}37)$$

with $y_i = (x - \xi)_i$ and $\delta_{ij}$ the Kronecker delta symbol. Equation (3-37) is obtained by differentiating $F(x, \xi) = -y_j\, n_j(x)/2\pi r^2$ as follows:

$$H(x, \xi) = \frac{-\kappa}{2\pi}\, n_i(x)\, n_j(\xi)\frac{\partial(y_i/r^2)}{\partial \xi_j} = \frac{-\kappa}{2\pi}\, n_i(x)\, n_j(\xi)\left(\frac{1}{r^2}\frac{\partial y_i}{\partial \xi_j} - \frac{2y_i}{r^3}\frac{\partial r}{\partial \xi_j}\right)$$

but $\partial y_i/\partial \xi_j = -\delta_{ij}$ and since $r^2 = y_k\, y_k$, $\partial r/\partial y_k = y_k/r$. Therefore,

$$\frac{\partial r}{\partial \xi_j} = \frac{\partial r}{\partial y_k}\frac{\partial y_k}{\partial \xi_j} = -\frac{y_k}{r}\delta_{jk} = -\frac{y_j}{r}$$

and
$$H(x, \xi) = \frac{\kappa}{2\pi r^2}\, n_i(x)\, n_j(\xi)\left(\frac{2y_i\, y_j}{r^2} - \delta_{ij}\right) \tag{3-38}$$

Results of this kind are very tedious to derive without using indicial notation since Eq. (3-38) contains eight distinct product terms when written out fully (two of which will be zero). The reader is therefore urged to master the few simple rules involved (Appendix A) at this stage, in preparation for the rather more extensive use of the notation which arises in elasticity, etc.


### 3-5-1 Discretization of the Surface and Volume Integrals and Formation of the System Matrices

The following development is very closely similar to that detailed in Sec. 3-4-1 for the indirect method and therefore only the main steps will be mentioned.

The key problem is the solution of Eq. (3-35) using the specified boundary conditions to calculate the remaining, initially unknown, boundary values of $p$ and $u$.

We shall again consider the simplest linear boundary element discretization scheme with constant distributions of variables over the elements and constant distribution of $\psi$ over each individual internal cell. If we discretize the boundary $S$ into $N$ such elements and the internal region into $M$ cells, of which the $q$th and $l$th respectively are typical members, then the identity (3-35) can be written for the $p$th element on the boundary as

$$\tfrac{1}{2}p(\xi_o^p) = \sum_{q=1}^{N} p(x)^q \int_{\Delta S} F(x^q, \xi_o^p)\, dS(x^q) - \sum_{q=1}^{N} u(x)^q \int_{\Delta S} G(x^q, \xi_o^p)\, dS(x^q)$$

$$+ \sum_{l=1}^{M} \psi(z^l)\, G(z^l, \xi_o^p)\, dA(z^l) \tag{3-39}$$

In Eq. (3-39) we have reintroduced coordinates $z$ to identify the location of internal area cells as in Sec. 3-4 in order to preserve $\xi$ unambiguously for surface points. The matrix equivalent of Eq. (3-39) is then

$$\tfrac{1}{2}p^p = \left(\int_{\Delta S} F^{qp}\, dS\right)\mathbf{p}^q - \left(\int_{\Delta S} G^{qp}\, dS\right)\mathbf{u}^q + \left(\int_{\Delta A} G^{lp}\, dA\right)\boldsymbol{\psi}^l \tag{3-40}$$

where $\mathbf{p}^q$ and $\mathbf{u}^q$ are $N \times 1$ columns, $\boldsymbol{\psi}^l$ an $M \times 1$ column, and the bracketted terms compatible row vectors. If Eq. (3-40) is compared with Eqs (3-15) and (3-16) then it will be seen that the order of the indices $pq$, $pl$ in the indirect BEM is reversed in direct BEM to $qp$, $lp$. This important change is the reflection in the matrix equations of the reversal of the order of the arguments discussed at length in Sec. 3-5. Consequently, from the symmetry of $G(x, \xi)$, $G^{pq} = G^{qp}$ whereas $F^{pq}$ and $F^{qp}$ are quite different from each other.

Again the integral terms in the row vectors may be evaluated either numerically, in all cases, or analytically for the simpler $F$ and $G$ functions (Sec. 3-6). After carrying out the summations the matrix equation can then be written

$$\tfrac{1}{2}p^p = (\bar{\mathbf{F}}^{qp})\,\mathbf{p}^q - (\mathbf{G}^{qp})\,\mathbf{u}^q + (\mathbf{G}^{lp})\,\mathbf{\psi}^l \qquad (3\text{-}41)$$

or $\qquad\qquad (\mathbf{F}^{qp})\,\mathbf{p}^q - (\mathbf{G}^{qp})\,\mathbf{u}^q + (\mathbf{G}^{lp})\,\mathbf{\psi}^l = 0 \qquad (3\text{-}42)$

if the $\tfrac{1}{2}p^p$ term is absorbed into the $p = q$ position of the $F$ matrix. By allowing $p$ to range over $1, ..., N$ we obtain the complete set of all equations similar to (3-42) as

$$\mathbf{F}^S\,\mathbf{p} - \mathbf{G}^S\,\mathbf{u} + \mathbf{G}^A\,\mathbf{\psi} = 0 \qquad (3\text{-}43)$$
$$(N \times N) \ (N \times N) \ (N \times M)$$

Again the superfix $S$ denotes a matrix derived from line integrals of $F$ and $A$, superfix $A$ a matrix derived from the integral of $G$ over area elements, and $\mathbf{p}$, $\mathbf{u}$ are vectors of the boundary potentials and fluxes identical to $\mathbf{p}^q$ and $\mathbf{u}^q$ respectively.

The simplest boundary-value problems will have either the potentials specified at all boundary points (hence $\mathbf{p}$ is known) or the fluxes specified similarly (in which case $\mathbf{u}$ is known). In either case the unspecified boundary information, $\mathbf{u}$ or $\mathbf{p}$ respectively, can be calculated directly from Eq. (3-43). In the general, mixed boundary-value problem, the potential will be specified on $S^p$ of $S$ and the flux over $S^u$ of $S$, where, in a well-posed problem, $(S^p + S^u) \equiv S$ always. Equation (3-43) should preferably be scaled by $\alpha \simeq \kappa$, to ensure that the terms in the $\mathbf{p}$ and $\mathbf{u}$ matrices are of comparable magnitude: viz.,

$$-\left(\frac{1}{\alpha}\mathbf{F}^S\right)(\alpha\mathbf{p}) + \mathbf{G}^S\,\mathbf{u} = \mathbf{G}^A\,\mathbf{\psi} \qquad (3\text{-}44)$$

Such a scaling can be achieved automatically by working throughout in dimensionless variables.

It is then a very simple operation to rearrange (3-44) so that the known $p$ and $u$ values form one vector $\mathbf{Y}$ $(N \times 1)$ and the unknown $\mathbf{p}$ and $\mathbf{u}$ values another $N \times 1$ vector $\mathbf{X}$, whence (3-44) can be re-written as

$$\mathbf{KX} + \mathbf{LY} = \mathbf{G}^A\,\mathbf{\psi} \qquad (3\text{-}45)$$

and solved for $\mathbf{X}$ with the result that all $u$ and $p$ components (i.e., both $p$ and $u$ at every boundary element) are now known around $S$.

As mentioned previously if the boundary does not have a unique tangent at any point where $F(x, \xi_o)$ has to be evaluated, then the coefficient of the free term in

Eq. (3-30), etc., will not be $+\frac{1}{2}$ but an unknown quantity, $\beta$ say. Equation (3-42) would then be

$$\beta p^p = (\bar{\mathbf{F}}^{qp})\,\mathbf{p}^q - (\mathbf{G}^{qp})\,\mathbf{u}^q + (\mathbf{G}^{lp})\,\psi^l \qquad (3\text{-}46)$$

and if the $\beta p^p$ term is absorbed into the leading diagonal terms of $\bar{\mathbf{F}}^{qp}$ we arrive at Eq. (3-43) once more, from which the value of $\beta$ can be deduced[12] directly by considering the effect of a uniform unit potential, $\mathbf{p} = \mathbf{I}$, applied around the whole boundary $S$, with $\psi = 0$. Clearly this produces zero boundary flux ($\mathbf{u}^q = 0$) and Eq. (3-43) then reduces to $\mathbf{F}^S\,\mathbf{I} = 0$ which implies that the sum of the components in each individual row of $\mathbf{F}^s$ has to be zero [see, for example, Eq. (2-26b)]. Therefore $\beta$ can be found directly from this requirement, which involves only the off-diagonal terms of $\bar{\mathbf{F}}^{qp}$ and therefore, if round-off errors are not significant, eliminates the need to consider the corner integrals separately.

### 3-5-2 Calculation of Internal Potentials and Velocities

Once all the components of $\mathbf{p}$ and $\mathbf{u}$ (on the boundary) are known then backsubstitution into Eqs (3-30) or (3-37) will yield the potential $p(\xi)$ or $n$ direction flux $u(\xi)$ at any subsequently selected point $\xi_i$ within $A$. These equations are again most usefully expressed in a discretized form. Following the discretization procedure explained previously the two equations then become, for, say, a selected internal point at $\xi^r$,

$$p(\xi^r) = \sum_{q=1}^{N} p(x^q) \int_{\Delta S} F(x^q, \xi^r)\,dS(x)^q - \sum_{q=1}^{N} u(x^q) \int_{\Delta S} G(x^q, \xi^r)\,dS(x)^q$$

$$+ \sum_{l=1}^{M} \psi(z)^l\, G(z^l, \xi^r)\,dA(z)^l$$

that is,

$$\mathbf{p}^r = (\mathbf{F}^{qr})\,\mathbf{p}^q - (\mathbf{G}^{qr})\,\mathbf{u}^q + (\mathbf{G}^{lr})\,\psi^l \qquad (3\text{-}47)$$

where $\mathbf{p}^q \equiv \mathbf{p}$ and $\mathbf{u}^q \equiv \mathbf{u}$ and all the components of the $F$ and $G$ matrices can be evaluated by normal summation since the only singularity which arises is a weak one involving the component of $(\mathbf{G}^{lr})$ for the cell in which $\xi^r$ lies.

When treated similarly Eq. (3-37) becomes

$$\mathbf{u}^r = (\mathbf{H}^{rq})\,\mathbf{p}^q - (\mathbf{F}^{rq})\,\mathbf{u}^q + (\mathbf{F}^{rl})\,\psi^l \qquad (3\text{-}48)$$

Again the term involving $\mathbf{F}^{rl}$ for the body cell encompassing $\xi^r$ will only give rise to a weak singularity, of order $1/r$ within an area integral, and therefore no principal-value integrals arise in connection with any of the $(\mathbf{F})$ or $(\mathbf{H})$ vectors in Eq. (3-48).

We have now completed the direct BEM development for a typical homogeneous two-dimensional potential flow problem and the reader is advised to follow through both the direct and indirect solution methods in parallel. By doing this it will probably be concluded, correctly, that the computational effort to arrive at $\phi$ in indirect BEM is virtually identical to that involved in determining the initially unknown boundary information in the direct method. Subsequently, due to the additional (**H**) matrix compilation which arises in direct BEM, rather more computational effort is involved in calculating internal point values of $\mathbf{u}^r(x)$ in this method which can be offset against the extra manipulation of the $\phi$ vector needed in indirect BEM to derive the remaining boundary information.

A further point of interest arises from Eq. (3-44) which can be written, in the absence of any $\Psi$ sources, as

$$\mathbf{u} = [\mathbf{G}^S]^{-1}\mathbf{F}^S\mathbf{p} = [\mathbf{K}]\mathbf{p} \tag{3-49}$$

Equation (3-49) then relates boundary fluxes and potentials in a way closely similar to a finite element formulation although we now have just one 'super-element' representing the whole of one homogeneous region of any shape. We shall show, in Sec. 3-8, how to assemble such zonal superelements together to solve problems involving 'piecewise' homogeneous bodies by BEM.

However, before doing this we shall establish, in Sec. 3-6, the equivalence of the direct and indirect formulations of BEM followed, in Sec. 3-7, by a complete set of all the intermediate integrals of the type $\int_{\Delta S} G^{pq}\,dS$, etc., which have to be evaluated to utilize Eqs (3-15) to (3-22) and (3-40) to (3-49). These are analytical solutions for linear or planar elements with sources uniformly distributed over them.

## 3-6 THE EQUIVALENCE OF INDIRECT AND DIRECT BOUNDARY ELEMENT METHOD ANALYSES

For convenience we repeat below the basic equation (3-30) from Sec. 3-5, which is the DBEM statement for the potential $p(x)$ at any point $x_i$ interior to $A$, but we now interchange the $x$ and $\xi$ symbols: viz.,

$$\int_A p(\xi)\,\delta(\xi, x)\,dA(\xi) = p(x)$$

$$= \int_S [p(\xi)\,F(\xi, x) - u(\xi)\,G(\xi, x)]\,dS(\xi)$$

$$+ \int_A \psi(\xi)\,G(\xi, x)\,dA(\xi) \tag{3-30a}$$

Consider the region $\bar{A}$, bounded by $S$, but exterior to $A$, within which there are no sources and assume $\bar{p}(x)$ to be a solution to the Laplace equation $\partial^2 \bar{p}/\partial x_i\,\partial x_i = 0$ within $\bar{A}$. An exact repetition of the analysis leading up to Eq. (3-31), carrying out integrations over $S$ and $\bar{A}$ but still with the field point at $x_i$

within $A$, develops an equivalent equation

$$0 = \int_S [-\bar{p}(\xi) F(\xi, x) - \bar{u}(\xi) G(\xi, x)] \, dS(\xi)$$

where, in comparison to Eq. (3-30),

1. The last term is zero, since $\bar{\psi}(x) = 0$.
2. The sign of the term involving $F(\xi, x)$ is reversed since the sense of the outward normal to $\bar{A}$ is opposite to that of $A$.
3. The left-hand-side term is zero, since $x_i$ is now exterior to $\bar{A}$ [that is, $\int_A \bar{p}(\xi) \delta(\xi, x) \, dA(\xi) = 0$].

If $\bar{p}(x)$ is specified to be that solution, in $\bar{A}$, which establishes on $S$ exactly the same boundary potentials as those in our initial interior region problem [that is, $\bar{p}(\xi) \equiv p(\xi), \xi \in S$], then substitution of $p(\xi)$ for $\bar{p}(\xi)$ in our second equation and adding it to (3-30) produces

$$p(x) = -\int_S [u(\xi) + \bar{u}(\xi)] G(\xi, x) \, dS(\xi) + \int_A \psi(\xi) G(\xi, x) \, dA(\xi)$$

Therefore,

$$p(x) = \int_S \phi(\xi) G(x, \xi) \, dS(\xi) + \int_A \psi(\xi) G(x, \xi) \, dA(\xi) \qquad (3\text{-}7a)$$

where

$$\phi(\xi) = -[u(\xi) + \bar{u}(\xi)]$$

which, apart from the arbitrary constant $C$, is exactly the IBEM statement, Eq. (3-7). All the subsequent IBEM operations to calculate fluxes, taking the field point to the boundary, etc., formally follow from Eq. (3-7) and we have therefore now established the IBEM formulation as rigorously as that for DBEM. The interested reader may care to refer to Lamb,[13] where the above argument is developed almost identically.

Throughout the book we shall develop our IBEM statements in the simple, physically satisfying way used in Chapters 2 and 3, although in each and every case they can also be established formally as above. It is interesting to note that if we adopt $\bar{u}$ as that solution in $\bar{A}$ which establishes, on $S, \bar{u} = u$ then we can obtain a second indirect formulation

$$p(x) = \int_S F(x, \xi) \mu(\xi) \, dS + \int_A \psi(\xi) G(x, \xi) \, dA(\xi) \qquad (3\text{-}30b)$$

where

$$\mu(\xi) = p(\xi) + \bar{p}(\xi)$$

Unfortunately the subsequent numerical treatment of (3-30b) presents considerable difficulties; therefore this alternative indirect formulation will not be considered further.

## 3-7 INTERMEDIATE INTEGRALS OVER BOUNDARY ELEMENTS AND INTERNAL CELLS

Even when both the boundary $S$ and the internal region of any problem have been discretized, ready for a matrix approximation to the governing integral equations, the fundamental solution $G$ and its derivatives $F$, $H$ have still to be integrated along individual boundary elements and over internal body cells. The following intermediate integrals of this kind will be required:

1. Indirect BEM [Eqs (3-15) and (3-16)]:

$$\int_{\Delta S} G^{pq}\, dS, \quad \int_{\Delta S} F^{pq}\, dS, \quad \int_{\Delta A} G^{pl}\, dA, \quad \int_{\Delta A} F^{pl}\, dA$$

2. Direct BEM [Eqs (3-41) and (3-48)]:

$$\int_{\Delta S} G^{qp}\, dS, \quad \int_{\Delta S} F^{qp}\, dS, \quad \int_{\Delta A} G^{lp}\, dA, \quad \int_{\Delta S} H^{qr}\, dS$$

Of these the two line integrals involving $G$ are identical, as are the two $(G)$ area integrals, in both cases due to the symmetry of $G$. The integrals $\int F\, dS$ and $\int H\, dS$ have strong singularities when $x_i \equiv \xi_i$, whereas all others are only weakly singular at such points.

For the simpler constitutive equations and simple discretization schemes, particularly those incorporating uniformly distributed potentials and sources, etc., over linear boundary elements and triangular body cells, all the above integrals can be evaluated analytically. They can, of course, also be approximated to any degree of accuracy by numerical integration methods, as can integrals involving very much more complicated singular solutions distributed non-uniformly over curved elements. We shall discuss numerical quadrature, isoparametric elements, etc., in detail later in the book.

Meanwhile, the $F$, $G$, $H$ functions which occur in potential flow are particularly amenable to direct integration when distributed over line and triangle elements and, as the analytical solutions generated also help to develop an understanding of the problems associated with singular functions, we shall now derive the complete set of intermediate integrals listed above.

In all cases we shall set up a local coordinate system (Fig. 3-6a), usually with our field point $x^p$ at the origin. Whenever a unit vector $n_i(x)$ is introduced we assume that it will have been transformed from the global to the local coordinate system.

(a) $\displaystyle\int_{\Delta S^q} G(x^p, \xi^q)\, dS(\xi^q)$

From Eq. (3-4),

$$G(x, \xi) = -\frac{1}{2\pi\kappa} \ln\left|\frac{r}{r_o}\right|$$

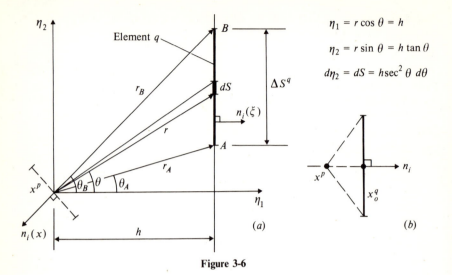

$$\eta_1 = r \cos \theta = h$$
$$\eta_2 = r \sin \theta = h \tan \theta$$
$$d\eta_2 = dS = h \sec^2 \theta \, d\theta$$

**Figure 3-6**

In local coordinates we therefore have

$$\int_{\Delta S} G(x,\xi) \, dS \Rightarrow \int_{\Delta S} G(0,\eta) \, d\eta_2 = \int_{\theta_A}^{\theta_B} -\frac{h}{2\pi\kappa} \ln \left| \frac{h \sec \theta}{r_o} \right| \sec^2 \theta \, d\theta$$

$$= -\frac{h}{2\pi\kappa} \left[ \tan \theta \left( \ln \left| \frac{h \sec \theta}{r_o} \right| - 1 \right) + \theta \right]_{\theta_A}^{\theta_B}$$

Thus

$$\int_{\Delta S} G(x^p, \xi^q) \, dS^q = \int_{\Delta S} G(\xi^q, x^p) \, dS^q = -\frac{1}{2\pi\kappa} \left[ r \sin \theta \left( \ln \left| \frac{r}{r_o} \right| - 1 \right) + \theta h \right]_{\theta_A, r_A}^{\theta_B, r_B} \tag{3-50}$$

Equation (3-50) provides the value of the potential generated at any field point $(x^p)$ by a uniformly distributed source of unit intensity along $AB$. The field point could be $(x_o^p)$, the midpoint of, say, the $p$th boundary element as required in Eq. (3-15). The summations over all boundary elements always involve the situation when $p = q$ (that is, $x^p$ approaches the midpoint of the $q$th boundary element $x^p \to x_o^p$) and it is always such diagonal components of the $[G]$ matrices, etc., which have to be evaluated with care because all the kernel functions are singular when $x_i = \xi_i$.

By approaching $(x_o^q)$ from inside $A$ (Fig. 3-6b) we have $h \to 0$, $\theta_B = -\theta_A \to \pi/2$, $r_B = r_A \to \Delta S/2 = b$ (say), when Eq. (3-50) becomes

$$\int_{\Delta S} G(x_o^q, \xi^q) \, dS \Rightarrow -\frac{1}{2\pi\kappa} \left[ 2b \left( \ln \frac{b}{r_o} - 1 \right) \right] = -\frac{b}{\pi\kappa} \left( \ln \frac{b}{r_o} - 1 \right) \tag{3-51}$$

and because $G$ is only weakly singular under line integration we obtain the required value, Eq. (3-51), without difficulty.

**(b)** $\int_{\Delta S^q} F(x^p, \xi^q)\, dS(\xi^q)$

From Eq. (3-6),

$$F(x, \xi) = \frac{(x - \xi)_i\, n_i(x)}{2\pi r^2}$$

and therefore in our local coordinate system

$$\int_{\Delta S} F(x, \xi)\, dS \to \int_{\Delta S} F(0, \eta)\, d\eta_2 = -\frac{1}{2\pi} \int_{\theta_A}^{\theta_B} n_i(x)\, n_i\, \frac{h \sec^2 \theta}{h^2 \sec^2 \theta}\, d\theta$$

$$= -\frac{1}{2\pi} \int_{\theta_A}^{\theta_B} (n_1 + n_2 \tan \theta)\, d\theta$$

$$= -\frac{1}{2\pi} \left[ n_1 \theta - n_2 \ln \left| \frac{h}{r} \right| \right]_{\theta_A, r_A}^{\theta_B, r_B} \qquad (3\text{-}52)$$

Again, if we let $x^p$ approach $(x_o^q)$ from inside $D$, $r_A = r_B$, $n_2 \to 0$, $n_1 \to 1$, $\theta_B = -\theta_A \to \pi/2$, and

$$\int_{\Delta S} F(x_o^q, \xi^q)\, dS \Rightarrow -\frac{1}{2\pi} (n_1)(\theta_B - \theta_A) = -\tfrac{1}{2}$$

**(c)** $\int_{\Delta S^q} F(\xi^q, x^p)\, dS(\xi^q)$

Now, $F(\xi, x) = [(\xi - x)_i/2\pi r^2]\, n_i(\xi)$ and we have $n_1(\eta) = 1$, $n_2(\eta) = 0$. Therefore,

$$\int_{\Delta S} F(\xi, x)\, dS \to \int_{\Delta S} F(\eta, 0)\, d\eta_2 = \frac{1}{2\pi} \int_{\theta_A}^{\theta_B} \frac{h(h \sec^2 \theta)}{h^2 \sec^2 \theta}\, d\theta = \left[ \frac{\theta}{2\pi} \right]_{\theta_A}^{\theta_B} \qquad (3\text{-}53)$$

and

$$\int_{\Delta S} F(\xi^q, x_o^q)\, dS = \tfrac{1}{2}$$

as $x^p$ and $x_o^q$ coalesce. In this particular case, as in the corresponding one in (b) above, these are the only non-zero contributions arising from the improper integral $f$.

Sections (b) and (c) above should help to clarify the fundamental differences between the $F(x, \xi)$ and $F(\xi, x)$ integrals which result in Eqs (3-52) and (3-53) respectively.

**(d)** $\int_{\Delta S^q} H(\xi^q, x^r)\, dS(\xi^q)$

Here we have used $x^r$ for the field point to emphasize that this integral only arises in the calculation, by direct BEM, of fluxes at points not on the boundary $S$. From Eq. (3-39),

$$H(x, \xi) = \frac{\kappa}{2\pi} n_i(x) \left[ \frac{n_j(\xi)}{r^2} \left( \delta_{ij} - \frac{2y_i y_j}{r^2} \right) \right]$$

where $y_i = (x_i - \xi_i)$. In our local coordinate system, $n_1(\eta) = 1$, $n_2(\eta) = 0$, this equation becomes

$$H(0, \eta) = \frac{\kappa}{2\pi} \left[ n_1(x) \left( \frac{1}{r^2} \right) \left( 1 - \frac{h^2}{r^2} \right) - n_2(x) \left( \frac{1}{r^2} \right) \left( \frac{h\eta_2}{r^2} \right) \right]$$

Therefore, from Fig. 3-5$a$ and recollecting that $d\eta_2/r^2 = d\theta/h$, we have, writing $n_i(x) = n_i$,

$$\int_{\Delta S} H(x, \xi) \, dS \rightarrow \int_{\Delta S} H(0, \eta) \, d\eta_2 = \int_{\theta_A}^{\theta_B} \frac{\kappa}{2\pi h} \left[ n_1(1 - \cos^2 \theta) - n_2(\sin \theta \cos \theta) \right] d\theta$$

$$= \frac{\kappa}{8\pi h} \left[ n_1(2\theta - \sin 2\theta) + n_2 \cos 2\theta \right]_{\theta_A}^{\theta_B} \qquad (3\text{-}54)$$

Equation (3-54) presents no problems for any field point within $A$ which is what we require. However, it is of interest to note that the strong singularity in $H$ of order $1/r^2$ does mean that $\int_{\Delta S} H(x_o^q, \xi^q) \, dS$ cannot be evaluated. Thus as $x^r \rightarrow x_o^q$ the value of Eq. (3-54) approaches $\kappa/4h$ which becomes infinite as $h \rightarrow 0$. Therefore, near surface points this presents a major difficulty with the direct method. Equations (3-50) to (3-54) provide analytical expressions for all the

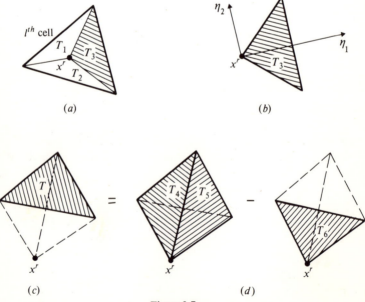

**Figure 3-7**

element line integrals required to compile all the matrices $(\mathbf{G}^{pq})$, $(\mathbf{F}^{pq})$ [Eqs (3-17) and (3-18)] and $(\mathbf{G}^{qp})$, $(\mathbf{F}^{qp})$ [Eqs (3-42) and (3-47)] and $(\mathbf{H}^{rq})$ [Eq. (3-48)]. To complete the set we require only integrals of $F$ and $G$ over internal area cells. It is important to calculate these integrals accurately, preferably analytically, for the cell in which the field point lies. These values will occupy the $r = l$ positions in the matrices in Eqs (3-47) and (3-48), etc., and although the following expressions can also be used to evaluate all the matrix components, those off the main diagonal are probably more conveniently obtained by numerical quadrature.

Figure 3-7a shows a typical $x^r$ point within the $l$th body cell over the area of which a uniform source $(z^l)$ is applied. Any such cell can be divided into three subtriangles (Fig. 3-7a) with $x^r$ at their common apex, so that if we analyse one of them (Fig. 3-7b) simple superposition will yield the answer we require. Furthermore, if $x^r$ lies outside any cell (Fig. 3-7c) or on the cell boundary, the same analytical result (Fig. 3-7b) can be used, in conjunction with the superposition scheme sketched in Fig. 3-7d, to solve this problem also. We can therefore obtain all we require to evaluate the components of $(\mathbf{G}^{pl})$, $(\mathbf{F}^{pl})$ [Eqs (3-17) and (3-18)] and $(\mathbf{G}^{lp})$ [Eqs (3-42) and (3-48)] by considering one subtriangle $(ABC)$ and a local $(\eta)$ coordinate system (Fig. 3-8).

(e) $\displaystyle \int_{\Delta A} G(x^r, z^l)\, dA(z^l) = \int_{\Delta A} G(z^l, x^r)\, dA(z^l)$

We have already [Eq. (3-50)] obtained the integral of $G$ over the strip $ab$ (Fig. 3-8) and we now merely have to integrate this expression with respect to $\eta$ $(0 \leqslant \eta_1 \leqslant h)$. Thus

$$\int_{\Delta A} G(x, z)\, dA \rightarrow \int_{\Delta A} G(0, \eta)\, d\eta_1\, d\eta_2$$

$$= \int_0^h -\frac{1}{2\pi\kappa}\left[ r\sin\theta\left( \ln\left|\frac{r}{r_o}\right| - 1 \right) + \eta_1\,\theta \right]_{r_a, \theta_A}^{r_b, \theta_B} d\eta_1$$

$$= \int_0^h -\frac{1}{2\pi\kappa}\left[ \eta_1 \tan\theta\left( \ln\left|\frac{\eta_1 \sec}{r_o}\right| - 1 \right) + \eta_1\,\theta \right]_{\theta_A}^{\theta_B} d\eta_1$$

$$= -\frac{1}{2\pi\kappa}\left[\left[ \tan\theta\frac{\eta_1^2}{2}\left( \ln\left|\frac{\eta_1 \sec\theta}{r_o}\right| - \frac{3}{2} \right) + \frac{\eta_1^2}{2}\theta \right]_{\theta_A}^{\theta_B}\right]_0^h$$

$$= -\frac{h^2}{4\pi\kappa}\left[ \tan\theta\left( \ln\left|\frac{r}{r_o}\right| - \frac{3}{2} \right) + \theta \right]_{r_A, \theta_A}^{r_B, \theta_B} \tag{3-55}$$

(f) $\displaystyle \int_{\Delta A} F(x^r, z^l)\, dA(z^l)$

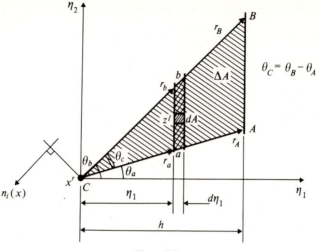

**Figure 3-8**

Again we simply integrate Eq. (3-52) w.r.t. $\eta_1$:

$$\int_{\Delta A} F(x, z)\, dA \rightarrow \int_{\Delta A} F(0, \eta)\, d\eta_1\, d\eta_2 = \int_0^h -\frac{1}{2\pi}\left[ n_1\, \theta - n_2 \ln\left|\frac{\eta_1}{r}\right| \right]_{r_a,\, \theta_A}^{r_b,\, \theta_B}$$

$$= -\frac{1}{2\pi}\int_0^h \left[ n_1\, \theta - n_2 \ln\left(\cos\theta\right) \right]_{\theta_A}^{\theta_B}\, d\eta_1$$

$$= -\frac{h}{2\pi}\left( n_1\, \theta_c - n_2 \ln\left|\frac{r_A}{r_B}\right| \right) \qquad (3\text{-}56)$$

[It is important to note that the $n_i(x)$ unit vector components have been transposed to suit the local axes chosen, with $\eta_2$ parallel to $AB$, and therefore they will take different numerical values for each subtriangle.]

Whereas $\int_{\Delta A} F(z^l, x^r)\, dA(z^l)$ would be, from Eq. (3-53),

$$\int_{\Delta A} F(z, x)\, dA \rightarrow \int_{\Delta A} F(\eta, 0)\, d\eta_1\, d\eta_2 = \int_0^h \frac{\theta_c}{2\pi}\, d\eta_1 = \frac{h\theta_c}{2\pi} \qquad (3\text{-}57)$$

We have now completed all the intermediate integrals required and, although the results obtained are quite simple and instructive in this case, it is very clear that any further complication, due to either the complexity of the singular solutions, or element geometry, or non-uniform distributions of sources and boundary conditions, will necessitate the introduction of numerical summation procedures to replace the corresponding analytic integrations. For potential flow problems the analytical results given above do provide the most convenient way of compiling the various matrix components; these are used in the illustrative examples in Sec. 3-10.

## 3-8 ZONED INHOMOGENEOUS BODIES

So far we have dealt entirely with problems involving one homogeneous zone of either isotropic or anisotropic material. In most practical situations the regions concerned contain a number of contiguous zones of materials each having different but homogeneous properties (i.e., they are zoned or piecewise homogeneous bodies). We shall now develop a very straightforward extension of the basic BEM algorithm to incorporate the solution of zoned body problems comprising any number of homogeneous zones.

Whereas the matrices which arise in the basic equations (3-22) and (3-45) are fully populated we shall find that zoned bodies lead to block-banded matrix systems with one block for each zone and overlaps between blocks where any zones have a common interface. In general there will be a number of zones $D^m$, $m = 1, 2, ...$, each enclosed by its surface boundary $S^m$. Where two regions have a common interface, e.g., regions 1, 2, then we have to ensure that corresponding points on the $S^1$ and $S^2$ boundaries are at the same potential and that there is continuity of flow across corresponding elements. Thus on the interface

$$\mathbf{p}_{12}^S - \mathbf{p}_{21}^S = 0 \quad \text{and} \quad \mathbf{u}_{12}^S + \mathbf{u}_{21}^S = 0 \qquad (3\text{-}58)$$

where the number of components in each of these interface potential and flux vectors is equal to the chosen number of interface boundary elements, say $R$.

For illustration we shall consider the two-zone problem (Fig. 3-9) where, for example, zone 1 will have in total $N_1$, $M_1$ boundary and internal cells respectively, of which $R_1 \equiv R_2$ will be interface elements. The known boundary potentials and velocities for zone 1 will be designated by $\mathbf{p}_1^S$, $\mathbf{u}_1^S$ respectively which therefore contain together a total of $(N_1 - R_1)$ elements.

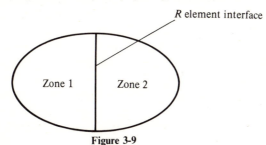

$R$ element interface

Zone 1    Zone 2

**Figure 3-9**

We can now write the basic solution set of equations (3-22) for an indirect BEM analysis of zone 1 as

$$\begin{Bmatrix} \alpha\mathbf{p}_1^S \\ \mathbf{u}_1^S \\ 0 \\ \mathbf{p}_{12} \\ \mathbf{u}_{12} \end{Bmatrix} = \begin{bmatrix} \mathbf{I}_1^o & \alpha\mathbf{G}_1^S \\ 0 & \mathbf{F}^S \\ 0 & \mathbf{b}_1^n \\ \mathbf{I}_{12}^o & \mathbf{G}_{12}^S \\ 0 & \mathbf{F}_{12}^S \end{bmatrix} \begin{Bmatrix} \alpha\mathbf{C}_1 \\ \boldsymbol{\phi}_1 \end{Bmatrix} + \begin{bmatrix} \alpha\mathbf{G}_1^A \\ \mathbf{F}_1^A \\ \mathbf{c}_1^n \\ \mathbf{G}_{12}^A \\ \mathbf{F}_{12}^A \end{bmatrix} \{\boldsymbol{\psi}_1\} \qquad (3\text{-}59a)$$

$(N_1 + R_1 + 1) \times 1$

or

$$\left\{\begin{array}{c} \mathbf{a}_1 \\ \mathbf{p}_{12} \\ \mathbf{u}_{12} \end{array}\right\} = \left[\begin{array}{c} \mathbf{A}_1 \\ \mathbf{B}_{12} \\ \mathbf{C}_{12} \end{array}\right] \{\boldsymbol{\phi}_1\} + \left[\begin{array}{c} \mathbf{J}_1 \\ \mathbf{K}_{12} \\ \mathbf{L}_{12} \end{array}\right] \{\boldsymbol{\psi}_1\} \qquad (3\text{-}59b)$$

where Eq. (3-59a) is merely (3-22) with the interface terms separated from the others and then, in (3-59b), the various matrices have been condensed with all the non-interface terms within the matrices $\mathbf{a}_1, \mathbf{A}_1, \mathbf{J}_1$ [$\mathbf{A}_1$ being $(N_1 - R_1 + 1) \times (N + 1)$], etc.

A precisely similar set of equations can be written for zone 2 as

$$\left\{\begin{array}{c} \mathbf{a}_2 \\ \mathbf{p}_{12} \\ \mathbf{u}_{21} \end{array}\right\} = \left[\begin{array}{c} \mathbf{A}_2 \\ \mathbf{B}_{21} \\ \mathbf{C}_{21} \end{array}\right] \{\boldsymbol{\phi}_2\} + \left[\begin{array}{c} \mathbf{J}_2 \\ \mathbf{K}_{21} \\ \mathbf{L}_{21} \end{array}\right] \{\boldsymbol{\psi}_2\} \qquad (3\text{-}60)$$

The interface compatibility equation (3-58) now allows us to combine (3-59b) and (3-60) to form

$$\left\{\begin{array}{c} \mathbf{a}_1 \\ \mathbf{0} \\ \mathbf{a}_2 \\ \mathbf{0} \end{array}\right\} = \left[\begin{array}{cc} \mathbf{A}_1 & \mathbf{0} \\ \mathbf{B}_{12} & -\mathbf{B}_{21} \\ \mathbf{0} & \mathbf{A}_2 \\ \mathbf{C}_{12} & \mathbf{C}_{21} \end{array}\right] \left\{\begin{array}{c} \boldsymbol{\phi}_1 \\ \boldsymbol{\phi}_2 \end{array}\right\} + \left[\begin{array}{cc} \mathbf{J}_1 & \mathbf{0} \\ \mathbf{K}_{12} & -\mathbf{K}_{21} \\ \mathbf{0} & \mathbf{J}_2 \\ \mathbf{L}_{12} & \mathbf{L}_{21} \end{array}\right] \left\{\begin{array}{c} \boldsymbol{\psi}_1 \\ \boldsymbol{\psi}_2 \end{array}\right\} \qquad (3\text{-}61)$$

a set of equations which will yield the only unknown $(\boldsymbol{\phi}_1 \, \boldsymbol{\phi}_2)^T$ vector, which is $(N_1 + N_2 + 2) \times 1$ since the constant $(C)$ terms have been absorbed into $\boldsymbol{\phi}$ in the condensation process (3-59a, b). Once all the $\boldsymbol{\phi}$ vectors have been found then each individual zone is treated subsequently as an absolutely independent region for which sufficient additional interface boundary information can be calculated [e.g., Eq. (3-60) can generate either $\mathbf{p}_{21}$ or $\mathbf{u}_{21}$ directly once $\boldsymbol{\phi}_2$ is known]. Thereafter potentials and fluxes at any point within each region can be calculated exactly as for a single-zone problem.

The location of the various submatrices within Eq. (3-61) follows a simple pattern if we note that the null matrices on the left-hand side are used to satisfy Eq. (3-58) with the rule that, for any two adjacent zones, the null associated with

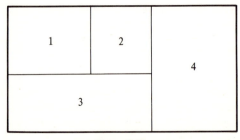

**Figure 3-10**

the first appearing zone equates potentials and the second null the fluxes. This procedure should be followed through for the four-zone problem illustrated in Fig. 3-10 which leads to the following self-explanatory set of equations for the complete set of fictitious potentials $\phi$.

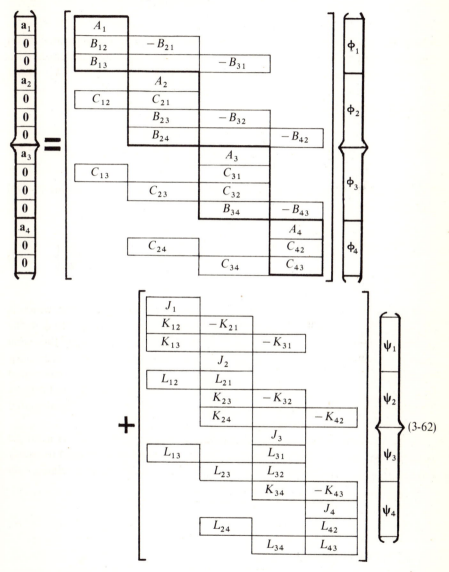

$$(3\text{-}62)$$

The block bandwidth is clearly governed by the maximum zone number difference of adjacent zones, and the overall dimension of the matrix to be inverted to obtain $\phi$ is $\sum_{i=1}^{4}(N+1)_i$ square where each zonal $N$ value includes both the interface and peripheral boundary elements.

The procedure for the direct BEM algorithm is very similar if we rewrite Eq. (3-43) in the form

$$\{p\} = [F^S]^{-1}([G^S]\{u\} - [G^A]\{\psi^A\})$$

or

$$\{p\} = [A^S]\{u\} - [D^A]\{\psi^A\} \tag{3-63}$$

Once more for the two-zone problem (Fig. 3-8) we can separate the peripheral fluxes and potentials $\{u_1\}$, $\{p_1\}$ for zone 1, say, from the interface values $\{u_{12}\}$, $\{p_{12}\}$, partition $A$ and $D$, and write

$$\begin{Bmatrix} p_1 \\ p_{12} \end{Bmatrix} = \begin{bmatrix} A_1 & A_{12} \\ B_{12} & C_{12} \end{bmatrix} \begin{Bmatrix} u_1 \\ u_{12} \end{Bmatrix} - \begin{bmatrix} D_1 \\ D_{12} \end{bmatrix} \{\psi_1\} \tag{3-64a}$$

and, for zone 2,

$$\begin{Bmatrix} p_2 \\ p_{21} \end{Bmatrix} = \begin{bmatrix} A_2 & A_{21} \\ B_{21} & C_{21} \end{bmatrix} \begin{Bmatrix} u_2 \\ u_{21} \end{Bmatrix} - \begin{bmatrix} D_2 \\ D_{21} \end{bmatrix} \{\psi_2\} \tag{3-64b}$$

The interface conditions (3-58) still apply and enable us to eliminate the interface potentials between Eqs (3-64a, b) which leads to

$$\begin{Bmatrix} p_1 \\ 0 \\ p_2 \end{Bmatrix} = \begin{bmatrix} A_1 & A_{12} & 0 \\ B_{12} & (C_{12} - C_{21}) & -B_{21} \\ 0 & A_{21} & A_2 \end{bmatrix} \begin{Bmatrix} u_1 \\ u_{12} \\ u_2 \end{Bmatrix} - \begin{bmatrix} D_1 & 0 \\ D_{12} & -D_{21} \\ 0 & D_2 \end{bmatrix} \begin{Bmatrix} \psi_1 \\ \psi_2 \end{Bmatrix} \tag{3-65}$$

If, for example, all the boundary potentials $(p_1, p_2)$ are specified, then Eq. (3-65) will allow us to calculate the boundary flux vector, including the interface $(u_{12})$ value. In a mixed boundary-value problem Eq. (3-65) has to be rearranged before it can be solved for the unspecified boundary values. Once $(u_{12})$ is known Eq. (3-64a) will provide $(p_{12})$ which completes the boundary information for zone 1 which is then treated as an entirely independent region for the determination of fluxes or potentials at points of interest within it using Eqs (3-47) and (3-48).

A four-zone problem (Fig. 3-10) leads to the following equations, from which the general procedure for assembling matrices for multizone problems should be quite clear:

The form of Eq. (3-66) is particularly convenient when all the boundary potentials are specified whereas when the boundary conditions are mixed the equation has to be rearranged so that all the specified boundary information is transposed to the left-hand side.


## 3-9 RELATED PROBLEMS


### 3-9-1 Free Surface Flow

Many practical groundwater flow problems involve free surface boundary conditions ($S_f$ in Fig. 3-11a). In such problems the free surface is a boundary

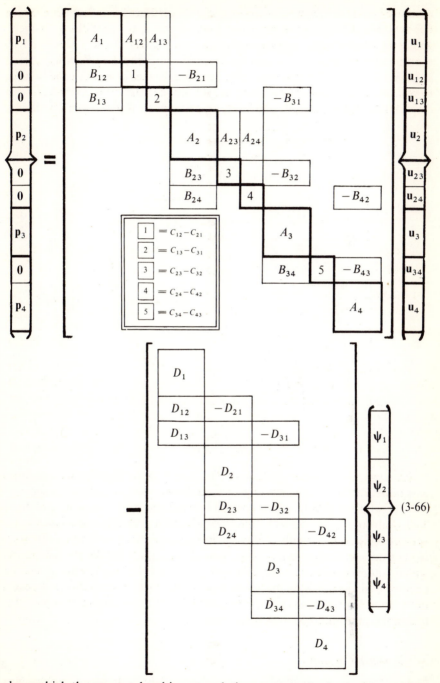

$$\begin{Bmatrix} \mathbf{p}_1 \\ 0 \\ 0 \\ \mathbf{p}_2 \\ 0 \\ 0 \\ \mathbf{p}_3 \\ 0 \\ \mathbf{p}_4 \end{Bmatrix} = [ \cdots ] \begin{Bmatrix} \mathbf{u}_1 \\ \mathbf{u}_{12} \\ \mathbf{u}_{13} \\ \mathbf{u}_2 \\ \mathbf{u}_{23} \\ \mathbf{u}_{24} \\ \mathbf{u}_3 \\ \mathbf{u}_{34} \\ \mathbf{u}_4 \end{Bmatrix}$$

$$\boxed{1} = C_{12} - C_{21}$$
$$\boxed{2} = C_{13} - C_{31}$$
$$\boxed{3} = C_{23} - C_{32}$$
$$\boxed{4} = C_{24} - C_{42}$$
$$\boxed{5} = C_{34} - C_{43}$$

$$- [ \cdots ] \begin{Bmatrix} \psi_1 \\ \psi_2 \\ \psi_3 \\ \psi_4 \end{Bmatrix} \qquad (3\text{-}66)$$

along which the pressure head is zero, relative to atmospheric pressure, and the potential at any point on $S_f$ is therefore simply its height ($h$) above an arbitrary datum (Fig. 3-11$a$). This figure shows a typical free surface for an earth dam which has two distinct sections with rather different boundary conditions on them. Along the whole of $S_f(1, 2, 3)$ we require $p(h) = h$ and, additionally, since $S_f(1, 2)$, the phreatic surface, is a streamline we have $u(h) = 0$ along $(1, 2)$. Section $(2, 3)$ is a seepage surface along which water emerges from the face of the dam and here we have $h$ specified unambiguously with $u(h) \neq 0$ along $(2, 3)$. The precise location of $S_f(1, 2)$ is not known initially and has to be determined as part of the solution by

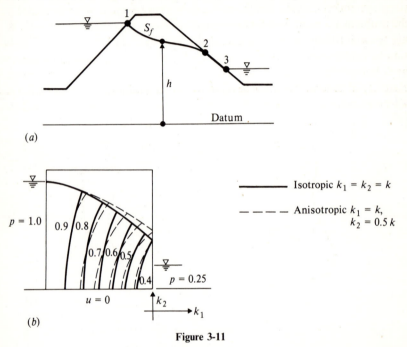

(a)

(b)

**Figure 3-11**

an iterative process. Niwa, Kobayashi, and Fukui[9] have solved such a problem using direct BEM in conjuction with the following simple iterative scheme.

The location of $S_f(1, 2, 3)$ is guessed initially and $p(h) = h$ assumed along it. This problem is then solved by direct BEM which generates automatically the values of $u(h)$ along $S_f$ (with, of course $u(h) \neq 0$ in general). These values of $u(h)$ are used as the specified boundary conditions along $S_f(1, 2)$ and the problem re-run with no other changes. The output is now $p(h) = h'$, say, along $S_f$ which generates an improved location $S'_f$ of the free surface defined by $h'$. The iteration is repeated until $h^n$ and $h^{n+1}$ agree as closely as required; Fig. 3-11$b$ shows an example of the free surface and equipotentials obtained in this way by Niwa and his colleagues.

## 3-9-2 Shaft Torsion

The application of 'boundary integral methods' to the shaft torsion problem was discussed in detail by Mendleson.[5] He considered its solution by 'indirect, semi-direct and direct' methods, using both warping functions and stress functions, and also extended his analysis from the purely elastic shaft to an elastoplastic one. Earlier Jaswon and Ponter[4] had presented solutions to the elastic torsion problem for a variety of solid and hollow shafts with different regular cross sections, again using a warping function in a 'direct' formulation.

The problem we shall consider here is the twisting of a uniform elastic shaft, of arbitrary cross section, by a torque applied as a specifically distributed shear stress system on its ends, which are free to warp. One approach would be to treat the problem as one in two-dimensional plane elasticity, which it obviously is, and use the algorithms detailed in Chapter 4. However, shaft torsion, as one of the simpler problems in elasticity, can be described by harmonic equations, as shown by St Venant, rather than the more complicated biharmonic ones which permeate elasticity in general. We shall develop the BEM solution in terms of a harmonic warping function $p(x)$, defined below, for an elastic shaft of cross-sectional area $A$,

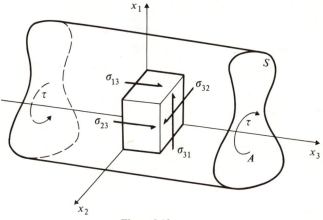

**Figure 3-12**

periphery $S$ (Fig. 3-12). A torque ($\tau$) acts on all cross sections to produce a twist $\alpha$ per unit shaft length where $\alpha = \tau/GJ$ with $G$ the shear modulus of the shaft material and $J$ the polar second moment of area of $A$ about a longitudinal axis through its centroid. The only non-zero stresses in the shaft are the shear stresses $\sigma_{13} = \sigma_{31}$ in the $(x_1, x_3)$ plane and $\sigma_{23} = \sigma_{32}$ in the $(x_2, x_3)$ plane (Fig. 3-12), where $x_3$ is directed along the shaft axis. If these stress components are specified in terms of $p(x)$ as

$$\sigma_{13} = \alpha G\left(\frac{\partial p}{\partial x_1} - x_2\right) \qquad \sigma_{23} = \alpha G\left(\frac{\partial p}{\partial x_2} + x_1\right) \tag{3-67}$$

then it can be shown[5] that $p(x_i)$ $(i = 1, 2)$ has to satisfy

$$\frac{\partial^2 p(x)}{\partial x_i \partial x_i} = 0 \qquad x_i \in A, S \tag{3-68a}$$

If we define $v_i(x) = -\partial p/\partial x_i$ and $u(x) = v_i(x)n_i(x)$, where $n_i(x)$ are the components of a specified unit vector at $x_i$, the value of $u(x_o)$, $x_o$ being on $S$, has to satisfy

$$u(x) = n_2 x_1 - n_1 x_2 \qquad x = x_o \in S \qquad (3\text{-}68b)$$

in which $n_1, n_2$ are the components of the outward unit normal vector $n_i(x_o)$ to $S$.

When formulated in this way we see immediately that Eqs (3-68$a, b$) correspond exactly to the two-dimensional potential flow problem with specified boundary fluxes, $\alpha = \tau/GJ$ being known. The solution for $p(x)$ then follows from either the direct or indirect BEM algorithms developed in this chapter. The stresses at any point involve $(\partial p/\partial x_i)$, the components of $v_i(x)$, which are again provided by the standard BEM analysis.

## 3-10 EXAMPLES OF SOLVED PROBLEMS

To demonstrate the general efficacy and precision of BEM for two-dimensional potential flow problems we conclude Chapter 3 with four examples illustrating solutions of increasing complexity.

These examples are taken from Tomlin[8] and the first of these is a test problem with a well-known analytical solution—that of the flow under an impermeable dam sitting on the surface of an isotropic material, across which there is a hydraulic potential difference of 100 units. The streamlines for this solution are elliptical and if we distort one of them by a 5/2 scale transformation, as shown in Fig. 3-13, and use it as an outer impermeable boundary, we have a very convenient anisotropic test problem ($k_1 = 4, k_2 = 25$, say) with a known solution. If, furthermore, we divide this region up arbitrarily into, say, five zones we can simultaneously check the precision of the zoned media algorithm explained in Sec. 3-8. Tomlin did exactly this and obtained the solutions shown in Fig. 3-14$a, b$ using indirect BEM and constant $\phi$ distribution along each element. For Fig. 3-14$a$ he used the potential at the midpoint of each field element to define the effect of each source element there, as explained in this chapter, whereas for Fig.

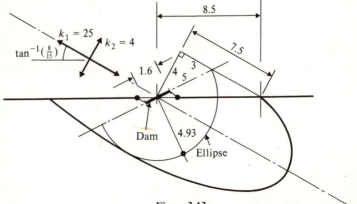

**Figure 3-13**

3-14*b* he elaborated the technique by using the average potential generated over the whole of each field element for this purpose. Inspection of the two figures will show that only a slight improvement in accuracy was obtained by doing this. Throughout most of the region the discrepancy between the calculated and analytical solutions for the potentials is less than 1 per cent of the total head across the dam with a maximum error of about 2 per cent under the dam base. Tomlin used eight boundary elements under the dam and a total number of 74 elements, of which 21 were on interfaces, leading to a final coefficient matrix of $100 \times 100$ and a solution time of about $100\,\text{CPUs}$ on an ICL 1907 computer utilizing 25k words core storage.

Figure 3-15 shows the distribution of potential under the base of the dam obtained from four numerical solutions compared with the analytical one. The four solutions are, in addition to the two BEM analyses already mentioned, a

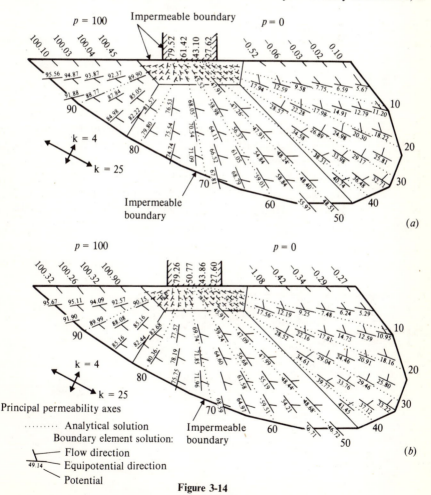

**Figure 3-14**

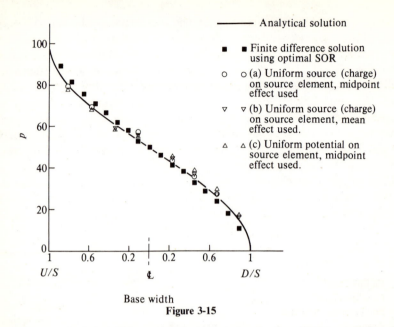

——— Analytical solution

■ ■ Finite difference solution using optimal SOR

○ ○ (a) Uniform source (charge) on source element, midpoint effect used

▽ ▽ (b) Uniform source (charge) on source element, mean effect used.

△ △ (c) Uniform potential on source element, midpoint effect used.

Base width

**Figure 3-15**

triangular mesh finite difference solution and a further BEM analysis of his incorporating the free-space solution for a uniformly distributed potential (as opposed to a uniformly distributed source) along each boundary element. Tomlin solved a number of problems in this way and found the distributed potential solution less convenient to use, and that it gave a slight decrease in precision. The third and fourth examples are rather similar, both of them relating to flow beneath surface dams underlain by complex beds of zoned anisotropic material. Figures 3-16 and 3-17 show the distribution of potential and the directions of the

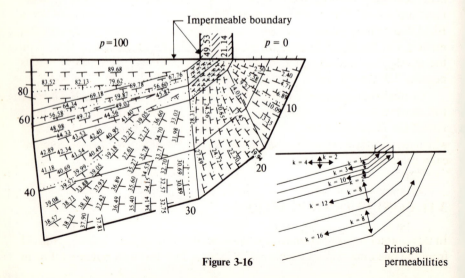

**Figure 3-16**

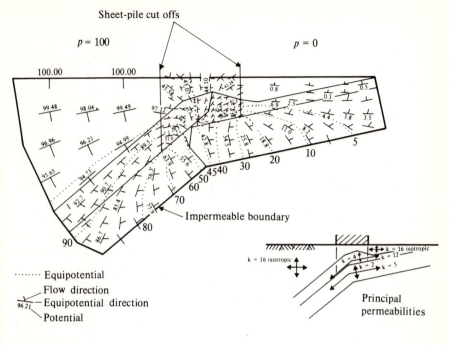

**Figure 3-17**

streamlines obtained from indirect BEM analyses for comparison with the dotted equi-potentials which were generated by Tomlin's triangular mesh finite difference programme. Again typical discrepancies between the two solutions are about 1 per cent of the total head across the dam, increasing to about 4 per cent near the singular points at the corners of the dam and the sheet piles. These are the most ambitious solutions to two-dimensional potential flow problems using BEM that we have yet seen, but nevertheless the computer requirements are quite modest. For the Fig. 3-16 problem he used 131 linear elements, including 77 on the interfaces of the 12 zones, which generated a $220 \times 220$ coefficient matrix needing 31k storage and a run time of about 150 CPUs on an ICL 1907.

Corresponding figures for the nine-zone 'cut-off' problem (Fig. 3-17) were 105 elements with 43 on the interfaces, a $157 \times 157$ matrix, 27k storage, and about 110 CPUs. Other examples can be found in Refs 3, 4, 6 to 9 and 14 to 17, all of which demonstrate accurate solutions which have been obtained very economically using BEM.

## 3-11 CONCLUDING REMARKS

In this chapter we have extended the ideas behind both indirect and direct BEM from the one-dimensional problems of Chapter 2 to deal with potential flow in two dimensions. One of the more remarkable features of the analyses is that the steps in the solution procedures remain virtually unchanged by the increase in dimensionality of the problems. Indeed, by using the tensor suffix notation introduced in this chapter we shall find that, in principle, the two-dimensional and three-dimensional potential flow algorithms remain identical (Chapter 5).

It is worth emphasizing again that the necessary 'free-space' unit solutions are well known for all the classical field equations, as are the relevant integral identities [Eq. (3-37)] which do not therefore have to be established before BEM can be used. In fact, once the technique has been thoroughly understood the solution procedure involves merely the systematic assembly of matrix equations such as (3-22) and (3-44), their solution, and back substitution into similar sets of equations [Eqs (3-7), (3-8) and (3-30), (3-38)] to obtain values of the solution variables at any subsequently selected points. The matrices to be reduced are not affected at all by changes in internal source distributions and the coefficients in the total sets of all equations such as (3-22), (3-44) and (3-62), (3-66) depend only on the geometry and material properties of the regions involved and are therefore independent of the specific boundary conditions stipulated. One consequence of this is that, in finite element terminology, the BEM algorithms establish a fully continuous 'stiffness matrix' for a homogeneous region of any arbitrary shape and, in this sense, each such region becomes one 'superelement' [see Eqs (3-49) and (3-19), (3-20)]. The coupling of BEM and FEM is discussed under 'hybrid methods' in Chapter 14.

We have taken particular care to bring out the 'two-point' nature of the singular solutions introduced and, at the expense of some of the equations looking rather cumbersome, persisted in tracing the roles of the two arguments through both the indirect and direct BEM procedures. Once the significance of the orderings of the arguments has been fully appreciated then the compact matrix form of the discretized integral equations, which is very simple indeed, can be used.

We have also incorporated closed-form integrations of the fundamental solutions over line elements and triangular cells as far as possible to illustrate both the integration techniques involved and the properties of the singular solutions themselves. These operations may, at first glance, appear to introduce some tedious algebra, but they do make the point that such subsidiary integrations are unavoidable, whether performed analytically or numerically, if accurate solutions are to be obtained efficiently. Every one of these integrals can, of course, be evaluated numerically and for the completely general procedure dealing with curvilinear elements numerical quadrature becomes quite unavoidable.

Some of the 'solved problem' examples in Sec. 3-10 are really quite ambitious; nevertheless, in spite of involving anisotropy, zoning, mixed boundary conditions, and even a phreatic surface (unspecified boundary location), in Sec. 3-9-2, the BEM solutions are very satisfactory in terms of both precision and computer requirements.

Finally, we would recommend very careful study of Chapters 2 and 3 before embarking upon Chapter 4 so that the underlying operations are quite clearly understood since they carry through identically into the solution of problems in elasticity. The slight additional complexity which arises in connection with the higher-rank kernel functions (stemming from the fourth-order differential equations of elasticity) will then present no real difficulty with the final matrix equation assembly procedures, being, in principle, identical to those already mastered.

## 3-12 REFERENCES

1. Kellog, O. D. (1929) *Foundations of Potential Theory*, Julius Springer, Berlin.
2. Jaswon, M. A. (1963) 'Integral equation methods in potential theory, I', *Proc. Roy. Soc., London*, **275**(A), 23–32.
3. Symm, G. T. (1963) 'Integral equation methods in potential theory, II', *Proc. Roy. Soc., London*, **275**(A), 33–46.
4. Jaswon, M. A., and Ponter, A. R. (1963) 'An integral equation solution of the torsion problem', *Proc. Roy. Soc., London*, **275**(A), 237–246.
5. Mendleson, A. (1973) 'Boundary integral methods in elasticity and plasticity', NASA Tech. Note T.N. D–7418, 36 pp.
6. Butterfield, R. (1972) 'The application of integral equation methods to continuum problems in soil mechanics' in *Stress strain behaviour of soils, Roscoe Meml Symp., Cambridge*, Foulis, pp. 573–587.
7. Butterfield, R., and Tomlin, G. R. (1972) 'Integral techniques for solving zoned anisotropic continuum problems', *Int. Conf. Var. Meth. in Engng*, Southampton University, pp. 9/31–9/51.
8. Tomlin, G. R. (1972) 'Numerical analysis of continuum problems in zoned anisotropic media', Ph.D. thesis, Southampton University.
9. Niwa, Y., Kobayashi, S., and Fukui, T. (1974) 'An application of the integral equation method to seepage problems', *Proc. Twenty-fourth Jap. Natnl Conf. for Appl. Mech.*, pp. 470–486.
10. Chang, Y. P., Kang, C. S., and Chen, D. J. (1973) 'The use of fundamental Green's functions for the solution of problems of heat conduction in anisotropic media', *Int. J. Heat and Mass Transfer*, **16**, 1905–1918.
11. Cruse, T. A., Snow, D. W., and Wilson, R. B. (1977) 'Numerical solutions in axi-symmetric elasticity', *Int. J. Solids and Structures*, **11**, 493–511.
12. Cruse, T. A. (1974) 'An improved boundary integral equation method for three-dimensional elastic stress analysis', *Int. J. Computers and Structs*, **4**, 741–754.
13. Lamb, H. (1932) *Hydrodynamics*, 6th ed., Dover, New York.
14. Christiansen, S., and Ramussen, J. (1976) 'Numerical solutions for two-dimensional annular electro-chemical machining problems', *J. Inst. Math. Applic*, **18**, 295–307.
15. Christiansen, S. (1978) 'A review of some integral equations for solving the Saint-Venant torsion problem', *J. Elasticity*, **8**(1), 1–20.
16. Liggett, J. (1977) 'Location of free surface in porous media', *J. Hydraul Div., ASCE*, **HY4**, 353–365.
17. Jaswon, M. A., and Symm, G. T. (1977) *Integral Equation Methods in Potential Theory and Elastostatics*, Academic Press, London.

# · FOUR
## TWO-DIMENSIONAL PROBLEMS OF ELASTOSTATICS

### 4-1 INTRODUCTION

This chapter describes the development and application of the BEM to the numerical solution of two-dimensional problems of small-strain elastostatics. Most of the theoretical background behind the derivations presented in this chapter was explored in Chapters 2 and 3. The development of BEM for elasticity problems follows[1,2] that of potential theory very closely. However, unlike the integral equations of potential theory, which are scalar equations, the resulting integral equations in elasticity are a coupled set of vector equations. The basic singular solutions for elasticity, as would be expected, are more complex than those of potential theory. Therefore, in order to introduce them in a compact and elegant manner we have made use of some elementary operations involving indicial notation. The reader unfamiliar with this notation is advised to read Appendix A, which explains the various symbolic operations which have been used.

### 4-2 GOVERNING EQUATIONS

Consider an isotropic elastic body referred to a cartesian coordinate system as shown in Fig. 4-1, with axes $x_1, x_2$. The governing differential equation of

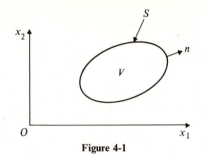

**Figure 4-1**

equilibrium for an element of the body can be written as

$$\frac{\partial \sigma_{ij}}{\partial x_j} + \psi_i = 0 \quad i,j = 1,2 \tag{4-1}$$

i.e., two equations of the form

$$\frac{\partial \sigma_{11}}{\partial x_1} + \frac{\partial \sigma_{12}}{\partial x_2} + \psi_1 = 0$$

in which $\sigma_{ij}$ are the stress components and $\psi_i$ the components of the body forces per unit volume. Hooke's law relating the stress and strain components in an isotropic elastic solid can be written as

$$\sigma_{ij} = \frac{2\mu\nu}{1-2\nu} \delta_{ij} \varepsilon_{kk} + 2\mu\varepsilon_{ij} \tag{4-2}$$

that is,

$$\sigma_{11} = \frac{2\mu\nu}{1-2\nu}(\varepsilon_{11} + \varepsilon_{22}) + 2\mu\varepsilon_{11} \quad \text{etc. for } \varepsilon_{33} = 0,$$

where $\mu$ and $\nu$ are elastic constants and $\delta_{ij}$ Kronecker's delta symbol. The strains and displacements are related by

$$\varepsilon_{ij} = \tfrac{1}{2}\left(\frac{\partial u_i}{\partial x_j} + \frac{\partial u_j}{\partial x_i}\right) \tag{4-3}$$

Substituting Eq. (4-2) in Eq. (4-1) and using Eq. (4-3) we obtain Navier's equations for equilibrium expressed in terms of the displacement components:

$$\frac{1}{1-2\nu}\frac{\partial^2 u_j}{\partial x_i \partial x_j} + \frac{\partial^2 u_i}{\partial x_j \partial x_j} + \frac{1}{\mu}\psi_i = 0 \tag{4-4}$$

Equation (4-4) is the governing differential equation for our problem which has to be solved subject to certain boundary conditions. For example, the displacement boundary-value problem assumes knowledge of the displacement over the boundary $S$, i.e.,

$$u_i(x) = g_i(x) \quad \text{on } S \tag{4-5}$$

The traction boundary-value problem requires

$$t_i(x) = \sigma_{ij}(x)\, n_j(x) = h_i(x) \qquad \text{on } S \tag{4-6}$$

that is, $g_i(x)$ and $h_i(x)$ are the specified conditions on the boundary.

## 4-3 THE SINGULAR SOLUTIONS

The fundamental singular solutions play the same important role in the BEM algorithms as did their counterparts in the previous potential flow problems. The classical result which evaluates the displacement field $u_i(x)$ due to a concentrated unti force $e_j(\xi)$ within an elastic body forms the basis of all the subsequent analysis. For plane strain problems[3]

$$u_i(x) = G_{ij}(x, \xi)\, e_j(\xi) \tag{4-7}$$

where

$$G_{ij}(x, \xi) = C_1\left(C_2\,\delta_{ij}\ln r - \frac{y_i\, y_j}{r^2}\right) + A_{ij}$$

$$C_1 = -\frac{1}{8\pi\mu(1-v)}$$

$$C_2 = 3 - 4v$$

$A_{ij}$ = arbitrary constant tensor, the value of which can be determined by specifying that at any distance from the load point the displacements are zero [i.e., Eq. (4-7) determines displacements relative to $u_i(r_0) = 0$ only]

$$y_i = x_i - \xi_i$$

$$y_j = x_j - \xi_j$$

$$r^2 = y_i\, y_i$$

The strains corresponding to the above displacement field can be obtained by substituting Eq. (4-7) into the strain displacement relations (4-3) as

$$\varepsilon_{ij}(x) = B_{ijk}(x, \xi)\, e_k(\xi) \tag{4-8}$$

where

$$B_{ijk}(x, \xi) = \frac{C_1}{r^2}\left[(1-2v)(\delta_{ik}\, y_j + \delta_{jk}\, y_i) + 2\frac{y_i\, y_j\, y_k}{r^2} - \delta_{ij}\, y_k\right]$$

The corresponding stresses can be deduced from the stress-strain relations as

$$\sigma_{ij}(x) = T_{ijk}(x, \xi)\, e_k(\xi) \tag{4-9}$$

where

$$T_{ijk}(x, \xi) = \left(\frac{C_3}{r^2}\right)\left[C_4(\delta_{ik}\,y_j + \delta_{jk}\,y_i - \delta_{ij}\,y_k) + \frac{2y_i\,y_j\,y_k}{r^2}\right]$$

$$C_3 = -\frac{1}{4\pi(1-v)}$$

$$C_4 = 1 - 2v$$

We shall also require the surface tractions $t_i(x)$ at a point $(x_i)$ on a surface with outward normal $n_j(x)$ which are calculated from

$$t_i(x) = \sigma_{ij}(x)\,n_j(x)$$

$$= F_{ik}(x, \xi)\,e_k(\xi) \tag{4-10}$$

where, in our case,

$$F_{ik} = \left(\frac{C_3}{r^2}\right)\left[C_4(n_k\,y_i - n_i\,y_k) + \left(C_4\,\delta_{ik} + \frac{2y_i\,y_k}{r^2}\right)y_j\,n_j\right]$$

Equations (4-9) to (4-10) provide all the displacement, stress, strain, and surface traction components of the unit solution that we shall require. Solutions for the corresponding plane stress problem can be obtained from those for the plane strain case given above by using an effective Poisson's ratio $\bar{v} = v/(1+v)$.

It is worth noting here that the various functions $G_{ij}$, $B_{ijk}$, $T_{ijk}$, $F_{ik}$ are singular when the load point and the field point coincide ($x_i = \xi_i$). The singularity in $G_{ij}$ is in the term $\ln r$ (weakly singular) and that in others in $1/r$ (strongly singular). Integrals involving weakly singular functions will always exist in the normal sense of integration even when $x_i = \xi_i$, but those involving strongly singular terms must be interpreted in the sense of a limiting value of the integral as the field point approaches the load point on the boundary. These features of the solution are exactly analogous to those already encountered in potential flow.

The reader might be dismayed by the apparently unnecessary complexity of Eq. (4-7) which could, after all, be equally well written in matrix notation as $\mathbf{u} = \mathbf{Ge}$, and since $\boldsymbol{\varepsilon}$, a strain vector, can also be expressed symbolically as $\boldsymbol{\varepsilon} = \mathbf{Lu}$, Eq. (4-8) can be stated simply as

$$\boldsymbol{\varepsilon} = \left\{\begin{array}{c}\varepsilon_{11} \\ \varepsilon_{22} \\ \gamma_{12}\end{array}\right\} = \begin{bmatrix}\dfrac{\partial}{\partial x_1} & 0 \\[2mm] 0 & \dfrac{\partial}{\partial x_2} \\[2mm] \dfrac{\partial}{\partial x_2} & \dfrac{\partial}{\partial x_1}\end{bmatrix}\left\{\begin{array}{c}u_1 \\ u_2\end{array}\right\} = \mathbf{Lu} = \mathbf{LGe}$$

However, in order to use this equation one has to evaluate separately each of the differential operations on every term in $\mathbf{G}$—not a simple matter when $\mathbf{G}$ is a quite complicated function. On the other hand, a brief study of the early part of

Appendix A will rectify any unfamiliarity with indicial notation following which the reader will appreciate how products like $B_{ijk}\, e_k$ are, in fact, a concise way of presenting the functions in a form convenient for computer coding.

## 4-4 INDIRECT BOUNDARY ELEMENT FORMULATION

### 4-4-1 Basic Formulation for a Homogeneous Isotropic Region

We consider a two-dimensional isotropic, linear, elastic, homogeneous region $V$ enclosed within a surface $S$ and apply a surface traction $\phi_j(\xi)$ distributed over $S$. The displacements $u_i(x)$ at any interior field point (see Fig. 4-2) $x_i$ due to the surface traction $\phi_j(\xi)$ and a known distribution of body forces $\psi_j(z)$ defined

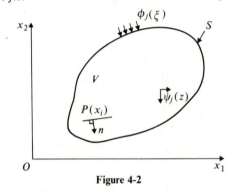

**Figure 4-2**

throughout the volume can be obtained by integrating the unit solutions related to $\phi$ and $\psi$ over $S$ and $V$ respectively as

$$u_i(x) = \int_S G_{ij}(x,\xi)\,\phi_j(\xi)\,ds(\xi) + \int_V G_{ij}(x,z)\,\psi_j(z)\,dv(z) + C_i \tag{4-11}$$

where $ds(\xi)$ and $dv(z)$ indicate that the variables of integration are $\xi$ and $z$ respectively. The $C_i$ are unknown rigid-body displacements which arise from the arbitrary nature of $r_0$ in Eq. (4-7) [i.e., the fact that this equation only provides relative values of $u_i(x)$—a problem similar to those we have already encountered in Chapters 2 and 3].

The strains can be obtained similarly from

$$\varepsilon_{ij}(x) = \int_S B_{ijk}(x,\xi)\,\phi_k(\xi)\,ds(\xi) + \int_V B_{ijk}(x,z)\,\psi_k(z)\,dv(z) \tag{4-12}$$

The corresponding stresses $\sigma_{ij}(x)$ at $(x_i)$ and the surface tractions $t_i(x)$ on a surface with an outward normal $n_j(x)$ passing through the point are given by

$$\sigma_{ij}(x) = \int_S T_{ijk}(x,\xi)\,\phi_k(\xi)\,ds(\xi) + \int_V T_{ijk}(x,z)\,\psi_k(z)\,dv(z) \tag{4-13}$$

whence, by using the relation $t_i(x) = \sigma_{ij}(x) n_j(x)$,

$$t_i(x) = \int_S F_{ik}(x, \xi)\, \phi_k(\xi)\, ds(\xi) + \int_V F_{ik}(x, z)\, \psi_k(z)\, ds(z) \tag{4-14}$$

Before proceeding further we might ask ourselves the following questions:

1. Do the integrals comprising Eqs (4-11), (4-12), (4-13), and (4-14) satisfy the governing differential equations of the problems everywhere within the region of interest?
2. Do they exist everywhere within $V$ and on $S$?

It is obvious that the surface integrals in Eqs (4-11) to (4-14) satisfy equations of equilibrium and compatibility everywhere within $V$ because the fundamental solutions on which they are based satisfy these conditions. These integrals also involve functions that are continuous and defined for all positions of $x_i$ within $V$. As far as the behaviour of the volume integrals within $V$ is concerned, they involve functions which become infinite at $x_i = z_i$ (i.e., when the load point and the field point coincide). Nevertheless, the integrals do exist in the normal sense because during the 'volume' integration the singularity vanishes (that is, $1/r$ terms are weakly singular under 'volumetric' integration). These integrals therefore also satisfy equations of equilibrium and compatibility within $V$. The conditions for uniqueness of the solution—that the effect of the distributed forces and tractions must not propagate to infinity—is satisfied by the integrals in Eqs (4-12), (4-13), and (4-14). Equation (4-11), which would not otherwise satisfy this condition, can be made to conform by requiring that the unique pair of rigid-body displacements $C_i$ take that specific value which exactly negates the cumulative effect on the infinite boundary of all the $\phi$ and $\psi$ forces.[4, 5] The $C_i$ values can be determined by using the augmented equations as described in Chapters 2 and 3. However, the existence of the integrals as $x_i$ approaches a boundary point $x_{oi}$, say, needs to be established.

By bringing the field point $x_i$ onto the surface we find that, whilst Eq. (4-11) provides displacement fields that are continuous everywhere, Eqs (4-12), (4-13), and (4-14) are not defined over the surface when the load point and the field point coincide. By using the standard methods of potential theory as described in Chapter 3, we can obtain, for example,

$$u_i(x_o) = \int_S G_{ij}(x_o, \xi)\, \phi_j(\xi)\, ds(\xi) + \int_V G_{ij}(x_o, z)\, \psi_j(z)\, dv(z) + C_i \tag{4-15}$$

$$t_i(x_o) = \pm \tfrac{1}{2}\delta_{ik}\, \phi_k(x_o) + \int_S F_{ik}(x_o, \xi)\, \phi_k(\xi)\, ds(\xi) + \int_V F_{ik}(x_o, z)\, \psi_k(z)\, dv(z) \tag{4-16}$$

provided the following conditions are not violated:[6, 7]

1. The point $x_o$ is not located at any edge or a corner (i.e., there must be a unique tangent plane at $x_o$).

2. The surface integral in Eq. (4-16) must be understood as a Cauchy principal-value integral.

The $\pm\frac{1}{2}\delta_{ik}\,\phi_k(x_o)$ term in Eq. (4-16) is positive if $x$ approaches the surface point $x_o$ from inside $S$ and negative if $x$ approaches $x_o$ from outside $S$. One has therefore to select the appropriate sign depending on whether it is the region inside or outside $S$ which is of interest.

We have now established that Eqs (4-15) and (4-16) are the two boundary integral equations governing the solution of any well-posed problem by the use of the indirect BEM. For example, if the displacements are specified on $S$ then the solution of (4-15) will provide the values of $\phi_j(\xi)$; on the other hand, if the tractions are specified on $S$ then Eq. (4-16) needs to be solved for $\phi_k(\xi)$. For a general mixed boundary-value problem Eq. (4-15) can be used for that portion of the boundary where displacements are specified and Eq. (4-16) can be used for the portion on which tractions are specified. The resulting equations in this case are then combined and solved together as described for the one-dimensional examples in Chapter 2.

### 4-4-2 Discretization of the Surface and Volume Integrals

Except for a few simple problems, closed-form solutions of Eqs (4-15) and (4-16) are not possible. Therefore a numerical method of solution has to be devised. We would emphasize that Eqs (4-15) and (4-16) are an exact formulation of the solution to the problem. Any numerical errors in the final result will have arisen solely from the discretization of the integrals and the subsequent solution of the algebraic equations. Whilst it is desirable to make the numerical algorithm as accurate as possible by using, for example, parametric representations for the system geometry and the functions $\phi$ and $\psi$ such as those developed in Chapter 8, we will in this chapter describe only the simplest possible algorithm which has been found to be efficient and usable for most practical problems. This utilizes linear boundary elements and triangular internal cells. For example, we can divide our two-dimensional region into a number $M$ of triangular cells and the

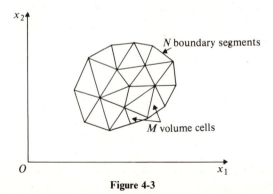

**Figure 4-3**

boundary of the region into $N$ straight line segments and assume the unknown function $\phi_j(\xi)$ and $\psi_j(z)$ to be uniform, or linearly varying, over the boundary elements or triangular cells. Figure 4-3 shows the details of such a system. It should be noted that although the interior cells have the appearance of elements in a finite element discretization scheme they are merely a convenient way of calculating the effects of $\psi_j(z)$ distributed throughout the volume. As pointed out previously the number of linear algebraic equations generated by numerical discretization schemes using BEM is governed only by the number of boundary elements used (i.e., the subdivision of $S$) and is quite unrelated to the number of internal cells employed (i.e., the subdivisions of $V$). Moreover, the volumetric cell discretization in BEM can be completely arbitrary in the sense that it need not match the surface discretization although, for simplicity, particularly from the point of view of data preparation, it is usually convenient to do so.

If we assume uniform distributions of $\phi$ and $\psi$ we can write for, say, the $p$th boundary element

$$u_i(x_o^p) = \sum_{q=1}^{N} \phi_j(\xi^q) \int_{\Delta S} G_{ij}(x_o^p, \xi^q)\, ds(\xi^q) + \sum_{l=1}^{M} \psi_j(z^l) \int_{\Delta V} G_{ij}(x_o^p, z^l)\, dv(z^l) + C_i$$

(4-17)

$$t_i(x_o^p) = \pm \tfrac{1}{2}\phi_i(x_o^p) + \sum_{q=1}^{N} \phi_k(\xi^q) \int_{\Delta S} F_{ik}(x_o^p, \xi^q)\, ds(\xi^q) + \sum_{l=1}^{M} \psi_k(z^l) \int_{\Delta V} F_{ik}(x_o^p, z^l)\, dv(z^l)$$

(4-18)

where    $x_o^p$ = coordinates of a representative field point on the $p$th boundary element, such as the centre of the element

$\Delta S$ = length of the $q$th boundary element

$\Delta V$ = area of the $l$th body cell

These equations can be written in a more convenient form by using matrix notation:

$$\mathbf{u}^p = \left( \int_{\Delta S} \mathbf{G}^{pq}\, ds \right)\boldsymbol{\phi}^q + \left( \int_{\Delta V} \mathbf{G}^{pl}\, dv \right)\boldsymbol{\psi}^l + \mathbf{IC} \qquad (4\text{-}19)$$

$$\mathbf{t}^p = \left( \int_{\Delta S} \mathbf{F}^{pq}\, ds \right)\boldsymbol{\phi}^q + \left( \int_{\Delta V} F^{pl}\, dv \right)\boldsymbol{\psi}^l \pm \tfrac{1}{2}\mathbf{I}\boldsymbol{\phi}^p \qquad (4\text{-}20)$$

where $q = 1, 2, ..., N$ and $l = 1, 2, ..., M$, which are written now as superfixes merely for convenience. If a linear variation of $\boldsymbol{\phi}$ and $\boldsymbol{\psi}$ needs to be considered then the quantities $\boldsymbol{\phi}^q$ and $\boldsymbol{\psi}^l$ cannot be taken outside the integral sign as was done in (4-19) and (4-20). However, by expressing $\boldsymbol{\phi}^q$ and $\boldsymbol{\psi}^l$ as a function of the nodal values using shape functions[8] $\mathbf{N}^q$ and $\mathbf{M}^l$, for the $q$th boundary element which has nodes $r$ and $s$ and for the $l$th cell which has nodes $r, s$, and $t$ as shown in

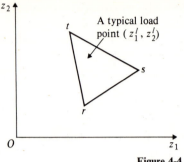

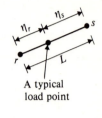

A typical
load point

**Figure 4-4**

Fig. 4-4, we can write

$$\phi^q = N^q \phi_{nodal} = N^q \phi_n$$

and

$$\psi^l = M^l \psi_{nodal} = M^l \psi_n$$

(4-21)

where

$$\phi_{nodal} = \phi_n = \left\{ \begin{matrix} \phi_r \\ \phi_s \end{matrix} \right\}, \text{ a } 4 \times 1 \text{ vector } \phi_r \text{ and } \phi_s \text{ being the nodal values of } \phi$$

$$N^q = \begin{bmatrix} N_1 & 0 & N_2 & 0 \\ 0 & N_1 & 0 & N_2 \end{bmatrix}, \text{ a } 2 \times 4 \text{ matrix}$$

$$N_1 = 1 - \frac{\eta_r}{L}$$

$$N_2 = 1 - \frac{\eta_s}{L}$$

$\eta_r, \eta_s$ = location of the instantaneous load point as shown above

$L$ = length of the boundary element

$$\psi_n = \left\{ \begin{matrix} \psi_r \\ \psi_s \\ \psi_t \end{matrix} \right\}, \text{ a } 6 \times 1 \text{ vector } \psi_r, \psi_s, \psi_t, \text{ being the nodal values of } \psi^l$$

$$M^l = \frac{1}{2\Delta} \begin{bmatrix} M_1 & 0 & M_2 & 0 & M_3 & 0 \\ 0 & M_1 & 0 & M_2 & 0 & M_3 \end{bmatrix}, \text{ a } 2 \times 6 \text{ matrix}$$

$\Delta$ = area of the $l$th cell

$$= \tfrac{1}{2} \times \text{determinant of} \begin{vmatrix} 1 & z_1^r & z_2^r \\ 1 & z_1^s & z_2^s \\ 1 & z_1^t & z_2^t \end{vmatrix}$$

$$= \tfrac{1}{2}[(z_1^s z_2^t - z_1^t z_2^s) - (z_1^r z_2^t - z_1^t z_2^r) + (z_1^r z_2^s - z_1^s z_2^r)]$$

$$M_1 = (z_1^s z_2^t - z_2^s z_1^t) + z_1^l(z_2^s - z_2^t) + z_2^l(z_1^t - z_1^s)$$

$$M_2 = (-z_1^r z_2^t + z_1^t z_2^r) + z_1^l(z_1^t - z_2^r) + z_2^l(z_1^r - z_1^t)$$

$$M_3 = (z_1^r z_2^s - z_1^s z_2^r) + z_1^l(z_2^s - z_2^r) + z_2^l(z_1^r - z_1^s)$$

Thus Eqs (4-19) and (4-20) can be rewritten for a typical boundary nodal point $x_o^p$ as

$$\mathbf{u}^p = \left[ \int_{\Delta s} \mathbf{G}^{pq} \mathbf{N}^q \, ds \right] \boldsymbol{\phi}_n + \left[ \int_{\Delta V} \mathbf{G}^{pl} \mathbf{M}^l \, dv \right] \boldsymbol{\psi}_n + \mathbf{IC} \qquad (4\text{-}22)$$

$$\mathbf{t}^p = \left[ \int_{\Delta s} \mathbf{F}^{pq} \mathbf{N}^q \, ds \right] \boldsymbol{\phi}_n + \left[ \int_{\Delta V} \mathbf{F}^{pl} \mathbf{M}^l \, dv \right] \boldsymbol{\psi}_n + \boldsymbol{\beta} \boldsymbol{\phi}^p \qquad (4\text{-}23)$$

where $\boldsymbol{\beta} = \frac{1}{2}\mathbf{I}$ if the nodal point is not located on a corner. Unfortunately, with a linear variation of $\phi$ on boundary elements, the most logical positions of field points $x_o^p$ are the nodal points, some of which may be located on corners. The value of $\boldsymbol{\beta}$ then becomes dependent on the solid angle subtended by the boundary elements as discussed in Chapter 3. The problem can be circumvented by taking the field point to be slightly away from the corner thus representing the corner by two separate nodes; hence

$$\mathbf{t}^p = \left[ \int_{\Delta s} \mathbf{F}^{pq} \mathbf{N}^q \, ds \right] \boldsymbol{\phi}_n + \left[ \int_{\Delta V} \mathbf{F}^{pl} \mathbf{M}^l \, dv \right] \boldsymbol{\psi}_n + \frac{1}{2} \boldsymbol{\phi}^p \qquad (4\text{-}23a)$$

This also takes account of the fact that, if the physical problem has corners, although displacements at a corner node can be specified unambiguously, the specified tractions can only be represented by considering two corner nodes indefinitely close to each other (typically 0.05 times the length of the local boundary element apart) representing the limiting endpoints of the two surfaces (see Fig. 4-5). For further details on edges and corners see Chapter 7.

**Figure 4-5**

Equations (4-22) and (4-23) can be used to construct the final system of algebraic equations. Before we can do this we need to evaluate the integrals within the square brackets. This can be done simply by using gaussian integration formulae (see Appendix C) provided the field point $x_o^p$ does not lie anywhere within the loaded boundary element because the integrals will then become singular.

If we look closely at the shape function matrix $\mathbf{N}^q$ we note that we can write

$$\mathbf{N}^q = \mathbf{N}_1^q + \mathbf{N}_2^q$$

where $N_1^q$ are the terms which are constant within the element ($=$ unity) and $N_2^q$ are the terms which contain the variable of integration. Hence we can write the surface integrals in (4-22) and (4-23) as

$$\int_{\Delta S} G^{pq} N^q \, ds = \left[ \int_{\Delta S} G^{pq} \, ds \right] N_1^q + \int_{\Delta S} G^{pq} N_2^q \, ds$$

and

$$\int_{\Delta S} F^{pq} N^q \, ds = \left[ \int_{\Delta S} F^{pq} \, ds \right] N_1^q + \int_{\Delta S} F^{pq} N_2^q \, ds$$

The second integrals on the right-hand side of the above equations can be calculated quite accurately by using gaussian integration formulae. The first integrals must be evaluated analytically by constructing a local coordinate system $(y_i)$ on the loaded element such that $y_1$ is normal to the element and $y_2$ is the positive tangential direction. Thus if the direction cosines of the local axes $y_1$ and $y_2$ with respect to the global axes are given by $e_{ij}$ then

$$\int_{\Delta S} G_{ij} \, ds = e_{ri} e_{sj} \int_{\Delta S} G'_{rs} \, dy_2 \tag{4-24}$$

$$\int_{\Delta S} F_{ik} \, ds = e_{ri} e_{sk} \int_{\Delta S} F'_{rs} \, dy_2 \tag{4-25}$$

where $G'_{rs}$ and $F'_{rs}$ are the expressions for $G_{rs}$ and $F_{rs}$ referred to the $y_1$, $y_2$ axes.

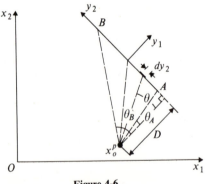

Figure 4-6

The integrals on the right-hand side of Eqs (4-24) and (4-25) can be easily calculated exactly by working with a polar coordinate system as shown in Fig. 4-6. Equations (4-24) and (4-25) can be written as

$$\int_{\Delta S} G_{ij} \, ds = e_{1i}(e_{1j} \Delta G'_{11} + e_{2j} \Delta G'_{12}) + e_{2i}(e_{1j} \Delta G'_{21} + e_{2j} \Delta G'_{22})$$

$$\int_{\Delta S} F_{ik} \, ds = e_{1i}(e_{1j} \Delta F'_{11} + e_{2j} \Delta F'_{12}) + e_{2i}(e_{1j} \Delta F'_{21} + e_{2j} \Delta F'_{22})$$

where $\Delta G'_{11}$, $\Delta F'_{11}$, $\Delta G'_{12}$, $\Delta F'_{12}$, etc., are the results of integrating $\int_{\Delta S} G'_{11}\,dy_2$, $\int_{\Delta S} F'_{11}\,dy_2$, etc.

In this way we obtain

$$\int_{\Delta S} G_{ij}\,ds = C_1 D[C_2\{\tan\theta(\log r - 1) + \theta\}\,\delta_{ij} - e_{1i}e_{1j}\theta$$

$$- (e_{1i}e_{2j} + e_{2i}e_{1j})\log r - e_{2i}e_{2j}(\tan\theta - \theta)]_{\theta_a, r_a}^{\theta_b, r_b} + A_{ij} \qquad (4\text{-}26)$$

$$\int_{\Delta S} F_{ik}\,ds = C_3[e_{1i}e_{1k}\{(C_4 + 1)\theta + \sin\theta\cos\theta\}\,\bar{n}_1 - e_{1i}e_{2k}\{C_4\ln r + \cos^2\theta\}\,\bar{n}_1$$

$$+ e_{1k}(e_{1i}\bar{n}_1 + e_{2i}\bar{n}_1)(C_4\ln r - \cos^2\theta)$$

$$+ e_{2i}e_{1k}\{(1 - C_4)\theta - \cos\theta\sin\theta\}\,\bar{n}_2 + e_{2k}(e_{1i}\bar{n}_2 + e_{2i}\bar{n}_1)$$

$$\times \{(C_4 + 1)\theta - \cos\theta\sin\theta\} + e_{2i}e_{2k}\{(C_4 + 2)\ln r + \cos^2\theta\}\,\bar{n}_2]_{\theta_a, r_a}^{\theta_b, r_b} \qquad (4\text{-}27)$$

where $\bar{n}_1 = e_{11}n_1 + e_{12}n_2$, $\bar{n}_2 = e_{21}n_1 + e_{22}n_2$, and Eqs (4-26) and (4-27) are valid for all field points $x_o^p$ except when the load point and the field point coincide

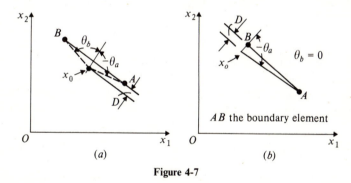

(a)      (b)

**Figure 4-7**

(i.e., as $D \to 0$ and $\theta_a$ and $\theta_b \to \mp\,\pi/2$ respectively, as in Fig. 4-7a). These can be evaluated either:

1. By taking the field point $x_o^p$ closely 'inside' the load point, as shown in Fig. 4-7 ($D = $ typically $0.01 \times$ length of the boundary element). In this case the $\pm\frac{1}{2}\phi_i(x_o)$ term should be discarded since it is being approximated directly by the proximity of $x_o^p$ and $x^p$.
2. By taking the field point $x_o^p$ onto the surface and then evaluating the integrals according to the principles outlined in Chapter 3.

The differences in the values of the coefficients found from a number of trial calculations using both methods 1 and 2 were quite negligible.

As far as the 'volume' integrals in Eqs (4-22) and (4-23) are concerned there are three classes of integrals to be evaluated:

1. Those involving the field point not located on a side or a vertex of the interior cell are integrals involving continuous functions; therefore these can be evaluated by using a gaussian integration formula.
2. Integrals involving the field point located on a vertex of an interior cell can be evaluated by using the four-point gaussian integration formula (see Fig. 4-8a).
3. Where the field point is located on a side of the triangular cell the integrals can be evaluated by splitting the triangular cell through the field point (see Fig. 4-8b) and applying the gaussian integration formula.

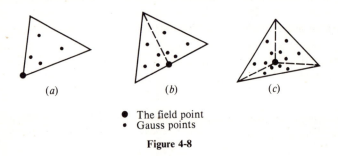

(a)                    (b)                    (c)

● The field point
• Gauss points

**Figure 4-8**

Equations (4-22) and (4-23) can then be applied to each boundary node in turn, which leads to a system of linear algebraic equations for the determination of the $\phi_n$ vectors on every node under a specified set of boundary conditions. This process is explained in detail below. Once all the $\phi_n$ values have been found, displacements, strains, and stresses at interior points can be evaluated from

$$\mathbf{u}(x) = \sum_{q=1}^{N}\left[\int_{\Delta S}\mathbf{G}(x,\xi^q)\mathbf{N}^q\,ds\right]\phi_n + \sum_{l=1}^{M}\left[\int_{\Delta V}\mathbf{G}(x,z^l)\mathbf{M}^l\,dv\right]\psi_n + \mathbf{IC} \quad (4\text{-}28)$$

$$\varepsilon(x) = \sum_{q=1}^{N}\left[\int_{\Delta S}\mathbf{B}(x,\xi^q)\mathbf{N}^q\,ds\right]\phi_n + \sum_{l=1}^{M}\left[\int_{\Delta V}\mathbf{B}(x,z^l)\mathbf{M}^l\,dv\right]\psi_n \quad (4\text{-}29)$$

$$\sigma(x) = \sum_{q=1}^{N}\left[\int_{\Delta S}\mathbf{T}(x,\xi^q)\mathbf{N}^q\,ds\right]\phi_n + \sum_{l=1}^{M}\left[\int_{\Delta V}\mathbf{T}(x,z^l)\mathbf{M}^l\,dv\right]\psi_n \quad (4\text{-}30)$$

The matrices $\mathbf{G}$, $\mathbf{B}$, and $\mathbf{T}$ are $2 \times 2$, $3 \times 2$, and $3 \times 2$ and the $\mathbf{u}$, $\varepsilon$, and $\sigma$ are $2 \times 1$, $3 \times 1$, and $3 \times 1$ vectors respectively, e.g.,

$$\varepsilon = \begin{Bmatrix} \varepsilon_{11} \\ \varepsilon_{22} \\ \varepsilon_{12} \end{Bmatrix} \quad \mathbf{B} = \begin{bmatrix} B_{111} & B_{112} \\ B_{221} & B_{222} \\ B_{121} & B_{122} \end{bmatrix} \quad \text{etc.}$$

Integrals in Eqs (4-28) to (4-30) can be evaluated using an integration scheme similar to that outlined above for the construction of the system equations.

One interesting feature of IBEM is that it is possible to calculate the stress tensor at an interior point by using Eq. (4-14) twice.[9] First, the kernel $F_{ij}$ is calculated with outward normal $n_i(x)$ as the unit vector in the direction of the global axis $x_1$ (that is, $n_i(x) = \delta_{i1}$). This gives

$$\sigma_{11}(x) = t_1(x) \qquad \sigma_{12}(x) = t_2(x)$$

Then, choosing $n_i(x) = \delta_{i2}$ (the unit normal in the direction of the global axis 2) we get

$$\sigma_{21}(x) = t_1(x) \qquad \sigma_{22} = t_2(x)$$

If it is necessary to calculate the stresses and strains at large numbers of interior points, it is more efficient to use Eq. (4-28), which is valid at all points including the boundary, to calculate accurate values of displacements at nodal points of interior cells. The stresses and strains can be obtained from them by using the same procedure as in the finite element or the finite difference method. Thus if $\mathbf{u}_n$ are the six components of the nodal displacement vectors for an interior triangular cell, the displacement vector at any point within the cell can be written as [see Eq. (4-21)]

$$\mathbf{u} = \mathbf{M}\mathbf{u}_n \tag{4-31}$$

The strains can be obtained by substituting Eq. (4-31) in Eq. (4-3). The result thus obtained can be written as

$$\boldsymbol{\varepsilon} = \bar{\mathbf{M}}\mathbf{u}_n \tag{4-32}$$

where $\bar{\mathbf{M}}$ is a $3 \times 6$ matrix. The stresses corresponding to these strains can then be obtained from

$$\boldsymbol{\sigma} = \mathbf{D}\bar{\mathbf{M}}\mathbf{u}_n \tag{4-33}$$

where $\mathbf{D}$ is the $3 \times 3$ elasticity matrix.

The major drawback of using such a procedure is that it is not possible to use Poisson's ratio of greater than 0.49 to represent an incompressible material ($v = 0.5$). The full BEM procedures Eqs (4-28) to (4-30) can, of course, cope with any admissible value of $v$.

### 4-4-3 Numerical Solution

The algorithm described above involves data relating to the surface geometry and the interior cell discretization scheme. The method of generating such quantities is outlined in Chapter 15, which deals with computer programming. The first step towards the solution is, as before, to assemble a specific system of linear algebraic equations incorporating the given boundary information for the problem in hand. Thus we can use Eqs (4-19) and (4-20) or (4-22) and (4-23) to generate the following system of equations for $N'$ boundary nodes:

$$\left\{ \begin{array}{c} \alpha\mathbf{u}^s \\ \mathbf{t}^s \\ 0 \end{array} \right\} = \left[ \begin{array}{cc} \mathbf{I}^o & \mathbf{G}^{ss}\alpha \\ 0 & \mathbf{F}^{ss} \\ 0 & \mathbf{b}_m \end{array} \right] \left\{ \begin{array}{c} \mathbf{C}\alpha \\ \boldsymbol{\phi} \end{array} \right\} + \left[ \begin{array}{c} \mathbf{G}^{sv}\alpha \\ \mathbf{F}^{sv} \\ \mathbf{c}_m \end{array} \right] \left\{ \boldsymbol{\psi} \right\} \tag{4-34}$$

$(2N'+2) \times 1 \quad (2N'+2) \times (2N'+2) \quad (2N'+2) \times 1 \quad (2N'+2) \times 2M' \quad 2M' \times 1$

where the superscripts $s$ and $v$ designate that these quantities are obtained by using either purely surface data ($ss$) or surface and volume data ($sv$), and

$\mathbf{u}^s = 2m \times 1$ vector corresponding to the numbers ($2m$) of given surface displacement components which usually comprise the known boundary displacements on the boundary nodes

$\mathbf{t}^s = 2n \times 1$ vector corresponding to the number ($2n$) of given surface traction components which may comprise the known tractions on $n$ boundary nodes

$\mathbf{0} = $ null matrices relating to the global equilibrium (auxiliary condition)

$$\mathbf{I}^o = 2m \times 2 \text{ matrix} = \begin{bmatrix} 1 & 0 \\ 0 & 1 \\ 1 & 0 \\ 0 & 1 \\ \cdots \\ \cdots \\ \cdots \end{bmatrix}$$

$$\mathbf{C} = 2 \times 1 \text{ vector} = \begin{Bmatrix} C_1 \\ C_2 \end{Bmatrix}$$

$\mathbf{G}^{ss}, \mathbf{F}^{ss}, \mathbf{G}^{sv}, \mathbf{F}^{sv}$ are $2m \times 2N'$, $2n \times 2N'$, $2m \times 2M'$, and $2n \times 2M'$ matrices of coefficients respectively

$\boldsymbol{\phi}, \boldsymbol{\psi}$ are $2N' \times 1, 2M' \times 1$ vectors of unknown tractions and known body forces within the volume respectively

$$b_m, c_m = \begin{bmatrix} b_1 & 0 & b_2 & 0 & \cdots \\ 0 & b_1 & 0 & b_2 & \cdots \end{bmatrix}, \begin{bmatrix} c_1 & 0 & c_2 & 0 & \cdots \\ 0 & c_1 & 0 & c_2 & \cdots \end{bmatrix},$$

which are $2 \times 2N'$ and $2 \times 2M'$ matrices involving weighting functions (length and area) of surface elements and volume cells respectively

$N', M' = $ total number of boundary nodes and volume nodes respectively

$\alpha = $ scale factor ($\mu$) by which equations for displacements are multiplied so that all the coefficients in the matrices are of the same order of magnitude

Solution of (4-38) will provide the values of $\boldsymbol{\phi}$ and $\mathbf{C}$ for specified admissible sets of values of $\mathbf{u}, \mathbf{t},$ and $\boldsymbol{\psi}$. Having obtained $\boldsymbol{\phi}, \mathbf{C}$ the solutions for interior values of the displacements, strains, and stresses can be calculated by using Eqs (4-28), (4-29), and (4-30) respectively.

## 4-5 DIRECT BOUNDARY ELEMENT FORMULATION

### 4-5-1 Basic Formulation for a Homogeneous Isotropic Region

The direct formulation of BEM is most conveniently approached from the reciprocal work theorem.[10-12] This theorem can be simply stated as follows: if two distinct elastic equilibrium states $(\psi_i^*, t_i^*, u_i^*)$, $(\psi_i, t_i, u_i)$ exist in a region $V$ bounded by the surface $S$ then the work done by the forces of the first system (*) on the displacements of the second is equal to the work done by the forces of the second system on the displacements of the first (*). Thus

$$\int_S t_i^*(x)\, u_i(x)\, ds(x) + \int_V \psi_i^*(z)\, u_i(z)\, dv(z) = \int_S t_i(x)\, u_i^*(x)\, ds(x) + \int_V \psi_i(z)\, u_i^*(z)\, dv(z)$$

$$(4\text{-}35)$$

where $x$ is a point on $S$ and $z$ is a point in $V$. Equation (4-35) is clearly a generalization of Eq. (2-31) relating to a simple elastic beam. If we choose the actual state of displacements, tractions, and body forces as $u_i$, $t_i$, and $\psi_i$ respectively and the (*) system as those corresponding to a unit force system in an infinite solid, as described in Sec. 4-3, we can write, from Eq. (4-35),

$$\int_S F_{ij}(x,\xi)\, u_i(x)\, ds(x) + \int_V \delta_{ij}\, \delta(z,\xi)\, u_i(z)\, dv(z) = \int_S t_i(x)\, G_{ij}(x,\xi)\, ds(x)$$

$$+ \int_V \psi_i(z)\, G_{ij}(z,\xi)\, dv(z) \qquad (4\text{-}36)$$

In writing this equation we have made use of Eq. (4-10), etc., $t_i^*(x) = F_{ij}(x,\xi)\, e_j^*(\xi)$, and the following manipulation of the unit force $\psi_i^*(z) = e_i^*(z)$ in the second left-hand-side term: viz.,

$$\int_V e_i^*(z)\, u_i(z)\, dv(z) = \int_V e_i^*(\xi)\, \delta(z,\xi)\, u_i(z)\, dv(z)$$

$$= \int_V e_j^*(\xi)\, \delta_{ij}\, \delta(z,\xi)\, u_i(z)\, dv(z)$$

The $e_j^*(\xi)$ term is now common to all the integrals and outside the integral signs, which allows us to write Eq. (4-36). This particular term can be simplified further by noting that

$$\int_V \delta_{ij}\, \delta(z-\xi)\, u_i(z)\, dv(z) = \int_V u_j(z)\, \delta(z,\xi)\, dv(z) = \beta u_j(\xi)$$

where $\beta = 1$ within $V$ and $\beta = 0$ outside $S$. Hence from Eq. (4-36) we arrive at

$$u_j(\xi) = \int_S [t_i(x)\, G_{ij}(x,\xi) - F_{ij}(x,\xi)\, u_i(x)]\, ds(x) + \int_V \psi_i(z)\, G_{ij}(z,\xi)\, dv(z) \qquad (4\text{-}37)$$

Equation (4-37) now provides the displacements $u_j(\xi)$ at an interior point $(\xi)$ due to any admissible combinations of $t_i$ and $u_i$ over $S$ and a given distribution of

$\psi_i$ within the volume—this equation is, in fact, well known as Somigliana's identity[11,12] for the displacement vector. The functions $G_{ij}$ and $F_{ij}$ are precisely those defined in Eqs (4-7) and (4-10) but, stemming from the use of the reciprocal theorem, three rather subtle changes have occurred in the significance of $(x, \xi)$ and $(i, j)$. A careful comparison of, say, Eqs (4-11) and (4-37) will show that Eq. (4-37) embodies an extension of the features discussed in connection with Eq. (3-29): viz.,

1. The first argument $(x)$ of the kernel function is now the load point (not $\xi$, which is now the field point coordinate) and the integration is carried through with respect to $x$.
2. Summation is over $i$, not $j$ (i.e., the roles of $i$ and $j$ are reversed).
3. The normal vector $n$ in $F_{ij}$ is now through the load point surface at $x$.

In functions such as $G_{ij}(x, \xi)$ which are symmetrical with respect to both $(i, j)$ and $(x, \xi)$ [see beneath Eq. (4-7)] these changes have no effect and all our previous results from exact integration will be valid. However, for $F_{ij}(x, \xi)$, which is antisymmetric in both $(i, j)$ and $(x, \xi)$ [see beneath Eq. (4-10)], the above changes lead to a totally different result from the integration.

By bringing the field point (now $\xi$) onto a point $x_o$ on the surface of the domain we obtain the following results (for $\lim \xi \to x_o$):

$$u_j(\xi) = u_j(x_o)$$

$$\int_S t_i(x) G_{ij}(x, \xi) \, ds(x) = \int_S t_i(x) G_{ij}(x, x_o) \, ds(x)$$

$$\oint_S F_{ij}(x, \xi) u_i(x) \, ds(x) = \alpha_{ij} u_i(x_o) + \int_S F_{ij}(x, x_o) u_i(x) \, ds(x)$$

where $\alpha_{ij} = -\frac{1}{2}\delta_{ij}$ for the 'interior' problem with $V$ inside a smooth boundary $S$. When the field point $x_o$ is at a corner with a subtended angle $\omega$ the discontinuity term $\alpha_{ij}(x_o)$ is given by[9,13]

$$\alpha_{11} = -1 - C_3 \left[ (C_4 + 1)\omega + \frac{\sin 2\omega}{2} \right]$$

$$\alpha_{22} = -1 - C_3 \left[ (C_4 + 1)\omega - \frac{\sin 2\omega}{2} \right]$$

$$\alpha_{12} = \alpha_{21} = -\sin^2\omega$$

Hence for a point $x_o$ on the smooth surface we can write Eq. (4-37) as

$$\frac{1}{2}u_j(x_o) = \int_S \left[ t_i(x) G_{ij}(x, x_o) - F_{ij}(x, x_o) u_i(x) \right] ds(x)$$

$$+ \int_V \psi_i(z) G_{ij}(z, x_o) \, dv(z) \tag{4-38}$$

Equation (4-38) is the required integral equation for the solution of any well-posed boundary-value problem. We can use Eq. (4-37) and the strain-displacement relationship to obtain

$$\varepsilon_{jk}(\xi) = \int_S [t_i(x) B^*_{ijk}(x, \xi) - C^*_{ijk}(x, \xi) u_i(x)] \, ds(x)$$

$$+ \int_V \psi_i(z) B^*_{ijk}(z, \xi) \, dv(z) \tag{4-39}$$

where

$$B^*_{ijk}(x, \xi) = \frac{1}{2} \left( \frac{\partial G_{ij}}{\partial \xi_k} + \frac{\partial G_{ik}}{\partial \xi_j} \right) \tag{4-39a}$$

$$C^*_{ijk}(x, \xi) = \frac{1}{2} \left( \frac{\partial F_{ij}}{\partial \xi_k} + \frac{\partial F_{ik}}{\partial \xi_j} \right) \tag{4-39b}$$

and the stresses, by using the stress-strain relationship, as

$$\sigma_{jk}(\xi) = \int_S [t_i(x) T^*_{ijk}(x, \xi) - E^*_{ijk}(x, \xi) u_i(x)] \, ds(x)$$

$$+ \int_V \psi_i(z) T^*_{ijk}(z, \xi) \, dv(z) \tag{4-40}$$

where

$$T^*_{ijk}(x, \xi) = \left[ \frac{2\mu v}{1 - 2v} \delta_{jk} \frac{\partial G_{im}}{\partial \xi_m} + \mu \left( \frac{\partial G_{ij}}{\partial \xi_k} + \frac{\partial G_{ik}}{\partial \xi_j} \right) \right] \tag{4-40a}$$

$$E^*_{ijk}(x, \xi) = \left[ \frac{2\mu v}{1 - 2v} \delta_{jk} \frac{\partial F_{im}}{\partial \xi_m} + \mu \left( \frac{\partial F_{ij}}{\partial \xi_k} + \frac{\partial F_{ik}}{\partial \xi_j} \right) \right] \tag{4-40b}$$

which simplify to

$$T^*_{ijk}(x, \xi) = \frac{a_1}{r} \left[ \frac{a_2}{r} (\delta_{ik} y_j + \delta_{jk} y_i - \delta_{ij} y_k) + \frac{2 y_i y_j y_k}{r^3} \right] \tag{4-41}$$

$$E^*_{ijk}(x, \xi) = \frac{a_3}{r^2} \left[ \frac{n_l y_l}{r^2} \left\{ 2a_2 \delta_{ij} y_k + 2v(\delta_{ik} y_j + \delta_{jk} y_i) - \frac{8 y_i y_j y_k}{r^2} \right\} \right.$$

$$+ n_i \left( 2v \frac{y_j y_k}{r^2} + a_2 \delta_{jk} \right) + n_j \left( 2v \frac{y_i y_k}{r^2} + a_2 \delta_{ik} \right)$$

$$+ \left. n_k \left( 2a_2 \frac{y_i y_j}{r^2} - a_4 \delta_{ij} \right) \right] \tag{4-42}$$

$a_1 = 1/[4\pi(1 - v)]$, $a_2 = 1 - 2v$, $a_3 = \mu/[2\pi(1 - v)]$, $a_4 = 1 - 4v$, and $y_i = x_i - \xi_i$.

## 4-5-2 Discretization of the Boundary and Volume Integrals

As before, if we divide a two-dimensional region into $M$ triangular cells and the boundary of the region into $N$ straight line segments, we can write Eq. (4-38) for the displacement vector of a representative nodal point on the $p$th boundary element with the assumption of constant $t_i$, $u_i$, and $\psi_i$ as

$$\tfrac{1}{2}u_j(x_o^p) = \sum_{q=1}^{N}\left[ t_i(x^q)\int_{\Delta S}G_{ij}(x^q,x_o^p)\,ds(x^q) - u_i(x^q)\int_{\Delta S}F_{ij}(x^q,x_o^p)\,ds(x^q)\right]$$
$$+ \sum_{l=1}^{M}\psi_i(z^l)\int_{\Delta S}G_{ij}(z^l,x_o^p)\,dv(z^l) \qquad (4\text{-}43)$$

This equation can be rewritten in matrix notation as

$$\tfrac{1}{2}\mathbf{I}\mathbf{u}^p = \sum_{q=1}^{N}\left\{\left[\int_{\Delta S}\mathbf{G}^{pq}\,ds\right]\mathbf{t}^q - \left[\int_{\Delta S}\mathbf{F}^{pq}\,ds\right]\mathbf{u}^q\right\} + \sum_{l=1}^{M}\left[\int_{\Delta V}\mathbf{G}^{pl}dv\right]\boldsymbol{\psi}^l \quad (4\text{-}44)$$

If we assume both the unknown and known values of tractions and displacements to vary linearly over the boundary segments and the specified body forces also to vary linearly within each volume cell the equipment form of Eq. (4-44) becomes

$$\tfrac{1}{2}\mathbf{I}\mathbf{u}^p = \sum_{q=1}^{N}\left\{\left[\int_{\Delta S}\mathbf{G}^{pq}\mathbf{N}^q\,ds\right]\mathbf{t}_n - \left[\int_{\Delta S}\mathbf{F}^{pq}\mathbf{N}^q\,ds\right]\mathbf{u}_n\right\} + \sum_{l=1}^{M}\left[\int_{\Delta V}\mathbf{G}^{pl}\mathbf{M}^l\,dv\right]\boldsymbol{\psi}_n$$
$$(4\text{-}45)$$

where $\mathbf{t}_n$, $\mathbf{u}_n$, and $\boldsymbol{\psi}_n$ are nodal values of tractions, displacements, and specified body forces respectively, $\mathbf{u}^p$ is the displacement vector of a representative point on the $p$th boundary element which is not located on a corner. $\mathbf{N}^q$ and $\mathbf{M}^l$ are the shape functions for the $q$th boundary element and the $l$th volume cell respectively.

If the physical problem has corners, Eq. (4-45) is not valid for a field point located on a corner node. Thus for a general case Eq. (4-45) can be written as

$$\boldsymbol{\beta}\mathbf{u}^p = \sum_{q=1}^{N}\left\{\left[\int_{\Delta S}\mathbf{G}^{pq}\mathbf{N}^q\,ds\right]\mathbf{t}_n - \left[\int_{\Delta S}\mathbf{F}^{pq}\mathbf{N}^q\,ds\right]\mathbf{u}_n\right\}$$
$$+ \sum_{l=1}^{M}\left[\int_{\Delta V}\mathbf{G}^{pl}\mathbf{M}^l\,dv\right]\boldsymbol{\psi}_n \qquad (4\text{-}45a)$$

where $\boldsymbol{\beta}$ is a $2\times 2$ matrix whose coefficients are functions of the solid angle subtended at the corner node.

Now, by taking the field point successively to all the nodal points on the boundary and absorbing the $\boldsymbol{\beta}$ matrix with the corresponding $2\times 2$ blocks of coefficients of $[\int\mathbf{F}^{pq}\mathbf{N}^q\,ds]$ we can write

$$\sum_{q=1}^{N}\left\{\left[\int_{\Delta S}\mathbf{G}^{pq}\mathbf{N}^q\,ds\right]\mathbf{t}_n - \left[\int_{\Delta S}\mathbf{F}^{pq}\mathbf{N}^q\,ds\right]\mathbf{u}_n\right\} + \sum_{l=1}^{M}\left[\int_{\Delta V}\mathbf{G}^{pl}\mathbf{M}^l\,dv\right]\boldsymbol{\psi}_n = 0$$
$$(4\text{-}46)$$

or more compactly

$$\mathbf{G}\mathbf{t} - \mathbf{F}\mathbf{u} + \mathbf{G}\boldsymbol{\psi} = 0 \qquad (4\text{-}47)$$

If we apply a rigid-body displacement to the body[14] (we can do this to any region of finite size) with specified body forces $\boldsymbol{\psi} = 0$, then this will generate no traction on the boundary (that is, $\mathbf{t} = 0$). Therefore we can write Eq. (4-47) as

$$\mathbf{Fu} = 0 \tag{4-48}$$

It is easy to see that for the validity of the above equation for any system of arbitrary rigid-body displacements each coefficient of the $2 \times 2$ on-diagonal blocks must be numerically equal to the sum of the corresponding coefficients of all off-diagonal blocks with a change in sign. Since the $2 \times 2$ diagonal blocks contained the terms involving $\beta$ as well as strongly singular integrals, the evaluation of which require a little more analytical effort, this method of determining the diagonal block components by utilizing the values of the off-diagonal blocks is a very attractive feature of the direct boundary element method.

Unfortunately, the method fails entirely for some problems where the off-diagonal blocks of the matrix $\mathbf{F}$ are zeros (e.g., the problem of a loaded half space). Watson[15] has recently developed a modification of the above procedure to cater for these semi-infinite exterior problems. Alternatively, we can use the method developed for the indirect formulation described earlier.

All volume integrals and the surface integrals involved in evaluating the off-diagonal blocks of Eq. (4-47) can be determined by using various numerical integration formulae discussed previously in Sec. 4-4-2. The on-diagonal blocks of $\mathbf{G}$, and in certain situations (as pointed out in the previous paragraph) those of $\mathbf{F}$, must be determined by separating the integrals into two parts as shown in Sec. 4-4-2. Integrals involving the constant part of the shape function can then be evaluated by constructing a local coordinate system as shown in Fig. 4-6 and by transforming the results of integration with respect to these local axes to global axes as described previously. Thus for a field point inside the region (the results for a field point indefinitely close to the boundary node can be obtained from these, as shown in Fig. 4-7) we have[13]

$$\int_{\Delta s} G_{ij}\, ds = C_1 D \left[ C_2 \delta_{ij}\{\tan\theta\,(\ln r - 1) + \theta\} - e_{1i}e_{1j}\theta \right.$$

$$\left. - (e_{1i}e_{2j} + e_{2i}e_{1j})\ln r - e_{2i}e_{2j}(\tan\theta - \theta) \right]_{\theta_a, r_a}^{\theta_b, r_b} + A_{ij} \tag{4-49}$$

$$\int_{\Delta s} F_{ij}\, ds = C_3 \left[ C_4 \delta_{ij}\theta + e_{1i}e_{1j}(\theta + \sin\theta\cos\theta) + (e_{1i}e_{2j} + e_{2i}e_{1j}) \right.$$

$$\left. \sin^2\theta + e_{2i}e_{2j}(\theta - \sin\theta\cos\theta) + C_4(e_{1i}e_{2j} - e_{2i}e_{1j})\ln r \right]_{\theta_a, r_a}^{\theta_b, r_b} \tag{4-50}$$

Note that, whereas Eq. (4-49) is identical to (4-26), Eq. (4-50) is entirely different from (4-27); in particular, the outward normal vectors ($n$) have been absorbed within the components of $e_{ij}$ because the normals now relate to the loaded elements.

Having obtained the unknown values of displacements and tractions, together with the specified values of tractions and displacements, the interior values of displacements, strains, and stresses can be calculated from

$$\mathbf{u}(\xi) = \sum_{q=1}^{N} \left\{ \left[ \int_{\Delta S} \mathbf{G}(x^q, \xi) \mathbf{N}^q \, ds \right] \mathbf{t}_n - \left[ \int_{\Delta S} \mathbf{F}(x^q, \xi) \mathbf{N}^q \, ds \right] \mathbf{u}_n \right\}$$
$$+ \sum_{l=1}^{M} \left[ \int_{\Delta V} \mathbf{G}(z^l, \xi) \mathbf{M}^l \, dv \right] \mathbf{\psi}_n \tag{4-51}$$

$$\mathbf{\varepsilon}(\xi) = \sum_{q=1}^{N} \left\{ \left[ \int_{\Delta S} \mathbf{B}^*(x^q, \xi) \mathbf{N}^q \, ds \right] \mathbf{t}_n - \left[ \int_{\Delta S} \mathbf{C}^*(x^q, \xi) \mathbf{N}^q \, ds \right] \mathbf{u}_n \right\}$$
$$+ \sum_{l=1}^{M} \left[ \int_{\Delta V} \mathbf{B}^*(z^l, \xi) \mathbf{M}^l \, dv \right] \mathbf{\psi}_n \tag{4-52}$$

$$\mathbf{\sigma}(\xi) = \sum_{q=1}^{N} \left\{ \left[ \int_{\Delta S} \mathbf{T}^*(x^q, \xi) \mathbf{N}^q \, ds \right] \mathbf{t}_n - \left[ \int_{\Delta S} \mathbf{E}^*(x^q, \xi) \mathbf{N}^q \, ds \right] \mathbf{u}_n \right\}$$
$$+ \sum_{l=1}^{M} \left[ \int_{\Delta V} \mathbf{T}^*(z^l, \xi) \mathbf{M}^l \, dv \right] \mathbf{\psi}_n \tag{4-53}$$

It is evident that the calculations for the stresses and strains at interior points involve much more computational effort than those necessary in the indirect method. In cases where the strains and stresses are required at a large number of interior points it is more efficient to calculate accurate values of displacements at a sufficient number of interior nodes and then obtain the strains and stresses from them by using the method outlined in Sec. 4-4-2.

### 4-5-3 Numerical Solution

By using the surface geometry and interior cell geometry data we can once again use Eqs (4-45) and (4-46) to generate the following system of equations relating the known and unknown components of tractions $t$ and displacements $u$ on the boundary of the region. Thus,

$$-[\mathbf{G}^{ss}]\{\mathbf{t}^s\} + [\mathbf{F}^{ss}]\{\mathbf{u}^s\} = [\mathbf{G}^{sv}]\{\mathbf{\psi}^v\} \tag{4-54}$$

where, as before, the superscripts $s$ and $v$ designate that these quantities have been obtained from surface and volume data respectively, and

$\{\mathbf{u}^s\}, \{\mathbf{t}^s\} = 2N' \times 1$ vectors of displacements and tractions respectively at $N'$ surface nodes

$\{\mathbf{\psi}^v\} = 2M' \times 1$ vector of known body forces defined at $M'$ interior nodal points

$[\mathbf{G}^{ss}], [\mathbf{F}^{ss}] = 2N' \times 2N'$ matrices of coefficients

$[\mathbf{G}^{sv}] = 2N' \times 2M'$ matrix of coefficients

and the free term involving $\mathbf{u}^p$ in Eq. (4-45) has been absorbed within the diagonal coefficients of the matrix $[\mathbf{F}^{ss}]$. If the tractions are specified on the boundaries then Eq. (4-54) can be written as

$$[\mathbf{F}^{ss}]\{\mathbf{u}^s\} = [\mathbf{G}^{sv}]\{\boldsymbol{\psi}^v\} + [\mathbf{G}^{ss}]\{\mathbf{t}^s\}$$

in which the right-hand side is known; hence $\mathbf{u}^s$ can be calculated. On the other hand, if the displacements are known, then

$$[\mathbf{G}^{ss}]\{\mathbf{t}^s\} = [\mathbf{F}^{ss}]\{\mathbf{u}^s\} - [\mathbf{G}^{sv}]\{\boldsymbol{\psi}^v\}$$

whence the tractions can be calculated. For a more general, mixed boundary-value problem it is convenient to write Eq. (4-54) in the form

$$-[\alpha\mathbf{G}^{ss}]\left\{\frac{1}{\alpha}\mathbf{t}^s\right\} + [\mathbf{F}^{ss}]\{\mathbf{u}^s\} = [\mathbf{G}^{sv}]\{\boldsymbol{\psi}^v\} \tag{4-55}$$

where $\alpha$ is a scale factor introduced to scale the coefficients of the matrix $\mathbf{G}$ so that they are of the same order of magnitude as those of matrix $\mathbf{F}$. By rearranging the coefficients of the matrices on the left-hand side of Eq. (4-55) we can write

$$[K]\{X\} + [L]\{Y\} = [\mathbf{G}^{sv}]\{\boldsymbol{\psi}^v\} \tag{4-56}$$

where     $[K] = 2N' \times 2N'$ matrix of coefficients

    $\{X\} = 2N' \times 1$ vector of unknown traction and displacement components on the boundary

    $[L] = 2N' \times 2N'$ matrix of coefficients

    $\{Y\} = 2N' \times 1$ vector of known traction and displacement components on the boundary

Solution of Eq. (4-56) now provides the unknown tractions and displacements. Thereafter the total set of prescribed and calculated values of the boundary tractions and displacements together with the specified body force distribution can be used to obtain displacements, strains, and stresses at any subsequently selected interior points $\xi_i$ by using Eqs (4-51), (4-52), and (4-53) respectively. Multizone problems can be analysed by using the multizone assembly procedure outlined in Chapter 3.

## 4-6 BODY FORCES

A large class of boundary-value problems in elastostatics involve body forces generated by either a steady state temperature of seepage gradient or a gravitational potential. For all such problems the body force terms $\psi_i$ can be expressed as

$$\psi_i = \frac{\partial p}{\partial x_i} \tag{4-57}$$

where $p$ is a scalar function which has to satisfy the following differential equation within $D$:

$$\frac{\partial^2 p}{\partial x_i \partial x_i} = Q \tag{4-58}$$

For the case of a steady state thermal stress analysis or a constant gravitational potential, $Q = 0$, and for centrifugal forces due to fixed axis rotation, $Q =$ constant.

The body force integral in (4-37) for this special class of body forces then becomes

$$\int_V G_{ij}(z, \xi)\,\psi_i(z)\,dv(z) = \int_V G_{ij}(z, \xi)\frac{\partial p(z)}{\partial z_i}\,dv(z) \tag{4-59}$$

which can be converted into surface integrals over $S$ by making use of the fact that $p$ satisfies (4-58). Rizzo[16, 17] has developed a technique for carrying out this operation which we follow here.

By applying the divergence theorem to the right-hand side of (4-59) we have

$$\int_V G_{ij}(z, \xi)\frac{\partial p(z)}{\partial z_i}\,dv(z) = \int_S p(x)\,G_{ij}(x, \xi)\,n_i(x)\,ds(x)$$

$$- \int_V p(z)\frac{\partial G_{ij}(z, \xi)}{\partial z_i}\,dv(z) \tag{4-60}$$

where $x$ is a point on $S$ and $z$ is in $V$. Noting that

$$\frac{\partial G_{ij}(z, \xi)}{\partial z_i} = \frac{2\mu}{\lambda + 3\mu}\frac{\partial}{\partial z_j}(\ln r) = -\frac{2\mu}{\lambda + 3\mu}\frac{\partial}{\partial \xi_j}(\ln r) \tag{4-61}$$

where $\lambda$ and $\mu$ are Lamé constants, we can express the volume integral in (4-60) as

$$-\int_V p(z)\frac{\partial G_{ij}(z, \xi)}{\partial z_i}\,dv(z) = \frac{2\mu}{\lambda + 3\mu}\int_V p(z)\frac{\partial}{\partial \xi_j}(\ln r)\,dv(z)$$

$$= \frac{2\mu}{\lambda + 3\mu}\frac{\partial}{\partial \xi_j}\int_V p(z)\ln r\,dv(z) \tag{4-62}$$

At this point let us consider the property of the function $(r^2 \log r)$. We can show easily that

$$\nabla^2(r^2 \ln r) = \frac{\partial^2}{\partial z_k \partial z_k}(r^2 \ln r) = \frac{1}{r}\frac{\partial}{\partial r}\left(r\frac{\partial}{\partial r}\right)(r^2 \ln r)$$

$$= 4(\ln r + 1) \tag{4-63}$$

By choosing harmonic functions $\phi = p$ and $\phi^* = \frac{1}{4}(r^2 \ln r)$ in the integral identity of potential theory (see Chapter 3),

$$\int_V (\phi\nabla^2 \phi^* - \phi^*\nabla^2 \phi)\,dv(z) = \int_S \left(\phi\frac{\partial \phi^*}{\partial n} - \phi^*\frac{\partial \phi}{\partial n}\right)ds(x)$$

and using (4-58) we can obtain

$$\int_V [(\ln r + 1)p(z) - \tfrac{1}{4}r^2 \ln rQ]\, dv(z)$$

$$= \frac{1}{4}\int_S \left[ p(x)\frac{\partial}{\partial n(x)}(r^2 \ln r) - r^2 \ln r \frac{\partial p}{\partial n(x)} \right] ds(x)$$

or $\quad \displaystyle\int_V p(z)\ln r\, dv(z) = \frac{1}{4}\int_S \left[ p(x)\frac{\partial}{\partial n(x)}(r^2 \ln r) - r^2 \ln r \frac{\partial p}{\partial n(x)} \right] ds(x)$

$$- \int_V p(z)\,dv(z) + \frac{Q}{4}\int_V (r^2 \ln r)\,dv(z) \tag{4-64}$$

Using (4-64) in (4-62) leads to

$$\frac{\partial}{\partial \xi_j}\int_V p(z)\ln r\, dv(z) = \frac{1}{4}\frac{\partial}{\partial \xi_j}\int_S \left[ p(x)\frac{\partial}{\partial n(x)}(r^2 \ln r) - r^2 \ln r \frac{\partial p}{\partial n(x)} \right] ds(x)$$

$$- \frac{\partial}{\partial \xi_j}\int_V p(z)\,dv(z) + \frac{Q}{4}\frac{\partial}{\partial \xi_j}\int_v (r^2 \ln r)\,dv(z)$$

$$= \frac{1}{4}\int_S \left[ p(x)\frac{\partial}{\partial n(x)}\frac{\partial}{\partial \xi_j}(r^2 \ln r) - \frac{\partial p}{\partial n(x)}\frac{\partial}{\partial \xi_j}(r^2 \ln r) \right] ds(x)$$

$$- 0 - \frac{Q}{4}\int_V \frac{\partial}{\partial z_j}(r^2 \ln r)\,dv(z) \tag{4-65}$$

where $(\partial/\partial \xi_j)(r^2 \ln r) = -(\partial/\partial z_j)(r^2 \ln r)$ has been used in the last integral. Applying the divergence theorem to the last term of (4-65) we obtain

$$\int_V G_{ij}(z, \xi)\psi_i(z)\,dv(z) = \int_V G_{ij}(z, \xi)\frac{\partial p}{\partial z_i}\,dv(z)$$

$$= \int_S G_{ij}(x, \xi)\,n_i(x)\,p(x)\,ds(x)$$

$$+ \frac{\mu}{2(\lambda + 3\mu)} \left\{ \int_S \left[ p(x)\frac{\partial}{\partial n(x)}\frac{\partial}{\partial \xi_j}(r^2 \ln r) \right. \right.$$

$$\left. \left. - \frac{\partial p(x)}{\partial n(x)}\frac{\partial}{\partial \xi_j}(r^2 \ln r) - Qr^2 \ln r n_j(x) \right] ds(x) \right\} \tag{4-66}$$

where $\partial p/\partial n$ and $p$ over the boundary $S$ are known from a prior solution of (4-58). Thus the problem has been reduced to one involving boundary discretization only.

In some problems it is often possible to choose a simple polynomial solution for the body force field. The direct integral representation (4-38) for such a case can be written as

$$\tfrac{1}{2}\delta_{ij} u_j(x_o) = \int_S \left[ G_{ij}(x, x_o) t_i(x) - F_{ij}(x, x_o) u_j(x) \right] ds(x) + \bar{u}_i(x_o) \tag{4-67}$$

where $\bar{u}_i$ is the simple polynomial representation of the displacement field due to body forces arising from the presence of a gravitational, centrifugal force field, etc. As an example, the particular solution $\bar{u}_i$ due to a uniform stress field $\sigma_{11}^\infty$ and $\sigma_{22}^\infty$, etc., is given by[18]

$$u_i(P) = k_{ij} x_j(P) \qquad (4\text{-}67a)$$

where $u_j(P) = 0$ when $x_j(P) = 0$, $x_j(P)$ being the $j$th coordinate describing the position of the point $P$ and

$$k_{11} = \frac{1}{E'} \left[ \sigma_{11}^\infty - v' \sigma_{22}^\infty \right]$$

$$k_{22} = \frac{1}{E'} \left[ \sigma_{22}^\infty - v' \sigma_{11}^\infty \right]$$

$$k_{12} = k_{21} = \frac{2(1+v')}{E'} \sigma_{12}^\infty$$

$$E' = E, \quad v' = v \qquad \text{for plane stress}$$

$$E' = \frac{E}{1-v^2}, \quad v' = \frac{v}{1-v} \qquad \text{for plane strain}$$

## 4-7 ANISOTROPIC BODIES

### 4-7-1 Governing Equations

In the previous sections we have used the singular solution for an isotropic elastic body although in a great many practical problems the materials involved may have strongly anisotropic elastic properties (e.g., laminates, reinforced materials and most naturally occurring materials). Increasing anisotropy is reflected in a reduction of the symmetry of elastic properties and an increase in the number of elastic parameters needed to interrelate the stresses and strains at a point in such a body. The theory of general anisotropy involves 21 independent elastic constraints and materials in which no planes of elastic symmetry exist.[19,20] The form of Hooke's law which has to be used in this case is best illustrated if we write the six independent stress and strain components for the three-dimensional situation as vectors $\sigma$ and $\varepsilon$ and note that the most general linear relationship between them will involve an elastic compliance $6 \times 6$ matrix $[C]$ whence

$$\varepsilon = [C] \sigma \qquad (4\text{-}68)$$

It can be shown that there is no loss of generality when $[C]$ is symmetric[21] which has therefore a maximum of 21 independent components. We shall illustrate solutions for which $[C]$ has nine components only, which means that the material must possess three orthogonal planes of elastic symmetry (i.e., an orthotropic elastic continuum) (see Fig. 4-9a). Written out fully for this case, Eq. (4-68) takes

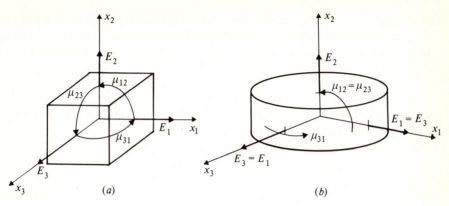

**Figure 4-9**

the form

$$
\begin{Bmatrix}
\varepsilon_{11} \\
\varepsilon_{22} \\
\varepsilon_{33} \\
\varepsilon_{12} \\
\varepsilon_{23} \\
\varepsilon_{31}
\end{Bmatrix}
=
\begin{bmatrix}
\dfrac{1}{E_1} & -\dfrac{v_{12}}{E_2} & -\dfrac{v_{13}}{E_3} & 0 & 0 & 0 \\[2mm]
-\dfrac{v_{21}}{E_1} & \dfrac{1}{E_2} & -\dfrac{v_{23}}{E_3} & 0 & 0 & 0 \\[2mm]
-\dfrac{v_{31}}{E_1} & -\dfrac{v_{32}}{E_2} & \dfrac{1}{E_3} & 0 & 0 & 0 \\[2mm]
0 & 0 & 0 & \dfrac{1}{2\mu_{12}} & 0 & 0 \\[2mm]
0 & 0 & 0 & 0 & \dfrac{1}{2\mu_{23}} & 0 \\[2mm]
0 & 0 & 0 & 0 & 0 & \dfrac{1}{2\mu_{31}}
\end{bmatrix}
\begin{Bmatrix}
\sigma_{11} \\
\sigma_{22} \\
\sigma_{33} \\
\sigma_{12} \\
\sigma_{23} \\
\sigma_{31}
\end{Bmatrix}
\quad (4\text{-}69)
$$

when the coordinate axes are perpendicular to the planes of elastic symmetry. In Eq. (4-69) $E_1$ and $\mu_{23}$ are the Young's modulus normal to, and shear modulus in, the $x_1 =$ constant planes of elastic symmetry and $v_{23}$ is the Poisson's ratio defining the extensional strain in the $x_2$ direction produced by unit compressive strain in the $x_3$ direction. The other components of $[C]$ are defined similarly by cyclic permutation of the subscripts. For $[C]$ to be symmetrical we must have $v_{13}/E_3 = v_{31}/E_1$, etc. (that is, $v_{(ij)}/E_j = v_{(ji)}/E_i$), which means that there are only nine independent compliances in Eq. (4-60). Furthermore, in order that the strain energy function $\boldsymbol{\sigma}^T \boldsymbol{\varepsilon}$ shall be positive definite the following restrictions have to be imposed on the individual elastic constants:[22]

$$
E_1, E_2, E_3, \mu_{12}, \mu_{23}, \mu_{31} > 0
$$
$$
(1 - v_{12} v_{21}), (1 - v_{23} v_{32}), (1 - v_{31} v_{13}) > 0 \qquad (4\text{-}70)
$$
$$
(1 - v_{12} v_{21} - v_{23} v_{32} - v_{31} v_{13} - 2 v_{12} v_{23} v_{31}) > 0
$$

These expressions serve as a useful check on the validity of both experimentally determined parameters or any adopted as arbitrary approximations and they also ensure that all evaluated stresses, strains, and expressions involving roots of functions of the compliances are real numbers.

The counterpart of Eqs (4-70) for an isotropic body is the well-known result that $E, \mu > 0$ and $-1 \leqslant \nu \leqslant \frac{1}{2}$, and it is worth noting that it is quite possible to have orthotropic materials with Poisson's ratios well outside this range. Lempriere[22] refers to a cross-ply composite with one $\nu$ component equal to 1.97. Equation (4-69) clearly generates a set of equations equivalent to (4-2), the isotropic case, when $E_1 = E_2 = E_3 = E$, $\mu_{12} = \mu_{23} = \mu_{12} = \mu_{23}$ all the Poisson's ratios are identical ($\nu$), and these three parameters are reduced to two independent ones by the relationship $E = 2\mu(1 + \nu)$. Alternatively, if we merely impose rotational symmetry on the elastic parameters (Fig. 4-9b) by requiring only that, say, $E_1 = E_3$, $\mu_{12} = \mu_{23}$, then Eq. (4-69) will interrelate the stress and strain components in a transversely isotropic material element. It is easily seen that the symmetry of $[C]$ then leads to

$$\frac{E_1}{E_2} = \frac{\nu_{21}}{\nu_{12}} = \frac{\nu_{23}}{\nu_{32}}, \qquad \nu_{13} = \nu_{31}, \quad \nu_{23} = \nu_{21} \tag{4-71}$$

$$\text{and} \quad E_1 = 2\mu_{31}(1 + \nu_{31}) \qquad \text{in the } (x_1, x_3) \text{ plane}$$

The result of these requirements is that a transversely isotropic elastic body can be completely specified by five elastic parameters (for example, $E_1$, $E_2$, $\mu_{12}$, $\nu_{12}$, and $\nu_{13}$).

The singular analytical solution for an infinitely long line load in an infinite orthotropic elastic space was developed by Tomlin,[4] and if this solution, rather than the simpler isotropic one, is used to generate the kernel functions in a BEM analysis we can solve two-dimensional problems in both orthotropic and transversely isotropic media. The only change required in the solution procedure which has been explained earlier in this chapter is that the new singular solution has to be used to calculate the components in all of the matrices $[\mathbf{F}]$, $[\mathbf{G}]$, etc.

### 4-7-2 The Singular Solution

The following solution for the stresses and displacements in an infinite orthotropic space deforming in either plane stress or plane strain due to an infinitely long line load was developed by Tomlin[4,23] from earlier work by Lekhnitskii.[20]

If, as in Fig. 4-9a, the coordinate planes are also the planes of elastic symmetry, then, for a case of plane stress, with the $x_3$ coordinate planes stress free, Eq. (4-69) with all of $\sigma_{33}, \sigma_{32}, \sigma_{31}, \sigma_{32},$ and $\sigma_{31}$ zero will relate the stress and strain components at any point. It is more convenient to write these equations as

$$\varepsilon_{(ii)} = C_{i\alpha} \sigma_{(\alpha\alpha)} \qquad i = 1, 2, 3; \; \alpha = 1, 2 \text{ with no indicial}$$
$$\text{summation implied} \tag{4-72}$$
$$\varepsilon_{12} = \frac{1}{2\mu_{12}} \sigma_{12}$$

It is then easily shown that for plane strain conditions the $C_{i\alpha}$ components only have to be replaced by $B_{i\alpha}$ which are related to them by

$$B_{i\alpha} = C_{i\alpha} - \frac{C_{i3} C_{\alpha 3}}{C_{33}} \tag{4-73}$$

The singular solution for a line load distributed with unit intensity along $x_3$ and acting in the $x_2$ direction provides the following stress and displacement components in the $(x_1, x_2)$ plane:

$$\frac{\sigma_{11}}{x_2} = \frac{[(b^2 + c)\sqrt{(B_{11} B_{22})} - 2(a^2 + c)b^2] x_1^2 + B_{22}(b^2 - c) x_2^2}{2\sqrt{(2)}\,\pi m b \sqrt{B_{11}}}$$

$$\frac{\sigma_{22}}{x_2} = \frac{\sigma_{12}}{x_1} = -\frac{\sqrt{B_{11}}(b^2 - c) x_1^2 + \sqrt{B_{22}}(b^2 + c) x_2^2}{2\sqrt{(2)}\,\pi m b} \tag{4-74}$$

in which $\quad c = \dfrac{1}{2\mu_{12}}$

$$a = [\sqrt{(B_{11} B_{22})} - B_{12} - c]^{1/2}$$

$$b = [\sqrt{(B_{11} B_{22})} + B_{12} + c]^{1/2}$$

and $\quad m = B_{11} x_1^4 + 2(B_{12} + c) x_1^2 x_2^2 + B_{22} x_2^4$

Whereas $b$ will always be real $a$ may take on real, imaginary, or zero values which leads to the following alternative values for the displacement components:

$a$ real: $\qquad u_1 = \dfrac{(a^2 b^2 + c^2)}{8\pi a b} \ln\left(\dfrac{1 + a l_1}{1 - a l_1}\right)$

$a$ zero: $\qquad u_1 = \dfrac{c^2 l_1}{4\pi b} \tag{4-75}$

$a$ imaginary: $\qquad u_1 = \dfrac{(a^2 b^2 + c^2)}{4\pi(ia) b} \tan^{-1}(ial_1)$

$a$ real: $\qquad u_2 = U - \dfrac{n}{4\sqrt{(2)}\,\pi a \sqrt{B_{11}}} \tan^{-1}(al_2)$

$a$ zero: $\qquad u_2 = U + \dfrac{c^2 l_2}{4\sqrt{(2)}\,\pi \sqrt{B_{11}}} \tag{4-76}$

$a$ imaginary: $\qquad u_2 = U - \dfrac{n}{8\sqrt{(2)}\,\pi b \sqrt{B_{11}}} \ln\left[\dfrac{1 + (ial_2)}{1 - (ial_2)}\right]$

in which
$$l_1 = \frac{\sqrt{(2)}\,x_1\,x_2}{\sqrt{(B_{11})}\,x_1^2 + \sqrt{(B_{22})}\,x_2^2}$$

$$l_2 = \frac{bx_2^2}{B_{11}\,x_1^2 + (B_{12} + c)\,x_2^2}$$

$$n = (b^2 - c)\,a^2 - (a^2 + c)\,c$$

and
$$U = -\frac{[(b^2 - c)\,c + (a^2 + c)\,b^2]\ln(m/B_{11}\,A^4)}{8\sqrt{(2)}\,\pi b\,\sqrt{B_{11}}}$$

Once again the displacements parallel to the line of action of the load, here $u_2$, are given as relative values only and the term involving $A$ in $U$, and hence in $u_2$, arises from the fact that Eqs (4-76) provide $u_2$ values relative to those at the arbitrary points $\pm(A, 0)$.

The stress and displacement components generated by a unit intensity line load along $x_3$ acting in the $x_1$ direction can be obtained by interchanging the subscripts 1 and 2 throughout all of equations (4-74) to (4-76).

Some care is needed in evaluating the above expressions involving inverse tangents; generally principal values $(-\pi/2$ to $+\pi/2)$ are used but it is possible for $l_2$ to become infinite and produce a discontinuity in $u_2$. Tomlin[4] overcomes this by adding $\pi$ to the value of $\tan^{-1}(al_2)$ whenever the denominator of $l_2$ is negative.

### 4-7-3 Numerical Solution

It is inevitable that as more sophisticated constitutive equations, anisotropy, etc., are introduced so the closed-form integration of equations such as (4-74) to (4-76) along line boundary elements[24] and internal areas becomes less attractive and numerical integration schemes have to be used.

Thus the solution of two-dimensional elastostatic problems for both orthotropic and transversely isotropic bodies, whether they be homogeneous or zoned, follows exactly the procedures outlined earlier incorporating numerical quadrature even down to the inclusion of a two-component rigid-body displacement vector in the indirect method to satisfy the infinite boundary radiation requirements. There are only two differences: (1) the unit solutions used are those in equations (4-74) to (4-76) rather than (4-7), (4-9); (2) in any particular zone the local coordinate axes $(x_1, x_2)$ are most conveniently directed along the axes of elastic symmetry for that zone. All boundary condition data should be transformed to these axes at the outset.

Tomlin used the above singular solution with the indirect BEM and uniform distribution of fictitious intensities along each element to solve a number of orthotropic media problems, two of which are illustrated in Sec. 4-8.

### 4-8 EXAMPLES OF TYPICAL APPLICATIONS

(a) **The problem of a thick cylinder** We now consider a long thick-walled circular cylinder, subjected to internal pressure, under plane strain and with a steady temperature gradient in the radial direction only. The problem is a very common one in heat-exchanger tubes and pressure vessels such as boiler drums and chemical reaction vessels. Figure 4-10 shows the problem and the discretization used in a DBEM computer programme with linear variations of $u_i$ and $t_i$. The boundary values of $T$ and $\partial T/\partial n$ were obtained from another DBEM analysis of the underlying potential flow problem using the same discretization but with uniform values of $T$ and $\partial T/\partial n$ over boundary elements.

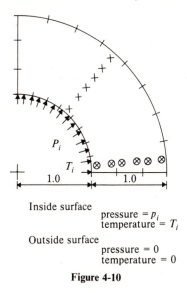

Inside surface
  pressure $= p_i$
  temperature $= T_i$

Outside surface
  pressure $= 0$
  temperature $= 0$

**Figure 4-10**

Here we show two solutions to the problem, under different boundary conditions, the first one involving zero thermal strain and the second one zero applied internal and external pressure. Figures 4-11 and 4-12 show the solutions for the stresses at the internal points marked in Fig. 4-10. The radial and circumferential stresses were calculated by simple transformation of the cartesian component computer output.

**(b) The problem of a loaded cantilever** Figure 4-13 shows the problem of an end-loaded cantilever incorporating artificial interfaces. The artificial subdivision of this long thin structure not only reduced the computer solution time but also improved solution accuracy by about 3 per cent over that obtained without the subdivisions.

The vertical displacements along the centreline of the beam are shown in Fig. 4-14 and the corresponding bending stresses are shown in Fig. 4-15. The stresses $\sigma_{11}$ have been plotted non-dimensionally as the ratio of $\sigma_{11}/\sigma_b$, where $\sigma_b$ is the bending stress from simple beam theory. The calculated stresses are typically 5 per cent lower than those given by beam theory. These discrepancies are unusually large and arise because the stresses being compared are those at

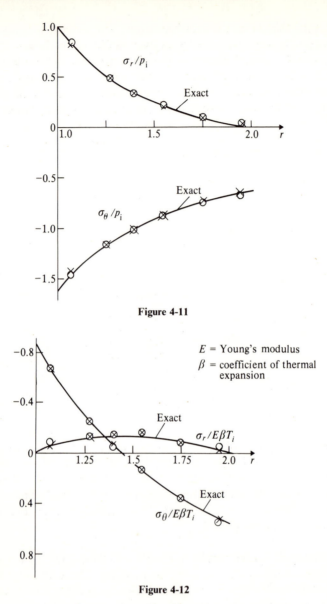

**Figure 4-11**

**Figure 4-12**

boundary points, where the errors in the numerical solution are always at a maximum. The cantilever problem, although trivial, is of interest because its geometry makes it particularly unamenable to solution using conventional, lower-order finite element discretization schemes.

**(c) The problem of an embankment resting on a stiffer layer** Figure 4-16 shows a typical embankment constructed on top of an underlying layer of stiffer material

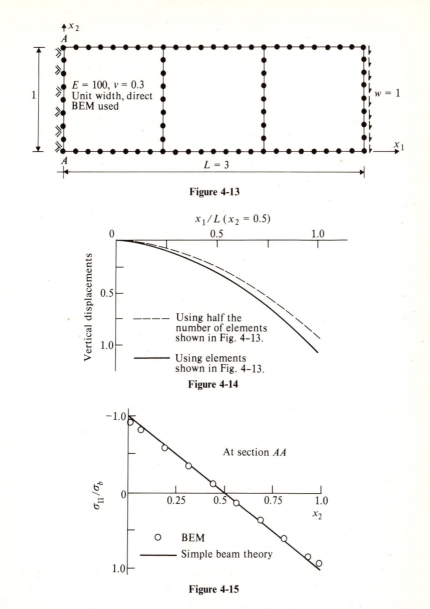

Figure 4-13

Figure 4-14

Figure 4-15

($\mu_2/\mu_1 = 2$). The calculated stresses and displacements at the original ground surface due to the construction of the embankment are shown in Fig. 4-17.

**(d) Problems involving zoned orthotropic media** The following two examples are taken from Ref. 23. The solutions for displacement vectors and principal stresses in Fig. 4-18a were obtained directly from the analytical solution[4] for a distributed surface load on an orthotropic half space. Figure 4-18b shows comparable

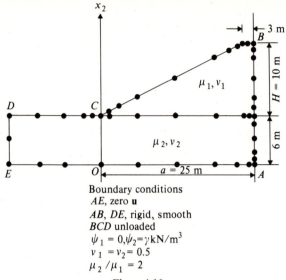

Boundary conditions
$AE$, zero $\mathbf{u}$
$AB$, $DE$, rigid, smooth
$BCD$ unloaded
$\psi_1 = 0, \psi_2 = \gamma\,\text{kN/m}^3$
$v_1 = v_2 = 0.5$
$\mu_2/\mu_1 = 2$

**Figure 4-16**

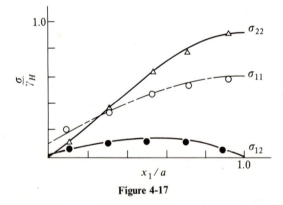

**Figure 4-17**

information from an indirect BEM solution in which, as a test case, the problem was treated as a two-zone system with each zone having identical elastic properties. The zonal interface was the small dashed rectangle shown in the figures and on the outer boundaries the displacements from the analytical solution were imposed. Figure 4-18*b* therefore shows the results calculated by BEM for a two-zone orthotropic material problem. The discretization scheme had five boundary elements under the half breadth of the footing and a total number of 42 elements including the zone interface subdivision. The agreement between the two solutions, apart from at shallow depths beneath the footing corners, was very close, as Table 4-1 shows.

**Table 4-1**

| Depth below edge of applied load ($b$ = half width of applied load) (1) | Displacement $\times b \times 10^{-6}$ | | Difference as a percentage (4) | Major principal stress | | Difference, as a percentage (7) |
|---|---|---|---|---|---|---|
| | BEM solution (2) | Half-space solution (3) | | BEM solutions (5) | Half-space solution (6) | |
| 0.5$b$ | 0.476 | 0.516 | 8 | 0.958 | 0.997 | 4 |
| 1.5$b$ | 0.304 | 0.317 | 4 | 0.548 | 0.566 | 3 |
| 2.5$b$ | 0.132 | 0.134 | 1.5 | 0.295 | 0.298 | 1 |

The second illustrative solution (Fig. 4-19) is included to illustrate a much more complex multizone problem. The properties of the different zones and the direction of the principal axes of elastic anisotropy are shown on the figure. The wide variety of boundary conditions incorporated is of particular interest.

**(e) A circular inclusion in an infinite elastic space**[18] Figure 4-20 shows the DBEM results for the problem of a circular inhomogeheity in an infinite elastic space compared with the analytical solution. DBEM results were obtained by modelling one quadrant using 20 boundary elements having a linear variation of **u** and **t** over them. In general the analytical and numerical results agreed to within four significant figures.

**(f) A typical mining problem**[18] Figure 4-21 shows a typical mining problem in which the elastic modulus of the ore-body ($E = 20\,\text{GPa}$) differs from that of the surrounding rock ($E = 35\,\text{GPa}$). Poisson's ratio is assumed to be 0.15 for both materials. The geostatic stresses measured at locations remote from the ore-body were

$$\sigma_{11}^{\infty} = 12.4\,\text{MPa} \qquad \sigma_{22}^{\infty} = 18.6\,\text{MPa} \qquad \text{and} \qquad \sigma_{12} = \sigma_{21} = 0$$

A DBEM analysis using 114 boundary nodes, giving a global system of equations involving 256 unknowns, was carried out and the results were compared with a finite element analysis using 855 constant strain quadrilateral elements.

The principal stresses along the lines $AA$, $BB$, $CC$, and $DD$ (see Fig. 4-21) obtained from both analyses are compared in Figs 4-22 to 4-25 respectively. Typically, the results agree to within 7 per cent.

**(g) Other applications** There are many other solutions in two-dimensional elastostatics reported in the literature.[24-35] Because of the continuous nature of

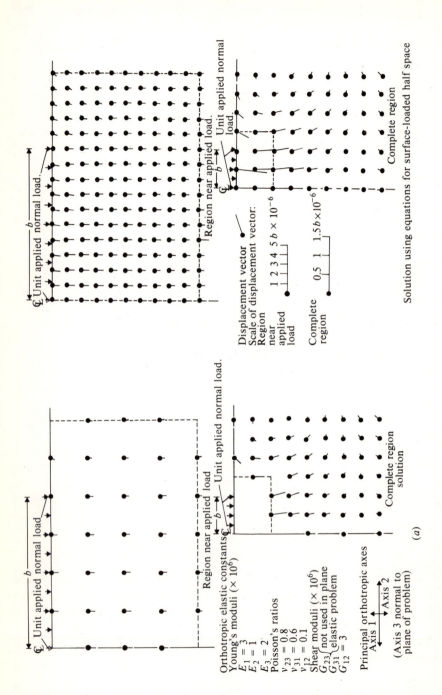

Unit applied normal load.

Region near applied load.

Unit applied normal load.

Displacement vector: ●

Scale of displacement vector:

Region near applied load: 1 2 3 4 5 $b \times 10^{-6}$

Complete region: 0.5 1 $1.5b \times 10^{-6}$

Complete region

Solution using equations for surface-loaded half space

Unit applied normal load.

Region near applied load.

Unit applied normal load.

Complete region solution

Orthotropic elastic constants

Young's moduli ($\times 10^6$)

$E_1 = 3$
$E_2 = 1$
$E_3 = 2$

Poisson's ratios

$v_{23} = 0.8$
$v_{31} = 0.6$
$v_{12} = 0.1$

Shear moduli ($\times 10^6$)

$G_{23}$ (not used in plane
$G_{31}$ elastic problem
$G_{12} = 3$

Principal orthotropic axes

Axis 1
Axis 2

(Axis 3 normal to
plane of problem)

(a)

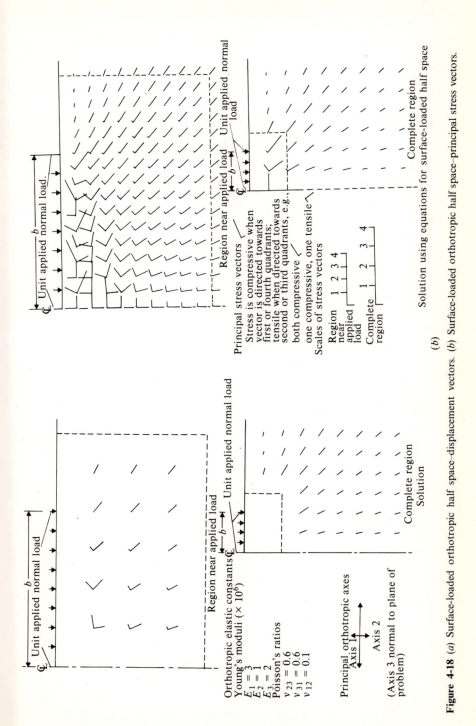

Orthotropic elastic constants
Young's moduli ($\times 10^6$)
$E_1 = 3$
$E_2 = 1$
$E_3 = 2$
Poisson's ratios
$v_{23} = 0.6$
$v_{31} = 0.6$
$v_{12} = 0.1$

Principal orthotropic axes

(Axis 3 normal to plane of problem)

Principal stress vectors

Stress is compressive when vector is directed towards first or fourth quadrants; tensile when directed towards second or third quadrants, e.g., both compressive

one compressive, one tensile

Scales of stress vectors

**Figure 4-18** (*a*) Surface-loaded orthotropic half space–displacement vectors. (*b*) Surface-loaded orthotropic half space–principal stress vectors.

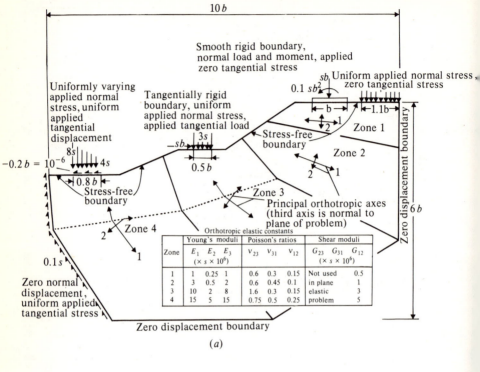

(a)

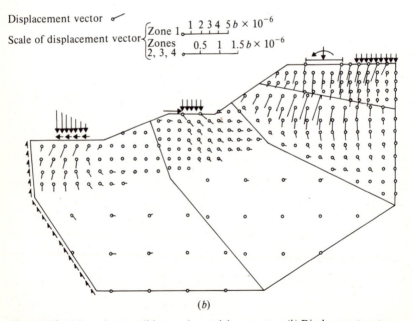

(b)

**Figure 4-19** (a) Boundary conditions and material parameters. (b) Displacement vectors.

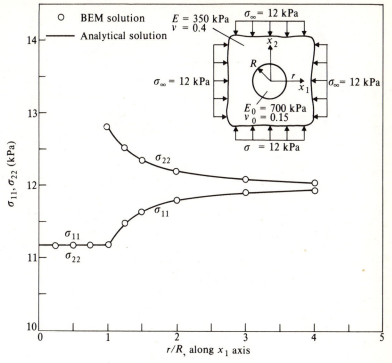

**Figure 4-20** Inclusion problem.

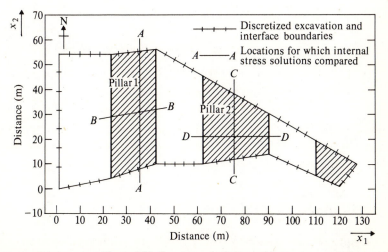

**Figure 4-21** Geometry of mining problem.

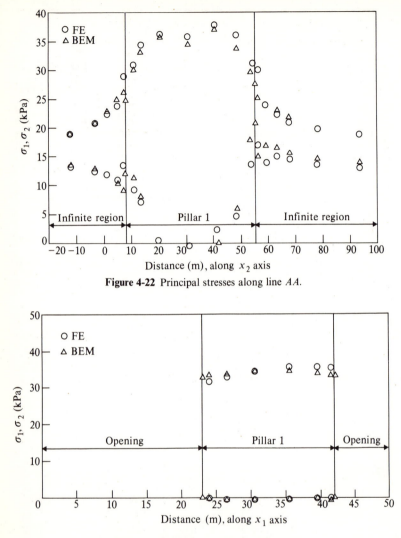

**Figure 4-22** Principal stresses along line *AA*.

**Figure 4-23** Principal stresses along line *BB*.

the solution the method has already become popular in fracture mechanics, and examples of the valuation of stress intensity factors around crack tips are described in Refs 24 to 28.

## 4-9 CONCLUDING REMARKS

In this chapter we have quite deliberately used the simplest possible numerical implementations of BEM which have been found to work successfully to

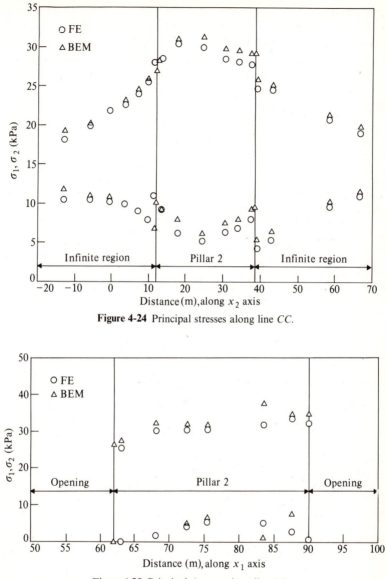

**Figure 4-24** Principal stresses along line $CC$.

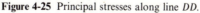

**Figure 4-25** Principal stresses along line $DD$.

produce routine solutions to engineering elasticity problems. One of the important features of these methods is that the numerical solution can be made as sophisticated as the analyst desires. For example, the surfaces and the functions can be varied by using parametric representations for them whereby problems can be modelled very much more accurately. These techniques will be discussed in Chapter 8. However, within the framework of the algorithm described here it is possible to improve the precision of the results by using weighted average displacements and tractions over the boundary elements, rather than calculating them solely at the nodal points of the boundary elements as we have done. This means that the boundary conditions over the element are satisfied in an average sense rather than merely at one selected point along their length (see Chapter 14).

## 4-10 REFERENCES

1. Lauricella, C. (1906) 'Atti della Reale Academia dei licei', **16**(1), 426–432.
2. Fredholm, I. (1905) 'Solution of fundamental problems in theory of elasticity', *Ark. Mat., Astr. och Fys.*, **2** (28), 1–8.
3. Sokolnikoff, I. (1956) *Mathematical Theory of Elasticity*, McGraw-Hill, New York.
4. Tomlin, G. R. (1972) 'Numerical analysis of continuum problems of zoned anisotropic media', Ph.D. thesis, Southampton University.
5. Banerjee, P. K., and Butterfield, R. (1977) 'Boundary element method in geomechanics', in G. Gudehus (ed.), *Finite Element in Geomechanics*, Chap. 16, Wiley, London.
6. Kupradze, V. D. (1963) *Potential Methods in Theory of Elasticity*, Davey, New York.
7. Mikhlin, S. G. (1965) *Multi-dimensional Singular Integrals and Integral Equations*, Pergamon Press, Oxford.
8. Zienkiewicz, O. C. (1971) *Finite Element Method in Engineering Science*, McGraw-Hill, London.
9. Mustoe, G. G. (1979) 'A combination of the finite element method and the boundary solution procedure for continuum problems', Ph.D. thesis, University of Wales, University College, Swansea.
10. Betti, E. (1872) 'Teoria dell elasticita', *Il Nuovo Ciemento*, **1872**, 7–10.
11. Cruse, T. A. (1969) 'Numerical solutions in three-dimensional elastostatics', *Int. J. Solids and Structs*, **5**, 1259–1274.
12. Rizzo, F. J. (1967) 'An integral equation approach to boundary value problems of classical elastostatics', *Q. Appl. Math.*, **25**, 83.
13. Ricardella, P. (1973) 'An implementation of the boundary integral techniques for plane problems in elasticity and elasto-plasticity', Ph.D. thesis, Carnegie Mellon University, Pittsburg.
14. Cruse, T. A. (1974) 'An improved boundary integral equation method for three-dimensional stress analysis', *Int. J. Computers and Structs*, **4**, 741–757.
15. Watson, J. O. (1979) Advanced implementation of the boundary element method in two- and three-dimensional elasto-statics', in P. K. Banerjee and R. Butterfield (eds), *Developments in Boundary Element Methods*, Chap. III, Applied Science Publishers, London.
16. Rizzo, F. J. (1979) Private communication.
17. Stippes, M., and Rizzo, F. J. (1977) 'A note on the body force integral of classical elastostatics', *ZAMP*, **28**, 339–341.
18. Wardle, L. J., and Crotty, J. M. (1978) 'Two-dimensional boundary integral equation analysis for nonhomogeneous mining applications', *Proc. Recent Dev. in Boundary Element Methods*, Southampton University.
19. Hearmon, R. F. S. (1961) *An Introduction to Applied Anisotropic Elasticity*, Oxford University Press.
20. Lekhnitskii, S. G. (1963) *Theory of Elasticity of an Anisotropic Elastic Body* (transl. P. Fern), Holden Day, San Francisco.
21. Love, A. E. H. (1927) *A Treatise on the Mathematical Theory of Elasticity*, Cambridge University Press.
22. Lempriere, B. M. (1968) 'Poisson's ratio in orthotropic materials', *J. Am. Inst. Aeronaut. and Astronaut.*, **6**, 2226–2227.
23. Tomlin, G. R., and Butterfield, R. (1974) 'Elastic analysis of zoned orthotropic continua', *Proc. ASCE*, **EM3**, 511–529.

24. Snyder, M. D., and Cruse, T. A. (1975) 'Boundary integral equation analysis of cracked anisotropic plates', *Int. J. Fracture Mech.*, **II**, 315–328.
25. Cruse, T. A., and Rizzo, F. (eds) (1975) *Proc. ASME Conf. on Boundary Integral Methods*, ASME, New York.
26. Cruse, T. A. (1973) 'Application of the boundary integral equation method for solid mechanics', *Proc. Var. Meth. in Engng*, Southampton University Press.
27. Cruse, T. A. (1978) 'Two-dimensional BIE fracture mechanics analysis', *Appl. Math. Modelling*, **3**, 287–293.
28. Cruse, T. A. (1979) 'Two and three-dimensional problems of fracture mechanics', in P. K. Banerjee and R. Butterfield (eds), *Developments in Boundary Element Methods*, Vol. I, Chap. V, Applied. Science Publishers, London.
29. Christiansen, S., and Hansen, E. (1975) 'A direct integral equation method for computing the hoop stress at holes in plane isotropic sheets', *J. Elasticity*, **5**(1), 1–14.
30. Rudolphi, T. J., and Ashbaugh, N. E. (1978) 'An integral equation solution for a bounded elastic body containing a crack: mode 1 deformation', *Int. J. Fracture*, **14**(5), 527–541.
31. Krenk, S. (1978) 'Stress concentration around holes in anisotropic sheets', Report No. 138, The Technical University of Denmark; also appeared in *Appl. Math. Modelling*, July 1978.
32. Heise, U. (1978) 'Application of the singularity method for the formulation of plane elasto-statical boundary value problems as integral equations', *Acta Mec.*, **31**, 33–69.
33. Brady, B. H. G., and Bray, J. W. (1978) 'The boundary element method for determining stresses and displacements around long openings in a triaxial stress field', *Int. J. Rock Mech.*, **15**, 21–28.
34. Brady, B. H. G., and Bray, J. W. (1978) 'The boundary element method for elastic analysis of tubular ore body extraction assuming complete plane strain', *Int. J. Rock Mech.*, **15**, 29–37.
35. Brady, B. H. G. (1979) 'A direct formulation of the boundary element method of stress analysis for complete plane strain', *Int. J. Rock Mech.*, **16**, 235–244.

# CHAPTER FIVE

# THREE-DIMENSIONAL PROBLEMS OF STEADY STATE POTENTIAL FLOW

## 5-1 INTRODUCTION

Many problems in engineering such as the steady state low-speed flow around a ship's hull, steady state groundwater flow, the field around an electromagnet, the electrostatic field around a porcelain insulator or buried electrical cables of variable cross section, etc., involve the solution of Laplace's or Poisson's equation in three dimensions, the governing differential equation with coordinate axes $(x_i)$ directed along the principal axes of 'conductivity' in a homogeneous medium being

$$\kappa_1 \frac{\partial^2 p(x)}{\partial x_1^2} + \kappa_2 \frac{\partial^2 p(x)}{\partial x_3^2} + \kappa_3 \frac{\partial^2 p(x)}{\partial x_3^2} = -\psi \qquad (5\text{-}1a)$$

or, for the generally anisotropic body,

$$\kappa_{ij} \frac{\partial^2 p(x)}{\partial x_i \partial x_j} = -\psi \qquad (5\text{-}1b)$$

Away from sources or sinks Eq. (5-1a) is clearly equivalent to Eq. (3-1). The related expressions for the flux vector, the normal boundary flux, and the transformation of the system into an equivalent isotropic body follow from the discussion in Sec. 3-2 except that for principal conductivities $\kappa(1, \alpha, \beta)$ the axis scaling ratios are, say, $(1, 1/\alpha^{\frac{1}{2}}, 1/\beta^{\frac{1}{2}})$ with a scaled isotropic permeability equal to $\kappa(\alpha\beta)^{\frac{1}{2}}$.

For the problems discussed in Chapters 3 and 4 the major advantages of the BEM are not apparent since standard finite element formulations can also provide routine solutions to such problems. However, this chapter and the following one on three-dimensional elastic stress analysis will demonstrate very clearly that for these cases BEM are probably the most realistic way of obtaining adequately precise results at a reasonable cost.

## 5-2 SINGULAR SOLUTIONS: THE INDIRECT AND DIRECT FORMULATIONS

The fundamental singular solution for the potential $p(x)$ at $x_i$, due to a unit source $e(\xi)$ at $\xi_i$ for an 'equivalent' isotropic body with 'conductivity' $\kappa$, may be written as

$$p(x) = G(x, \xi) e(\xi) \tag{5-2}$$

where

$$G(x, \xi) = \frac{1}{4\pi\kappa} \frac{1}{r}$$

$$r^2 = (x - \xi)_i (x - \xi)_i$$

By comparing Eqs (3-4) and (5-2) we note that whereas Eq. (3-4) provides the relative potential $p(x)$, that given by Eq. (5-2) is an absolute value and the behaviour at infinity is also satisfied automatically. Therefore in an indirect BEM formulation the augmented equations to satisfy the uniqueness condition will not be required.

The flux (velocity) vector corresponding to the potential $p(x)$ is given by

$$v_i(x) = -\kappa \frac{\partial p(x)}{\partial x_i} = \frac{1}{4\pi} \frac{1}{r^2} \frac{\partial r}{\partial x_i} e(\xi) = \frac{1}{4\pi} \frac{y_i}{r^3} e(\xi) \tag{5-3}$$

where $y_i = x_i - \xi_i$.

The corresponding normal velocity on the boundary is then

$$u(x) = v_i(x) n_i(x) = \frac{y_i n_i}{4\pi r^3} e(\xi)$$

or

$$u(x) = F(x, \xi) e(\xi) \tag{5-4}$$

with

$$F(x, \xi) = \frac{y_i n_i}{4\pi r^3}$$

The indirect formulation of BEM follows from these singular solutions as demonstrated in Eqs (3-7) to (3-12) in Chapter 3, except that for three-dimensional problems the constant $C$ in Eqs (3-7) and (3-9) is zero, as explained above.

Multiplying Eq. (5-1) by the function $G$ and integrating by parts, as shown in Chapter 3, we obtain the reciprocal identity (3-28) which can be used to develop the corresponding direct formulation represented by Eqs (3-37) and (3-38). The kernel function $H(x, \xi)$, which is necessary to obtain the velocity in any direction $n$ at an interior point $\xi$, is now given by

$$H(x, \xi) = -\kappa \frac{F(x, \xi)}{\partial \xi_i} n_i(\xi) = -\frac{\kappa}{4\pi} n_i(\xi) \, n_j(x) \frac{\partial}{\partial \xi_i} \frac{y_j}{r^3}$$

$$= -\frac{\kappa}{4\pi r^3} \left( \frac{3 y_i y_i}{r^2} - \delta_{ij} \right) n_i(\xi) \, n_j(x) \tag{5-5}$$

# 5-3 INTEGRABILITY OF THE KERNEL FUNCTIONS

If we compare the singularities of the kernel functions involved in the two-dimensional analysis with those derived above we find that the latter are at least one order higher. Therefore there may be some doubt as to the existence of the integrals involved for the case when the source point and the field point coincide. Mikhlin[1] discusses these problems with reference to integrals over $m$-dimensional space $E_m$ and shows that integrals with singularities of the order of $1/r^{m-1}$ are weak and therefore exist in the normal sense. But integrals with singularities of the order of $1/r^m$ exist only under certain conditions (most of which apply here) as Cauchy principal values. Singularities of orders higher than $1/r^m$ are not integrable.

On the basis of these results we can see that for $x = \xi$ the surface and the volume integrals involving the function $G(x, \xi)$ and, in the indirect formulation only, the volume integral involving the function $F(x, \xi)$ are weak, and therefore integrable in the normal sense. The surface integrals involving the function $F(x, \xi)$ exist only as a Cauchy principal value and those involving $H(x, \xi)$ do not exist at all as $\xi$ approaches $x$ on the boundary. It is worth noting that the behaviour of these integrals is identical to the corresponding ones discussed for the two-dimensional analysis of Chapter 3.

# 5-4 NUMERICAL SOLUTION

## 5-4-1 Local Coordinates

For the sake of algebraic simplicity we shall assume that for the present there are no sources specified in the interior of the region (that is, $\psi = 0$). The surface $S$ may be discretized as an assembly of flat triangular elements as shown in Fig. 5-1. This discretization scheme is essentially identical to the idealization of shells as an assembly of flat elements in the finite element method (see Zienkiewicz[2]).

If we consider the $q$th triangular element, with nodes $r, s$, and $t$, in isolation and construct an orthogonal local axis system $z_i$ such that $z_1$ coincides with the side $rs$ and $z_3$ is normal to the plane of the triangle (see Fig. 5-2), the component

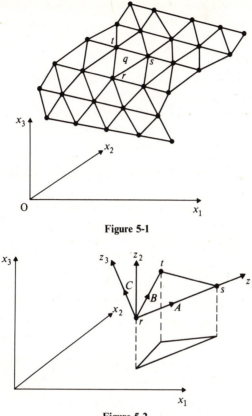

**Figure 5-1**

**Figure 5-2**

vectors $\mathbf{A}$ and $\mathbf{B}$ which define the sides $rs$ and $rt$ are then given by

$$\mathbf{A} = \begin{Bmatrix} x_1^s - x_1^r \\ x_1^s - x_2^r \\ x_3^s - x_3^r \end{Bmatrix} \qquad \text{and} \qquad \mathbf{B} = \begin{Bmatrix} x_1^t - x_1^r \\ x_2^t - x_2^r \\ x_3^t - x_3^r \end{Bmatrix} \qquad (5\text{-}6)$$

and their lengths by

$$l_a = \sqrt{(\mathbf{A}^T \cdot \mathbf{A})} \qquad \text{and} \qquad l_b = \sqrt{(\mathbf{B}^T \cdot \mathbf{B})} \qquad (5\text{-}7)$$

The components of the normal vector $\mathbf{C}$ are then given by the cross-product of $\mathbf{A}$ and $\mathbf{B}$ (i.e., symbolically):

$$\mathbf{C} = \mathbf{A} \times \mathbf{B}$$

The length of $\mathbf{C}$ can easily be shown to be equal to twice the area of the triangle (that is, $2\Delta$). The direction cosines of the normal, or the $z_3$ axis, are then

$$\mathbf{e}^c \qquad \text{or} \qquad \mathbf{n} = \frac{1}{2\Delta} \mathbf{C} \qquad (5\text{-}8)$$

Equations (5-6) to (5-8) allow us to construct a local coordinate system on the surface element (see also Chapter 8).

## 5-4-2 Shape Functions

In Chapter 3 we proposed a simple numerical solution scheme for potential flow problems by using a 'piecewise constant' distribution of $\phi$, $p$, and $u$ over the boundary elements. Whilst this simple approach allowed us to demonstrate the essential features of the solution technique, a more efficient algorithm is obtained if we consider at least a linear variation of these quantities over each boundary element. Indeed, for some problems of elasticity (such as the problem of bending of a beam) the 'piecewise constant' approximation does not provide the correct distribution of the shear stress over the cross section of the beam, and we therefore had to use a linear variation of $\mathbf{t}$ and $\mathbf{u}$ in Chapter 4.

If we construct the local coordinates $z_1$ and $z_2$ in the plane of the element, as shown in Fig. 5-2 for a linear variation of $\phi$ (for example) over this $q$th element, we have

$$\phi = \mathbf{N}^q \boldsymbol{\phi}_n$$

where      $\phi$ = fictitious potential at any interior point within the element

$\mathbf{N}^q = 1 \times 3$ matrix

$\boldsymbol{\phi}_n = 3 \times 1$ vector of nodal values of $\phi = \begin{Bmatrix} \phi_r \\ \phi_s \\ \phi_t \end{Bmatrix}$

The components $N_1$ $N_2$, and $N_3$ of the *shape function* matrix may be obtained from $M_1$, $M_2$, and $M_3$ respectively of Eq. (4-21) by allowing for the fact that the origin of the coordinate system now passes through node $r$ (see Figs 4-4 and 5-2 and Chapter 8).

We can therefore write the discretized form of the boundary integrals for the indirect formulation as (with $\psi = 0$)

$$p(x_o^p) = \sum_{q=1}^{N} \left( \int_{\Delta S^q} G(x_o^p, \xi) \, \mathbf{N}^q(\xi) \, dS \right) \boldsymbol{\phi}_n^q \tag{5-9}$$

and      $$u(x_o^p) = \beta \phi(x_o^p) + \sum_{q=1}^{N} \left( \int_{\Delta S^q} F(x_o^p, \xi) \, \mathbf{N}^q(x) \, dS \right) \boldsymbol{\phi}_n^q \tag{5-10}$$

where $\beta = \frac{1}{2}$, if the boundary point $x_o^p$ is not located on any edge or corner and the shape function matrix $\mathbf{N}^q$ is a function of the variable of integration $\xi$.

For the direct formulation the discretized boundary integral becomes

$$\beta p(\xi_o^p) = \sum_{q=1}^{N} \left\{ \left[ \int_{\Delta S^q} F(x, \xi_o^p) \, \mathbf{N}^q(x) \, dS \right] \mathbf{p}_n^q - \left[ \int_{\Delta S^q} G(x, \xi_o^p) \, \mathbf{N}^q(x) \, dS \right] \mathbf{u}_n^q \right\} \tag{5-11}$$

where the matrix $\mathbf{N}^q$ is now a function of the variable of integration $x$.

Equations (5-9) and (5-10) for the indirect BEM or Eq. (5-11) for the direct method can be used to construct the final system of equations exactly as discussed in Chapters 3 and 4.

### 5-4-3 Numerical Integration

The methods and strategies used for evaluating the various integrals have been discussed in Chapter 4. These involve a study of the character of the kernel-shape function products. If they remain bounded over the complete range of integration then they can be integrated numerically. Methods of numerical integration and the relevant quadrature formulas are described in Appendix C. In order to obtain an efficient algorithm it is essential to vary the number of sampling points depending upon the distance between the node $p$ and the element $q$.

### 5-4-4 Exact Integration

If the kernel-shape function product becomes infinite anywhere within the interval of integration numerical integration is not possible. Although it is sometimes possible to take the field point slightly inside the region so that the integrand remains bounded and the resulting continuous (rapidly varying) kernel-shape function product can be integrated by using a high-order quadrature formula, such procedures cannot be recommended. Ideally they should be integrated analytically.

We can do this by splitting the singular kernel-shape function product, as discussed in Chapter 4; the singular part is then given by the integrals evaluated below.

**(a) Evaluation of** $\int_{\Delta S} G(x_o, \xi)\, dS(\xi)$ **or** $\int_{\Delta S} G(x, \xi_o)\, dS(x)$
Because of the symmetry of the function $G$ with respect to the arguments $x$ and $\xi$ the results of integrating either of them will be the same and only one of them need be considered.

By constructing a local polar coordinate system on the $q$th element, as shown in Fig. 5-3, we can express the integral as

$$\int_{\Delta S} G(x_o, \xi)\, dS = \frac{1}{4\pi\kappa} \int_{\theta_A}^{\theta_B} \int_0^{r(\theta)} \frac{1}{r} r\, dr\, d\theta = \frac{1}{4\pi\kappa} \int_{\theta_A}^{\theta_B} r(\theta)\, d\theta$$

$$= \frac{1}{4\pi\kappa} \int_{\theta_A}^{\theta_B} D \sec\theta\, d\theta = \frac{D}{4\pi\kappa} \left[ \ln\left\{ \tan\left(\pi/4 + \theta/2\right) \right\} \right]_{\theta_A}^{\theta_B} \quad (5\text{-}12)$$

**(b) Evaluation of** $\int_{\Delta S} F(x_o, \xi)\, dS(\xi)$ **and** $\int_{\Delta S} F(x, \xi_o)\, dS(x)$
Both these integrals contain strong singularities and therefore only exist as Cauchy principal values. They can be evaluated by assuming an arbitrarily small

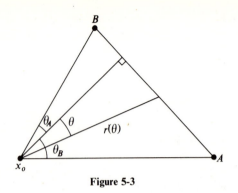

**Figure 5-3**

region of exclusion $\Delta S_\varepsilon$ around the singularity and taking the limit $\Delta S_\varepsilon \to 0$; i.e.,

$$\int_{\Delta S} F(x_o, \xi)\, dS = \int_{\Delta S - \Delta S_\varepsilon} F(x_o, \xi)\, dS + \int_{\Delta S_\varepsilon} F(x_o, \xi)\, dS \qquad \lim \Delta S_\varepsilon \to 0 \quad (5\text{-}13)$$

It can easily be shown that the first integral on the right-hand side of Eq. (5-12) contributes nothing to the system equations whereas the second integral contributes to the discontinuity term $\beta$ and is equal to $\omega/4\pi$, where $\omega$ is the solid angle subtended at the corner (see Chapter 3).

Alternatively, if we take the field point slightly inside in the node shown in Fig. 5-4, the integrals then exist in the normal sense and therefore $\beta = 0$ in Eq. (5-10) and $\beta = 1$ in Eq. (5-11) (since the field point is still inside the region, although very near to the surface). Once again we construct a local coordinate system $z_i$ through one nodal point $r$ of the element under consideration (Fig. 5-4) such that $z_1, z_2$ are as shown in Fig. 5-4 and $z_3$ is directed along the normal to the surface element. For a general field point at $(x_1, x_2, x_3)$ we can evaluate these integrals analytically. Note, first, that the integral

$$\Delta F = \int_{\Delta S} F(x, \xi)\, ds(\xi) = -\frac{1}{4\pi} \int_{\Delta S} \frac{\partial}{\partial x_i} \frac{1}{r}\, n_i\, ds(\xi) = -\frac{n_i}{4\pi} \int_{\Delta S} \frac{\partial}{\partial x_i} \frac{1}{r}\, ds(\xi)$$

$$= -\frac{n_i}{4\pi} \int_C \frac{1}{r}\, e_i\, dc \qquad (5\text{-}14)$$

where $e_i$ are the global components of the normal to the contour $C$ surrounding the area $\Delta S$.

Along the side $rs$ we have

$$\Delta F = -\frac{n_i}{4\pi} E_{1i} \left[ \ln(z_1 + r) \right]_r^s \qquad (5\text{-}15)$$

where

$$E_{1i} = \frac{\partial z_1}{\partial x_i}$$

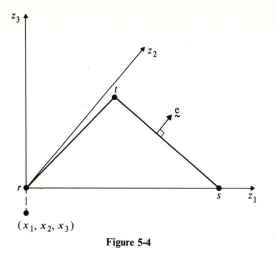

**Figure 5-4**

The total result is then obtained by reorienting the local coordinate system such that the $z_1$ axis coincides, in turn, with the $st$ and $tr$ sides of the triangle. The results are then expressed in terms of the coordinates of the field point and those of the corners. Clearly the result (5-15) can also be used to evaluate the integral over a flat quadrilateral element in a similar manner. Hess and Smith[3] also evaluated the contour integral in Eq. (5-14) and provided exact analytical expressions for $\Delta F$.

In order to evaluate the integral $\int_{\Delta S} F(x, \xi) \, ds(x)$ we note that:

1. Since the normal $n$ in this kernel function is at the integration point $x$, for flat elements it can be taken outside the integral sign.
2. By interchanging the arguments $x$ and $\xi$ the function changes its sign only.

Therefore,

$$\int_{\Delta S} F(x, \xi) \, ds(x) = -\int_{\Delta S} F(x, \xi) \, ds(\xi) \tag{5-16}$$

The result of integrating this simple expression is important. We shall see in Chapters 8 and 15 that even when high-order representations of geometries and boundary parameters are used we can always isolate the strongly singular part and use this solution. The remaining integrals involving the variations of curvature and of the functions (such as $\phi$, $u$, or $p$ over the boundary) can be evaluated numerically.

In the direct method it is, of course, possible to calculate the total contribution to the diagonal coefficients of the final system matrices by using the condition of uniform potential, as discussed in Chapters 3 and 4.

## 5-5 AXISYMMETRIC FLOW

### 5-5-1 General

Many axisymmetric problems of potential flow can best be represented by a cylindrical coordinate system $r, z$, as shown in Fig. 5-5. Although they can be solved by using the algorithms outlined in the previous sections it is usually more efficient to formulate them in cylindrical coordinates. The algorithm then becomes essentially a two-dimensional one.

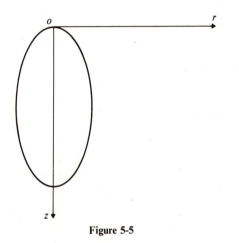

**Figure 5-5**

Such axisymmetric potential flow problems have been considered by Jaswon and Symm[4] and Lennon, Liu, and Liggett[5] in recent publications related to electrostatic fields and groundwater flow respectively. The following axisymmetric algorithm is based on their work.

### 5-5-2 Axisymmetric Singular Solution

The fundamental singular solution due to an axisymmetric source $e$ acting at $Q$ may be written as (see Fig. 5-6)

$$p(P) = G(P, Q) e(Q) \tag{5-17}$$

where the function $G(P, Q)$ is obtained by expressing Eq. (5-2) in $(r, \theta, z)$ coordinates and integrating the result with respect to $\theta$ between the limits 0 and $2\pi$. Thus we can write

$$G(P, Q) = \frac{1}{2\pi\kappa} \int_0^\pi \frac{d\theta}{\sqrt{(a - b\cos\theta)}} \tag{5-18}$$

where $a = r^2 + r_o^2 + (z - z_o)^2$ and $b = 2rr_o$. The integral in Eq. (5-18) cannot be

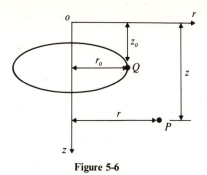

**Figure 5-6**

evaluated exactly and has to be dealt with as an elliptic integral, i.e.,

$$G(P,Q) = \frac{1}{\pi\kappa} \frac{2}{\sqrt{(a+b)}} K(m) \tag{5-19}$$

where $K(m)$ is a complete elliptic integral of the first kind of modulus $m$ and complementary modulus $m_1 = 1 - m$. Thus

$$K(m) = \int_0^\pi \frac{d\theta}{\sqrt{(1 - m\sin^2\theta)}}$$

$$m = \frac{4rr_o}{(r+r_o)^2 + (z-z_o)^2} \qquad 0 \leqslant m < 1 \tag{5-20}$$

$$m_1 = \frac{(r-r_o)^2 + (z-z_o)^2}{(r+r_o)^2 + (z-z_o)^2}$$

It is rather inconvenient to work with the complete integral (5-20) and therefore it is usually necessary to obtain a suitably close polynomial approximation (see Hastings[6] and Hart et al.[7]). Jaswon and Symm[4] discuss a number of possible approximations for different values of the modulus $m$. A typical one for $K(m)$ might be[8]

$$K(m) = \sum_{i=0}^{n} [a_i m_1^i + b_i m_1^i \ln(1/m_1)] + \varepsilon(m) \tag{5-21}$$

where, for $n = 4$, the error term $\varepsilon(m)$ is $\leqslant 2 \times 10^{-8}$, and $a_i$ and $b_i$ are constants.

Equation (5-21) shows that the singularity at $P = Q$ is similar to that of the two-dimensional problem (i.e., a weak logarithmic one). The normal 'velocity' in the direction $n$ at a point $P$ can be obtained from Eq. (5-19) as

$$u(P) = F(P,Q)e(Q) \tag{5-22}$$

where

$$F(P,Q) = -\kappa\left(\frac{\partial G}{\partial r} n_r + \frac{\partial G}{\partial z} n_z\right)$$

and $n_r$ and $n_z$ are the components of the vector $n$ in the direction of $r$ and $z$ axes respectively.

By using the identity

$$E(m) = m_1 \left[ 2m \frac{dK(m)}{dm} + K(m) \right]$$

where $E(m)$ is the complete elliptic integral of the second kind, we can express $F(P,Q)$ as[8]

$$F(P,Q) = \frac{n_r[E(m) - K(m)]}{\pi r \sqrt{(a+b)}} + \frac{n_r(r - r_o) + n_z(z - z_o)}{2\pi(a-b)\sqrt{(a+b)}} E(m) \tag{5-23}$$

and $E(m)$ in Eq. (5-23) may be written as the polynomial approximation

$$E(m) = 1 + \sum_{i=1}^{n} [c_i m_1^i + d_i m_1^i \ln(1/m_1)] \tag{5-24}$$

where $0 \leqslant m < 1$, for $n = 4$, $\varepsilon(m)$ is $< 2 \times 10^{-8}$, and $c_i$ and $d_i$ are constants.

### 5-5-3 Indirect and Direct Formulations

By distributing the source $\phi(Q)$ over the surface, the potential $p(p)$ at $P(r, z)$ may be obtained (ignoring for simplicity the presence of any internal sources) by reducing the integral for a three-dimensional problem to the axisymmetric form:

$$p(p) = \int_s r_o G(P,Q) \phi(Q) dr_o dz_o \tag{5-25}$$

The normal velocity at $p$ then becomes

$$u(p) = \int_s r_o F(P,Q) \phi(Q) dr_o dz_o \tag{5-26}$$

If we now bring the point $p$ to a surface point $p_o$ we can obtain as before our two basic integral equations for the solution of any boundary-value problem:

$$p(P_o) = \int_s r_o G(P_o, Q) \phi(Q) dr_o dz_o$$

$$u(P_o) = \beta\phi(P_o) + \int_s r_o F(P,Q) \phi(Q) dr_o dz_o \tag{5-27}$$

The boundary discretization for the axisymmetric problem is in fact identical to that for a two-dimensional one. For example, we divide the boundary along $r_o, z_o$ by straight line segments and assume a linear variation of $\phi$ over each segment (see Fig. 5-7), i.e.,

$$\phi(Q) = \mathbf{N}\boldsymbol{\phi}_n \tag{5-28}$$

where $\mathbf{N} = [N_1, N_2]$, $\boldsymbol{\phi}_n = \begin{Bmatrix} \phi_s \\ \phi_t \end{Bmatrix}$, $N_1 = 1 - \xi_s/L, N_2 = \xi_s/L$. We can then discretize the boundary integrals in the usual manner.

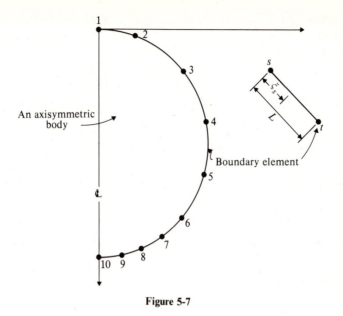

**Figure 5-7**

However, in view of the complicated form of the kernel functions it is always necessary to integrate the kernel-shape function products numerically by using quadrature formulae. This can be accomplished by using ordinary quadrature formulae for all cases except those contributing to the leading diagonal of the final system matrices. Integrals involving the function $G$ have a logarithmic singularity and can be evaluated accurately by using the special gaussian quadrature formulae described in Appendix C, but those involving the function $F$ must be evaluated analytically. We can do this by the method discussed in Sec. 5-4-4 (i.e., by considering the singular part of the integral together with the contribution arising from the discontinuity term). The function $F$ can be simplified for this particular case.

The corresponding direct boundary integral statement may be written as

$$p(Q) = \int_s r[F(P,Q)\,p(P) - G(P,Q)\,u(P)]\,dr\,dz \tag{5-29}$$

where $Q(r_o, z_o)$ is a point in the interior of the region.

For a point on the boundary we have

$$\beta p(Q_o) = \int_s r[F(P,Q_o)\,p(P) - G(P,Q)\,u(P)]\,dr\,dz \tag{5-30}$$

An equation which can be discretized and the various integrals evaluated in the manner as described above to provide a numerical solution for any well-posed problem.

## 5-6 EXAMPLES

In view of the inherent advantage of BEM for obtaining numerical solutions to problems involving low-speed flow regimes[9, 10] there are already very many published papers which explore the flow around a ship's hull, around an aircraft or around a screw propeller, and, of course, a variety of problems related to groundwater flow.[3, 5, 11 – 21]

If a body moves slowly through a fluid of low viscosity then the resulting flow may be considered irrotational, since the vorticity in such a case convects with the fluid as if it were attached to fluid particles. The vorticity 'diffuses' through the fluid in a manner which mathematically resembles the diffusion equation with 'conductivity' equal to the kinematic viscosity. Therefore if a body moves through a fluid otherwise at rest, it encourages no vorticity and the flow can be idealized by the potential theory[9, 10] set out above.

Hess and Smith[3] were the first to apply this method to full-scale practical problems. Figure 5-8 shows a typical comparison between analytical and boundary element results for velocities at surface points on an ellipsoid using flat

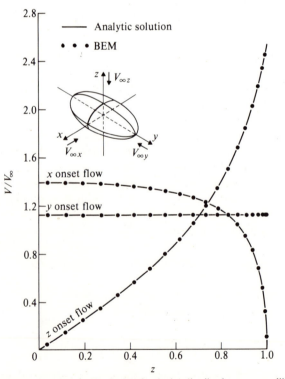

**Figure 5-8** Comparison of analytic and calculated velocity distributions on an ellipsoid with axes ratios $1:2:\frac{1}{2}$ in $x, y, z$ directions. (*a*) Velocities in $x, z$ plane. (*b*) Velocities in $y, z$ plane. (*c*) Velocities in $x, y$ plane.

quadrilateral elements with constant source values ($\phi$) over each element. The total number of equations were reduced by taking advantage of quadrantal symmetry (i.e., coefficients for elements having the same values of $\phi$ were summed). The numerical results are clearly in excellent agreement with the exact solution. They also compared the calculated and experimental pressure distributions on two delta wings. One such comparison is shown in Fig. 5-9 which clearly demonstrates the usefulness of a BEM analysis applied to a real low-speed flight problem.

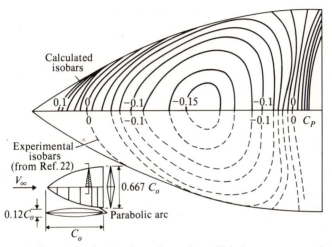

**Figure 5-9** Comparison of calculated and experimental zero-lift isobars on a symmetric delta wing of Gothic planform.

The foregoing solutions, obtained by using IBEM, are only applicable to problems of non-lifting flow. Hess[15] extended the analysis to deal approximately with problems of lifting flow by introducing vorticity (in addition to the surface source) strips over both the lifting portion boundaries (see Fig. 5-10) and in the trailing vortex 'wake'. He also introduced the effect of a boundary layer by using a displacement approximation due to Lighthill[23] which allows for the diffusion of the vorticity outwards into a thin boundary layer near the body, from the viscosity of the fluid, and its convection into the 'wake' with the fluid passing the surface. He found that this additional vorticity could be usefully distributed at a constant strength obtained from the 'Kutta' circulation condition, over the individual lifting strips and that the presence of a trailing vortex made little difference to the values of the lift coefficients. Using this modified form of IBEM he obtained many useful solutions for low-speed flow regimes around aircraft, one example of which is shown in Figs 5-11 and 5-12. The boundary element pattern of Fig. 5-11 was used to calculate the lift coefficients on a wing, shown in Figs. 5-12, which are compared with experimental data for a 7 degree angle of attack: a Mach number of 0.5 and a Reynolds number of 6.25 million. The

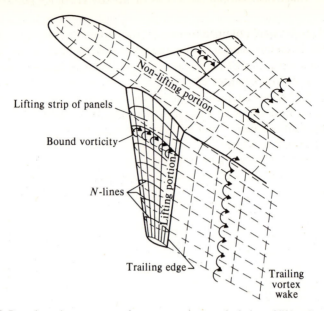

Lifting strip of panels

Bound vorticity

$N$-lines

Trailing edge

Non-lifting portion

Lifting portion

Trailing vortex wake

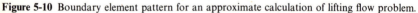

**Figure 5-10** Boundary element pattern for an approximate calculation of lifting flow problem.

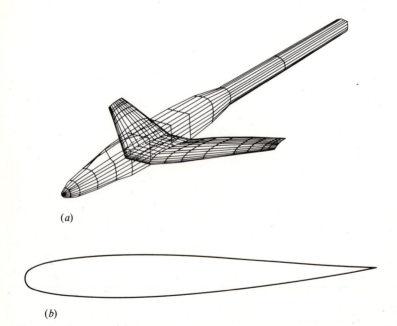

(a)

(b)

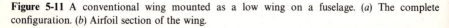

**Figure 5-11** A conventional wing mounted as a low wing on a fuselage. (a) The complete configuration. (b) Airfoil section of the wing.

numerical results were obtained by using eight vorticity strips on the wing and a total of 391 boundary elements over the wing and the fuselage.

Other examples of BEM applications to problems of fluid flow can be found in Refs 17 to 21 and Chapter 13.

Figure 5-13 shows a typical example[16] of an axisymmetric quick draw-down problem, with radial flow to a pumped well, solved by DBEM. The difference between the finite element[24] and BEM solutions is negligible. Similar examples are also described in Kipp.[26,27]

More recently, Rizzo and Shippy[28] have developed boundary integral equation solutions for axisymmetric bodies under very general boundary conditions.

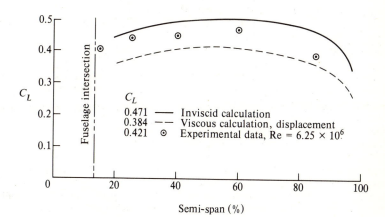

**Figure 5-12** Comparison of calculated and experimental spanwise distributions of section lift coefficient for a conventional swept wing mounted low on a fuselage at 6.9° angle of attack.

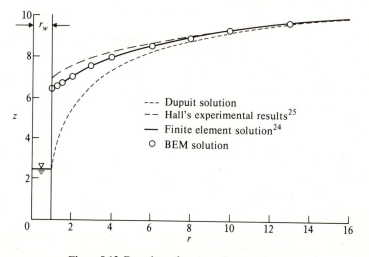

**Figure 5-13** Drawdown from an axisymmetric well.

## 5-7 CONCLUDING REMARKS

The examples discussed above clearly demonstrate the versatility and appeal of BEM for three-dimensional problems. Although we have discretized the surface of our body using triangular elements, the algorithm remains essentially unchanged for quadrilateral ones which can, in certain cases, lead to more accurate and efficient solutions.

## 5-8 REFERENCES

1. Mikhlin, S. G. (1965) *Multi-dimensional Singular Integral Equations*, Pergamon Press, Oxford.
2. Zienkiewicz, O. C. (1978) *The Finite Element Method*, 3rd ed., McGraw-Hill, London.
3. Hess, J. L., and Smith, A.M.O. (1964) 'Calculations of nonlifting potential flow about arbitrary three-dimensional bodies', *J. Ship Res.*, **8**(2), 22–44.
4. Jaswon, M. A., and Symm, G. T. (1977) *Integral Equation Methods in Potential Theory and Elastostatics*, Academic Press, London.
5. Lennon, G. P., Liu, P. L. F., and Liggett, J. A. (1979) 'Boundary integral equation solution to axisymmetrical potential flows: I, basic formulations' Unpublished work, Department of Civil and Environmental Engineering, Cornell University, Ithaca, N.Y.
6. Hastings, C. (1955) *Approximations for Digital Computers*, Princeton University Press.
7. Hart, J. F., et al. (1968) *Computer Approximations*, Wiley, New York.
8. Abramowitz, M., and Stegun, I. A. (1974) *Handbook of Mathematical Functions*, Dover, New York.
9. Milne-Thomson, L. M. (1976) *Theoretical Hydrodynamics*, 5th ed., Macmillan, London.
10. Lamb, H. (1932) *Hydrodynamics*, Cambridge University Press.
11. Hess, J. L., and Smith, A. M. O. (1966) 'Calculations of potential flow about arbitrary bodies', in *Progress in Aeronautical Sciences*, Vol. 8, pp. 1–138, Pergamon Press, New York.
12. Hess, J. L. (1974) 'The problem of three-dimensional lifting potential flow and its solution by means of surface singularity distributions, *Computer Meth. in Appl. Mech. Engng*, **4**, 283–319.
13. Hess, J. L. (1975) 'Improved solution for potential flow about arbitrary axisymmetric bodies by the use of a higher order surface source method', *Computer Meth. in Appl. Mech. Engng*, **5**, 297–308.
14. Hess, J. L. (1975) 'Consistent velocity and potential expansions for higher order surface singularity method', Report No. MDC J6911, McDonnell Douglas Corporation.
15. Hess, J. L. (1977) 'A fully automated combined potential flow boundary layer procedure for calculating viscous effects on lifts and pressure distribution of arbitrary three-dimensional configuration', Report No. MDC J7491, McDonnell Douglas Corporation.
16. Lennon, G. P., Liu, P. L. F., and Liggett, J. A. (1979) 'Boundary integral equation solution to axisymmetric potential flows: II, recharge and well problems in porous media', Unpublished work, Cornell University, Ithaca, N.Y.
17. Luu, T. S., Coulmy, G., and Corniglion, J. (1969) 'Technique effets elementaires de singularites dans la resolution des problems d'hydro et d'aerodynamique', *Ass. Technq. Marit. et Aeronaut.*, Session 1969, Paris. 'Etude des ecoutements instationnaires autour des ambes passantes par une theorie non lineaire', *Ass. Technq. Marit. et Aeronaut.*, Session 1971, Paris.
18. Luu, T. S., Coulmy, G., and Sagnard, J. (1971) 'Calcul non lineaire de l'ecoulement a potentiel autour d'une aile d'envergure finie de forme arbitraire', *Ass. Technq. Marit. et Aeronaut.*, Session 1971, Paris.
19. Luu, T. S., Coulmy, G., and Dulieu, A. (1972) 'Calcul de l'ecoulement transsonique autour d'un profil eu admettant la loi de compressibilite exacte', *Ass. Technq. Marit. et Aeronaut.*, Session 1972, Paris.

20. Luu, T. S., and Coulmy, G. (1975) 'Calcul de l'ecoulement transsonique avec choc a traverse une goille d'anbes', *Ass. Technq. Marit. et Aeronaut.*, Session 1975, Paris.
21. Luu, R. S., and Dulieu, A. (1977) 'Calcul de l'helice fouctionnaut en arriere d'une corps a symetrie axiale', *Ass. Technq. Marit. et Aeronaut.*, Session 1977, Paris.
22. Peckham, D. H. (1961) 'Low speed wind-tunnel tests on a series of uncambered slender pointed wings with sharp edges', British R and M 3186.
23. Lighthill, M. J. (1958) 'On displacement thickness', *J. Fluid Mech.*, part 4.
24. Neuman, S. P., and Witherspoon, P. A. (1971) 'Analyses of non-steady flow with a free surface using the finite element method', *Wat. Resour. Res.*, 7(3), 611–623.
25. Hall, H. P. (1955) 'An investigation of steady flow toward a gravity well', *La Houille Blanche*, 10(8).
26. Kipp, K. L. (1973) 'Unsteady flow to a partially penetrating, finite radius well into an unconfined aquifer', *Wat. Resour. Res.*, 9(2), 448–462.
27. Kipp, K. L. (1971) 'Unsteady flow to a partially penetrating finite-radius well in an unconfined aquifer', Ph.D. thesis, University of Washington, Seattle.
28. Rizzo, F. J., and Shippy, D. J. (1979) 'A boundary integral approach to potential and elasticity problems for axi-symmetric bodies with arbitrary boundary conditions' Unpublished report of Dept Engng Mechs, University of Kentucky.

# CHAPTER SIX

# THREE-DIMENSIONAL PROBLEMS IN ELASTICITY

## 6-1 INTRODUCTION

The development of algorithms for three-dimensional problems in elasticity follows closely that of two-dimensional elasticity and three-dimensional potential flow described in Chapters 4 and 5, although the kernel functions involved in the various integral equations are of course different.

The great potential of BEM in three-dimensional elasticity has already been exploited by a number of research workers who have published stress analyses of bulky three-dimensional solids using both IBEM and DBEM[1-11] and a number of commercial computer programmes based on these algorithms are currently in use.

## 6-2 SINGULAR SOLUTIONS

### 6-2-1 Isotropic Point Force Solution

For three-dimensional homogeneous, isotropic, elastic media the point force solution required is that due to Kelvin:[12]

$$u_i(x) = G_{ij}(x, \xi) e_j(\xi) \tag{6-1}$$

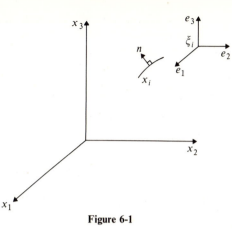

**Figure 6-1**

where $e_j(\xi)$ are the unit forces acting at a point $\xi_i$ with reference to a cartesian coordinate system $x_i$ as shown in Fig. 6-1, and

$$G_{ij}(x, \xi) = \frac{1}{16\pi\mu(1-v)} \frac{1}{r}\left[(3-4v)\,\delta_{ij} + \frac{y_i y_j}{r^2}\right] \tag{6-2}$$

with $\mu$ and $v$ the shear modulus and Poisson's ratio of the medium respectively, and

$$y_i = (x_i - \xi_i) \qquad r^2 = y_i y_i$$

The strain $\varepsilon_{ij}(x)$ is given by

$$\varepsilon_{ij}(x) = B_{ijk}(x, \xi)\,e_k(\xi) \tag{6-3}$$

where

$$B_{ijk}(x, \xi) = -\frac{1}{16\pi\mu(1-v)} \frac{1}{r^2}\left[(1-2v)\left(\delta_{ik}\frac{y_j}{r} + \delta_{jk}\frac{y_i}{r}\right) - \delta_{ij}\frac{y_k}{r} + \frac{3 y_i y_j y_k}{r^3}\right] \tag{6-4}$$

and the corresponding stresses

$$\sigma_{ij}(x) = T_{ijk}(x, \xi)\,e_k(\xi) \tag{6-5}$$

where

$$T_{ijk}(x, \xi) = -\frac{1}{8\pi(1-v)} \frac{1}{r^2}\left[\left(\delta_{ik}\frac{y_j}{r} + \delta_{jk}\frac{y_i}{r} - \delta_{ij}\frac{y_k}{r}\right)(1-2v) + \frac{3 y_i y_j y_k}{r^3}\right]$$

The tractions on a surface through $x$ with outward normal $n_i$ can be calculated from the standard expression $t_i(x) = \sigma_{ij}(x)\,n_j(x)$, whence

$$t_i(x) = F_{ij}(x, \xi)\,e_j(\xi) \tag{6-6}$$

with

$$F_{ij}(x, \xi) = -\frac{1}{8\pi(1-v)} \frac{1}{r^2}\left\{(1-2v)\left(n_j\frac{y_i}{r} - n_i\frac{y_j}{r}\right) + \left[\frac{3 y_i y_j}{r^2} + (1-2v)\,\delta_{ij}\right]\frac{y_k n_k}{r}\right\} \tag{6-7}$$

Equations (6-1) to (6-7) provide us with all the components and derivatives, etc., of the singular, or fundamental, solution that we require to solve, in principle, any isotropic three-dimensional elasticity problem.

## 6-2-2 Anisotropic Point Force Solution

Although there are a number of formal solutions for a point force within anisotropic solids the only available closed-form analytical solution is that for a point force within a transversely isotropic body.[13-15] Most of the other solutions are not suitable for a general algorithm although they may be useful in specific cases.

The displacement kernel function $G_{ij}(x, \xi)$ can be expressed for a general anisotropic solid as[16]

$$G_{ij}(x, \xi) = \frac{1}{8\pi^2 r} \oint_{|\lambda| = 1} K_{ij}^{-1}(\lambda)\, ds \tag{6-8}$$

where the line integral is taken on a unit circle in the plane normal to $r$ and passing through $x$. The function $K_{ij}^{-1}$ is given by

$$K_{ij}^{-1}(\lambda) = [C_{ijkm} \lambda_k \lambda_m]^{-1} \tag{6-9}$$

where $C_{ijkm}$ are the elastic constants.

It is possible to evaluate the displacement kernel by a series expansion of (6-8) or a direct numerical method,[17,18] but both of these procedures are unsatisfactory for routine numerical use. Wilson and Cruse[16] presented a very efficient and elegant evaluation of the contour integral in Eq. (6-8). They defined a modulation function

$$M_{ij}(v_1, v_2) = \oint_{|\lambda| = 1} K_{ij}^{-1}(\lambda)\, ds \tag{6-10}$$

where $v_1$ and $v_2$ describe the orientation of the vector **r**. Equation (6-8) can then be written as

$$G_{ij}(x, \xi) = \frac{1}{8\pi^2 r} M_{ij}(v_1, v_2) \tag{6-11}$$

The function $M_{ij}(v_1, v_2)$ which is independent of $r$ (the distance between the load point and the field point) has no singularities when the load point and field point coincide, and is continuously differentiable.

In order to obtain the strain at a point $x$ we need to differentiate the function $G_{ij}$ to produce

$$\frac{\partial G_{ij}}{\partial x_k} = -\frac{1}{8\pi^2 r^2} \frac{y_k}{r} M_{ij} + \frac{1}{8\pi^2 r} \frac{\partial M_{ij}}{\partial v_\alpha} \frac{\partial v_\alpha}{\partial x_k} \tag{6-12}$$

where the derivatives of $M_{ij}$ are non-singular and those of $v_\alpha$ can be obtained as closed-form expressions.

By using Eq. (6-12) we can calculate all the necessary components of the point force solution that are required, although it is apparently important to avoid any numerical differentiation in the evaluation $M_{ij,\alpha}$. Further details on this can be found in Wilson and Cruse.[16]

## 6-3 BASIC INTEGRAL FORMULATIONS

The basic direct and indirect integral formulations developed in Chapter 4 apply equally to three-dimensional stress analysis provided that we use the appropriate three-dimensional kernel functions $G, F, T$, etc., introduced in Sec. 6-2. Furthermore, exactly as in the case of three-dimensional potential flow, the function $G$ satisfies the infinite boundary conditions automatically and therefore the rigid-body displacement vector $\mathbf{C}$ introduced in the two-dimensional IBEM formulation will not be required.

The order of singularity in the function $G$ is $1/r$ and that for functions $F, B$, and $T$ is $1/r^2$. The existence and integrability of singularities such as these was discussed in Sec. 5-3.

## 6-4 BODY FORCES

The presence of body forces of a general nature requires an integration scheme throughout the volume of the body. However, if the body forces are generated by a steady state field of temperature gradients, seepage forces, or centrifugal forces the body force integrals can be reduced to equivalent surface integrals. The problem is then once more one of boundary integration only.

### 6-4-1 Thermal Strains or Seepage Gradients

In many situations it is necessary to carry out a three-dimensional stress analysis of bodies subjected to a distribution of temperature or a hydraulic potential which satisfies the following steady state potential flow equation:

$$\frac{\partial^2 p}{\partial x_i \partial x_i} = -q \tag{6-13}$$

where $q$ is either zero or constant throughout the region. The total stresses can then be written as[19-21]

$$\sigma_{ij} = \sigma'_{ij} - \gamma \delta_{ij} p \tag{6-14}$$

where

$$\sigma'_{ij} = \frac{2\mu v}{1-2v} \delta_{ij} \frac{\partial u_m}{\partial x_m} + \mu \left( \frac{\partial u_i}{\partial x_j} + \frac{\partial u_j}{\partial x_i} \right)$$

and $\mu$ and $v$ are again the shear modulus and the Poisson's ratio of the solid (in the case of a porous body these are the equivalent properties of the skeleton and $\sigma'_{ij}$ is the effective stress tensor), $\gamma = -1$ for the seepage problem and $\alpha E/(1-2v)$ for thermal stress analysis problems, $\alpha$ being the coefficient of thermal expansion and $E$ Young's modulus.

The total surface traction can then be written as

$$t_i = \sigma_{ij} n_j = \sigma'_{ij} n_j - \gamma p n_i$$
$$= t'_i - \gamma p n_i \tag{6-15}$$

Equilibrium requires

$$\frac{\partial \sigma_{ij}}{\partial x_j} = \frac{\partial \sigma'_{ij}}{\partial x_j} - \gamma \frac{\partial p}{\partial x_i} = 0 \tag{6-16}$$

We can now use these equations to construct both the direct and the indirect integral formulations for the solution of boundary-value problems. If we note that the second term in Eq. (6-16) is an equivalent body force we can write the direct integral representation for the displacements at any interior point as

$$u_j(\xi) = \int_S [t'_i(x) G_{ij}(x, \xi) - F_{ij}(x, \xi) u_i(x)] \, ds - \int_V \gamma \frac{\partial p(x)}{\partial x_i} G_{ij}(x, \xi) \, dv \tag{6-17}$$

where

$$t'_i(x) = \sigma'_{ij}(x) n_j(x) = t_i + \gamma p n_i$$

Using the divergence theorem the volume integral in Eq. (6-17) can be rewritten as

$$-\int_V \gamma \frac{\partial p(x)}{\partial x_i} G_{ij}(x, \xi) \, dv = -\int_S \gamma p n_i(x) G_{ij}(x, \xi) \, dv + \int_V \gamma p(x) \frac{\partial G_{ij}(x, \xi)}{\partial x_i} \, dv \tag{6-18}$$

Substituting Eq. (6-18) into (6-17) and using (6-15) to eliminate $t'_i$ (the modified traction) we arrive at

$$u_j(\xi) = \int_S [t_i(x) G_{ij}(x, \xi) - F_{ij}(x, \xi) u_i(x)] \, ds + \int_V \gamma p(x) \frac{\partial G_{ij}(x, \xi)}{\partial x_i} \, dv \tag{6-19}$$

For the analogous thermal problem, substituting for the displacement kernel function $G_{ij}$ from (6-2), we can write the volume integral as

$$\int_V \gamma p(x) \frac{\partial G_{ij}(x, \xi)}{\partial x_i} \, dv = \frac{\alpha(1+v)}{4\pi(1-v)} \int_V \frac{\partial}{\partial x_j} \frac{1}{r} p(x) \, dv \tag{6-20}$$

where $r$ is the distance between $x$ and $\xi$.

In order to transform the volume integrals in (6-19) and (6-20) we note the following property of the laplacian operator in three dimensions:[22]

$$\nabla^2 r = \frac{2}{r} \tag{6-21}$$

By using both Eq. (6-13) with $q = 0$ and Eq. (6-21), the volume integral in (6-20) can be written as[22]

$$\int_v \frac{\partial}{\partial x_j} \frac{1}{r} p(x)\, dv = \frac{1}{2} \int_v \left[ \nabla^2 \frac{\partial r}{\partial x_j} p(x) - \frac{\partial r}{\partial x_j} \nabla^2 p(x) \right] dv \qquad (6\text{-}22)$$

The divergence theorem applied to the right-hand side of (6-22) produces

$$\int_v \frac{\partial}{\partial x_j} \frac{1}{r} p(x)\, dv = \frac{1}{2} \int_S \left[ \frac{1}{r} \left( n_j - \frac{\partial r}{\partial x_j} \frac{\partial r}{\partial n} \right) p(x) - \frac{\partial r}{\partial x_j} \frac{\partial p(x)}{\partial n} \right] ds$$

and therefore Eq. (6-20) can be reduced to

$$u_j(\xi) = \int_S \left[ t_i(x) G_{ij}(x,\xi) - F_{ij}(x,\xi) u_i(x) \right] ds$$

$$+ \frac{\alpha(1+\nu)}{8\pi(1-\nu)} \int_S \left\{ \frac{1}{r} \left[ n_j(x) - \frac{\partial r}{\partial x_j} \frac{\partial r}{\partial n} \right] p(x) - \frac{\partial r}{\partial x_j} \frac{\partial p(x)}{\partial n} \right\} ds \qquad (6\text{-}23)$$

in which all the volume integrals have been transformed into equivalent boundary integrals.

Equation (6-23) can now be written for a point $\xi$ on the boundary in the usual manner. Its solution requires knowledge of both the potential $p(x)$ and its normal derivative $\partial p(x)/\partial n$ over the boundaries. These can be obtained from a prior solution of the underlying potential flow problem as discussed in Chapter 5. These comments also apply to the case of a porous body and only the constants outside the second surface integral in (6-23) would be different.

Clearly the derivatives of the displacement field given by Eq. (6-23) when substituted in Eqs (6-14) will give us the stresses $\sigma_{ij}$[22,23] as

$$\sigma_{ij}(\xi) = \sigma'_{ij}(\xi) - \gamma \delta_{ij} p(\xi)$$

$$= \int_S \left[ D_{kij}(x,\xi) t_k(x) - S_{kij}(x,\xi) u_k(x) \right] ds$$

$$+ \int_S \left[ S^*_{ij}(x,\xi) p(x) - V^*_{ij}(x,\xi) \frac{\partial p(x)}{\partial n} \right] ds - \gamma \delta_{ij} p(\xi) \qquad (6\text{-}24)$$

where

$$D_{kij} = \frac{1}{r^{\alpha_0}} \frac{(1-2\nu)(\delta_{ki} r_{,j} + \delta_{kj} r_{,i} - \delta_{ij} r_{,k}) + \beta r_{,i} r_{,j} r_{,k}}{4\pi\alpha_0(1-\nu)}$$

$$S_{kij} = \frac{2\mu}{r^\beta} \left[ \beta \frac{dr}{dn} \{ (1-2\nu)\delta_{ij} r_{,k} + \nu(\delta_{ik} r_{,j} + \delta_{jk} r_{,i}) - \psi r_{,i} r_{,j} r_{,k} \} \right.$$

$$+ \beta\nu(n_i r_{,j} r_{,k} + n_j r_{,i} r_{,k}) + (1-2\nu)(\beta n_k r_{,i} r_{,j} + n_j \delta_{ik} + n_i \delta_{jk})$$

$$\left. - (1-4\nu) n_k \delta_{ij} \right] \Big/ [4\alpha_0 \pi(1-\nu)]$$

Then $\alpha_0 = 2$, $\beta = 3$, and $\psi = 5$ for the three-dimensional case discussed here whereas $\alpha_0 = 1$, $\beta = 2$, and $\psi = 4$ apply to the two-dimensional elastic case discussed in Chapter 4. For the three-dimensional case,

$$S_{ij}^* = \frac{\alpha E}{8\pi(1-v)} \frac{1}{r^2} \left[ \frac{1}{1-2v} \delta_{ij} \frac{\partial r}{\partial n} + \left( n_i r_{,j} + n_j r_{,i} - 3r_{,i} r_{,j} \frac{\partial r}{\partial n} \right) \right]$$

$$V_{ij}^* = -\frac{\alpha E}{8\pi(1-v)} \frac{1}{r} \left( \frac{1}{1-2v} \delta_{ij} + r_{,i} r_{,j} \right)$$

In these equations the implied differentiations are with respect to the variable $\xi$ and the normal is evaluated at $x$.

Following the equivalence between the direct and indirect BEM formulations discussed in Sec. 3-6, we can write the equivalent displacement integral for an indirect formulation from Eq. (6-19) as

$$u_i(x) = \int_S G_{ij}(x, \xi)\, \phi_j(\xi)\, ds + \int_V \gamma p(\xi) \frac{\partial G_{ij}(x, \xi)}{\partial \xi_i}\, dv \tag{6-25}$$

where the volume integral can be converted into a surface integral in the same manner as shown earlier. Whence we arrive at

$$u_i(x) = \int_S G_{ij}(x, \xi)\, \phi_j(\xi)\, ds + \frac{\alpha(1+v)}{8\pi(1-v)}$$
$$\times \int_S \left\{ \frac{1}{r} \left[ n_j(\xi) - \frac{\partial r}{\partial \xi_j} \frac{\partial r}{\partial n} \right] p(\xi) - \frac{\partial r}{\partial \xi_j} \frac{\partial p(\xi)}{\partial n} \right\} ds \tag{6-26}$$

By using (6-26) in the strain displacement equations together with the stress–strain relations we can obtain the 'pseudo' stresses as

$$\sigma_{ij}'(x) = \int_S T_{ijk}(x, \xi)\, \phi_k(\xi)\, ds + \int_S \left[ S_{ij}'(x, \xi) p(\xi) - V_{ij}'(x, \xi) \frac{\partial p(\xi)}{\partial n} \right] ds \tag{6-27}$$

where $S_{ij}'(x, \xi)$ and $V_{ij}'(x, \xi)$ are the same as $S_{ij}^*$ and $V_{ij}^*$ defined above except that the implied differentiations are with respect to the variable $x$ and the normal is evaluated at $\xi$.

The total traction $t_i(x)$ on a surface through $x$ with an outward normal $n(x)$ is given by

$$t_i = \sigma_{ij}(x) n_j(x)$$
$$= \sigma_{ij}'(x) n_j(x) - \gamma p(x) n_i(x)$$

or $\quad t_i(x) = \int_S F_{ij}(x, \xi)\, \phi_j(\xi)\, ds + \int_S \left[ S_{ij}'(x, \xi) n_j(x) p(\xi) - V_{ij}'(x, \xi) n_j(x) \frac{\partial p(\xi)}{\partial n} \right] ds$

$$- \gamma p(x) n_i(x) \tag{6-28}$$

By taking the point $x$ to a boundary point $x_o$ in Eqs (6-26) and (6-28), as shown in Chapter 4, we can obtain the two boundary integral equations necessary for the solution of any well-posed boundary-value problem by the indirect BEM.

## 6-4-2 Mechanical Body Forces

Recently Rizzo and Shippy[7, 8, 24] extended the analysis to include body forces that are gradients of a scalar potential and presented a DBEM formulation which included the combined effects of a temperature field and mechanical body forces.

If the temperature field satisfies the equation

$$\frac{\partial^2 p(x)}{\partial x_i \partial x_i} = 0 \tag{6-29}$$

and the mechanical body forces can be described by

$$\psi_i(x) = \frac{\partial f(x)}{\partial x_i} \quad \text{with} \quad \frac{\partial^2 f(x)}{\partial x_i \partial x_i} = q \tag{6-30}$$

where $q$ is a constant for the entire region, we can use either of the boundary element approaches to solve Eqs (6-29) and (6-30) for the surface values of $(p, \partial p/\partial n)$ and $(f, \partial f/\partial n)$ respectively. The DBEM development follows that explained in the preceding section and Chapter 4, with the displacements at any interior point now given by[8]

$$u_j(\xi) = \int_S \{[t_i(x) - f(x) n_i(x)] G_{ij}(x, \xi) - u_i(x) F_{ij}(x, \xi)\} \, ds$$

$$+ \int_S \left\{ [f(x) + \gamma p(x)] \frac{\partial w'(x, \xi)}{\partial x_j} - \frac{\partial}{\partial n} [f(x) + \gamma p(x)] \frac{\partial w(x, \xi)}{\partial x_j} \right\} ds$$

$$+ q \int_S n_j(x) w(x, \xi) \, ds \tag{6-31}$$

where, once more, only surface integrals are involved.

For two-dimensional problems (plane strain),

$$w = -\frac{(1-2v)(1+v)}{8\pi E(1-v)} r^2 (\ln r - 1) \qquad w' = \frac{\partial w}{\partial n}$$

whereas in three dimensions,

$$w = \frac{(1-2v)(1+v)}{8\pi E(1-v)} r \qquad w' = \frac{\partial w}{\partial n}$$

To solve any well-posed boundary-value problem, Eq. (6-31) is taken to the boundary point $\xi_o$ as before.

## 6-5 INITIAL STRESSES AND INITIAL STRAINS

Initial stresses or initial strains may be present in a body due to any number of effects. For example, a fluid pressure within a porous elastic solid can be dealt with as an initial stress, and lack of fit, thermal strains, creep, etc., may be considered as initial strains. The concept of initial stress ('eigenspannungen') was first introduced by Reisner.[25]

If the initial strains $(\varepsilon_{ij}^o)$ or the initial stresses $(\sigma_{ij}^o)$ are known, then the correct stresses are given by

$$\sigma_{ij} = D_{ijkl}(\varepsilon_{kl} - \varepsilon_{kl}^o) \tag{6-32a}$$

or $$\sigma_{ij} = D_{ijkl}\varepsilon_{kl} - \sigma_{ij}^o \tag{6-32b}$$

Equations (6-32a, b) are clearly very similar to Eqs (6-14) and the analysis developed in Sec. 6-4-1 can be applied [see Eqs (6-14) to (6-19)] to arrive at both the DBEM and IBEM statements. However, it is instructive to introduce an alternative approach[26,27] using the principle of virtual work.

It is well known[21] that the displacement field $(u_i)$ which solves any problem can be constructed as

$$u_i = u_i' + u_i'' \tag{6-33}$$

where $u_i'$ is the solution that satisfies the homogeneous form of the governing differential equation (without the body forces or initial stress gradients, etc.) and the boundary conditions and $u_i''$ is a solution that satisfies the inhomogeneous differential equation (complete with the body forces and the initial stress gradient terms).

By analogy with the standard method for solving inhomogeneous differential equations, we can regard $u_i'$ as the complementary function and $u_i''$ as a particular integral. Hence

$$u_j'(\xi) = \int_S [t_i(x)G_{ij}(x, \xi) - F_{ij}(x, \xi)u_i(x)]\,ds \tag{6-34}$$

In order to obtain the particular integral we consider the virtual work equation

$$\int_S t_i^*(x)u_i''(x)\,ds + \int_S f_i^*(x)u_i''(x)\,dv = \int_V \sigma_{ij}^*(x)\varepsilon_{ij}^0(x)\,dv \tag{6-35}$$

where $t_i^*(x), f_i^*(x)$, and $\sigma_{ij}^*(x)$ belong to a virtual state which is quite unrelated to the real state $u_i''(x)$ and $\varepsilon_{ij}^o(x)$.

If we choose the (*) system to be that caused by a point body force within an infinite solid and $u_i''(x), \varepsilon_{ij}^o(x)$ are the real displacements and initial strains, then Eq. (6-35) gives

$$u_k''(\xi) = \int_V T_{ijk}(x, \xi)\varepsilon_{ij}^o(x)\,dv \tag{6-36}$$

where $T_{ijk}(x, \xi)$ generates the stresses $\sigma_{ij}^*(x)$ due to a unit force vector $e_k(\xi)$ [Eq. (6-5)]. Equation (6-36) is in fact a generalized version of Maysel's well-known[28,29] expression for displacements caused by a steady state temperature field in which the initial strain is simply $\varepsilon_{ij}^o = \delta_{ij}\alpha p$, $\alpha$ being the coefficient of thermal expansion (note that plane stress conditions can be handled by using a modified $\alpha$) and $p$ the temperature field. The reader may verify that for the thermal strain problem the right-hand side of Eqs (6-20) and (6-36) are identical.

By using the relationships

$$\int_V \sigma_{ij}^*(x)\,\varepsilon_{ij}^0(x)\,dv = \int_V D_{ijkl}\,\varepsilon_{kl}^*(x)\,\varepsilon_{ij}^0(x)\,dv = \int_V \varepsilon_{kl}^*(x)\,\sigma_{kl}^0(x)\,dv$$

we can write Eq. (6-36) equivalently as

$$u_k''(\xi) = \int_V B_{ijk}(x,\xi)\,\sigma_{ij}^0(x)\,dv \tag{6-37}$$

where $B_{ijk}(x,\xi)$ generates the strains $\varepsilon_{ij}^*(x)$ due to a unit force vector $e_k(\xi)$ [Eq. (6-3)].

Equations (6-36) and (6-37) provide the particular integrals we require for initial strain or initial stress problems respectively. Thus for specified initial stresses the displacements at any interior point $\xi$ can be calculated from

$$u_j(\xi) = \int_S [t_i(x)\,G_{ij}(x,\xi) - F_{ij}(x,\xi)\,u_i(x)]\,ds + \int_V B_{ikj}(x,\xi)\sigma_{ik}^0(x)\,dv \tag{6-38}$$

The boundary constraint equations can be constructed from Eq. (6-38) in the usual manner.

The corresponding indirect formulation for displacements is

$$u_i(x) = \int_S G_{ij}(x,\xi)\,\phi_j(\xi)\,ds + \int_V B_{jki}(\xi,x)\,\sigma_{jk}^0(\xi)\,dv \tag{6-39}$$

and the correct stresses $\sigma_{mn}$ are given by

$$\sigma_{mn}(x) = \sigma_{mn}'(x) - \sigma_{mn}^0(x)$$

$$= \int_S T_{mnk}(x,\xi)\,\phi_k(\xi)\,ds + \int_V L_{jkmn}(\xi,x)\,\sigma_{jk}^0(\xi)\,dv - \sigma_{mn}^0(x) \tag{6-40}$$

where

$$L_{jkmn}(\xi,x) = \frac{2\mu v}{1-2v}\,\delta_{mn}\frac{\partial}{\partial x_i}\int_V B_{jki}(\xi,x)\,dv$$

$$+ \mu\frac{\partial}{\partial x_n}\int_V B_{jkm}(\xi,x)\,dv + \mu\frac{\partial}{\partial x_m}\int_V B_{jkn}(\xi,x)\,dv \tag{6-41}$$

The reason for keeping the derivative signs outside the various integrals becomes clear if we look at the function $B_{jki}$ of Eq. (6-39). The order of the singularities of the function $B_{jki}$ is $1/r$ for two-dimensional problems and $1/r^2$ for three-dimensional ones; therefore the volume integral exists in the ordinary sense. However, if we differentiate under the integral sign (admissible since the variable of integration is $\xi$) the resulting singularities, when $x$ and $\xi$ coincide, will be of the order of $1/r^2$ and $1/r^3$ in two and three dimensions respectively. The volume integral in Eq. (6-40) then ceases to have any meaning. It is therefore necessary to integrate the volume integrals in Eq. (6-41) analytically when $x = \xi$, as shown in Chapter 3, and then evaluate the derivatives. This problem is not peculiar to the indirect formulation since interior stress calculations based on Eq. (6-40) would also involve identical difficulties. They can be circumvented by calculating the contributions to the displacement field arising from the volume integrals in Eqs (6-39) and (6-40) and using a local finite difference approximation for Eq. (6-41) around the singular point.

The surface traction $t_i(x)$ on a surface through $x$ with an outward normal $n_j(x)$ can be calculated from Eq. (6-40) via

$$t_i(x) = \int_S F_{ij}(x, \xi)\, \phi_j(\xi)\, ds + \int_V M_{jki}(\xi, x)\, \sigma^o_{jk}(\xi)\, dv - \sigma^o_{ij}(x)\, n_j(x)$$

where

$$M_{jki} = L_{jkim}(\xi, x)\, n_m(x) \tag{6-42}$$

Equations (6-39) and (6-42) can again be written for a boundary point $x_o$ and used for the solution of any boundary-value problem.


## 6-6 DISCRETIZATION


### 6-6-1 General

We can discretize the surface of the body with flat triangular elements, construct a local axis system as described in Chapter 5 (Sec. 5-4-1), and assume linear variations of the parameters $u_i$, $t_i$, and $\phi_i$ over each element. The basic algorithm is therefore identical to that described in Chapter 5 for the three-dimensional potential flow problem. For simplicity we shall assume that body forces, initial stresses, etc., are absent.


### 6-6-2 Linear Shape Functions

If we assume a linear variation of the unknown fictitious intensities over the $q$th boundary element (for example) we have[30]

$$\boldsymbol{\phi}(\xi) = \mathbf{N}^q(\xi)\boldsymbol{\phi}_n \tag{6-43}$$

where    $\boldsymbol{\phi}(\xi)$ = the fictitious tractions at any point within the element

$$= \left\{ \begin{array}{c} \phi_1(\xi) \\ \phi_2(\xi) \\ \phi_3(\xi) \end{array} \right\}, \text{ the vector of the three components of } \phi \text{ in the direction of the } x_1, x_2, \text{ and } x_3 \text{ axes}$$

$\mathbf{N}^q(\xi) = 3 \times 9$ matrix

$$= \begin{bmatrix} N_1 & 0 & 0 & N_2 & 0 & 0 & N_3 & 0 & 0 \\ 0 & N_1 & 0 & 0 & N_2 & 0 & 0 & N_3 & 0 \\ 0 & 0 & N_1 & 0 & 0 & N_2 & 0 & 0 & N_3 \end{bmatrix}$$

$\boldsymbol{\phi}_n = 9 \times 1$ vector of the nodal values of $\phi$

The discretized forms of the relevant boundary integrals can then be expressed in the manner discussed in Chapter 4.

### 6-6-3 Integration of the Kernel Shape Function Products

The integration schemes for the singular and non-singular contribution will be identical to those discussed in earlier chapters, namely, when the kernel shape function products remain finite throughout the interval of integration we adopt numerical integration formulas but the singular case (i.e., when the field point is at a node where the shape function tends to unity) would be isolated and integrated exactly (see Secs 4-4-2 and 5-4-4).

**(a) Evaluations of** $\int_{\Delta S} G_{ij}(x_o, \xi)\, ds(\xi)$ **or** $\int_{\Delta S} G_{ij}(x, \xi_o)\, ds(x)$

By considering a local coordinate system $z_1$ and $z_2$ in the plane of the element and $z_3$ in the direction normal through the node under consideration[5] (see Fig. 6-2),

$$\frac{y_i}{r} = \frac{\partial r}{\partial x_i} = \sin\theta \frac{\partial z_1}{\partial x_i} + \cos\theta \frac{\partial z_2}{\partial x_i} = \sin\theta\, e_{1i} + \cos\theta\, e_{2i}$$

when $e_{1i}$ and $e_{2i}$ are the direction cosines of the $z_1$ and $z_2$ axes in the global coordinate system.

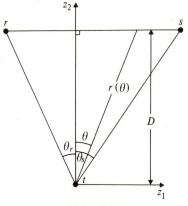

**Figure 6-2**

We can then recast the integral as

$$\int_{\Delta s} G_{ij}(x, \xi) \, ds(x)$$

$$= \int_{-\theta_r}^{\theta_s} \int_0^{r(\theta)} A \frac{1}{r} \left[ \delta_{ij}(3 - 4v) + \frac{y_i y_j}{r} \right] r \, dr \, d\theta \qquad \text{where } A = \frac{1}{16\pi\mu(1 - v)}$$

$$= \int_{-\theta_r}^{\theta_s} Ar(\theta) [\delta_{ij}(3 - 4v) + (\sin \theta e_{1i} + \cos \theta e_{2i})(\sin \theta e_{1j} + \cos \theta e_{2j})] \, d\theta$$

where $r(\theta) = D/\cos \theta$. Thus

$$\int_{\Delta s} G_{ij}(x, \xi) \, ds(x)$$

$$= \int_{\Delta s} G_{ij}(x, \xi) \, ds(\xi)$$

$$= DA[\delta_{ij}(3 - 4v) \ln(\tan \theta + \sec \theta) + a_{ij}\{\ln(\tan \theta + \sec \theta) - \sin \theta\}$$

$$+ b_{ij} \cos \theta + c_{ij} \sin \theta]_{-\theta_r}^{\theta_s} \qquad (6\text{-}44)$$

where     $a_{ij} = e_{1i} e_{1j}$

$$b_{ij} = -(e_{2i} e_{1j} + e_{1i} e_{2j})$$

$$c_{ij} = e_{2i} e_{2j}$$

**(b) Evaluations of $\int_{\Delta s} F_{ij}(x, \xi) \, ds(\xi)$ or $\int_{\Delta s} F_{ij}(x, \xi) \, ds(x)$**
Cruse[5] presented a method for evaluating these integrals exactly which we follow here. For a flat element $y_k n_k = 0$; therefore the function $F_{ij}(x, \xi)$ reduces to

$$F_{ij}(x, \xi) = B \frac{1}{r^2} \left[ n_j \frac{y_i}{r} - n_i \frac{y_j}{r} \right] \qquad \text{where } B = -\frac{1 - 2v}{8\pi(1 - v)}$$

By using the identities

$$\varepsilon_{ijk} \varepsilon_{rsk} n_r \frac{\partial}{\partial x_s} \frac{1}{r} = \frac{1}{r^2} \left( n_j \frac{\partial r}{\partial x_i} - \frac{\partial r}{\partial x_j} n_i \right) = \frac{1}{r^2} \left( n_j \frac{y_i}{r} - n_i \frac{y_j}{r} \right)$$

where

$\varepsilon_{ijk}$ is the permutation tensor (see Appendix A)

$= 0$   when $i = j$ or $j = k$ or $k = i$

$= 1$   when $i, j, k$ are cyclic

$= -1$   when $i, j, k$ are anticyclic

we can express the integral $\int_{\Delta S} F_{ij}(x, \xi)\, ds(x)$ as

$$\Delta F_{ij} = \int_{\Delta S} F_{ij}(x, \xi)\, ds(x) = B\varepsilon_{ijk} \int_{\Delta S} \varepsilon_{rsk}\, n_r \frac{\partial}{\partial x_s} \frac{1}{r}\, ds(x)$$

Stoke's theorem can be used to convert this into a line integral, i.e.,

$$\Delta F_{ij} = B\varepsilon_{ijk} \oint \frac{1}{r}\, dx_k \tag{6-45}$$

In terms of the local variables

$$dx_k = e_{1k}\, dz_1 + e_{2k}\, dz_2$$

Along the side $rs$ we have

$$\Delta F_{ij} = B\varepsilon_{ij}\, e_{1k}[\ln(z_1 + r)]_r^s \tag{6-46}$$

The total result is then obtained by reorienting the local coordinate system such that the $z_1$ axis coincides, in turn, with the $st$ and $tr$ sides of the triangle. The results are then expressed in terms of the coordinates of the field point and those of the corners. Clearly Eq. (6-46) can also be used to evaluate the integral over a flat quadrilateral element in a similar manner.

Since the normal $n$ does not change direction over our flat element (i.e., it is immaterial whether the normal is over a load point or a field point) and $\partial r/\partial x_i = -\partial r/\partial \xi_i$ it is easy to show that, for this local element

$$\int_{\Delta S} F_{ij}(x, \xi)\, ds(x) = -\int_{\Delta S} F_{ij}(x, \xi)\, ds(\xi)$$

Therefore for an IBEM element we use the same result with a change of sign.

If in Eq. (6-46) the field point is taken slightly inside the region the discontinuity terms resulting from the treatment of this integral as an improper one are included automatically.

The reader should note that the integrals in Eqs. (5-14) and (6-45) are similar.

## 6-7 AXISYMMETRIC STRESS ANALYSIS

In many practical situations involving elastic stress analysis of three-dimensional bodies the geometry and loading involved are such that they can be reasonably approximated by an axisymmetric three-dimensional system. There are, of course, also many problems which are truly axisymmetric, e.g., that shown in Fig. 6-3. This represents an axially loaded insert (a pile whose base diameter is larger than the shaft diameter) embedded within an elastic solid. The boundary discretization would consist of a number of ring elements as shown in the figure. An axisymmetric idealization would therefore result in a substantial reduction in analytical effort compared to that in a conventional three-dimensional analysis. Problems such as these were analysed by the authors[31] nearly a decade ago using

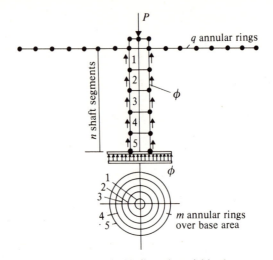

**Figure 6-3** A single pile under axial load.

IBEM. Kermanidis[32,33] has published a general indirect formulation for axisymmetric problems whereas Cruse, Snow, and Wilson[34] presented a completely general DBEM analysis including both thermal and centrifugal loadings.

### 6-7-1 Fundamental Solutions

The fundamental singular solutions due to point loads required for the analysis could be obtained by re-casting the general three-dimensional case described in Sec. 6-2, in a polar coordinate system (see Sec. 5-5-2), and integrating the various functions along a ring as shown in Fig. 6-4. This approach has been used by some workers.[31,33]

Alternatively, Cruse, Snow, and Wilson[34] used a Galerkin vector representation of a point force in a cylindrical coordinate system. In order to maintain continuity with the axisymmetric analysis outlined in Chapter 5 we shall use the former approach (i.e., direct integration of the three-dimensional point force solution described in Sec. 6-2).

The displacement field due to a *radially* loaded ring source of intensity $2\pi r_o\, e_r$ can be obtained from (6-1) as[33]

$$u_r = \frac{e_r}{4\pi\mu(1-v)}\left[\frac{4(1-v)(\rho^2+\bar{z}^2)-\rho^2}{2rR}\,K\!\left(\frac{\pi}{2},m\right)\right.$$
$$\left.-\left\{\frac{3.5-4v}{2r}\,R-\frac{e^4-\bar{z}^4}{4rR^3(1-m^2)}\right\}E\!\left(\frac{\pi}{2},m\right)\right] \qquad (6\text{-}47a)$$

$$u_z = \frac{e_r\,\bar{z}}{4\pi\mu(1-v)}\left[\frac{(e^2+\bar{z}^2)\,E(\pi/2,m)}{2R^3(1-m^2)}-\frac{K(\pi/2,m)}{2R}\right] \qquad (6\text{-}47b)$$

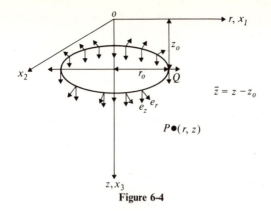

**Figure 6-4**

The corresponding field due to an axially loaded ring source of intensity $2\pi r_o\, e_z$ is given by

$$u_r = \frac{e_z r_o \bar{z}}{4\pi\mu(1-v)}\left[\frac{e^2-\bar{z}^2}{2rR^3(1-m^2)}E\left(\frac{\pi}{2},m\right)+\frac{K(\pi/2,m)}{2rR}\right] \qquad (6\text{-}47c)$$

$$u_z = \frac{e_z r_o}{4\pi\mu(1-v)}\left[\frac{(3-4v)}{R}K\left(\frac{\pi}{2},m\right)+\frac{\bar{z}^2\,E(\pi/2,m)}{R^3(1-m^2)}\right] \qquad (6\text{-}47d)$$

where $\rho^2 = r^2 + r_o^2$, $e^2 = r^2 - r_o^2$, $R = \sqrt{[(r+r_o)^2+\bar{z}^2]}$, and $\bar{z} = z - z_o$. $K$ and $E$ are complete elliptic integrals of the first and second kind, modulus $m = \sqrt{(4rr_o/R^2)}$, $0 \leqslant m^2 \leqslant 1$, and the complementary modulus $m_1 = \sqrt{(1-m^2)}$. Such functions can be represented by polynomial approximations as mentioned in Chapter 5. We can write Eqs (6-47a–d) in matrix form as

$$\begin{Bmatrix} u_r \\ u_z \end{Bmatrix} = \begin{bmatrix} G_{rr} & G_{rz} \\ G_{zr} & G_{zz} \end{bmatrix} \begin{Bmatrix} e_r \\ e_z \end{Bmatrix} \quad \text{or} \quad \mathbf{u} = \mathbf{Ge} \qquad (6\text{-}48)$$

The partial derivatives of elliptic functions which are necessary in order to calculate the strains, stresses, and tractions are given by

$$\frac{\partial E}{\partial a} = \frac{E-K}{2a} \quad \text{and} \quad \frac{\partial K}{\partial a} = \frac{E}{2a(1-a)} - \frac{K}{2a} \qquad (6\text{-}49)$$

where $a = m^2$. Using (6-48) and (6-49) the strains can be calculated from

$$\begin{Bmatrix} \varepsilon_r \\ \varepsilon_\theta \\ \varepsilon_z \\ \gamma_{rz} \end{Bmatrix} = \begin{bmatrix} \dfrac{\partial G_{rr}}{\partial r} & \dfrac{\partial G_{rz}}{\partial r} \\[2mm] \dfrac{G_{rr}}{r} & \dfrac{G_{rz}}{r} \\[2mm] \dfrac{\partial G_{zr}}{\partial z} & \dfrac{\partial G_{zz}}{\partial z} \\[2mm] \dfrac{\partial G_{rr}}{\partial z}+\dfrac{\partial G_{zr}}{\partial r} & \dfrac{\partial G_{rz}}{\partial z}+\dfrac{\partial G_{zz}}{\partial r} \end{bmatrix} \begin{Bmatrix} e_r \\ e_z \end{Bmatrix} \quad \text{or} \quad \boldsymbol{\varepsilon} = \mathbf{Be} \qquad (6\text{-}50)$$

and from the stress–strain relationships the stresses can be obtained as

$$
\begin{Bmatrix} \sigma_r \\ \sigma_\theta \\ \sigma_z \\ \tau_{rz} \end{Bmatrix} = \frac{2\mu}{1-2v} \begin{bmatrix} 1 & \dfrac{v}{1-v} & \dfrac{v}{1-v} & 0 \\ & 1 & \dfrac{v}{1-v} & 0 \\ & & 1 & 0 \\ \text{symm} & & & \dfrac{1-2v}{2(1-v)} \end{bmatrix} \begin{Bmatrix} \varepsilon_r \\ \varepsilon_\theta \\ \varepsilon_z \\ \gamma_{rz} \end{Bmatrix} \quad \text{or} \quad \boldsymbol{\sigma} = \mathbf{D}\boldsymbol{\varepsilon}
$$

(6-51)

The tractions on a surface through $(r, z)$ having an outward normal $n$ (with global components $n_r$ and $n_z$ referred to the $r$ and $z$ axes) may be calculated from[34]

$$
\begin{Bmatrix} t_r \\ t_z \end{Bmatrix} = \begin{bmatrix} F_{rr} & F_{rz} \\ F_{zr} & F_{zz} \end{bmatrix} \begin{Bmatrix} e_r \\ e_z \end{Bmatrix} \quad \text{or} \quad \mathbf{t} = \mathbf{Fe}
$$

(6-52)

where

$$
F_{rr} = 2\mu \left\{ \left[ c\frac{\partial G_{rr}}{\partial r} + d\left( \frac{G_{rr}}{r} + \frac{\partial G_{zr}}{\partial z} \right) \right] n_r + \frac{1}{2}\left( \frac{\partial G_{rr}}{\partial z} + \frac{\partial G_{zr}}{\partial r} \right) n_z \right\}
$$

$$
F_{rz} = 2\mu \left\{ \left[ c\frac{\partial G_{zr}}{\partial z} + d\left( \frac{G_{rr}}{r} + \frac{\partial G_{rr}}{\partial r} \right) \right] n_z + \frac{1}{2}\left( \frac{\partial G_{rr}}{\partial z} + \frac{\partial G_{zr}}{\partial r} \right) n_r \right\}
$$

$$
F_{zr} = 2\mu \left\{ \left[ c\frac{\partial G_{rz}}{\partial r} + d\left( \frac{G_{rz}}{r} + \frac{\partial G_{zz}}{\partial z} \right) \right] n_r + \frac{1}{2}\left( \frac{\partial G_{rz}}{\partial z} + \frac{\partial G_{zz}}{\partial r} \right) n_z \right\}
$$

$$
F_{zz} = 2\mu \left\{ \left[ c\frac{\partial G_{zz}}{\partial z} + d\left( \frac{G_{rz}}{r} + \frac{\partial G_{rz}}{\partial r} \right) \right] n_r + \frac{1}{2}\left( \frac{\partial G_{rz}}{\partial z} + \frac{\partial G_{zz}}{\partial r} \right) n_z \right\}
$$

$$
c = \frac{1-v}{1-2v} \quad \text{and} \quad d = \frac{v}{1-2v}
$$

Equations (6-48) to (6-52) provide us with all the components of the fundamental singular solution which are needed.

## 6-7-2 Direct and Indirect Formulations

Both the direct and indirect BEM statements follow once again from the point force solution. For example, an indirect formulation for an interior point $P$ can be written as [note that the coordinates of $P$ are now $(r, z)$ and those of $Q$ are $(r_o, z_o)$]

$$
\mathbf{u}(P) = \int_S \mathbf{G}(P, Q)\boldsymbol{\phi}(Q)\, dr_o\, dz_o
$$

(6-53)

and

$$
\mathbf{t}(P) = \int_S \mathbf{F}(P, Q)\boldsymbol{\phi}(Q)\, dr_o\, dz_o
$$

(6-54)

where $\boldsymbol{\phi}(Q) = \begin{Bmatrix} \phi_r \\ \phi_z \end{Bmatrix}$ and the normal for $\mathbf{F}$ is at $P$.

Similarly, the direct formulation for an interior point $Q(r_o, z_o)$ becomes

$$\mathbf{u}(Q) = \int_S [\mathbf{F}^T(P, Q)\mathbf{u}(P) - \mathbf{G}^T(P, Q)\mathbf{t}(P)] \, dr \, dz \qquad (6\text{-}55)$$

where $\mathbf{F}^T, \mathbf{G}^T$ are $\mathbf{F}$ and $\mathbf{G}$ transposed.

We can again use Eqs (6-53) and (6-54) or (6-55) to construct a numerical algorithm for the solution of boundary-value problems by taking the field point $P$ in Eqs (6-53) and (6-54), or $Q$ in Eq. (6-55), to a boundary point. The development of this algorithm is essentially identical to that discussed in both Sec. 5-5-3 and Refs 31 and 34.

### 6-7-3 Body Forces

In a great many problems of mechanical engineering, three-dimensional axisymmetric stress analyses are required for components in which steady state thermal and centrifugal forces are present; similar problems arise in other branches of engineering. The simpler approach is again to convert the integrals derived in Sec. 6-4 to their equivalent axisymmetric forms following the above pattern. (The corresponding axisymmetric Galerkin vector analysis will be found in Cruse, Snow, and Wilson.[34])

## 6-8 EXAMPLES

Because of the inherent attractions of BEM for three-dimensional stress analysis in engineering there is a considerable published literature which amply demonstrates the general usefulness of the method in routine analysis. Indeed, for bulky solids it is probably the only reliable method currently available for obtaining detailed results at reasonable cost. In addition BEM are able to take advantage of theoretical developments in singular solutions, representation of boundary geometry, anisotropy, etc., with each step placing them yet further ahead of competing analyses. This section contains a selection of solved problems together with some assessment of the accuracy of the results.

**(a) The problem of a loaded cube**[11]  The boundary discretization and the applied loading are shown in Fig. 6-5*a–c*. For problem 1 a linear variation of displacement $u_x$ was specified at the end of the cantilever (Fig. 6-5*c*). Figure 6-6*a* shows the stresses $\sigma_x$ at various locations including interior and boundary points with the calculated bending stresses in excellent agreement with the exact solution.

Figure 6-6*b* shows the bending and shear stress distributions along the vertical centreline of the specimen, at mid-length, for problems 2 and 3. The results of an analysis (three-dimensional) using uniform variations of tractions

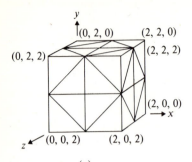

(a)

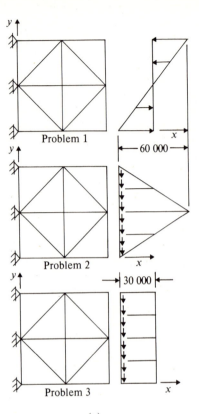

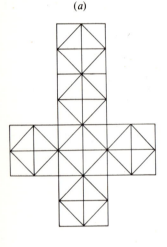

(b)

(c)

**Figure 6-5** (a) Test problem geometry. (b) Test problem boundary segments. (c) Beam boundary conditions for test problems.

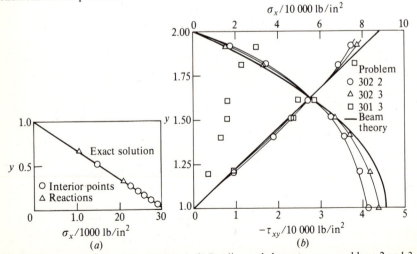

(a)

(b)

**Figure 6-6** (a) Bending stresses—problem 1. (b) Bending and shear stresses—problems 2 and 3.

and displacements for problem 3 are also shown for comparison. The three-dimensional analysis is quite clearly unsatisfactory and one can conclude that, for bodies in which bending is dominant over the boundary, it is necessary to consider at least a linear variation of functions over the boundary.

**(b) A three-dimensional mine structure problem**[10] A room-and-pillar structure representing roof support components in a coal mine was selected to provide a direct comparison of finite element FE and BEM in three-dimensional stress analysis. The geometry of the problem, the finite element discretization, and two boundary element meshes are shown in Fig. 6-7a–d. Figure 6-7a illustrates a symmetric portion of the entire structure where the $xy$ and $xz$ planes are planes of symmetry; faces 1, 2, and 3 have zero normal displacement and zero shear; and face 4 is loaded by a normal pressure.

Cruse[10] used DBEM with a piecewise constant approximation for $u_i$ and $t_i$ to solve this problem and investigate the following:

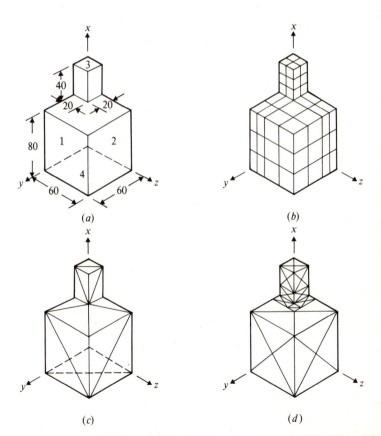

(a)   (b)

(c)   (d)

**Figure 6-7** (a) Mine structure geometry. (b) FE model of the mine structure. (c) BEM-001 model of the mine structure. (d) BEM-003 model of the mine structure.

1. Do FEM and BEM give comparable values of the internal stresses?
2. What are the relative problem sizes and computing times for the two methods?
3. How well do the two methods resolve the stress concentration problem at the room-pillar intersection?

The finite element analysis was carried out using modified Irons–Zienkiewicz hexahedra. He found that the finite element stresses at all interior points were generally in agreement to within 10 to 15 per cent of those obtained using BEM-001 (Fig. 6-7c) and within 5 per cent of those obtained using BEM-003 (Fig. 6-7d).

Table 6-1 summarizes the computational statistics using an early version of Cruse's BEM programme. The stresses computed using BEM for points approaching the corners of the room-pillar intersection are shown in Fig. 6-8. The finite element discretization was too crude to provide such detailed stresses; the finite element results are clearly not acceptable and a much finer discretization in the neighbourhood of the room-pillar interaction would be required with a parallel increase in computer time. The overall comparison is therefore not as precise as one might wish.

**Table 6-1 FEM and BEM results in three-dimensional mine structure**

|  | FEM | BEM-001 | BEM-003 |
| --- | --- | --- | --- |
| Problem size | 92 bandwidth<br>274 degrees of<br>freedom<br>24 208 coefficient array | 16 boundary segments<br>48 degrees of<br>freedom<br>2304 coefficient array | 44 boundary segments<br>132 degrees of<br>freedom<br>19 044 coefficient array |
| CPU times† | 50 s total | 34 s boundary solution<br><br>7 s each interior<br>point | 315 s boundary<br>solution<br>19 s each interior<br>point |

† CPU times obtained on a CDC 6400 computer.

**(c) Three-dimensional analysis of pile groups**[35]  Figure 6-9a shows the geometry of a $3 \times 3$ pile group embedded in a multilayered soil together with Figure 6-9b, the discretization scheme adopted. The average vertical stress over the pile cross sections is shown in Fig. 6-10, illustrating clearly that the central pile carries the least load; the four corner piles support nearly half the load and the percentage of the load carried directly by the pile cap is negligible.

Figure 6-11 summarizes results obtained from a three-dimensional analysis of the load displacement behaviour of groups of rigid piles with a rigid cap both resting on isotropic elastic ground and clear of it. (The settlement ratio is defined as the ratio of the settlement of an $N$ pile group to that of a single pile when all of the piles carry identical average loads.)

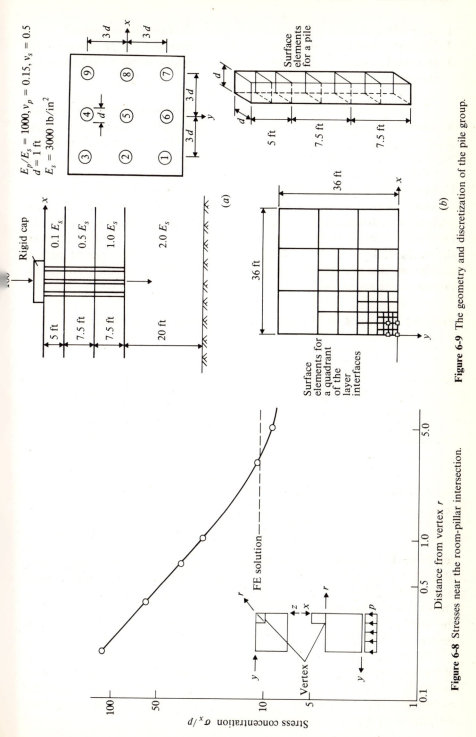

**Figure 6-9** The geometry and discretization of the pile group.

**Figure 6-8** Stresses near the room-pillar intersection.

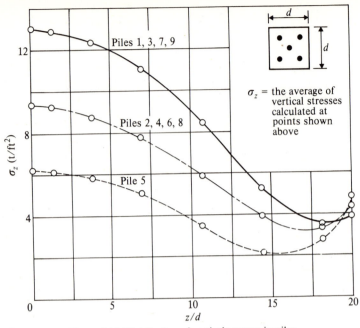

**Figure 6-10** Distribution of vertical stresses in piles.

This problem was solved by IBEM (Butterfield and Banerjee[36]) utilizing Mindlin's solution[37] for a point load in the interior of a semi-infinite solid to generate the kernel functions. Since this particular point force solution automatically satisfies the stress-free boundary conditions on the surface of the half space, only the cap-soil and the pile-soil interfaces had to be discretized. An advanced version[38] of this program (PGROUP) which incorporates compressible raked piles, lateral loading, inhomogeneity, and slip at the pile-soil interface is commercially available through DoT (UK).

**(d) Axisymmetric stress analysis of three-dimensional bodies** Cruse, Snow, and Wilson[34] carried out three-dimensional axisymmetric stress analysis of discs subjected to boundary, steady state thermal, and centrifugal loading. Since the axisymmetric kernel functions are complicated and expensive to evaluate they adopted a simple linear and circular arc element representation of the boundary geometry. The various functions were assumed to vary parabolically over each element as shown in Fig. 6-12.

Figure 6-13 shows the boundary discretization for a disc of 1 in thickness, 2 in inner radius, and 10 in outer radius, which was analysed for three test cases:

1. An outer radius, radial rim load, $\sigma_r = 300$
2. Centrifugal loading
3. A steady non-uniform temperature distribution

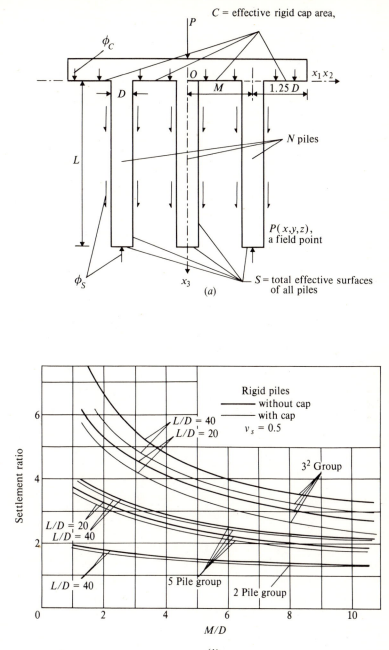

(a)

(b)

**Figure 6-11** Three-dimensional elasticity problems. (a) Geometry of rigid pile group-cap system. (b) Calculated settlement ratios for different pile groups.

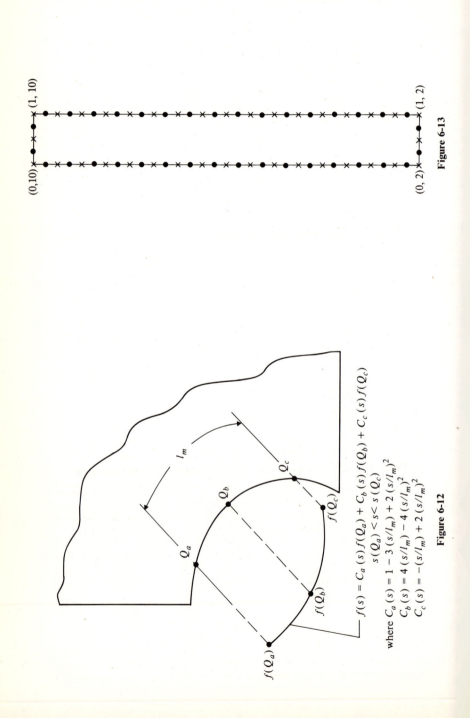

$f(s) = C_a(s)f(Q_a) + C_b(s)f(Q_b) + C_c(s)f(Q_c)$
$$s(Q_a) < s < s(Q_c)$$

where $C_a(s) = 1 - 3(s/l_m) + 2(s/l_m)^2$
$C_b(s) = 4(s/l_m) - 4(s/l_m)^2$
$C_c(s) = -(s/l_m) + 2(s/l_m)^2$

Figure 6-12

Figure 6-13

## Table 6-2 Disc with rim load

| | $u_r$ | | $\sigma_r$ | | $\sigma_\theta$ | |
|---|---|---|---|---|---|---|
| $r$ | BIE($\times 10^{-4}$) | Exact($\times 10^{-4}$) | BIE | Exact | BIE | Exact |
| 2 | 0.417 | 0.417 | 4 | 0 | 626 | 625 |
| 3 | 0.408 | 0.408 | 176 | 174 | 452 | 451 |
| 4 | 0.443 | 0.443 | 235 | 234 | 391 | 391 |
| 5 | 0.495 | 0.495 | 263 | 263 | 363 | 363 |
| 6 | 0.556 | 0.556 | 278 | 278 | 347 | 347 |
| 7 | 0.621 | 0.621 | 287 | 287 | 338 | 338 |
| 8 | 0.690 | 0.690 | 293 | 293 | 332 | 332 |
| 9 | 0.761 | 0.761 | 297 | 297 | 328 | 328 |
| 10 | 0.833 | 0.833 | 300 | 300 | 325 | 325 |

## Table 6-3 Disc with centrifugal load

| | $u_r$ | | $\sigma_r$ | | $\sigma_\theta$ | |
|---|---|---|---|---|---|---|
| $r$ | BIE($\times 10^{-4}$) | Exact($\times 10^{-4}$) | BIE | Exact | BIE | Exact |
| 2 | 0.547 | 0.547 | 5 | 0 | 821 | 820 |
| 3 | 0.532 | 0.532 | 209 | 205 | 584 | 583 |
| 4 | 0.566 | 0.567 | 257 | 256 | 489 | 489 |
| 5 | 0.614 | 0.615 | 256 | 256 | 433 | 433 |
| 6 | 0.662 | 0.662 | 231 | 231 | 389 | 389 |
| 7 | 0.701 | 0.702 | 190 | 190 | 348 | 348 |
| 8 | 0.729 | 0.730 | 137 | 137 | 308 | 308 |
| 9 | 0.740 | 0.741 | 73 | 73 | 265 | 265 |
| 10 | 0.732 | 0.733 | 0 | 0 | 220 | 220 |

## Table 6-4 Disc with one-dimensional steady state temperature

| | $u_r$ | | $\sigma_r$ | | $\sigma_\theta$ | |
|---|---|---|---|---|---|---|
| $r$ | BIE | Exact | BIE | Exact | BIE | Exact |
| 2 | $0.577 \times 10^{-2}$ | $0.588 \times 10^{-2}$ | 921 | 0 | 86827 | 88237 |
| 3 | $0.627 \times 10^{-2}$ | $0.636 \times 10^{-2}$ | 21046 | 19722 | 37552 | 38105 |
| 4 | $0.818 \times 10^{-2}$ | $0.824 \times 10^{-2}$ | 21719 | 21158 | 14702 | 15093 |
| 5 | $0.108 \times 10^{-1}$ | $0.109 \times 10^{-1}$ | 18851 | 18449 | 893 | 1067 |
| 6 | $0.139 \times 10^{-1}$ | $0.140 \times 10^{-1}$ | 14996 | 14685 | $-8920$ | $-8844$ |
| 7 | $0.174 \times 10^{-1}$ | $0.175 \times 10^{-1}$ | 11013 | 10758 | $-16500$ | $-16478$ |
| 8 | $0.212 \times 10^{-1}$ | $0.212 \times 10^{-1}$ | 7136 | 6953 | $-22727$ | $-22688$ |
| 9 | $0.252 \times 10^{-1}$ | $0.252 \times 10^{-1}$ | 3497 | 3361 | $-27970$ | $-27930$ |
| 10 | $0.294 \times 10^{-1}$ | $0.294 \times 10^{-1}$ | 88 | 0 | $-32562$ | $-32471$ |

Tables 6-2, 6-3, and 6-4 compare the calculated results with the exact solutions from Timoshenko and Goodier.[39] Agreement is generally very good with maximum errors less than 2 per cent.

They also analysed a round notched bar, commonly used for fatigue testing. The BEM discretization is shown in Fig. 6-14 with a large number of elements used to provide details of the stress concentration at the notch. The nominal stress concentration factor $K_T$ varies rapidly with changes in notch radius and therefore high analytical precision is required around the re-entrant notch tip.

They also carried out a finite element analysis of the disc using the NASTRAN package with linear strain, ring elements. The discretization used (Fig. 6-15) involves 750 nodes and 300 elements compared with 70 nodes and 35 boundary elements shown in Fig. 6-14. Computation times for the FEM and BEM analyses were 3 min and 1 min respectively on an IBM 360/168 computer.

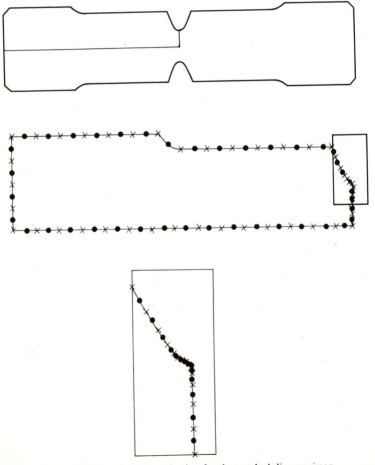

**Figure 6-14** Boundary discretization for the notched disc specimen.

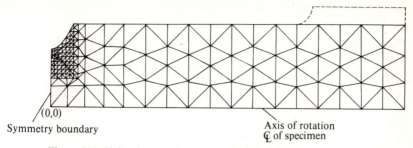

Symmetry boundary

Axis of rotation
₵ of specimen

**Figure 6-15** Finite element discretization for the notched disc specimen.

**Table 6-5 Fatigue specimen stress concentration factors**

| Nominal geometry | NASTRAN | BEM |
|---|---|---|
| $K_T = 2$ specimen | 2.18 | 2.15 |
| $K_T = 3$ specimen | 3.30 | 3.36 |
| $K_T = 4$ specimen | 4.43 | 4.45 |

Table 6-5 shows the stress concentration factors obtained by both the methods for different notch geometries. It is worth emphasizing also that the data preparation time for the BEM analysis was considerably lower than that required by NASTRAN. Other examples can be seen in Refs. 1 to 11, 22 to 24, 31 to 36, 38, and 40 to 47.

## 6-9 CONCLUDING REMARKS

Examples presented in this chapter will, we hope, help to convince the reader of the general utility and versatility of BEM as a problem-solving tool. Since the numerical schemes outlined here are those developed for the first generations of BEM computer programmes the computational efficiencies will certainly be improved upon. Nevertheless, the algorithms presented in Chapters 2 to 6 do illustrate all the essential steps involved in the solution process which need to be mastered.

BEM solutions are algebraically complicated; it is probable that more efficient implementation will be achieved by using higher-order representations of both geometries and functions so that the number of elements required to describe the boundary data, and over which the integrations are performed, is minimized. Such higher-order schemes will be described in Chapter 8 for we believe that the future success of BEM lies ultimately in accurate modelling of boundary information.

## 6-10 REFERENCES

1. Banerjee, P. K. (1976) 'Integral equation methods for analysis of piece-wise non-homogeneous three-dimensional elastic solids of arbitrary shape', *Int. J. Mech. Sci.*, **18**, 293–303.
2. Banerjee, P. K., and Davies, T. G. (1978) 'The behaviour and axially and laterally loaded single piles embedded in non-homogeneous soils', *Géotechnq.*, **28**(3), 309–326.
3. Cruse, T. A. (1979) 'Two and three-dimensional problems of fracture mechanics', in P. K. Banerjee and R. Butterfield (eds), *Developments in Boundary Element Methods*, Vol. I, Chap. V, Applied Science Publishers, London.
4. Banerjee, P. K. (1971) 'Foundations within a finite elastic layer—application of the integral equation method', *Civ. Engng*, **1971**, 1197–1202.
5. Cruse, T. A. (1969) 'Numerical solutions in three-dimensional elasto-statics', *Int. J. Solids and Structs*, **5**, 1259–1274.
6. Cruse, T. A., and Rizzo, F. J. (eds) (1975) 'Boundary integral equation method: computational applications', *Proc. ASME Conf. on Boundary Integral Equation Meth.*, AMD–11, ASME, New York.
7. Rizzo, F. J., and Shippy, D. J. (1977) 'An advanced boundary integral equation method for three-dimensional thermo-elasticity', *Int. J. Num. Meth. in Engng*, **11**, 1753.
8. Rizzo, F. J., and Shippy, D. J. (1979) 'Recent advances of the boundary element method in thermoelasticity', in P. K. Banerjee and R. Butterfield (eds), *Developments in Boundary Element Methods*, Vol. I, Chap. VI, Applied Science Publishers, London.
9. Lachat, J. C., and Watson, J. O. (1976) 'Effective numerical treatment of boundary integral equations: a formulation for three-dimensional elasto-statics', *Int. J. Num. Meth. in Engng*, **10**, 991–1005.
10. Cruse, T. A. (1973) 'Application of the boundary integral equation method to three-dimensional stress analysis', *Int. J. Computer and Structs*, **3**, 509–527.
11. Cruse, T. A. (1974) 'An improved boundary integral equation method for three-dimensional elastic stress analysis', *Computer and Structs*, **4**, 741–754.
12. Love, A. E. H. (1944) *A Treatise on the Mathematical Theory of Elasticity*, Dover, New York.
13. Willis, J. R. (1965) 'The elastic interaction energy of dislocation loops in anisotropic media', *Q. Mech. and Appl. Math.*, **18**, 419–433.
14. Lifschitz, I. M., and Rozenweig, L. N. (1974) *J. Exp. Theor. Physics*, **17**, 783.
15. Pan, Y. C., and Chou, T. W. (1976) 'Point force solution for an infinite transversely isotropic solid', *J. Appl. Mech. Trans. ASME*, **98**(E), 608–612.
16. Wilson, R. B., and Cruse, T. A. (1978) 'Efficient implementation of anisotropic three-dimensional boundary integral equation stress analysis', *Int. J. Num. Meth. in Engng*, **12**, 1383–1397.
17. Kinoshita, M., and Mura, T. (1975) 'Green's function for anisotropic elasticity', AEC contract Report C00–2034–5, North-Western University, Illinois.
18. Vogel, S. M., and Rizzo, F. J. (1973) 'An integral equation formulation of three-dimensional anisotropic elastostatic boundary value problems', *J. Elasticity*, **3**, 203–216.
19. Goodier, J. N. (1937) 'Integration of thermo-elastic equations', *Phil. Mag.*, **23**.
20. Terzaghi, K. (1943) *Theoretical Soil Mechanics*, Wiley, New York.
21. Fung, Y. C. (1965) *Foundations of Solid Mechanics*, Prentice-Hall, New Jersey.
22. Cruse, T. A. (1975) 'Boundary integral equation method for three-dimensional elastic fracture mechanics', AR OSR–TR–75–0813, ADA 011660, Pratt and Whitney Aircraft, Connecticut.
23. Cruse, T. A. (1975) 'Mathematical foundations of the boundary integral equation method in solid mechanics', AF OSR–TR–77–1002, Pratt and Whitney Aircraft, Connecticut.
24. Rizzo, F. J., and Shippy, D. J. (1976) 'Thermomechanical stress analysis of an advanced turbine blade cooling configuration', US–AFOSR Interim Science Report.
25. Reisner, H. (1931) 'Initial stresses and sources of initial stresses' (in German), *ZAMP*, **11**, 1–8.
26. Banerjee, P. K., and Mustoe, G. G. W. (1978) 'Boundary element methods for two-dimensional problems of elasto-plasticity', *Proc. Int. Conf. on Recent Developments of Boundary Element Methods*, Southampton University, pp. 283–300, Pentec Press, London.

27. Banerjee, P. K., Cathie, D. N., and Davies, T. G. (1979) 'Two and three-dimensional problems of elasto-plasticity', in P. K. Banerjee and R. Butterfield (eds), *Developments in Boundary Element Methods*, Vol. I, Chap. IV, Applied Science Publishers, London.
28. Lin, T. Y. (1967) 'Reciprocal theorem for displacements in inelastic bodies', *J. Composite Matter*, **1**, 144–151.
29. Lin, T. Y. (1968) *Theory of Inelastic Structures*, Wiley, New York.
30. Zienkiewicz, O. C. (1977) *The Finite Element Method*, McGraw-Hill, London.
31. Butterfield, R., and Banerjee, P. K. (1971) 'The elastic analysis of compressible piles and pile groups', *Géotechnq.*, **21**(1), 43–60.
32. Kermanidis, Th. (1973) 'Eine integral gleichungs methode Zure lösung des Umdrehungskörpers', *Acta Mech.*, **16**, 175.
33. Kermanidis, Th. (1975) 'A numerical solution for axially symmetrical elasticity problems', *Int. J. Solids and Structs*, **11**, 493–500.
34. Cruse, T. A., Snow, D. W., and Wilson, R. B. (1977) 'Numerical solutions in axi-symmetric elasticity', *Computers and Structs*, **7**, 445–451.
35. Banerjee, P. K., and Butterfield, R. (1977) 'Boundary element methods in geomechanics', in G. Gudehus (ed.), *Finite Elements in Geomechanics*, Chap. 16, Wiley, London.
36. Butterfield, R., and Banerjee, P. K. (1971) 'The problem of pile cap, pile group interaction', *Géotechnq.*, **21**(2), 135–141.
37. Mindlin, R. D. (1935) 'A point force in the interior of a semi-infinite solid', *Physics*, **7**, 195–202.
38. Banerjee, P. K., Driscoll, R. M. C., and Davies, T. G. (1978) 'A computer programme for the analysis of pile groups of any geometry and subjected to any boundary conditions', Program PGROUP, Highways Engineering Computer Division, Department of Transport, London.
39. Timoshenko, S., and Goodier, J. N. (1951) *Theory of Elasticity*, 2nd ed., McGraw-Hill, New York.
40. Cruse, T. A., and Van Bauren, W. (1971) 'Three-dimensional elastic stress analysis of fracture specimen with an edge crack', *Int. J. Fracture Mech.*, **7**, 1–15.
41. Cruse, T. A., and Meyers, G. J. (1977) 'Three-dimensional elastic fracture mechanics analysis', *J. of Struct. Div.*, *ASCE*, **103**(ST2), 309–320.
42. Cruse, T. A. (1972) 'Some classical elastic sphere problems solved numerically by integral equations', *J. Appl. Mech.*, **March**, 272–274.
43. Dominguez, J. A. (1977) 'Calculations of stresses due to an anchor by elements on contours', Ph.D. thesis, University of Sevilla, Spain.
44. Paris, F. C. (1979) 'A method of element on contours for elasticity and potential theory', Ph.D. thesis, Technical University of Madrid, Spain.
45. Alarcon, E., Brebbia, C., and Dominguez, J. (1978) 'The boundary element method in elasticity', *Int. J. Mech, Sci.*, **September 1978**.
46. Davies, T. G. (1979) 'Linear and nonlinear analysis of pile groups', Ph.D. thesis, University of Wales, Cardiff.
47. Mustoe, G. G. W. (1979) 'A combination of the finite element and boundary solution procedure for continuum problems', Ph.D. thesis, University of Wales, Swansea.

# SEVEN

## PROBLEMS OF EDGES AND CORNERS

### 7-1 INTRODUCTION

In previous chapters we have quite deliberately avoided any detailed treatment of discontinuities in the geometry and the boundary conditions although most practical problems do have these features in the form of edges and corners.

One obvious way to tackle such problems would be to round off edges and corners. We would expect such an analysis to produce reasonable answers away from the rounded regions but less good results near them. This was in fact the approach used in the first generation of computer programmes developed using the indirect formulation (see Jaswon and Symm[1]).

Although it is not always necessary to obtain detailed results at or near a boundary discontinuity the 'rounding-off' procedure cannot be considered a satisfactory solution to the problem since the results, even at some distance away from the rounded edges or corners, must be affected. Moreover, there is a large class of problems involving re-entrant corners, etc., where results at or near the geometric discontinuity form the most important part of the solution.

In this chapter we shall discuss the various procedures which have been developed to date to simulate such discontinuities using both the direct and indirect methods.

### 7-2 DIRECT METHODS

#### 7-2-1 Definition of the Problem

In the case of a potential flow problem the potential is uniquely defined but its normal derivatives are multivalued at a corner node. Similarly, for an elasticity problem the displacements are uniquely defined but the surface tractions are multivalued at a corner node. Thus if we wish to write the equation (for an elasticity problem)

$$\beta\mathbf{u}^p = \sum_{q=1}^{N}\left[\left(\int_{\Delta S} \mathbf{G}^{pq}\mathbf{N}^q\,ds\right)\mathbf{t}_n - \left(\int_{\Delta S} \mathbf{F}^{pq}\mathbf{N}^q\,ds\right)\mathbf{u}_n\right] \qquad (7\text{-}1)$$

for $m$ boundary nodes including one true corner node, the resulting final system of equations will be

$$\mathbf{Gt} - \mathbf{Fu} = 0 \qquad (7\text{-}2)$$

where     $\mathbf{F} = 2m \times 2m$ or $3m \times 3m$ matrix for two- and
                three-dimensional problems respectively
and      $\mathbf{G} = 2m \times (2m+2)$ or $3m \times (3m+4)$ matrix
                for two- and three-dimensional problems respectively

The additional columns in **G** arise from the multivalued tractions defined at the corner node. If these tractions are all specifically prescribed then the solution of (7-2) presents no difficulty. A suitable mixture of tractions and displacement boundary conditions (Fig. 7-1a, b) at the corner also presents no difficulty if the final system matrix involving all the unknowns is square and of order $2m \times 2m$ for two-dimensional problems and $3m \times 3m$ for the three-dimensional ones. If the displacements alone are specified at the corner (Fig. 7-1c) it is not possible to solve Eq. (7-2) and an alternative approach must be found.

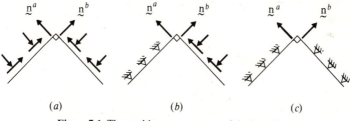

*(a)*                 *(b)*                 *(c)*

**Figure 7-1** The problems at a corner of the boundary.

It should be noted that the preceding remarks only apply to a sharp corner that actually exists in a specific problem. The discretization of a smooth surface using flat boundary elements also results in boundary discontinuities, but these must be treated as if the boundary were continuous if correct results are to be obtained.

## 7-2-2 Single-node Representation

One obvious way to circumvent this problem when either the potential or the displacement is specified at a corner node is to assume that the corresponding (unknown) multivalued normal derivatives of the potential or the unknown multivalued tractions are equal. For example, if $\mathbf{t}^a$ and $\mathbf{t}^b$ are the limiting values of tractions at a corner node (Fig. 7-1), then

$$t_i^a = \sigma_{ij} n_j^a = t_i^b = \sigma_{ij} n_j^b \tag{7-3}$$

where $\mathbf{n}^a$ and $\mathbf{n}^b$ are the outward normal vectors to the surfaces meeting at the corner node. Lachat and Watson[2-4] used this inherently simple method and found that the errors involved were mainly confined to the corner and were not significant even at points close to it.

### 7-2-3 Multiple Independent Node Concept

In order to eliminate the ambiguities involved at a corner node Riccardella[5] introduced the double-node concept (for two-dimensional problems). This involves writing equations such as (7-1) for multiple nodes defined at a corner (at two or three nodes respectively, depending on whether the problem is two dimensional or three dimensional). This would result in the matrices $\mathbf{G}$ and $\mathbf{F}$ in the final system of Eqs (7-2) remaining square for the case illustrated in Fig. 7-1c.

The major drawback of this technique when used with the direct method is that it is a potential source of instability unless a sufficient gap is left between the corner nodes so that equations such as (7-1) written for them are really independent.

### 7-2-4 Multiple-node Concept with Auxiliary Relationships

A somewhat different multiple-node concept was proposed by Chaudonneret[6] for the elasticity problem and Alarcon, Martin and Paris[7] for potential flow problems in which they derived an additional set of equations, to supplement (7-1) written for a corner node, when the boundary conditions are of the type shown in Fig. 7-1c.

**Potential flow problems** At a corner node the normal derivatives of the potential $p$ can be written as

$$u^a = \frac{\partial p}{\partial x_i} n_i^a = u \cos \alpha^a$$

$$u^b = \frac{\partial p}{\partial x_i} n_i^b = u \cos \alpha^b \tag{7-4}$$

where $u$ is the modulus of the gradient at the corner and $\alpha^a$, $\alpha^b$ are the angles between $u$ and the respective normals to the boundary.

The direction cosines $(m)$ of this 'dummy' gradient are obviously

$$\mathbf{m} = \left[ \frac{\partial p/\partial x_1}{\sqrt{[(\partial p/\partial x_1)^2 + (\partial p/\partial x_2)^2]}}, \frac{\partial p/\partial x_2}{\sqrt{[(\partial p/\partial x_1)^2 + (\partial p/\partial x_2)^2]}} \right] \tag{7-5}$$

From the known distribution of the potential $p$ we can calculate the quantities on the right-hand side of Eq. (7-5) by using, for example, a finite difference formula.

The value of (**m**) and the known boundary normals $\mathbf{n}^a$ and $\mathbf{n}^b$ enable us to evaluate $\cos \alpha^a$ and $\cos \alpha^b$, leaving a determinate problem involving only one unknown dummy variable $u$ at the corner.

**Elastostatic problems** Chaudonneret[6] derived two auxiliary relationships for points in the neighbourhood of a corner by considering the invariance of the stress tensor and the invariance of the trace of the strain tensor.

The surface tractions $t_i^a$ and $t_i^b$ corresponding to the unique value of stress field $\sigma_{ij}$ in the neighbourhood of a corner are given by

$$t_i^a = \sigma_{ij} n_j^a$$
$$t_i^b = \sigma_{ij} n_j^b$$

(7-6)

These equations can be solved for the stress components $\sigma_{ij}$ and the solution can be expressed in terms of $t_i^a$, $t_i^b$, $n_j^a$, and $n_j^b$.

Considering the symmetry of the stress tensor we can write

$$n_1^b t_1^a - n_1^a t_1^b = n_2^a t_2^b - n_2^b t_2^a$$

(7-7)

To derive the second auxiliary relationship let us consider orthogonal axes (Fig. 7-2) with the origin at the corner. From the invariance of the trace of the strain

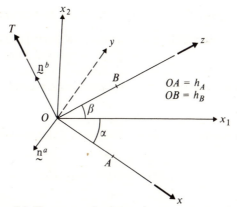

**Figure 7-2** The system of orthogonal axes at a corner.

tensor at the corner we have

$$\varepsilon_{xx} + \varepsilon_{yy} = \varepsilon_{zz} + \varepsilon_{TT}$$

(7-8)

The displacements $u_1$ and $u_2$, referred to the axes $OX_1$ and $OX_2$ (the reference axes), are assumed to vary linearly along $OA$ and $OB$. Therefore we have

$$u_i = u_i(O) + \frac{u_i(A) - u_i(O)}{h_A}(x - x_O)$$

Its component along $OX$ is

$$u_x = u_1 \cos \alpha + u_2 \sin \alpha$$

Therefore

$$\varepsilon_{xx} = \frac{\partial u_x}{\partial x} = \frac{u_1(A) - u_1(O)}{h_A} \cos \alpha + \frac{u_2(A) - u_2(O)}{h_A} \sin \alpha \qquad (7\text{-}9)$$

Similarly we can obtain

$$\varepsilon_{zz} = \frac{\partial u_z}{\partial z} = \frac{u_1(B) - u_1(O)}{h_B} \cos \beta + \frac{u_2(B) - u_2(O)}{h_B} \sin \beta \qquad (7\text{-}10)$$

Considering axes $OY$ and $OT$ we have

$$\sigma_{yy} = t_1^a \sin \alpha - t_2^a \cos \alpha$$

$$\sigma_{TT} = t_2^b \cos \beta - t_1^b \sin \beta$$

and, from Hooke's law,

$$\sigma_{yy} = (a + \mu) \varepsilon_{yy} + (a - \mu) \varepsilon_{xx}$$

$$\sigma_{TT} = (a + \mu) \varepsilon_{TT} + (a - \mu) \varepsilon_{zz}$$

where

$$a = \lambda + \mu$$

From these we can derive

$$
\begin{aligned}
(a + \mu) \varepsilon_{yy} + (a - \mu) \varepsilon_{xx} &= t_1^a \sin \alpha - t_2^a \cos \alpha \\
(a + \mu) \varepsilon_{TT} + (a - \mu) \varepsilon_{zz} &= t_2^b \cos \beta - t_1^b \sin \beta
\end{aligned}
\qquad (7\text{-}11)
$$

Now, by substituting from (7-8), (7-9), and (7-10) in (7-11), we have

$$t_1^a \sin \alpha + t_1^b \sin \beta - t_2^a \cos \alpha - t_2^b \cos \beta$$

$$= -\frac{2\mu}{h_A} \cos \alpha\, u_1(A) + 2\mu \left( \frac{\cos \alpha}{h_A} - \frac{\cos \beta}{h_B} \right) u_1(O) + \frac{2\mu}{h_B} \cos \beta\, u_1(B)$$

$$-\frac{2\mu}{h_A} \sin \alpha\, u_2(A) + 2\mu \left( \frac{\sin \alpha}{h_A} - \frac{\sin \beta}{h_B} \right) u_2(O) + \frac{2\mu}{h_B} \sin \beta\, u_2(B) \qquad (7\text{-}12)$$

Using Eqs (7-7) and (7-12) we can now eliminate one set of tractions at the corner node thereby converting the problem into a well-posed one.

The procedure outlined above can be extended to the three-dimensional case by considering a triple point at the corner to represent the traction discontinuities. For the displacement boundary-value problem at a corner node we would have three equations involving nine unknowns (as opposed to two equations with four unknowns for the two-dimensional case discussed above). Of the six additional relationships that are required, three equations can be obtained by considering the symmetry of the stress tensor (that is $\sigma_{12} = \sigma_{21}$, $\sigma_{13} = \sigma_{31}$, and $\sigma_{23} = \sigma_{32}$) and the remaining three are provided by the invariance of trace of the strain tensor and the relationships between the strain and displacements of the surface elements meeting at the corner node.

A much simpler approach using basically the same principle was recently developed by Mustoe[8] who considered a polynomial interpolation for the displacement field within a triangular region (for two-dimensional problems) containing the two adjoining boundary elements. Thus for a point within this triangle

$$\mathbf{u} = \mathbf{M}\mathbf{u}^n \tag{7-13}$$

where    $\mathbf{M}$ = the shape function
         $\mathbf{u}^n$ = displacements at the local boundary nodes

The strains at a point within this triangular region are then

$$\boldsymbol{\varepsilon} = \mathbf{L}\mathbf{u} = \mathbf{L}\mathbf{M}\mathbf{u}^n \tag{7-14}$$

where

$$\mathbf{L} = \begin{bmatrix} \dfrac{\partial}{\partial x_1} & 0 \\[2ex] 0 & \dfrac{\partial}{\partial x_2} \\[2ex] \dfrac{\partial}{\partial x_2} & \dfrac{\partial}{\partial x_1} \end{bmatrix}$$

and the stresses via Hooke's law

$$\boldsymbol{\sigma} = \mathbf{D}\boldsymbol{\varepsilon} = \mathbf{D}\mathbf{L}\mathbf{M}\mathbf{u}^n \tag{7-15}$$

The surface traction $\mathbf{t}$ at any point with the outward normal $\mathbf{n}$ is given by

$$\mathbf{t} = \mathbf{T}\boldsymbol{\sigma} \tag{7-16}$$

where

$$\mathbf{T} = \begin{bmatrix} n_1 & 0 & n_2 \\ 0 & n_2 & n_1 \end{bmatrix}$$

Therefore we can express tractions $\mathbf{t}$ in terms of the nodal displacements $\mathbf{u}^n$ in the form

$$\mathbf{t} = \mathbf{T}\mathbf{D}\mathbf{L}\mathbf{M}\mathbf{u}^n \tag{7-17}$$

Equation (7-17) can be used at a corner to provide two additional relationships. The normal vector $\mathbf{n}$ used in (7-17) can be chosen to be either $\mathbf{n}^a$ or $\mathbf{n}^b$. Further details can be found in Ref. 8.

## 7-3 INDIRECT METHODS

### 7-3-1 Multiple Independent Node Concept

The major source of difficulty in the indirect method is that the magnitude of the source distribution $\phi$ becomes infinite at a corner node; therefore the corners have to be represented by using the multiple independent node concept outlined in Sec. 7-2-3. This must be used for all types of boundary condition, but when used at corners where either $p$ (for potential flow problems) or $u_i$ (in elasticity problems) are specified the numerical results for $\partial p/\partial n$ or $\sigma_{ij}$ at points close to the corner are generally unsatisfactory.

The source distribution $\phi_j$ should be discontinuous at a smooth boundary node, if there is a known stress discontinuity, e.g., the crack tip in an edge crack problem.

### 7-3-2 Other Methods

Some workers [9-11] have suggested using a separate set of kernel functions which satisfy the governing differential equations and certain boundary conditions at the corner 'a priori'. For example, in the vicinity of the re-entrant of the L-shaped domain shown in Fig. 7-3 we have

$$p(r, \theta) = B_0 + B_1 r^{\pi/\alpha} \cos\left(\frac{\pi}{\alpha}\theta\right) + B_2 r^{2\pi/\alpha} \cos\left(\frac{2\pi}{\alpha}\theta\right) + \cdots \qquad (7\text{-}18)$$

where $\alpha$ is the interior angle subtended at the corner which is equal to $3\pi/2$ in the case shown.

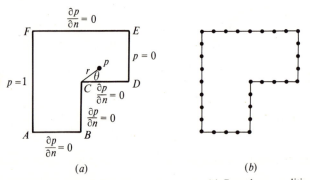

Figure 7-3 An L-shaped domain with a reentrant corner. (a) Boundary conditions. (b) BEM discretization.

Each of the terms in Eq. (7-18) satisfies the governing differential equation $\partial^2 p/\partial x_i \partial x_i = 0$ and therefore (7-18) may be used to supplement an indirect formulation which, if used in isolation, may produce unacceptable results near a singular corner.

An analysis of this type can be performed by writing the integral representation for the potential $p$ as

$$p(x) = \int_S G(x, \xi)\, \phi(\xi)\, ds + p^* \qquad (7\text{-}19)$$

where $p^*$ is the supplementary solution (7-18) at the corner node at which the fictitious density $\phi(\xi)$ is assumed to be zero.

## 7-4 MULTIZONE PROBLEMS

If the corners are represented by multiple independent nodes then the assembly process for multizone problems is identical to those discussed in Chapter 3. However, the results near the corners of zones would be inaccurate, but probably not unacceptably so, unless specific information at the junction is the objective of the analysis.

On the other hand, if the concept of multiple nodes with auxiliary relationships are used then some additional care is necessary. Let us consider the case of an interface between two regions as shown in Fig. 7-4a, where double nodes are used to represent the discontinuous tractions. After the interface equilibrium and compatibilities are applied six unknowns remain at the corner. Only four independent equations (two per region) are contributed to the global system by the discretized version of the boundary integrals for the two regions.

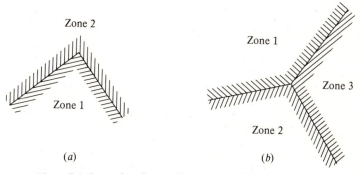

Zone 2

Zone 1

Zone 1

Zone 3

Zone 2

(*a*)

(*b*)

**Figure 7-4** Corner interface problem between different regions.

Equations (7-7) and (7-12) provide the two extra equations we require of which Eq. (7-7) can be written in terms of the tractions and normals for either region but Eq. (7-12) cannot hold simultaneously for both regions unless they both have the same shear modulus $\mu$. In the case illustrated in Fig. 7-4a, condition (7-12) should be written[12] for zone 1, since the stress tensor at the reentrant corner for zone 2 is likely to exhibit singular behaviour at the corner.

When three regions intersect at a point, as shown in Fig. 7-4b, after applying the usual interface equilibrium and compatibility relations eight boundary variables remain: the two displacement components and six traction components corresponding to the three distinct boundary normals involved at the point. Six independent equations are contributed to the global system for the three regions involved. The two additional equations are once again provided by Eqs (7-7) and (7-12) which can now only be written for one of the regions. On intuitive grounds the region with the largest included angle less than 180° is probably the region for which Eqs (7-7) and (7-12) should be written (zone 1 in Fig. 7-4b).[12]

## 7-5 CONCLUDING REMARKS

In this chapter we have attempted to highlight some of the difficulties involved in modelling the edges and corners that often arise in practical problems and discussed the current state of development in this area. Obviously a great deal of work still remains to be done, particularly for the indirect method, before all problems associated with edges and corners can be solved definitively.

## 7-6 REFERENCES

1. Jaswon, M. A., and Symm, G. T. (1977) *Integral Equation Methods in Potential Theory and Elasto-statics*, Academic Press, London.
2. Lachat, J. C. (1975) 'Further development of boundary integral technique for elasto-statics', Ph.D. thesis, Southampton University.
3. Lachat, J. C., and Watson, J. O. (1975) 'A second generation boundary integral equation program for three-dimensional stress analysis', in T. A. Cruse and F. J. Rizzo (eds), *Proc. ASME Conf. on Boundary Integral Equation*, AMD–11, ASME, New York.
4. Lachat, J. C., and Watson, J. O. (1976) 'Effective numerical treatment of boundary integral equation', *Int. J. Num. Meth. in Engng*, **10**, 991–1005.
5. Riccardella, P. (1973) 'An implementation of the boundary integral techniques for plane problems in elasticity and elasto-plasticity', Ph.D. thesis, Carnegie Mellon University, Pittsburgh.
6. Chaudonneret, M. (1977) 'Resolution of traction discontinuity problem in boundary integral equation method applied to stress analysis' (in French), *C. r., Acad. de Sci., Ser. A, Math.*, **284**(8), 463–466.
7. Alarcon, A., Martin, A., and Paris, F. (1978) 'Improved boundary elements in torsion problems', *Proc. Int. Conf. on Recent Advances in Boundary Element Method*, Southampton University, pp. 149–165.
8. Mustoe, G. G. W. (1980) 'A combination of the finite element method and boundary integral procedure for continuum problems', Ph.D. thesis, University of Wales, University College, Swansea.
9. Kermanidis, T. (1975) 'Kupradze functional equation for the torsion problem of prismatic bars—part 2', *Comp. Meth. in Appl. Mech. Engng*, **7**, 249–259.
10. Symm, G. T. (1973) 'Treatment of singularities in the solution of Laplace equation', NPL Report NAC31, Teddington.
11. Kelly, D. W., Mustoe, G. G. W., and Zienkiewicz, O. C. (1978) 'On a hierarchical order for trial functions in numerical procedures based on satisfaction of the governing differential equations', *Proc. Int. Conf. on Recent Advances in Boundary Element Methods*, Southampton University, pp. 359–373.
12. Wardle, L. J., and Crotty, J. M. (1978) 'Two dimensional boundary integral equation analysis for non-homogeneous mining applications', *Proc. Int. Conf. on Recent Advances in Boundary Element Methods*, Southampton University, pp. 233–251.

# CHAPTER EIGHT

# PARAMETRIC REPRESENTATION OF FUNCTIONS AND GEOMETRY

## 8-1 INTRODUCTION

Throughout boundary element analysis we are faced with the evaluation of boundary integral equations which might be typified in any number of spatial dimensions by Eq. (4-38); viz., for elastostatic problems,

$$\beta_{ij}\, u_i(x_o) = \int_S [t_i(x)\, G_{ij}(x, x_o) - u_i(x)\, F_{ij}(x, x_o)]\, ds(x) + \int_V \psi_i(z)\, G_{ij}(z, x_o)\, dv(z)$$

(8-1)

where $\beta_{ij}$ is the discontinuity term at a general boundary node $x_o$. Since the corresponding expression for potential flow [Eq. (3-37)] is simpler than (8-1) we can use the above elasticity equation to illustrate the general case and to explain the purpose of Chapter 8.

Because we are not able to solve (8-1) analytically we subdivide $S$ and $V$ into, say, $n$ and $m$ elements respectively and write an equivalent matrix equation for the vector of boundary displacements at a boundary node $p$ as

$$\boldsymbol{\beta}\mathbf{u}^p = \sum_{q=1}^{n} \left[ \left( \int_{\Delta S_q} \mathbf{G}^{pq}\, \mathbf{t}^q\, ds \right) - \left( \int_{\Delta S_q} \mathbf{F}^{pq}\, \mathbf{u}^q\, ds \right) \right] + \sum_{l=1}^{m} \left( \int_{\Delta V_l} \mathbf{G}^{pl}\, \boldsymbol{\psi}^l\, dv \right)$$ (8-2)

Evaluation of the terms exemplified by

$$\int_{\Delta S_q} \mathbf{F}^{pq}\, \mathbf{u}^q\, ds \quad \text{and} \quad \int_{\Delta V_l} \mathbf{G}^{pl}\, \boldsymbol{\psi}^l\, dv$$

are the nub of our problem.

The special cases in which the $\mathbf{u}^q$ terms in each element can be represented by constant or linearly varying values have been dealt with previously. We now need to discuss more complicated distributions of $\mathbf{u}$ (or $\mathbf{t}$ or $\boldsymbol{\psi}$) over elements with, typically, second-order (quadratic) or third-order (cubic) variation. This can be achieved by using multiple nodes on each element, say $\rho$ on each, and adopting some general relationship between the displacement components $u_i(x)$ at any point within the element and the complete set of nodal displacements for each element $U_{i\alpha}$ referred to global axes $x_i$. Here $U_{i\alpha}$ is the general term of the nodal displacement matrix, with components specified at *function nodes*, and $\alpha = 1 \cdots \rho$ the number of such nodes per element which will, in this chapter, always be specified by Greek suffixes. Italic letters $i, j, k$, etc., will be used to denote the dimensions of local (curvilinear) and global (cartesian) axes. Thus we shall have in general

$$u_i = U_{i\alpha} N_\alpha \qquad (8\text{-}3)$$

in which the components of the vector $N_\alpha$, known as *shape functions*, are expressed in terms of local (element) coordinates, $\zeta_i$ say. Clearly a scalar quantity $\psi$ could be treated similarly, resulting in a row matrix of nodal $\Psi_\alpha$ values multiplying $N_\alpha$.

It might be noted here that in previous chapters, following earlier work, Eq. (8-3) has been written differently, but equivalently, in terms of the entire element nodal displacement component vector $\mathbf{u}_n$ as

$$\mathbf{u} = \mathbf{N}\mathbf{u}_n$$

where, for example, in Eq. (6-43) there are three shape functions $N_1, N_2, N_3$ for the three-noded triangle corresponding to $N_\alpha$ ($\rho = 3$).

Once we have embarked upon non-linear variations of $\mathbf{u}$, $\mathbf{t}$, $\boldsymbol{\psi}$, etc., over elements, it is convenient to investigate curvilinear element geometries at the same time. The reason for this becomes clear if we consider also describing the geometry of our elements by multiple nodes (*geometric nodes*), $\tau$ per element, for example, characterized by the general term of a nodal coordinate matrix, $X_{i\beta}$ ($\beta = 1 \cdots \tau$). We shall find that the global coordinates $x_i$ of any point within an element can be expressed, in terms of $X_{i\beta}$, as

$$x_i = X_{i\beta} M_\beta \qquad (8\text{-}4)$$

The $M_\beta$ vector components are again *shape functions*, which now interrelate the global $x_i$ coordinates of any point within an element to the nodal coordinate matrix $X_{i\beta}$.

If the geometrical and functional nodes coincide and, conveniently, the sets of shape functions are identical (that is, $N \equiv M$) such elements are called *isoparametric* (Zienkiewicz[1]).

The terminology introduced above will be recognized by readers who have studied finite element analysis; indeed, the shape functions we shall discuss are common to both FEM and BEM. Our presentation of the material does, however, differ in some respects and we hope that it may prove to be both interesting and helpful, even to those already familiar with the basic ideas.

## 8-2 GEOMETRICAL TRANSFORMATIONS

We can perhaps best start our discussion of curvilinear segments, patches, cells, etc., by developing some geometrical ideas based on coordinate transformations. The philosophy underlying all the subsequent analysis can be appreciated from the diagrams in Fig. 8-1, some of which are closely similar to those in D'Arcy Thompson's famous book on 'growth and form'.[2] He was intrigued by how different species of an animal (e.g., the fish in Fig. 8-1$a, b$) can apparently be reproduced by simple, regular mapping from an orthogonal cartesian coordinate system $(x_i)$ to a curvilinear system $(\eta_i)$. The point of particular interest to us is to note how, between these two figures, the curved $(\eta_i)$ coordinate lines in Fig. 8-1$a$, and areas bounded by them, map as straight lines and rectilinear areas respectively in the $\eta_i$ plane in Fig. 8-1$b$ (clearly such mappings are equally possible in three dimensions with $i$ ranging over 1, 2, 3). Rather than adopt arbitrary, simple $\eta_i$ lines such as the approximately elliptic and hyperbolic ones shown here we could, in principle, find a much more elaborate $\zeta_i$ coordinate system (Fig. 8-1$c$) which would map the *boundary* of our fish into a rectangle (Fig. 8-1$d$) and simultaneously transform sets of curved internal cells into squares. It is now evident that mathematical operations on the straight boundaries and rectilinear cells of Fig. 8-1$d$ will be simpler than those on the curvilinear elements of Fig. 8-1$c$. It will therefore be advantageous to generate shape functions, etc., in Fig. 8-1$d$ in terms of $\zeta_i$, providing that we can always transform them (via the inverse $\zeta_i \rightarrow x_i$ mapping) back into $x$.

The set of diagrams in Fig. 8-1 is worth studying since it does illustrate clearly how, for example, straight lines in one mapping may become curved when transformed, and vice versa; area (and volume) mappings may change both shape and size; and when the $\eta_i$ or $\zeta_i$ coordinate lines are mapped as orthogonal cartesian axes (Fig. 8-1$b, d$) the $x_i$ coordinate lines then become curvilinear.

Since we shall hardly ever have a convenient transformation function for the whole of our 'fish' we are obliged to operate piecemeal on small, arbitrary, 'elements' of our system within which the curved boundaries are approximated locally by linear, quadratic, or cubic, etc., shape functions.

If we consider, initially, points located by $x_i$ in a global cartesian space $(X)$ and by $\zeta_i$ in local curvilinear coordinates associated with a space $(Z)$ of the same dimensionality, the $x_i$ will be expressible as a function of the $\zeta_i$, $x_i = f_i(\zeta_1, \zeta_2, \zeta_3)$ and conversely the $\zeta_i = g_i(x_1, x_2, x_3)$. Such transformation equations are usually written in the abbreviated form $x_i \equiv x_i(\zeta)$ and $\zeta_i \equiv \zeta_i(x)$, whence differential components of line elements in $X$ and $Z$ will be interrelated by

$$dx_i = \frac{\partial f_i}{\partial \zeta_j} d\zeta_j = \frac{\partial x_i}{\partial \zeta_j} d\zeta_j$$

$$(8\text{-}5)$$

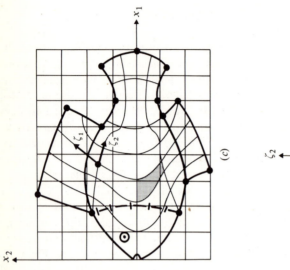

(a)

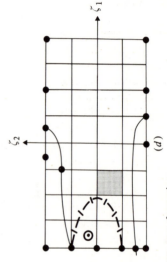

(c)

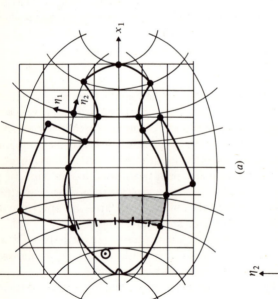

(b)

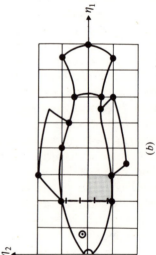

(d)

**Figure 8-1** Curvilinear coordinate transformations.

or
$$dx = J \, d\zeta$$

The matrix $\mathbf{J}$ is known as the jacobian of the transformation

$$\mathbf{J} = \left[ \frac{\partial x_i}{\partial \zeta_j} \right] \tag{8-6}$$

The inverse operation, if it exists, will be defined by

$$d\zeta = \mathbf{J}^{-1} \, dx$$

with
$$d\zeta_i = \frac{\partial \zeta_i}{\partial x_j} \, dx_j \tag{8-7}$$

where
$$\mathbf{J}^{-1} = \left[ \frac{\partial \zeta_i}{\partial x_j} \right] \tag{8-8}$$

The determinant of $\mathbf{J}$, $J = ||\mathbf{J}||$, is known as the *jacobian* (determinant) of the transformation. One could use, equally validly, either transformation (8-5) or (8-7) to define a jacobian and when reading other texts (e.g., Fung[3]) it is important to note their starting point.

In order that any transformation $x \to \zeta$ is reversible (has an inverse) and that $(x_i, \zeta_i)$ points are in one-to-one correspondence in some region of interest $(R)$ (i.e., that any set of numbers $x_i$ defines a unique set of numbers $\zeta_i$ in $R$, and vice versa) it is sufficient that:[3]

1. The $f_i$ are single-valued, continuous functions of $x_i$ with continuous first partial derivatives in $R$.
2. $J$ must be finite (that is, $1/J \neq 0$) at any point in $R$.

These conditions alone ensure that the transformation is *admissible*. Since we also require that right-handed coordinate systems shall remain right handed when transformed, our transformation must also be *proper*, which requires that $J$ be *positive* everywhere (e.g., for simple transformations between orthogonal cartesian systems $J = +1$). There are a few basic transformation operations which we shall use subsequently; these are given below, with primed symbols referring to functions in $Z$ and unprimed ones in $X$.

**(a) Scalar fields** A scalar field $\psi(x)$ in $X$ transforms identically to $\psi'(\zeta)$ in $Z$ at corresponding $(x_i, \zeta_i)$ points. Thus,

$$\psi(x_1, x_2, x_3) = \psi'(\zeta_1, \zeta_2, \zeta_3)$$

or
$$\psi(x) = \psi'(\zeta) \qquad \text{or} \qquad \psi = \psi' \tag{8-9}$$

**(b) Vector fields** The gradient of a scalar field $(\partial \psi / \partial x_i) = v_i$, say, in $X$ transforms by the differentiation chain rule as

$$\frac{\partial \psi}{\partial x_i} = \frac{\partial \psi}{\partial \zeta_j} \frac{\partial \zeta_j}{\partial x_i}$$

that is,

$$\mathbf{v} = \mathbf{J}^{-1}\mathbf{v}' \qquad \mathbf{v}' = \mathbf{J}\mathbf{v} \tag{8-10}$$

whereas infinitesimal displacement field components, $u_i(x)$ transform via

$$u_i(x) = \frac{\partial x_i}{\partial \zeta_j} u_j'(\zeta)$$

that is,

$$\mathbf{u} = \mathbf{J}\mathbf{u}' \qquad \mathbf{u}' = \mathbf{J}^{-1}\mathbf{u} \tag{8-11}$$

Such transformation rules can be extended systematically to higher-rank tensor quantities, a very lucid account of which will be found in Fung[3] (see also Appendix A). The jacobian matrix of a transformation is seen to be fundamental to both geometrical mappings and the manner in which tensor components change between the $x_i$ and $\zeta_i$ coordinate systems.

## 8-3 TRANSFORMATION OF DIFFERENTIAL VOLUME, AREA, AND LINE ELEMENTS

### 8-3-1 Internal Cells

Figure 8-2 shows a number of differential length, area, and volume elements mapped between $X$ and $Z$. Since they all take simpler geometrical forms in $Z$ we shall find it more convenient to evaluate the quantities $dV(x)$ [and $dA(x)$, $dS(x)$,

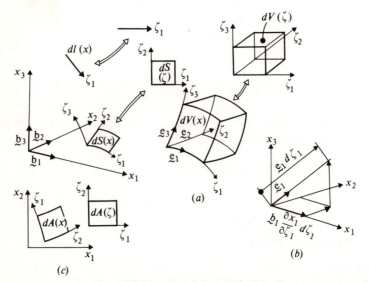

**Figure 8-2** Mapping of elements in $X$ to $Z$.

etc.] appearing in the basic BEM statements in terms of their mappings in $Z$, which can all be identical 'unit' elements irrespective of their size in $X$. Although the surface patches (in three dimensions) and boundary line segments (in two dimensions) are the major features of BEM it is simpler to deal with them after the transformation of volume (in three dimensions) and area (in two dimensions) cells. Consider a differential volume element $(d\zeta_1\, d\zeta_2\, d\zeta_3)$ in $Z$ (Fig. 8-2a) which transforms to the curvilinear elemental cell of volume $dV(x)$ in $X$. From Fig. 8-2b,

$$\mathbf{e}_1 = \mathbf{b}_1 \frac{\partial x_1}{\partial \zeta_1} + \mathbf{b}_2 \frac{\partial x_2}{\partial \zeta_1} + \mathbf{b}_3 \frac{\partial x_3}{\partial \zeta_1} \tag{8-12}$$

where $\mathbf{e}$ and $\mathbf{b}$ are unit base vectors in $Z$ and $X$ respectively. Elementary vector algebra provides the volume of the elemental cell $dV(x)$ in $X$ as

$$dV(x) = (\mathbf{e}_2 \times \mathbf{e}_3) \cdot \mathbf{e}_1 (d\zeta_1\, d\zeta_2\, d\zeta_3)$$

Substituting from Eq. (8-12) and equivalent expressions for $\mathbf{e}_2$ and $\mathbf{e}_3$ leads to

$$dV(x) = \begin{Vmatrix} \mathbf{b}_1 & \mathbf{b}_2 & \mathbf{b}_3 \\ \dfrac{\partial x_1}{\partial \zeta_2} & \dfrac{\partial x_2}{\partial \zeta_2} & \dfrac{\partial x_3}{\partial \zeta_2} \\ \dfrac{\partial x_1}{\partial \zeta_3} & \dfrac{\partial x_2}{\partial \zeta_3} & \dfrac{\partial x_3}{\partial \zeta_3} \end{Vmatrix} \cdot \mathbf{e}_1 (d\zeta_1\, d\zeta_2\, d\zeta_3)$$

that is,

$$dV(x) = \begin{Vmatrix} \dfrac{\partial x_1}{\partial \zeta_1} & \dfrac{\partial x_1}{\partial \zeta_2} & \dfrac{\partial x_1}{\partial \zeta_3} \\ \dfrac{\partial x_2}{\partial \zeta_1} & \dfrac{\partial x_2}{\partial \zeta_2} & \dfrac{\partial x_2}{\partial \zeta_3} \\ \dfrac{\partial x_3}{\partial \zeta_1} & \dfrac{\partial x_3}{\partial \zeta_2} & \dfrac{\partial x_3}{\partial \zeta_3} \end{Vmatrix} (d\zeta_1\, d\zeta_2\, d\zeta_3) = J_V(d\zeta_1\, d\zeta_2\, d\zeta_3) \tag{8-13}$$

where the $V$ suffix has been added to the jacobian determinant $J$ to indicate its relevance to volume transformation. If we are considering a flat, two-dimensional elemental cell of area $(d\zeta_1\, d\zeta_2)$ in $Z$ (Fig. 8-2c), then the corresponding transformation rule is easily shown to be

$$dA(x) = \begin{Vmatrix} \dfrac{\partial x_i}{\partial \zeta_i} \end{Vmatrix} (d\zeta_1\, d\zeta_2) = J_A(d\zeta_1\, d\zeta_2) \qquad i,j = 1,2 \tag{8-14}$$

[i.e., only the range of the suffixes has changed between Eqs (8-13) and (8-14)].

## 8-3-2 Boundary Patches

A boundary patch (in three dimensions) of elemental area $dS(x)$ must be carefully distinguished from a two-dimensional cell $[dA(x)]$, for $dS(x)$ is *not* flat in $X$

although it *is* flat and bounded by $(\zeta_1, \zeta_2)$ in $Z$ (Fig. 8-2c). Again we use the well-known vector algebra equation that

$$dS(x) = |\mathbf{e}_1 \times \mathbf{e}_2|(d\zeta_1 \, d\zeta_2)$$

which from (8-13) leads to

$$dS(x) = \sqrt{[(d_1)^2 + (d_2)^2 + (d_3)^2]}(d\zeta_1 \, d\zeta_2) = J_S(d\zeta_1 \, d\zeta_2) \tag{8-15}$$

where

$$d_1 = \left( \frac{\partial x_2}{\partial \zeta_3} \frac{\partial x_3}{\partial \zeta_2} - \frac{\partial x_3}{\partial \zeta_3} \frac{\partial x_2}{\partial \zeta_2} \right)$$

and $d_2, d_3$ are the other lower minors of the basic jacobian matrix $[\mathbf{J}]$ (i.e., those which involve $\zeta_1, \zeta_2$ only).

Equation (8-15) is clearly very different from (8-14) and only reduces to it when all derivatives related to $x_3$ are zero and $dS(x) = d_3(d\zeta_1 \, d\zeta_2) = J_A(d\zeta_1 \, d\zeta_2)$ again.

### 8-3-3 Line Segments

The transformation of a general length element, $dl(x)$, from $X$ to $Z$ follows immediately from Eq. (8-5):

$$dl(x) = J_l \, dl(\zeta) \tag{8-16}$$

By analogy with the development of (8-15) we need to evaluate line integrals involving $dl(x)$ within which the kernel and shape functions will be expressed in terms of $\zeta$. Therefore, since $dl(\zeta)$ will be directed along $\zeta_1$, say,

$$J_l = \left[ \left( \frac{\partial x_1}{\partial \zeta_1} \right)^2 + \left( \frac{\partial x_2}{\partial \zeta_1} \right)^2 \right]^{1/2} \tag{8-17}$$

## 8-4 'LINEAR' CELLS AND BOUNDARY PATCHES

We are now in a position to apply the foregoing ideas to transformations from an orthogonal cartesian system $(X)$ to a skew cartesian one $(Z)$ which will allow us to define shape functions for a whole range of elements along the boundaries of which parameters vary linearly—so-called 'linear' elements.

In order to emphasize how simply the foregoing ideas can be applied to 'linear' elements we shall start by considering a two-dimensional example involving skew cartesian $(\zeta_i)$ axes $(i = 1, 2)$ defined by the three points $\mathbf{X}_1, \mathbf{X}_2, \mathbf{X}_3$ in Fig. 8-3.

A general point $P(x_1, x_2)$ or $(\zeta_1, \zeta_2)$ (Fig. 8-3a) allows us to establish the $\zeta_i \rightarrow x_i$ transformation rules by simple trigonometry as

$$x_1 - X_{13} = \zeta_1(\cos \alpha_1) + \zeta_2(\cos \alpha_2) = \zeta_1 \frac{X_{11} - X_{13}}{l_1} + \zeta_2 \frac{X_{12} - X_{13}}{l_2}$$

$$x_2 - X_{23} = \zeta_1(\sin \alpha_1) + \zeta_2(\sin \alpha_2) = \zeta_1 \frac{X_{21} - X_{23}}{l_1} + \zeta_2 \frac{X_{22} - X_{23}}{l_2} \tag{8-18}$$

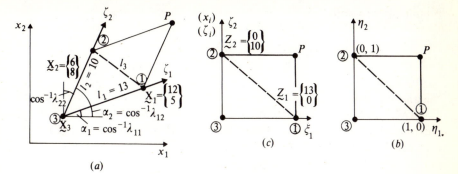

**Figure 8-3** A skew cartesian mapping.

or, alternatively, in terms of the direction cosines of the $\zeta_i$ axes ($\lambda_{ij} = \cos \widehat{x_i \zeta_j}$),

$$\mathbf{x} = [\lambda]\boldsymbol{\zeta} + \mathbf{X}_3 \qquad (8\text{-}19)$$

Therefore for $d\mathbf{x} = \mathbf{J}\,d\boldsymbol{\zeta}$ we have $[J] = [\lambda]$, the components of which are provided by Eq. (8-18) in terms of the $(X, l)$ parameters. This latter set will be of particular interest to us as we shall always be defining both boundary and internal elements in terms of nodal coordinates. With this in mind it is helpful to introduce $\eta_i$ as *normalized, local* curvilinear coordinates, where by *local* we mean that the space is particular to each element considered, with $\eta_i$ directed favourably for the study of any element, and by *normalized* we mean that the transformed element is also scaled so that its dimensions are arithmetically convenient (e.g., lines of unit length, triangles of unit side, squares $2 \times 2$, etc.).

By defining $\eta_i = \zeta_{(i)}/l_{(i)}$ we can simpilfy (and generalize) Eq. (8-18) to

$$\begin{Bmatrix} x_1 \\ x_2 \end{Bmatrix} = [\mathbf{X}_1, \mathbf{X}_2] \begin{Bmatrix} \eta_1 \\ \eta_2 \end{Bmatrix} + (1 - \eta_1 - \eta_2)\mathbf{X}_3 \qquad (8\text{-}20)$$

with a corresponding

$$[J] = [(\mathbf{X}_1 - \mathbf{X}_3), (\mathbf{X}_2 - \mathbf{X}_3)] \qquad (8\text{-}21)$$

and the mapping of the $(1, 2, 3)$ triangle in $\eta_i$ now has unit sides along $(\eta_1 \, \eta_2)$ (Fig. 8-3c).

Alternatively, we can develop more symmetrical equations by adopting *homogeneous* coordinates, a well-established technique in analytical geometry,[4] whereby a single extra (dependent) coordinate is introduced as $x_3$ and $\eta_3$ in both $x$ and $\eta$. We can take $x_3 = 1$ and $\eta_3 = 1 - \eta_1 - \eta_2$ (that is, $\eta_1 + \eta_2 + \eta_3 = 1$), whence (8-20) becomes

$$\begin{Bmatrix} x_1 \\ x_2 \\ 1 \end{Bmatrix} = \begin{bmatrix} \mathbf{X}_1 & \mathbf{X}_2 & \mathbf{X}_3 \\ 1 & 1 & 1 \end{bmatrix} \begin{Bmatrix} \eta_1 \\ \eta_2 \\ \eta_3 \end{Bmatrix} = [J]\eta_x \qquad (8\text{-}22)$$

Note that if we write just

$$\begin{Bmatrix} x_1 \\ x_2 \end{Bmatrix} = [\mathbf{X}_1, \mathbf{X}_2, \mathbf{X}_3] \begin{Bmatrix} \eta_1 \\ \eta_2 \\ \eta_3 \end{Bmatrix}$$

the transformation matrix is $2 \times 3$, does not have an inverse, and $\mathbf{J}^{-1}$ is undefined.

An interesting minor extension of the previous section will cover the corresponding three-dimensional tetrahedral cell (Fig. 8-4$a$). It is easy to prove that, for the node numbering shown, Eqs (8-20) to (8-22) become, equivalently,

$$\mathbf{x} = [\lambda]\boldsymbol{\zeta} + \mathbf{X}_4 \tag{8-23}$$

or

$$\begin{Bmatrix} x_1 \\ x_2 \\ x_3 \end{Bmatrix} = [\mathbf{X}_1, \mathbf{X}_2, \mathbf{X}_3] \begin{Bmatrix} \eta_1 \\ \eta_2 \\ \eta_3 \end{Bmatrix} + (1 - \eta_1 - \eta_2 - \eta_3)\mathbf{X}_4 \tag{8-24}$$

and

$$[J] = [(\mathbf{X}_1 - \mathbf{X}_4), (\mathbf{X}_2 - \mathbf{X}_4), (\mathbf{X}_3 - \mathbf{X}_4)] \tag{8-25}$$

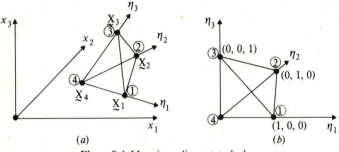

**Figure 8-4** Mapping a linear tetrahedron.

If we introduce homogeneous coordinates, as above, with $x_4 = 1$ and $\eta_1 + \eta_2 + \eta_3 + \eta_4 = 1$, we obtain either

$$\mathbf{x} = \begin{Bmatrix} x_1 \\ x_2 \\ x_3 \end{Bmatrix} = [\mathbf{X}_1 \, \mathbf{X}_2 \, \mathbf{X}_3 \, \mathbf{X}_4] \begin{Bmatrix} \eta_1 \\ \eta_2 \\ \eta_3 \\ \eta_4 \end{Bmatrix} \tag{8-26}$$

or, preferably,

$$\begin{Bmatrix} x_1 \\ x_2 \\ x_3 \\ 1 \end{Bmatrix} = \begin{bmatrix} \mathbf{X}_1 & \mathbf{X}_2 & \mathbf{X}_3 & \mathbf{X}_4 \\ 1 & 1 & 1 & 1 \end{bmatrix} \begin{Bmatrix} \eta_1 \\ \eta_2 \\ \eta_3 \\ \eta_4 \end{Bmatrix} = [J]\eta_\alpha \tag{8-27}$$

The normalized coordinates $\eta_i$ ($i = 1, 2, 3$) plot as in Fig. 8-4$b$.

An alternative interpretation of the symmetrical, homogeneous $\eta_\alpha$ ($\alpha = 1, 2, 3$ in two dimensions and $\alpha = 1, 2, 3, 4$ in three dimensions) is shown in Fig. 8-5a, b.

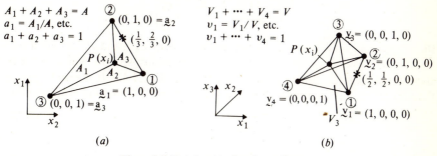

Figure 8-5 'Area' and 'volume' coordinates.

The first diagram defines what are usually called 'area' coordinates for a triangle $a_\alpha$ ($\alpha = 1, 2, 3$) where $a_1 = A_1/A$ with $A$ the total triangle area and $A_1$ the area of the subtriangle opposite node 1. We necessarily have $a_1 + a_2 + a_3 = 1$ and any point $x_i$ can be defined unambiguously in terms of $a_\alpha$. By requiring $\mathbf{X}_1$ to transform to $(1, 0, 0)^T$ in terms of $a_\alpha$ we obtain a linearly varying $x_i$ field given by

$$x_i = \begin{Bmatrix} x_1 \\ x_2 \end{Bmatrix} = [\mathbf{X}_1 \, \mathbf{X}_2 \, \mathbf{X}_3] \begin{Bmatrix} a_1 \\ a_2 \\ a_3 \end{Bmatrix} = X_{i\alpha} a_\alpha \qquad (8\text{-}28)$$

Therefore, $a_\alpha$ and $\eta_\alpha$ are identical.

If we define 'volume' coordinates on the tetrahedron (Fig. 8-5b) $v_\alpha$ ($\alpha = 1, 2, 3, 4$) with $v_1 = V_1/V$ by analogy with the triangle, the outer terms in Eq. (8-28) still equate with $v_\alpha$ replacing $a_\alpha$ and $i = 1, 2, 3$. Therefore for the 'linear' tetrahedron $v_\alpha = \eta_\alpha$.

This idea could equally well be applied to a line element defined by $l_1 + l_2 = 1$ with $x = [X_1 \, X_2] \begin{Bmatrix} l_1 \\ l_2 \end{Bmatrix}$ or extended to specify a rectangle by four triangular area coordinates (see, for example, Fig. 8-7c) with $a_1 + a_3 = 1$, $a_2 + a_4 = 1$, and $a_1 = 2A_1/A$, etc., for the parallelogram shown.

It is now evident that we have progressed from a differential area element and are really discussing, in Fig. 8-3, the transformation of a plane triangular cell $(1, 2, 3)$ from the $x$ to the $\eta$ coordinate system, and Eq. (8-28) could equally well be written, to correspond with (8-4), as $x_i = X_{i\beta} M_\beta$ where the geometrical shape functions are simply $\eta_\beta$. We can similarly define a linearly varying displacement field over the triangle by, say, $U_{i\alpha}$ (a nodal displacement matrix), leading to $u_i = U_{i\alpha} N_\alpha$ [Eq. (8-3)] or some scalar variable $\psi$ (value $\Psi_\alpha$ at the nodes 1, 2, 3), whence $\psi = \Psi_\alpha N_\alpha$. Quite obviously, $N_\alpha = M_\alpha = \eta_\alpha$.

The final, and most common, 'linear' element in BEM is the triangular surface patch on a three-dimensional body. Figure 8-6 shows one such

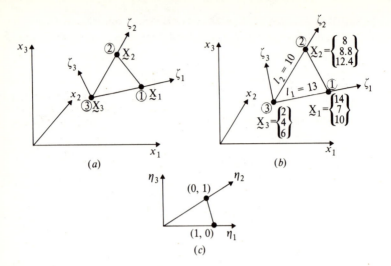

**Figure 8-6** Mapping boundary elements.

boundary element which in fact corresponds to triangle $(1, 2, 4)$ in Fig. 8-4$a$ apart from modified node numbering (node 3 coinciding with and replacing 4).

The geometrical transformation equations follow from (8-23), (8-24):

$$\mathbf{x} = [(\mathbf{X}_1 - \mathbf{X}_3), (\mathbf{X}_2 - \mathbf{X}_3)] \begin{Bmatrix} \eta_1 \\ \eta_2 \end{Bmatrix} + \mathbf{X}_3 \tag{8-29}$$

or from (8-28):

$$x_i = \begin{Bmatrix} x_1 \\ x_2 \\ x_3 \end{Bmatrix} = [\mathbf{X}_1 \, \mathbf{X}_2 \, \mathbf{X}_3] \begin{Bmatrix} a_1 \\ a_2 \\ a_3 \end{Bmatrix} \tag{8-30}$$

which differs from (8-28) only in that all the $\mathbf{x}, \mathbf{X}$ vectors now have three components.

From all the foregoing we see that for 'linear' elements (with triangular faces) defined by nodal coordinates the jacobian transformation matrices have simple constant elements and the associated shape functions are closely related to the homogeneous coordinates of the elements.

## 8-5 INTERPOLATION FUNCTIONS

Interpolation functions enable us to generate useful curvilinear shape functions in a systematic way. Again the simplest introduction is by considering linear internal cells, this time parallelograms (two dimensional) and parallelipipeds (three dimensional).

Consider a simple parallelogram cell (Fig. 8-7a). We could seek a set of functions $f(\Phi_\alpha)$ where $\Phi_\alpha$ are the values of some function $\phi$ at each of the nodes $\alpha = 1, 2, 3, 4$, in $X$, which map $\phi$ over the square in $\eta$ through

$$\phi = f_0 + f_1\eta_1 + f_2\eta_2 + f_3\eta_1\eta_2 \tag{8-31}$$

$\phi$ will thereby vary linearly along each $\eta_i = \pm 1$ boundary and we need to determine the four $(f)$ functions to satisfy $\phi = \Phi_\alpha$ at each node $\alpha$.

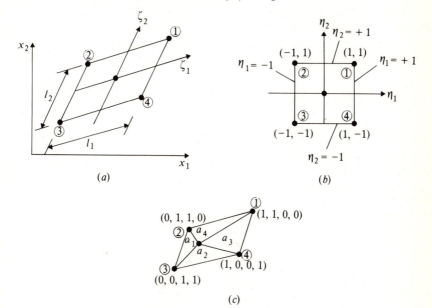

(a)

(b)

(c)

**Figure 8-7** A parallelogram cell.

This *interpolation function* can be used to produce shape functions for so-called 'linear' rectangular elements. Higher-order expressions provide a means of generating most of the commoner shape functions used in finite element analysis.[1,5,6] Since we have one nodal $\Phi_\alpha$ value at each of the $\eta_i$ nodes ($\Phi_1$ at node 1, etc.; Fig. 8-7b) we can deduce the four functions $(f)$ from equations such as, at node 1, $\Phi_1 = f_0 + f_1 + f_2 + f_3$, etc. Having found the $f = f(\Phi_\alpha)$, Eq. (8-31) can be rewritten as

$$\phi = \phi(\eta) = \Phi_\alpha N_\alpha \tag{8-32}$$

where each of the shape functions $N_\alpha(\eta)$ is now a function of $\eta_i$. The result we obtain from these operations is

$$\left. \begin{aligned} 4N_1 &= (1+\eta_1)(1+\eta_2) \\ 4N_2 &= (1-\eta_1)(1+\eta_2) \\ 4N_3 &= (1-\eta_1)(1-\eta_2) \\ 4N_4 &= (1+\eta_1)(1-\eta_2) \end{aligned} \right\} \quad \text{or} \quad 4N_\alpha = (1+s_{\alpha 1}\eta_1)(1+s_{\alpha 2}\eta_2) \tag{8-33}$$

where the second, composite expression is a most useful and convenient way of writing the whole set of $N_\alpha$ functions; $s_{\alpha 1}$ (always $\pm 1$) takes the *sign* of the $\eta_1$ coordinate at node $\alpha$, etc. (Fig. 8-7b). *No sum* over $\alpha$ is implied in any equations involving $s_{\alpha i}$. We note two important features of (8-33) which are common to all such shape functions:

1. Each function $N_\alpha$ has unit value *at node* $\alpha$ and is zero at all other nodes (i.e., the effects of all the $\Phi_\alpha$ are uncoupled one from the other).
2. The sum of the $N_\alpha$ values $\sum_{\alpha=1}^{\alpha} N_\alpha = 1$ at *any* point. This ensures that the interpolation function contains a 'complete' polynomial in $\eta_i$ which is a necessary condition for such a shape function transformation to be admissible.

If, instead of $\phi$, we have a displacement field $u_i$, specified by $\mathbf{U}_\alpha$ at each node,

$$u_i = u_i(x) = U_{i\alpha} N_\alpha \qquad \mathbf{u} = [\mathbf{U}]\mathbf{N} \tag{8-34}$$

we now see our way forward to more sophisticated curvilinear elements in $X$ for we can, in principle, use more complicated complete polynomials in (8-31), match the number of terms ($\alpha$) with nodal values $\Phi_\alpha$, and arrive at a shape function equation corresponding to (8-34) in any number of spatial dimensions.

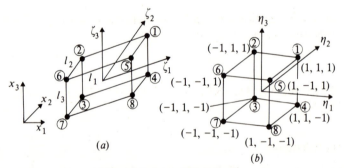

**Figure 8-8** A parallelipiped cell.

The equations corresponding to (8-31) and (8-33) for the three-dimensional parallelipiped transformation shown in Fig. 8-8a, b and the associated 'linear' shape functions can be derived easily from the following interpolation function which contains the requisite eight terms:

$$\phi = f_0 + f_1\eta_1 + f_2\eta_2 + f_3\eta_3 + f_4\eta_1\eta_2 + f_5\eta_2\eta_3 + f_6\eta_3\eta_1 + f_7\eta_1\eta_2\eta_3 \tag{8-35}$$

The shape functions then follow as

$$8N_\alpha = (1 + s_{\alpha 1}\eta_1)(1 + s_{\alpha 2}\eta_2)(1 + s_{\alpha 3}\eta_3) \tag{8-36}$$

where the $s_{\alpha i}$ signs agree with the $\eta_i$ coordinate component signs in the nodal numbering scheme adopted (Fig. 8-8b) [for example $(-1, +1, +1)$ for node 2, etc.]. Once again $N_\alpha = 1$ at node $\alpha$, is zero at other nodes and $\sum_{\alpha=1}^{8} N_\alpha = 1$ always.

## 8-6 RECAPITULATION

Before we move on to curvilinear elements it may be helpful to recapitulate briefly how the techniques which we have developed in this chapter are useful in enabling us to calculate components for each boundary element in the final matrix equations such as (8-2), (4-19), (4-20), etc.

As stated in Sec. 8-1, the basic requirement is the integration, over possibly a triangular surface patch $\Delta S(x)$, of a set of known functions $\mathbf{F}(x, x_0)$ multiplied by a vector of known (or unknown) components $\mathbf{u}(x)$, say. We can now write this as

$$\int_{\Delta S(x)} \mathbf{F}(x, x_0) \mathbf{u}(x)\, dS(x) = \left[ \int_{\Delta S(\eta)} \mathbf{F}(x(\eta), x_0) \mathbf{N}(\eta)\, J_s\, dS(\eta) \right] \mathbf{U} \qquad (8\text{-}37)$$

in which $\mathbf{U}$ are constant nodal displacement components related to $\mathbf{u}(x)$ by the shape functions $\mathbf{N}(\eta)$. $J_s$ is the specific jacobian determinant which transforms $dS(\eta)$ to $dS(x)$ calculated [Eq. (8-15)] from the terms of the jacobian matrix $\mathbf{J}$. The terms of $\mathbf{J}$ for isoparametric elements are obviously given by

$$J_{ij} = \frac{\partial x_i}{\partial \eta_j} = X_{i\alpha} \frac{\partial N_\alpha}{\partial \eta_j} \qquad (8\text{-}38)$$

and the limits of integration for $\eta_1, \eta_2$ in the right-hand-side integral will be over unit intervals (see Chapter 15).

## 8-7 CURVILINEAR TRANSFORMATIONS AND SHAPE FUNCTIONS

### 8-7-1 Line Elements

We shall take as our paradigm the transformation of a vector of coordinates $x_i (i = 1, 2)$ specified by its components $\mathbf{X}_\alpha = X_{i\alpha}$ at a number of geometric nodes $(\alpha)$ along each line element sufficient to describe the order of variation required by a complete polynomial interpolation function. For any line element, all this means is that we must have two nodes per element for a linear variation of each $x_i$ component; three for quadratic; four for cubic; five for quartic; etc. Thus for $\zeta_i$ along the element (Fig. 8-9a) and quadratic variation, a suitable interpolation function would be $\phi = f_0 + f_1 \zeta_1 + f_2 \zeta_1^2$, augmented by $f_3 \zeta_1^3$ for third order, etc. Following the same pattern of calculation that led to Eq. (8-32) we arrive at the following sets of shape functions $M_\alpha$ (that is, $x_i = X_{i\alpha} M_\alpha$; $\mathbf{x} = [\mathbf{X}]\mathbf{M}$). These are given in terms of both normalized coordinates $\eta_1 = \zeta_1/l$ and homogeneous length coordinates (Sec. 8-4), $l_1 + l_2 = 1$, for equally spaced nodes along the element length $(l)$, numbered as shown in Fig. 8-9 with a central origin in $\eta$.

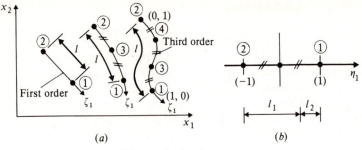

**Figure 8-9** Line elements.

**Linear variation** $(\alpha = 1, 2)$

$$\left.\begin{array}{l} M_1 = \frac{1}{2}(1+\eta_1) = l_1 \\ M_2 = \frac{1}{2}(1-\eta_1) = l_2 \end{array}\right\} \qquad M_\alpha = \frac{1}{2}(1+s_{\alpha 1}\eta_1) \tag{8-39}$$

**Quadratic (second-order) variation** $(\alpha = 1, 2, 3)$

$$\left.\begin{array}{l} M_1 = \frac{1}{2}\eta_1(1+\eta_1) = l_1(l_1 - l_2) \\ M_2 = -\frac{1}{2}\eta_1(1-\eta_1) = l_2(l_2 - l_1) \end{array}\right\} = \frac{1}{2}s_{\alpha 1}\eta_1(1 + s_{\alpha 1}\eta_1)$$
$$M_3 = (1+\eta_1)(1-\eta_1) = 1-\eta_1^2 = 4l_1 l_2 \tag{8-40}$$

**Cubic (third-order) variation** $(\alpha = 1, 2, 3, 4)$

$$\left.\begin{array}{l} M_1 = \frac{1}{16}(1+\eta_1)(1+3\eta_1)(3\eta_1 - 1) \\ \quad = \frac{1}{2}l_1(2l_2 - l_1)(l_2 - 2l_1) \\ M_2 = \frac{1}{16}(1-\eta_1)(1+3\eta_1)(3\eta_1 - 1) \\ \quad = \frac{1}{2}l_2(2l_1 - l_2)(l_1 - 2l_2) \end{array}\right\} = \frac{1}{16}(9\eta_1^2 - 1)(1 + s_{\alpha 1}\eta_1)$$

$$\left.\begin{array}{l} M_3 = \frac{9}{16}(1-\eta_1)(1-\eta_1)(1+3\eta_1) \\ \quad = \frac{9}{2}l_1 l_2(2l_1 - l_2) \\ M_4 = \frac{9}{16}(1+\eta_1)(1-\eta_1)(1-3\eta_1) \\ \quad = \frac{9}{2}l_1 l_2(2l_2 - l_1) \end{array}\right\} = \frac{9}{16}(1-\eta_1)(1+3s_{\alpha 1}\eta_1) \tag{8-41}$$

The jacobian matrix can be calculated for any of these transformations, although for curved elements the matrix components will no longer be constants. If homogeneous length coordinates $(l_1 = \eta_1, l_2 = 1-\eta_1)$ are used then the components of $J$ are given by

$$J_{ij} = X_{i\alpha}\left(\frac{\partial M_\alpha}{\partial l_r}\frac{\partial l_r}{\partial \eta_j}\right) \qquad r = 1, 2 \text{ and } j = 1 \tag{8-42}$$

### 8-7-2 Plane Triangular Cells

Since these are all minor extensions of the foregoing argument the results will be simply listed in terms of area coordinates ($a_1 = \eta_1, a_2 = \eta_2$, and $a_3 = 1 - \eta_1 - \eta_2$).

**Linear variation** (Sec. 8-4) ($\alpha = 1, 2, 3$)

$$M_\alpha = a_\alpha \qquad (8\text{-}43)$$

**Quadratic variation** The complete polynomial interpolation function now has six terms and we therefore require the six nodes shown in Fig. 8-10a:

$$\phi = f_0 + f_1 \eta_1 + f_2 \eta_2 + f_3 \eta_1 \eta_2 + f_4 \eta_1^2 + f_5 \eta_2^2$$

which leads to (for $\alpha = 1, 2, ..., 6$)

$$\mathbf{M}^T = \{a_1(2a_1 - 1), a_2(2a_2 - 1), a_3(2a_3 - 1), 4a_1 a_2, 4a_2 a_3, 4a_3 a_1\} \qquad (8\text{-}44)$$

**Cubic variation** The complete cubic polynomial has ten terms and therefore the four nodes per triangle side (to allow cubic variation) have to be augmented by a tenth (centroidal) node (Fig. 8-10a). The shape functions are

$$\begin{aligned}
M_1 &= \tfrac{1}{2} a_1(3a_1 - 1)(3a_1 - 2) & M_2, M_3 \text{ similarly} \\
M_4 &= \tfrac{9}{2} a_1 a_2(3a_1 - 1) \Big\} & \\
M_5 &= \tfrac{9}{2} a_1 a_2(3a_2 - 1) \Big\} & M_6, M_7; M_8, M_9 \text{ similarly} \qquad (8\text{-}45) \\
M_{10} &= 27 a_1 a_2 a_3 &
\end{aligned}$$

All the above are taken from Zienkiewicz[1] where a recurrence relationship is deduced for 'triangular' elements relating the $M_\alpha^{n+1}$ [for the $(n+1)$th-order

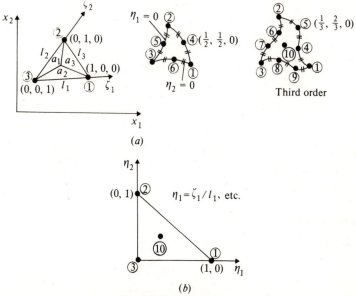

$(a)$

$\eta_1 = \zeta_1 / l_1$, etc.

$(b)$

**Figure 8-10** Planar cells.

element] to $M_\alpha^n$ (for the $n$th-order element) which enables shape functions to be developed for 'triangles' of any order.

## 8-7-3 Plane Parallelipiped Cells

Although internal cells are less important than boundary patches in BEM it is simpler, once started, to complete the treatment of cells without a break.

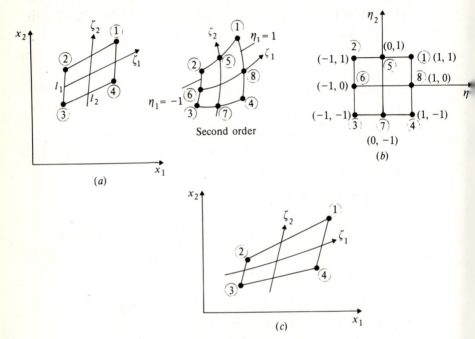

Figure 8-11 Quadrilateral cells.

Our second-order variation nodal pattern (Fig. 8-11$a$) requires three nodes per side (totalling eight) and therefore the interpolation function has to expand to, say,

$$\phi = f_0 + f_1\eta_1 + f_2\eta_2 + f_3\eta_1\eta_2 + f_4\eta_1^2 + f_5\eta_2^2 + f_6\eta_1^2\eta_2 + f_7\eta_2^2\eta_1$$

whence shape functions, calculated as before for equally divided rectangle sides, are as follows.

**Linear variation** ($\alpha = 1, 2, 3, 4$)

$$\left.\begin{aligned}
M_1 &= \tfrac{1}{4}(1+\eta_1)(1+\eta_2) = a_1 a_2 \\
M_2 &= \tfrac{1}{4}(1-\eta_1)(1+\eta_2) = a_2 a_3 \\
M_3 &= \tfrac{1}{4}(1-\eta_1)(1-\eta_2) = a_3 a_4 \\
M_4 &= \tfrac{1}{4}(1+\eta_1)(1-\eta_2) = a_4 a_1
\end{aligned}\right\} \qquad M_\alpha = \tfrac{1}{4}(1+s_{\alpha 1}\eta_1)(1+s_{\alpha 2}\eta_2) \qquad (8\text{-}46)$$

where the area coordinates $(a_\alpha)$ for a parallelogram are defined as in Sec. 8-4 $(a_2 + a_4 = 1; a_1 + a_3 = 1; \eta_1 = 1 - 2a_4; \eta_2 = 1 - 2a_1)$. It should be noted that a general quadrilateral (Fig. 8-11c) can also be transformed into a square in $\eta$ by (8-46) and therefore we are not confined to parallelogram cells.

**Quadratic variation** $(\alpha = 1, 2, ..., 8)$

$$\left.\begin{aligned}
M_1 &= \tfrac{1}{4}(1+\eta_1)(1+\eta_2)(\eta_1+\eta_2-1) = a_1 a_2(2a_1 + 2a_2 - 3) \\
M_2 &= \tfrac{1}{4}(1-\eta_1)(1+\eta_2)(-\eta_1+\eta_2-1) = a_2 a_3(2a_2 + 2a_3 - 3) \\
M_\alpha &= \tfrac{1}{4}(1+s_{\alpha 1}\eta_1)(1+s_{\alpha 2}\eta_2)(s_{\alpha 1}\eta_1 + s_{\alpha 2}\eta_2 - 1)
\end{aligned}\right\} M_3, M_4 \text{ similarly}$$

$$(8\text{-}47)$$

$$\left.\begin{aligned}
M_5 &= \tfrac{1}{2}(1-\eta_1)(1+\eta_1)(1+\eta_2) \\
&= 4a_1 a_2 a_3 = \tfrac{1}{2}(1-\eta_1^2)(1+s_{\alpha 2}\eta_2) \\
M_6 &= \tfrac{1}{2}(1-\eta_1)(1+\eta_2)(1-\eta_2) \\
&= 4a_2 a_3 a_4 = \tfrac{1}{2}(1-\eta_2^2)(1+s_{\alpha 1}\eta_1)
\end{aligned}\right\} M_7, M_8 \text{ similarly}$$

**Cubic variation** $(\alpha = 1, 2, ..., 12)$

$$\left.\begin{aligned}
M_1 &= \tfrac{1}{32}(1+\eta_1)(1+\eta_2)[9(\eta_1^2+\eta_2^2)-10] \\
&= a_1 a_2[1+\tfrac{9}{2}(a_4 a_2 - a_3 a_1)] \\
M_2 &= \tfrac{1}{32}(1-\eta_1)(1+\eta_2)[9(\eta_1^2+\eta_2^2)-10] \\
&= a_2 a_3[1+\tfrac{9}{2}(a_4 a_2 - a_3 a_1)] \\
M_\alpha &= \tfrac{1}{32}(1+s_{\alpha 2}\eta_1)(1+s_{\alpha 2}\eta_2)[9(\eta_1^2+\eta_2^2)-10]
\end{aligned}\right\} M_3, M_4 \text{ similarly}$$

$$(8\text{-}48)$$

$$\begin{aligned}
M_5 &= \tfrac{9}{32}(1+\eta_1)(1-\eta_1)(1+\eta_2)(1+3\eta_1) = \tfrac{9}{2} a_2 a_3 a_4(1-3a_4) \\
M_6 &= \tfrac{9}{32}(1+\eta_1)(1-\eta_1)(1+\eta_2)(1-3\eta_1) = \tfrac{9}{2} a_2 a_3 a_4(3a_4-1)
\end{aligned}$$

Here

$$M_\alpha = \tfrac{9}{32}(1-\eta_1^2)(1+s_{\alpha 2}\eta_2)(1+s_{\alpha 1}\eta_1)$$

and $M_7, ..., M_{12}$ follow a corresponding pattern.

## 8-7-4 Three-dimensional Cells

**(a) Tetrahedral cells** These are a straightforward extension of the curvilinear triangle in the same way that the straight-sided tetrahedral transformation in Sec. 8-4 followed on from the triangle. Thus, in volume coordinates $(v_1 + v_2 + v_3 + v_4 = 1)$ (Sec. 8-4, Fig. 8-5b) with equal divisions of the 'tetrahedral' edges we have (Fig. 8-12) the following variations.

*Linear variation* $(\alpha = 1, 2, 3, 4)$

$$M_\alpha = v_\alpha \qquad (8\text{-}49)$$

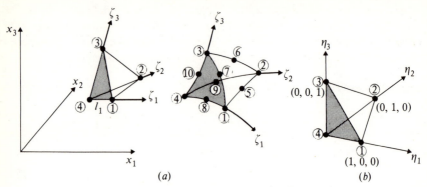

**Figure 8-12** Three-dimensional cells.

***Quadratic variation*** ($\alpha = 1, 2, ..., 10$)

$$M_1 = v_1(2v_1 - 1) \qquad M_2, M_3, M_4 \text{ similarly}$$
$$\left.\begin{array}{l} M_5 = 4v_1 v_2 \\ M_6 = 4v_2 v_3 \end{array}\right\} \qquad M_7, ..., M_{10} \text{ similarly} \tag{8-50}$$

***Cubic variation*** ($\alpha = 1, 2, ..., 20$)

$$M_1 = \tfrac{1}{2}v_1(3v_1 - 1)(3v_1 - 2) \qquad M_2, M_3, M_4 \text{ similarly}$$
$$\left.\begin{array}{l} M_5 = 9/2v_1 v_2(3v_1 - 1) \\ M_6 = 9/2v_1 v_2(3v_2 - 1) \end{array}\right\} \begin{array}{l} M_7, M_8; M_9, M_{10} \text{ and all other mid-side nodes} \\ \text{similarly} \end{array}$$
$$M_{11} = 27v_1 v_2 v_3 \text{ and the other three 'centroidal' nodes similarly}$$

**(b) Hexahedral cells** These are an extension of the 'quadrilateral' cells of Sec. 8-7-3. Nodes are identified as shown in Fig. 8-13.

***Linear case*** ($\alpha = 1, 2, ..., 8$)

$$M_\alpha = \tfrac{1}{8}(1 + s_{\alpha 1}\eta_1)(1 + s_{\alpha 2}\eta_2)(1 + s_{\alpha 3}\eta_3) \tag{8-51}$$

***Quadratic case*** ($\alpha = 1, 2, ..., 20$)

Nodes $1, 2, ..., 8$:

$$M_\alpha = \tfrac{1}{8}(1 + s_{\alpha 1}\eta_1)(1 + s_{\alpha 2}\eta_2)(1 + s_{\alpha 3}\eta_3)$$
$$\times (s_{\alpha 1}\eta_1 + s_{\alpha 2}\eta_2 + s_{\alpha 3}\eta_3 - 2)$$

Nodes $9, 10, 11, 12$: $\qquad M_\alpha = \tfrac{1}{4}(1 - \eta_1^2)(1 + s_{\alpha 2}\eta_2)(1 + s_{\alpha 3}\eta_3) \qquad$ (8-52)

Nodes $13, 14, 15, 16$: $\qquad M_\alpha = \tfrac{1}{4}(1 - \eta_2^2)(1 + s_{\alpha 1}\eta_1)(1 + s_{\alpha 3}\eta_3)$

Nodes $17, 18, 19, 20$: $\qquad M_\alpha = \tfrac{1}{4}(1 - \eta_3^2)(1 + s_{\alpha 1}\eta_1)(1 + s_{\alpha 2}\eta_2)$

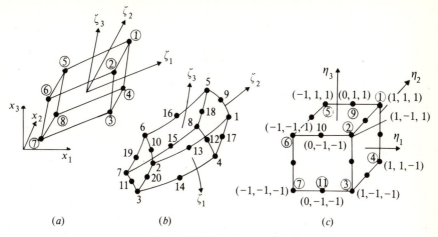

**Figure 8-13** Hexahedral cells.

### 8-7-5 General Comments on Cell-Shape Functions

All the foregoing shape functions have been listed here for convenient reference although one might note:

1. By substituting $s_{\alpha 3}\eta_3 = \eta_3 = 1$ in the hexahedral shape functions [(5-51), (8-52)] the 'quadrilateral' values are recovered [(8-46), (8-47)] and, also, replacement of $v_\alpha$ by $a_\alpha$ in the tetrahedral set [(8-49), (8-50), etc.] produces those for the 'triangles' [(8-43), (8-44), (8-45)].
2. Other forms of contraction from higher- to lower-dimensional elements are possible. Watson,[7] for example, points out that by collapsing nodes 2, 3, 6 for the second-order quadrilateral (Fig. 8-11a) to a single node, quadratic triangular shape functions will be generated. The reference axes will, of course, remain the $\zeta_i$ set of Fig. 8-11a, which will not be aligned along the 'triangle' sides. Similar operations on hexahedra (Fig. 8-13) could be used to generate 'triangular' prisms, although it would probably be more convenient to combine triangular shape functions in the $(\zeta_1, \zeta_3)$ plane with, say, functions from a hexahedral cell along $(\zeta_2)$. For example (Fig. 8-14), the prism of linear triangle cross section and quadratic variation along its length will have for

Nodes $1, 2, ..., 6$:  $\quad M_\alpha = \frac{1}{2}a_\alpha(1 + s_{\alpha 2}\eta_2)s_{\alpha 2}\eta_2$

Nodes $7, 8, 9$:  $\quad M_\alpha = \frac{1}{2}a_\alpha(1 - \eta_2^2)$  $\qquad\qquad$ (8-53)

All our previous remarks concerning the freedom of choice of functional and geometric nodes, $N_\alpha$ and $M_\alpha$, apply equally well to all the shape functions presented in this chapter.

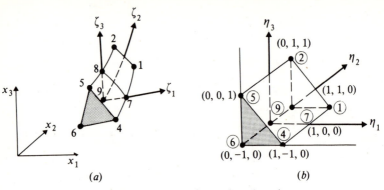

**Figure 8-14** Mapping a triangular prism.

## 8-8 CURVILINEAR BOUNDARY ELEMENTS

Most of the above material relating to shape functions has been taken from finite element sources[1] in which the emphasis is on interior (cell) rather than surface discretization schemes. We have already discussed curvilinear boundaries in two-dimensional problems (Sec. 8-7-1) and all that remains to be explained is the three-dimensional counterpart of curved planar boundary patches (elements), as shown in Fig. 8-15.

We could develop the shape functions for the doubly curved quadratic triangle of Fig. 8-15a by contracting the second-order tetrahedron (Fig. 8-12a) as explained above. Nodes 3, 10, 4 will all have identical coordinates and coalesce into, say, node 3′. The shape function $M'_3$ for this node will then be the sum of $M_3 + M_{10} + M_4$ with $\eta_3 = v_3 = 0$. Recollecting that $v_1 = \eta_1$, $v_2 = \eta_2$, $v_3 = \eta_3$ but $v_4 = 1 - \eta_1 - \eta_2 - \eta_3$, and using $M_\alpha$ from Eqs (8-50), we obtain, with $\eta_3 = 0$,

$$M'_3 = (1 - \eta_1 - \eta_2)(1 - 2\eta_1 - 2\eta_2)$$

$M_1, M_2$ and $M_5 = M'_4$ are unaffected, but $M'_5 = M_6 + M_9$ and $M'_6 = M_7 + M_8$. If we evaluate these and note that on the curvilinear surface element $\eta_\alpha = a_\alpha$ ($\alpha = 1, 2, 3$) with $a_1 + a_2 + a_3 = 1$, the shape functions we require (Fig. 8-15a) are found to be *identical* to those for a second-order triangle in two dimensions (8-44)! On reflection, this must be so, and to deal with such curvilinear surface elements we need to notice only that (1) in $x_i = X_{i\alpha} M_\alpha$ summation is over $i = 1, 2, 3$ (i.e., all three coordinate components are operated on by $M_\alpha$); (2) the jacobian for transforming $dS(x) = J_S \, dS(\eta)$ has to be that discussed in Sec. 8-3, Eq. (8-15).

Curvilinear quadrilateral elements can obviously be handled in an exactly similar way. The shape functions we require are therefore those listed in Secs 8-7-2 and 8-7-3.

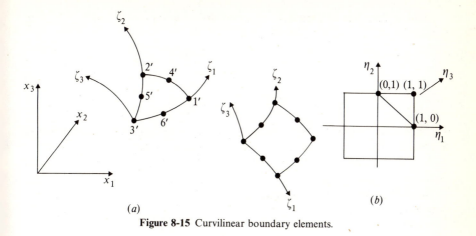

**Figure 8-15** Curvilinear boundary elements.

## 8-9 INFINITE BOUNDARY ELEMENTS

Although BEM are able to account for boundaries which are at infinity in their entirety without any discretization, for surfaces which extend from a region of major interest to infinity the discretization has to be curtailed at some arbitrary distance. It is then assumed that, beyond this point, the region under consideration remains connected to the infinite space.

In order to eliminate these difficulties Watson[7] developed the idea of infinite boundary elements. A typical example of the use of such elements is illustrated in Fig. 8-16,[7] which shows the discretization for the problem of a half space loaded uniformly over a rectangular area. Three finite-size boundary elements and two infinite boundary elements have been used to model a quadrant of the surface.

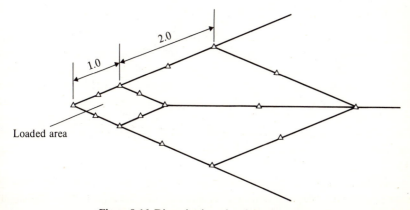

**Figure 8-16** Discretization of surface of half space.

For infinite line elements he assumed the variations of $u_i$ and $t_i$ to be (Fig. 8-17a)

$$u_i(\eta) = u_i^1 \left[\frac{r(z, x_o)}{r(x, x_o)}\right]$$

$$t_i(\eta) = t_i^1 \left[\frac{r(z, x_o)}{r(x, x_o)}\right]^2$$

(8-54)

where $z$ is the location of node 1, $r(z, x_o)$ is the distance from node 1 to an arbitrary reference point $x_o$, $r(x, x_o)$ is the distance from $x$ and $x_o$, and $u_i^1$, $t_i^1$ are $u_i$ and $t_i$ at node 1. Equations (8-54) clearly indicate quantities which decay asymptotically to zero at infinity.

Similarly, over an infinite surface element (Fig. 8-17b) the following variations were assumed:

$$u_i(\eta_1, \eta_2) = u_i(1, \eta_2) \left[\frac{r(z, x_o)}{r(x, x_o)}\right]$$

$$t_i(\eta_1, \eta_2) = t_i(1, \eta_2) \left[\frac{r(z, x_o)}{r(x, x_o)}\right]^2$$

(8-55)

where $z$ is the point $(1, \eta_2)$.

The terms within the square brackets play the role of shape functions for the displacements and tractions over these elements.

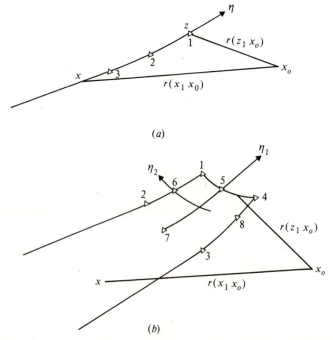

(a)

(b)

**Figure 8-17** Asymptotic functions over infinite element. (a) Plane problems. (b) Three-dimensional problem.

It is necessary to transform these infinite elements so that quadrature formulae for the finite interval $(-1, 1)$ can be used to evaluate them. This can be accomplished by[7] transformation into the $\xi$ plane via

$$\eta = \frac{3\xi - 1}{1 + \xi} \tag{8-56}$$

for line integrals. The jacobian is given by

$$J_l = \frac{d\eta}{d\xi} = \frac{4}{(1 + \xi)^2} \tag{8-57}$$

For an infinite surface element in the three-dimensional case the transformation is

$$\eta_1 = \frac{3\xi_1 - 1}{(1 + \xi_1)^2} \qquad \eta_2 = \xi_2 \tag{8-58}$$

with the corresponding jacobian

$$J_s = \frac{d\eta}{d\xi} = \frac{4}{(1 + \xi_1)^2} \tag{8-59}$$

Equations (8-56) to (8-59) enable us to complete the mapping. For applications involving IBEM the variations of $\phi_i$ over these infinite boundary elements would be identical to those of $t_i$ in Eqs (8-54) and (8-55).

## 8-10 INTEGRATION OF THE KERNEL SHAPE FUNCTION PRODUCTS

Whatever problem we tackle, no matter how elaborately or simply we represent the variation of geometry and other distributed parameters, we shall be faced with the solution of an equation similar to (8-1). What we have developed in this chapter is a means of reducing both the surface and volume integrals involved to forms such as

$$\int_{\Delta S(x)} u_i(x) F_{ij}(x, x_o) \, dS(x) = U_{i\alpha} \int_{\Delta S(\eta)} \underline{F_{ij}(\eta, x_o) N_\alpha(\eta) J_s(\eta) \, dS(\eta)} \tag{8-60}$$

where

1. The values $U_{i\alpha}$ at the functional nodes can now be taken outside the integral.
2. The $N_\alpha$ are the shape functions introduced in this chapter.
3. $J_S$ is the relevant jacobian for the geometric shape functions $(M_\alpha)$ which has to relate to line, area, or volume transformations as explained previously.
4. The limits of integration in $\eta$ are now very simple $(0, 1$ or $-1, +1,$ etc.$)$.

In order to evaluate these integrals it is necessary to examine the behaviour of the underlined term in (8-60) which has to be integrated over the element. If this remains bounded within the interval it can be integrated numerically; if not, then it must be obtained either in an indirect manner by using the rigid-body displacement modes as discussed in Chapters 3 and 4 and Refs 7 to 11 or by splitting the integrand into a singular part which has to be integrated analytically and a continuous part which can be evaluated by numerical quadrature. A detailed discussion of this will be found in Chapter 15.

## 8-11 EXAMPLES OF APPLICATION

During the last five years a series of second-generation BEM computer pro-grammes have been developed by a number of workers. These show very clearly that a great deal of the potential of BEM has now been realized. The examples presented below will, we hope, serve to illustrate this point.

(a) **The test specimen CT.15**[9, 12] A test specimen is shown in Fig. 8-18 in which a load of 20 kN was applied through the hole. A state of plane strain was assumed to exist in the specimen for which Young's modulus = $210 \, \text{kN/mm}^2$ and Poisson's ratio = 0.3.

The stress intensity factor $K_I$ for mode I in plane strain elastic problems can be obtained from[12]

$$K_I = \frac{P}{B\sqrt{W}} Y\left(\frac{a}{W}\right) \tag{8-61}$$

where

$$Y\left(\frac{a}{W}\right) = \frac{B}{P}\left[\frac{E}{1-v^2} \frac{\partial U}{\partial(a/W)}\right]^{1/2}$$

$U$ = potential energy per unit thickness

$P$ = load

Figure 8-19a, b shows a finite element discretization involving 223 trian-gular elements (six-noded isoparametric) and a BEM discretization involving 28 cubic boundary line elements. Both the data preparation time and the computer cost for the BEM analysis were considerably lower than those for the finite element analyses. The numerical output for the variation of $U$, the potential energy per unit thickness, from both analyses is shown in Fig. 8-20. The results are essentially the same over a range of $(a/W)$ values.

(b) **Analysis of a specimen used in low-cycle fatigue testing**[13, 14] The cross section of the specimen is shown in Fig. 8-21. The presence of curved surfaces, a stress concentration, and three-dimensional effects make the problem representative of those encountered in practical analysis. The specimen is loaded through pins at $A'$ and $B$. The solid line shows the simplified shape used for both the BEM

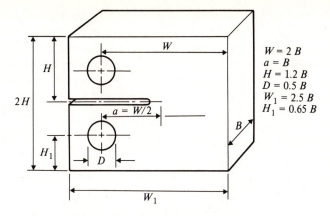

$W = 2\,B$
$a = B$
$H = 1.2\,B$
$D = 0.5\,B$
$W_1 = 2.5\,B$
$H_1 = 0.65\,B$

**Figure 8-18** A fracture test specimen CT.15.

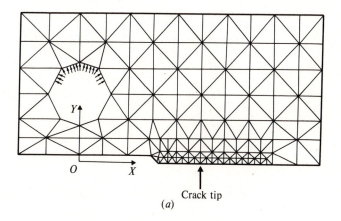

Crack tip

(a)

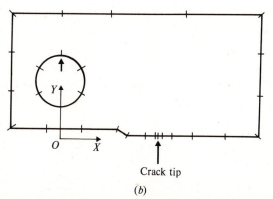

Crack tip

(b)

**Figure 8-19** (a) The finite element network. (b) The BEM analysis network.

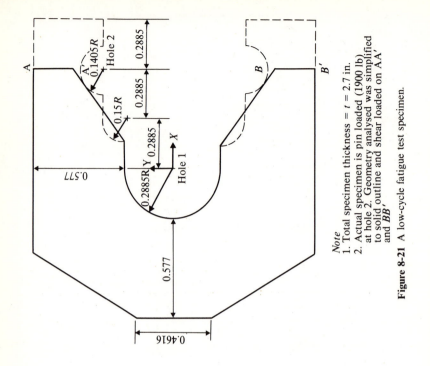

*Note*
1. Total specimen thickness = *t* = 2.7 in.
2. Actual specimen is pin loaded (1900 lb) at hole 2. Geometry analysed was simplified to solid outline and shear loaded on *AA'* and *BB'*

**Figure 8-21** A low-cycle fatigue test specimen.

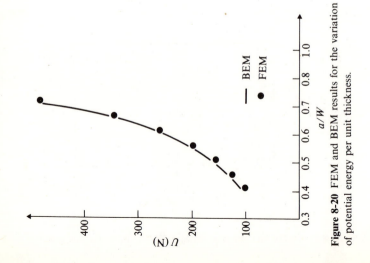

**Figure 8-20** FEM and BEM results for the variation of potential energy per unit thickness.

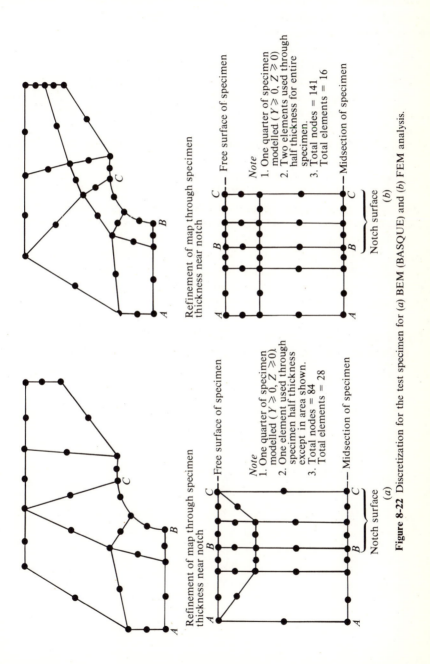

Refinement of map through specimen thickness near notch

Free surface of specimen

*Note*
1. One quarter of specimen modelled ($Y \geqslant 0, Z \geqslant 0$)
2. Two elements used through half thickness for entire specimen.
3. Total nodes = 141
   Total elements = 16

Midsection of specimen

Notch surface

(b)

Refinement of map through specimen thickness near notch

Free surface of specimen

*Note*
1. One quarter of specimen modelled ($Y \geqslant 0, Z \geqslant 0$).
2. One element used through specimen half thickness except in area shown.
3. Total nodes = 84
   Total elements = 28

Midsection of specimen

Notch surface

(a)

**Figure 8-22** Discretization for the test specimen for (a) BEM (BASQUE) and (b) FEM analysis.

analysis (using the program BASQUE[9]) and the finite element analysis using NASTRAN. A very refined (BASQUE) BEM analysis was carried out to establish a 'baseline solution' used to determine the minimum cost, less refined, BASQUE analysis which would produce satisfactory results.

The discretization schemes used for the two analyses are shown in Fig. 8-22a, b. It is worth noting that, due to the large surface area-to-volume ratio of the specimen, the problem is not a very favourable one for BEM.

The BEM map shown in Fig. 8-22a was used on the free surface and in the midplane of symmetry together with only one layer of surface elements across the thickness except in the immediate neighbourhood of the notch. The discretization consisted of 28 boundary elements with quadratic variation of both geometry and functions resulting in 84 nodes. The computing time was 65 CPU on an IBM 360/168—the 'baseline solution' time was, of course, considerably higher.

For the NASTRAN analysis two layers of elements were used through the thickness of the specimen. The discretization consisted of 16 HEX 20 elements and 141 nodes and took 48 CPU. However, the NASTRAN results were very much less accurate than the BASQUE ones, as shown in Figs 8-23 and 8-24. A second NASTRAN analysis using four layers of elements through the Fig. 8-22b was also performed. This involved 32 elements and 245 nodes and took 85 CPU to run. The 'peak' results did not improve significantly but the response through the thickness of the body was somewhat better.

**(c) A double-edge notch specimen**[13] The BEM discretization for a double-edge notch specimen designed for low-cycle fatigue testing of various anisotropic

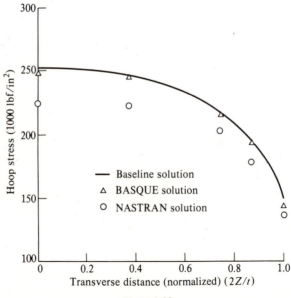

Figure 8-23

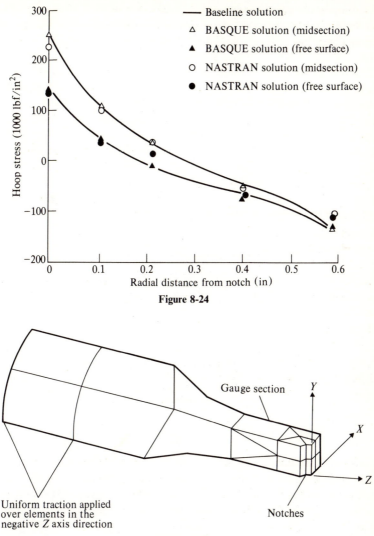

**Figure 8-24**

**Figure 8-25**

alloys used in turbines is shown in Fig. 8-25. A one-eighth section of the specimen was modelled using 26 surface elements with quadratic variation of geometry and functions over them which generated 93 nodes. The analysis was carried out for both isotropic and transversely isotropic materials loaded in the direction of the $Z$ axis.

Figure 8-26 shows the variation of $\sigma_Z$ at the bottom of the notch for the two materials. In both cases the maximum stress is at the centre of the specimen and varies only slightly over the central 50 to 75 per cent of its thickness, with a rapid drop as the free surface is approached. The anisotropic analysis was carried out using a numerically constructed fundamental solution, as outlined in Chapter 6, due to Wilson and Cruse.[15, 16] The corresponding plane strain results are also shown in Fig. 8-26.

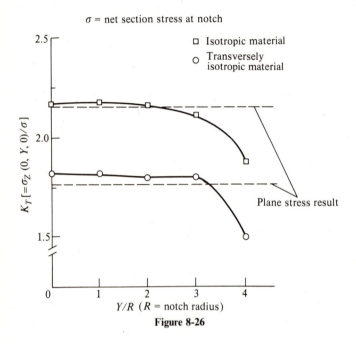

Figure 8-26

**(d) Analysis of a turbine disc rim-slot**[14] Turbine disc rim-slots, which retain blades in a gas turbine rotor assembly, produce strain concentrations in the disc rim under the inertia loading of the disc and the attached blades. To supply cooling air to the turbine blades, cooling holes are machined into the disc rim and exit through the disc rim-slots. The stress field at the intersection of the disc rim-slot with a cooling hole has to be deduced to predict the turbine disc cycled loading fatigue life and also to optimize the geometry of the cooling holes and the rim-slots. In such problems three-dimensional effects in the vicinity of the rim-slot–cooling hole intersections are of key importance.

Figure 8-27 shows a rectilinear model (six times actual size) of the problem, with views parallel to the rim-slots, of the cross section, and of the geometry of the cooling holes. Since, in the rectilinear model, a blade pull does not give rise to transverse stress, independent loads were applied in the radial and tangential directions using hydraulic jacks to enable the actual disc load state to be modelled using superposition. Both sides of the centre rim-slot and cooling hole were instrumented and the measured results averaged, the model being symmetrical.

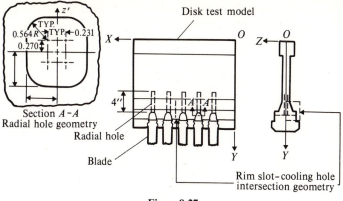

**Figure 8-27**

The three-dimensional finite element discretization for one-quarter of the rim-slot–cooling hole geometry using eight noded isoparametric elements is shown in Fig. 8-28. This provided only a very crude idealization of the problem and required 1·5 h of computation time for an IBM 370/168 computer. In order to resolve the detailed behaviour in the rim-slot and cooling hole, BEM discretizations involving the region $ABCD$ (see Fig. 8-28) were constructed. The first of which used 436 plane triangular elements with linear variations of tractions and displacements over them (BINTEQ) whilst the second utilized 97 isoparametric surface elements with quadratic variations (BASQUE). The displacements from the finite element analysis were applied as boundary conditions to the top and bottom surfaces of the BEM models.

A typical set of results for the strain concentration factor (defined as local strain/nominal strain) in the cooling hole as obtained from the test and the BEM modelling of Fig. 8-29 is shown in Fig. 8-30. The peak strain predicted by BINTEQ is about 8 per cent lower than the test data while that given by the BASQUE program is only 1 per cent lower.

The computer time for the BINTEQ analysis was about 1 h compared to 11 min using BASQUE, which demonstrates clearly the advantages of higher-order modelling.

**(e) The field in a transformer core**[17]  The current density $J$ within a coil is a source of magnetic potential $p$, which satisfies Poisson's equation

$$\nabla^2 p = -\mu_o J$$

throughout the coil area, where $\mu_o$ is the permeability of free space. In the air surrounding the coil we have

$$\nabla^2 p = 0$$

with core permeability assumed to be infinite.

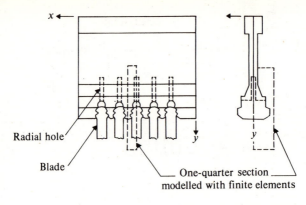

Symmetrical one-quarter section of disk text model

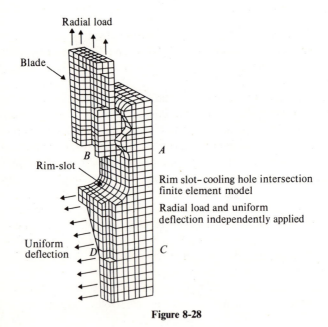

**Figure 8-28**

Because of the symmetry of the problem about both axes (see Fig. 8-31) we need to consider only the one quadrant shown in Fig. 8-32, where the regions $A_1, A_2$ are coil areas with current densities $J_1$ and $J_2$ respectively, subject to the conditions $J_1 A_1 + J_2 A_2 = 0$, since the nett current flow is zero. The region $A_3$ is the air area. Thus the problem may be posed as the solution of,

$$\nabla^2 p = 0 \qquad \text{in } A_3$$

$$\nabla^2 p = -1 \qquad \text{in } A_1$$

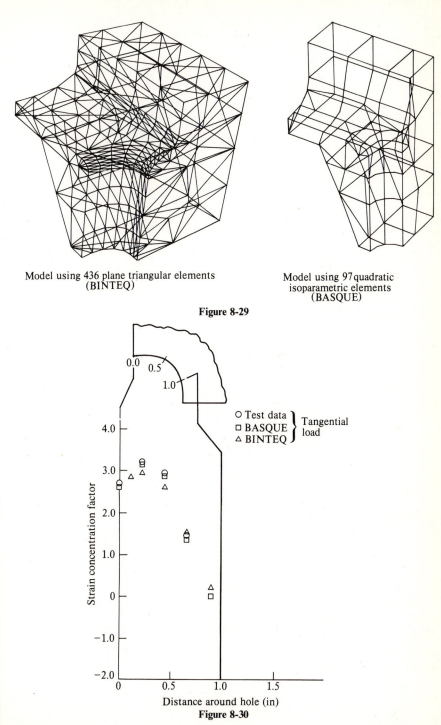

Model using 436 plane triangular elements
(BINTEQ)

Model using 97 quadratic
isoparametric elements
(BASQUE)

**Figure 8-29**

**Figure 8-30**

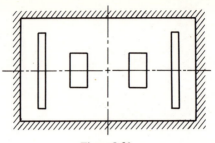

**Figure 8-31**

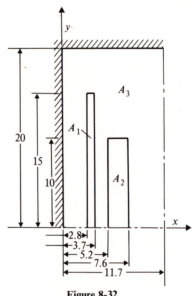

**Figure 8-32**

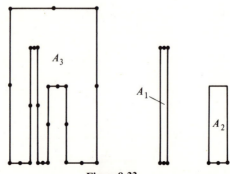

**Figure 8-33**

and $$\nabla^2 = \frac{A_1}{A_2} \quad \text{in } A_2$$

subject to the boundary condition $\partial p/\partial n = 0$ over the outer rectangular boundary. Across the interfaces between the regions the usual continuity of $p$ and $\partial p/\partial n$ must be enforced.

This problem has been solved by three different BEM discretizations. BEM-1 and BEM-2 had quadratic boundary and interface elements with 26 nodes and 38 nodes respectively, whilst the BEM-C discretization had 47 boundary and interface elements with a piecewise constant assumption imposed on them. The BEM-1 and BEM-C discretizations are shown in Figs 8-33 and 8-34 respectively; BEM-2 had 12 more nodes than BEM-1, two on each of the vertical sides of Fig. 8-33.

Table 8-1 shows a comparison between the numerical and analytical solution[18] for $x$ and $y$ components of the force $F$ (per unit length in the $Z$ direction) acting on each conductor. The components of the force may be obtained by a straightforward application of Green's theorem on the $xy$ plane to the 'force-grad $p$' relations. Thus for the coil area $A_K$ we have

$$(F_x)_K = J_K \int_{S_K} p(Q)\, n_x(Q)\, dS(Q) \qquad (F_y)_K = J_K \int_{S_K} p(Q)\, n_y(Q)\, dS(Q)$$

where $S_K$ denotes the boundary of $A_K$.

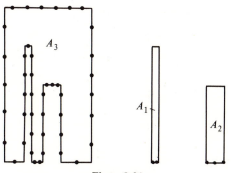

Figure 8-34

**Table 8-1  A comparison of different BEM analysis and analytical results**

|  | BEM-1 | BEM-2 | BEM-C | Analytical |
|---|---|---|---|---|
| Maximum error in $p$ at nodes, % | 13.7 | 2.4 | 4.5 | ... |
| Average error in $p$ at nodes, % | 5.5 | 0.7 | 2.0 | ... |
| $F_x$ in $A_1$, N/m | 4.83 | 4.43 | 4.46 | 4.46 |
| $F_y$ in $A_1$, N/m | 1.17 | 1.41 | 1.34 | 1.42 |
| $F_x$ in $A_2$, N/m | 5.50 | 4.84 | 4.71 | 4.76 |
| $F_y$ in $A_2$, N/m | 2.29 | 2.55 | 2.48 | 2.54 |

The values of the force $F$ for all three discretizations are in remarkably good agreement with the analytical data. Although we have not discussed the application of BEM to electromagnetic field problems specifically in this book there is a well-established literature summarized in Lean, Freidman, and Wexler.[19]

Other examples of advanced BEM applications will be found in Refs 20 to 31.


## 8-12 CONCLUDING REMARKS

We have demonstrated that, by using a parametric representation along the boundary, we can develop very elegant algorithms to cater for continuously varying geometry, boundary conditions, and body forces. Whereas fewer, more sophisticated, boundary elements would appear to be the way ahead for the ultimate commercial competitiveness of BEM it is crucial that numerical quadrature formulae be developed which are specifically tailored to handle the types of kernel function which arise in BEM. Although the examples presented in this chapter might well convince a reader that BEM are, even now, useful, powerful, and cost effective, the use of conventional gaussian quadrature formulae is perhaps the major current limitation on their efficiency.


## 8-13 REFERENCES

1. Zienkiewicz, O. C. (1971) *The Finite Element Method in Engineering Science*, McGraw-Hill, London.
2. Thompson, D'Arcy, W. (1917) *Growth and Form*, Cambridge University Press; 2nd ed., 1942.
3. Fung, Y. C. (1965) *Foundations of Solid Mechanics*, Prentice-Hall, Englewood Cliffs, N.-J.
4. McCrea, W. H. (1953) *Analytical Geometry of Three-dimensions*, Oliver and Boyd, Edinburgh and London, and Interscience, New York.
5. Ergatoudis, J. G. (1968) 'Isoparametric finite elements in two and three-dimensional stress analysis', Ph.D. thesis, University of Wales, University College, Swansea; also M.Sc. thesis, University of Wales.
6. Norrie, H. N., and Vires, G. de (1973) *The Finite Element Method*, Academic Press, London.
7. Watson, J. O. (1979) 'Advanced implementation of the boundary element method for two and three-dimensional elasto-statics', in P. K. Banerjee and R. Butterfield (eds), *Developments in Boundary Element Methods*, Chap. III, Applied Science Publishers, London.
8. Lachat, J. C. (1975) 'Further developments of the boundary integral technique for elasto-statics', Ph.D. thesis, Southampton University.
9. Lachat, J. C., and Watson, J. O. (1976) 'Effective numerical treatment of boundary integral equations: a formulation for three-dimensional elasto-statics', *Int. J. Num. Meth. in Engng*, **10**, 991–1005.
10. Cruse, T. A. (1974) 'An improved boundary integral equation method for three-dimensional elastic stress analysis', *Computers and Structs*, **4**, 741–754.
11. Rizzo, F. J., and Shippy, D. J. (1977) 'An advanced boundary integral equation method for three-dimensional thermo-elasticity', *Int. J. Num. Meth. in Engng*, **11**, 1753–1768.
12. Boissenot, J. M., Lachat, J. C., and Watson, J. O. (1974) 'Etude par equations integrals d'une éprouvette C.T.15', *Revue de Physique Appliq.*, **9**, 611–651.

13. Wilson, R. B., Potter, R. G., and Cruse, T. A. (1978) 'Calculations of three-dimensional concentrated stress fields by boundary integral method', Paper presented to a symposium at University of Connecticut, USA.
14. Wilson, R. B., Potter, R. G., and Wong, J. K. (1978) 'Boundary integral equation analysis of an advanced turbine disk rim-slot', Paper No. 14, *Proc. AGARD Conf.*, USA.
15. Cruse, T. A., and Wilson, R. B. (1978) 'Advanced applications of boundary integral equation methods', *Nucl. Engng and Des.*, **46**, 223–234.
16. Wilson, R. B., and Cruse, T. A. (1978) 'Efficient implementation of anisotropic three-dimensional boundary integral equation stress analysis', *Int. J. Num. Meth. in Engng*, **12**, 1383–1397.
17. Wu, Y. S., Rizzo, F. J., Shippy, D. J., and Wagner, J. A. (1977) 'An advanced boundary integral equation method for two-dimensional electro-magnetic field problems', *Electromech.*, **1**, 301–303.
18. Binns, K. J., and Lawrenson, P. L. (1963) *Analysis and Computation of Electric and Magnetic Field Problems*, Macmillan, New York.
19. Lean, M. H., Freidman, M., and Wexler, A. (1979). 'Advances in application of the boundary element method in electrical engineering problems', in P. K. Banerjee and R. Butterfield (eds), *Developments in Boundary Element Methods*, Chap. IX, Applied Science Publishers, London.
20. Patterson, R. E. (1963) *Stress Concentration Design Factors*, Wiley, New York.
21. Hess, J. L. (1973) 'Higher order numerical solutions of the integral equation for the two-dimensional Neumann problem', *Comp. Meth. in Appl. Mech. Engng*, **2**, 1–15.
22. Hess, J. L. (1975) 'The use of higher order surface singularity distributions to obtain improved potential flow solutions for two-dimensional lifting airfoils', *Comp. Meth. in Appl. Mech. Engng*, **5**, 11–35.
23. Hess, J. L. (1975) 'Improved solutions for potential flow about arbitrary axi-symmetric bodies by the use of a higher order surface source method', *Comp. Meth. in Appl. Mech. Engng*, **5**, 297–308.
24. Hess, J. L. (1975) 'Consistent velocity and potential expansions for higher order surface singularity method', Report MDC J6911, McDonnel Douglas Aircraft Corporation, Long Beach, California.
25. Banerjee, P. K., and Cathie, D. N. (1981) 'An advanced boundary element algorithm for two-dimensional elasticity and elasto-plasticity' (to be published).
26. Banerjee, P. K., and Davies, T. G. (1981) 'An advanced boundary element algorithm for three-dimensional elasticity and elasto-plasticity (to be published).
27. Rizzo, F. J., and Shippy, D. J. (1979) 'Recent advances of the boundary element method in thermo-elasticity', in P. K. Banerjee and R. Butterfield (eds), *Developments in Boundary Element Methods*, Chap. VI, Applied Science Publishers, London.
28. Grodtjaer, E. (1973) 'A direct integral equation method for potential flow about arbitrary bodies', *Int. J. Num. Meth. in Engng*, **6**, 253–264.
29. Nedelec, J. C. (1976) 'Curved finite element method for the solution of singular integral equations on surfaces in $R^3$', *Comp. Meth. in Appl. Mech. Engng*, **8**, 61–80.
30. Argyris, J. H., and Scharpf, D. W. (1969) 'Two and three-dimensional potential flow by the method of singularities', *Aero. J. of the Roy. Acro. Soc.*, **73**(11), 959–961.
31. Baratanow, T., and Speheit, T. (1977) 'Hydrodynamics of ice resistance, part 1: foundation of the method of analysis', Technical Report ENG 76–00354, National Science Foundarion, Grantro.

# CHAPTER NINE

# TRANSIENT POTENTIAL FLOW (DIFFUSION) PROBLEMS

## 9-1 INTRODUCTION

Previous chapters have all been concerned with 'steady state' systems—those in which neither the problem variables nor boundary conditions change with time. However, a great many problems of practical importance do involve 'transient' phenomena, the simplest of which are a large group governed by the linear 'diffusion' equation. In addition to the classical diffusion of gases and liquids the topics of most interest to the engineering analyst might be the heating and cooling of bodies, electrical and hydraulic diffusion phenomena, and the 'consolidation' of soil-like materials under load.

The major reference source for analytical solutions, Green's functions, etc., for the diffusion equation (in the terminology of heat transfer) is Carslaw and Jaeger's well-known book.[1] There is also an extensive numerical-solution literature which can be categorized by the manner in which the time-dependent terms in the equation are dealt with, irrespective of the underlying method of analysis (BEM, FEM, finite differences, etc.). The two basic methods used are either (1) a 'time-marching' process in which, step by step, the solution is evaluated at successive time intervals following an initially specified state, or (2) Laplace transformation of the time variable under which the (parabolic) diffusion equation becomes an elliptic one, resembling Poisson's equation, which can be solved in the transform space by the techniques described in Chapters 3 and 5.

Previously published work on the application of boundary integral equations to the solution of the diffusion equation,[2-5, 9-11]—of which the most ambitious by far was Tomlin,[2] who solved by IBEM problems involving the consolidation of general, anisotropic, multizone soil systems—has been restricted mainly to problems in which there were no distributed time-dependent sources throughout the body. This chapter presents an extension of earlier analyses of ours which encompassed all such effects.[6, 11]

## 9-2 GOVERNING EQUATIONS

The equation we are to solve can be written as

$$C_{ij} \frac{\partial^2 H(y, \bar{t})}{\partial y_i \partial y_j} = \frac{\partial H(y, \bar{t})}{\partial \bar{t}} - Q(y, \bar{t}) \tag{9-1}$$

where, in a heat-transfer problem for example, $H(y, \bar{t})$ would be the scalar temperature field at any point $y_i$ in a region $V$ at time $\bar{t}$, $C_{ij}$ the thermal 'diffusivity' tensor, and $Q(y, \bar{t})$ the specified time-dependent sources or sinks, distributed throughout the region. As explained previously (Sec. 5-1), Eq. (9-1) can be reduced to an equivalent isotropic equation (diffusivity $C$) if the $y_i$ axes are directed along the principal axes of $C_{ij}$ and the problem geometry suitably scaled. It is also always helpful to express the governing equation in dimensionless form which, in addition to generalizing the solution, allows the dimensionless variables to take on values that improve the conditioning of the various matrices by reducing the range of the magnitudes of their elements. In this case we would use, say, $p = H/H_o$ ($H_o$ being an arbitrary value of $H$) and divide our transformed coordinates by some reference length $L$ to produce the dimensionless coordinates $x_i$. The dimensionless time then becomes $t = C\bar{t}/L^2$. Our notation now corresponds with that used in the steady state flow chapters and (9-1) can be written as

$$\frac{\partial^2 p(x, t)}{\partial x_i \partial x_i} - \frac{\partial p(x, t)}{\partial t} = -\psi(x, t) \tag{9-2}$$

or

$$\nabla^2 p = \frac{\partial p}{\partial t} - \psi$$

The flux vector components $v_i$ are, at any point,

$$v_i(x, t) = -\frac{\partial p(x, t)}{\partial x_i} \tag{9-3}$$

and therefore the flux ($u$) across any boundary defined by its outward normal $n_i(x)$ is

$$u(x, t) = v_i n_i = -\frac{\partial p}{\partial x_i} n_i \tag{9-4}$$

In a well-posed problem the boundary conditions will be

1. $p(x, o) = f(x)$, say, specified throughout $V$, at $t = 0$.
2. $p(x_o, t) = g(x_o, t)$ specified over part $(S_1)$ of the boundary $S$ at all times, $(x_o \in S)$.
3. $u(x_o, t) = h(x_o, t)$ specified over the remainder $(S_2)$ of $S$.

Alternatively, a linear combination of $u$ and $p$, the so-called convection boundary condition, might be specified over part or all of $S$, as

$$Ap(x_o, t) + Bu(x_o, t) = q(x_o, t) \qquad (9\text{-}5)$$

Clearly, conditions 2 and 3 are special cases of (9-5).

Our problem is to develop BEM algorithms to solve (9-2) throughout $V$ at all times $t$ subject to the above boundary conditions.

## 9-3 THE FUNDAMENTAL SINGULAR SOLUTION

The fundamental solution which we require is that for the effect at $x_i$, at time $t$, of a unit point source applied at point $\xi_i$ and time $\tau$ in an infinite region. The 'instantaneous' unit point source is again a Dirac impulse function now written, *in extenso*, as $\delta(x, t; \xi, \tau)$, a notation which, although at first sight cumbersome, is worth persevering with in order to follow the roles played by the various arguments in the theoretical developments below. If we again call our free space Green's function $G(x, t; \xi, \tau)$ then, from (9-2), it has to be the solution of

$$\frac{\partial^2 G}{\partial x_i \partial x_i} - \frac{\partial G}{\partial t} = -\delta(x, t; \xi, \tau) \qquad (9\text{-}6)$$

or

$$\nabla^2 G = \frac{\partial G}{\partial t} - \delta$$

which is well known to be[1]

$$G(x, t; \xi, \tau) = \frac{\exp[-r^2/4(t-\tau)]}{[4\pi(t-\tau)]^m} \qquad (9\text{-}7)$$

where $r = z_i z_i$, $z_i = (x_i - \xi_i)$, and $2m$ is the number of spatial dimensions of the problem. We shall deal mainly with planar examples $(i = 1, 2; m = 1)$ although the analysis presented is equally valid for any range of $i$.

The 'directed' flux $(F)$ associated with $G$ is given by

$$F(x, t; \xi, \tau) = -\frac{\partial G}{\partial x_i} n_i = \frac{G}{2} \frac{z_i n_i}{(t-\tau)} \qquad (9\text{-}8)$$

Integrations of the above solution along lines and over triangular areas will be found useful; these are summarized in Sec. 9-7.

## 9-4 DIRECT BOUNDARY ELEMENT FORMULATION

Although the DBEM statement can be derived directly using Green's identity,[11, 12] it is probably more instructive to use a minor extension of the integration by parts analysis presented in Sec. 3-5 so as to clarify the new time integration operations involved.

We are to multiply (9-2) throughout by $G$ and integrate by parts, twice with respect to $x$ and once with respect to time ($\tau$). The spatial integration is exactly as before and leads, for the planar case illustrated in Sec. 3-5, to Eq. (3-28) again, which, using (9-4) and (9-8), is

$$\int_A G\nabla^2 p\, dA = \int_S (pF - Gu)\, dS + \int_A p\nabla^2 G\, dA$$

and, substituting for $\nabla^2 p$ and $\nabla^2 G$ from (9-2) and (9-6), we have

$$\int_A G\left(\frac{\partial p}{\partial t} - \psi\right) dA = \int_S (pF - Gu)\, dS + \int_A p\left(\frac{\partial G}{\partial t} - \delta\right) dA \tag{9-9}$$

One should note in this equation that terms like $\int_S pF\, dS$ are shorthand for $\int_S p(x, t)\, F(x, t;\, \xi, \tau)\, ds(x)$. We might also note the close similarity between the way that terms such as $(x - \xi)_i$ and $(t - \tau)$ appear in $G$ [Eq. (9-7)] and $F$. In fact, we could introduce additional coordinates, say $(x_4 = t, \xi_4 = \tau)$, in three dimensions and absorb the time variables simply as an increase in the dimensionality of our basic problem, providing we always bear in mind that sources applied at time $\tau$ can *only* affect events at later times (i.e., discretized events along the time axis are *not* retroactive). It is then clear that our DBEM statement will be generated by integrating (9-9) once more—now with respect to time. Reference to Fig. 9-1 will

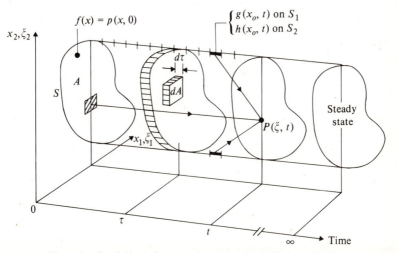

**Figure 9-1** Spatial and time domains for a plane diffusion problem.

show why, in order to sum the effects of instantaneous sources applied at times $(0 \leqslant \tau < \infty)$, we have to consider $t = \tau$ in all terms involving $p(x,\tau)$, $u(x,\tau)$ and $\partial p/\partial \tau \, (x,\tau)$ in (9-9) and then integrate with respect to $\tau$ from $\tau = 0$ to $\tau = t$.

A convenient notation to adopt for this operation is that of 'Riemann convolution integrals', which are written as[29]

$$\int_0^t \phi(x, t-\tau)\, \chi(x,\tau)\, d\tau \equiv (\phi * \chi)(x,t) \tag{9-10}$$

if $\phi$ is defined for $0 < t \leqslant \infty$ and $\phi(x, t^-) = 0$. This condition is met by $G$ and $F$ by approaching the $t$ boundary, to $(\tau = t^-)$ and $(\tau = 0^+)$, from 'inside' the region as before. The 'composition' product $(\phi * \chi)$ possesses all the group properties of commutativity, associativity, and distributivity.

Using this notation, integrating Eq. (9-9) with respect to $\tau$ leads to

$$\int_A \left\{ [Gp]_0^t - \int_0^t p \frac{\partial G}{\partial \tau}\, d\tau - (G * \psi) \right\} dA$$

$$= \int_S (F * p - G * u)\, dS + \int_0^t \int_A p \frac{\partial G}{\partial \tau}\, dA\, d\tau - \alpha p(\xi, t)$$

The final term results from the property of the delta function that

$$\int_0^t \int_A p(x,\tau)\, \delta(x, t; \, \xi, \tau)\, dA\, d\tau = \alpha p(\xi, t)$$

where $\alpha = 1$ for $\xi_i$ inside $V$ and $\alpha = 0$ if $\xi_i$ is outside $V$. Since $G(x, t; \, \xi, t^-) = 0$, as mentioned above, $p(x,0) = f(x)$, and $\partial G/\partial t = -\partial G/\partial \tau$, we have, finally,

$$\alpha p(\xi, t) = \int_S (F * p - G * u)\, dS + \int_A (G * \psi + fG)\, dA \tag{9-11}$$

where the $G$ term in the non-convoluted $(fG)$ product is $G(x, t; \, \xi, 0)$.

Equation (9-11) provides us with the potential $p(\xi, t)$, at any time $t$, due to initial sources $f(x)$, the time-dependent sources $\psi(x, t)$ throughout $V$, and all (both known and unknown) boundary potentials and fluxes over $S$. We note that the kernel functions involving $G$ and $F$ are singular when $\xi \to x$ and $\tau \to t$. Nevertheless, as discussed in Chapter 7, etc., the integrals involving them exist in the ordinary and Cauchy principal-value senses respectively. The kernel functions are also well behaved on all infinite boundaries and therefore, in contradistinction to planar steady state problems, we shall not need to introduce auxiliary terms. The initially unknown boundary data can be calculated by the now standard device of taking $\xi$ to the boundary at $\xi_o$, whence (9-11) becomes

$$\alpha p(\xi_o, t) = \frac{\omega}{4\pi} p(\xi_o, t) + \int_S (F * p - G * u)\, dS + \int_A (G * \psi + fG)\, dA \tag{9-12}$$

where $\omega$ is the solid angle enclosed by the boundary at $\xi_o$. When the boundary is smooth (i.e., has a unique tangent plane) $\omega = 2\pi$.

Equations (9-11) and (9-12) are the complete statement of the DBEM solution to the diffusion equation. The development is clearly equally applicable to a region $V$ bounded by a surface $S$ by merely increasing the spatial dimensions of $A$ and $S$ in the equations.

## 9-5 INDIRECT BOUNDARY ELEMENT FORMULATION

The IBEM formulation can be deduced from the direct one, and thereby established rigorously, by using a device originally due to Lamb.[8]

Consider a complementary region $(V^+)$ exterior to $V$ with a common boundary $S$, no internal sources $(\psi^+ = 0)$, and zero initial $p$ values (that is, $f^+ = 0$) (Fig. 9-2). If the boundary distributions of $(p, u)$ over the common surface

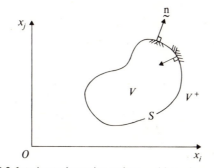

**Figure 9-2** Interior and exterior regions and boundary normals.

$(S)$ of $V^+$ are prescribed as $(p^+, u^+)$, say, then from (9-11), in $V^+$,

$$\alpha^+ p(\xi, t) = \int_S (-F * p^+ - G * u^+) \, dS \qquad (9\text{-}13)$$

where the sign has changed on $F$ since the outward normal to $S$ of $V^+$ is opposed to that of $S$ of $V$ (Fig. 9-2). Writing both (9-13) and (9-11) for any point within $V$ leads to

$$0 = \int_S (-F * p^+ - G * u^+) \, dS$$

$$\tag{9-14}$$

and $\qquad p(\xi, t) = \int_S (F * p - G * u) \, dS + \int_V (G * \psi + fG) \, dV$

We may choose $(p^+, u^+)$ such that $p = p^+$ and define $(u + u^+) = -\phi$, a new boundary source density distribution, which produces, when Eqs (9-14) are added together,

$$p(\xi, t) = \int_S (G * \phi) \, dS + \int_V (G * \psi + fG) \, dV \tag{9-15}$$

Since $G$ is symmetric in $x$ and $\xi$ we can interchange them, which means that integration is now carried out with respect to $\xi$ and $\tau$ and (9-15), when written out fully, becomes

$$p(x, t) = \int_S [G(x, t; \xi, \tau) * \phi(\xi, \tau)] \, ds$$
$$+ \int_V [G(x, t; \xi, \tau) * \psi(\xi, \tau) + f(\xi, \tau) G(x, t; \xi, 0)] \, dV \tag{9-16}$$

Differentiating (9-16) with respect to $x$ and using (9-4) we can calculate the $n$ direction flux at $x_i$ as

$$u(x, t) = \int_S (F * \phi) \, dS + \int_V (F * \psi + fF) \, dV \tag{9-17}$$

with arguments identical to (9-16).

Finally, the IBEM formulation is generated by taking $x$ to $x_o$ on $S$, from inside $V$, as

$$p(x_o, t) = \int_S (G * \phi) \, dS + \int_V (G * \psi + fG) \, dV$$
$$u(x_o, t) = \frac{\omega}{4\pi} \phi(x_o, t) + \int_S (F * \phi) \, dS + \int_V (F * \psi + fF) \, dV \tag{9-18}$$

In both the direct and indirect BEM statements the only changes from the steady state formulations are:

1. One additional dimension (time) of integration is involved [expressed here by the convolutions $(F * \phi)$, etc.].
2. There is a corresponding increased dimensionality of the singular solution.
3. Additional contributions occur on the right-hand side of the final equations generated by the initially specified values of the potential $p(x, o) = f(x)$ throughout the region.

In principle the solution now proceeds as before with the exception that the time-dependent contribution, which has converted the governing equation from elliptic to parabolic, may be treated in a number of different ways.

## 9-6 SOLUTION OF THE DIRECT AND INDIRECT BOUNDARY ELEMENT EQUATIONS

### 9-6-1 Solution by Laplace Transformation

The most powerful analytical methods for solving the diffusion equation (and, in fact, other classes of problem involving convolution integrals such as those of viscoelasticity, Chapter 10) are based on Laplace transformation of the time coordinate.[1,13] Some authors, principally Rizzo and Shippy,[5,9] have explored the use of this technique in conjunction with BEM and, although we are of the opinion that it is unlikely to lead to any substantial improvement over alternative numerical solution procedures, the main features of the method are worth mentioning.

The key attraction is that when the time coordinate is so transformed the dimensionality of the differential equation is effectively reduced by one to produce an equivalent elliptic equation. We have already established that BEM solve elliptic (steady state) problems very efficiently and it is therefore worth investigating whether or not there are advantages in combining BEM and Laplace transform techniques.

The major steps involved require:

1. *The Laplace transform of a function $p(x,t)$, which is defined as $p^*(x,s)$ where*

$$p^*(x,s) \equiv \mathscr{L}[p(x,s)] = \int_0^\infty p(x,t)\,e^{-st}\,dt \qquad (9\text{-}19)$$

By operating on the diffusion equation (9-2) and the boundary conditions (9-5) by $\mathscr{L}$ we obtain

$$\frac{\partial^2 p^*(x,s)}{\partial x_i\,\partial x_i} - sp^*(x,s) = \psi^*(x,s) \qquad (9\text{-}20a)$$

and 
$$Ap^* + Bu^* = C^* \qquad (9\text{-}20b)$$

It may well prove to be very difficult to evaluate the transform of $\psi(x,t)$ and Rizzo and Shippy only consider the $\psi = 0$ case.

2. *The fundamental solution $G^*$ of (9-20a), which is, in fact, $G^* = \mathscr{L}(G)$ with $G$ defined by (9-7), leading to*

$$G^* = \tfrac{1}{2}\pi K_o\!\left(\frac{\sqrt{s}}{r}\right) \qquad (9\text{-}21)$$

where $r^2 = (x-\xi)_i(x-\xi)_i$ and $K_o$ is a zero-order, modified, Bessel function of the second kind.

3. *A solution for $p^*$ throughout V, which can be obtained by using $G^*$ (and the associated $F^*$) in the DBEM solution of (9-20a), as*

$$p^*(\xi,s) = \int_S (p^* F^* - u^* G^*)\,dS$$

The major difficulty lies in calculating $p(\xi,t)$ from $p^*(\xi,s)$ (i.e., obtaining the inverse transform of $p^*$). There is, as yet, no universal numerical method for doing this and those in use require the analysis to be carried out for a series of real values of the parameter $(s)$; $p(\xi,t)$ can then be calculated from the sequence of $p^*(\xi,s)$ values thus obtained.[5] Since the transformed kernel functions have become more complicated and $\psi$ has been excluded from the analysis it seems to us much more promising to use the BEM statements [Eqs (9-11), (9-12), and (9-16), (9-17), (9-18)] together with incremental solutions directly in $t$ rather than $s$ (i.e., a time-marching process).

## 9-6-2 Time-marching Processes

The essence of such processes is to 'march', from $t = 0$, step by step using time increments $\Delta\tau$ to any specified time $\tau = N\Delta\tau$ in $N$ such steps. One fundamental question is immediately obvious. Are there limits on the magnitude of $\Delta\tau$, since, from the point of view of computational efficiency, a few large steps would be preferable to many small ones?

The published information on admissible $\Delta\tau$ values (dimensionless time) relates mainly to finite difference methods (Crandall[14] gives a particularly good account of the basic problem) and finite element methods for which Smith[15] provides very interesting information on the use of higher-order approximations to the essentially exponential variation of $p$ with time which allows larger time steps to be used. Using whole-body discretization schemes and the conventional explicit forward difference marching process the maximum value of the ratio $\Delta\tau/(\Delta x)^2 = \frac{1}{2}$ to ensure convergence, $\Delta x$ being the mesh or element dimension.[28] Thus, for, typically, $\Delta x = L/10$, say, $\Delta\tau \not> 0.005$. Although, as mentioned previously, this field is one in which BEM have not been studied in great detail, and Tomlin[2] used such a marching scheme successfully with $\Delta\tau = 0.04$ to 0.005 for a rectangular region and $\Delta t = 0.05$ for a hollow cylinder (see examples 1 and 2 below in Sec. 9-8) and Shaw[16] used $\Delta\tau = 0.05$ to 0.002 in a study of the transient cooling of a plane region.

However, an examination of Eqs (9-12) and (9-18) will show that the BEM formulations are implicit in time (i.e., the various quantities at time $t = N\Delta\tau$ are calculated from the boundary integrals involving the known and unknown boundary values and known sources in the interior up to time $t$). Therefore, if in the discretized system we do not introduce gross approximations in the time domain the criteria governing stability should be less stringent than that discussed above.

Two basically different time-marching processes can be used, both of which lead to systems of equations that are simultaneous in space but successive in time.

**Method 1** In both the direct and indirect formulations established above the treatment of the time variable is virtually identical to that of the spatial variables. Therefore, as depicted in Fig. 9-1, it should be possible to treat a planar diffusion problem as one in 'three dimensions' (the third being time) and proceed directly to a solution at time $t$. The time axis 'boundaries' would have been discretized into elements but these could, presumably, become larger (possibly logarithmically[17]) with time as the steady state solution was approached. The basic boundary element philosophy would lead us to discretize the whole spatial and temporal boundary (Fig. 9-1) and determine all the unknown boundary information, from which any $p(\xi, t)$, etc., could be calculated, bearing in mind that sources introduced at $\tau > t$ have no retroactive effect at $\tau \leqslant t$.

We shall demonstrate all the important features of this algorithm by solving a general one-dimensional problem using DBEM. Consider the $(x, t)$ plane of Fig. 9-3 and a uniform one-dimensional field ($L = 1$) extending from $x = 0$ to $x = 1$ with $f(x, 0)$ the initially specified potential along it. The other boundaries of our

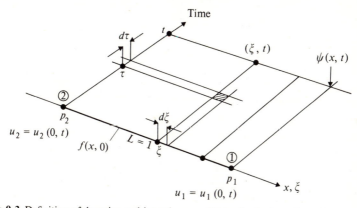

**Figure 9-3** Definition of domains and boundary conditions for the 'one-dimensional' example.

problem are lines parallel to the time axis and, as an example, we shall consider the potentials $(p_1, p_2)$ to be constant and specified along them. We realize immediately that the boundary fluxes $(u_1, u_2)$ are thereby both (unknown) functions of time. By analogy with Eq. (2-24) the DBEM statement (9-11) now becomes (noting that, in one dimension, $u = v$)

$$p(\xi, t) = -[G * u - F * p]_0^L + \int_0^L (fG + G * \psi)\, dx \qquad (9\text{-}22)$$

and, from (9-7), (9-8),

$$G = \frac{\exp[-r^2/4(t - \tau)]}{2\sqrt{\pi}\,(t - \tau)^{1/2}} \qquad F = \frac{G}{2(t - \tau)} r\, \text{sgn}\, r \qquad r = x - \xi \qquad (9\text{-}23)$$

Since $(p_1, p_2)$ are constant we can evaluate the $(F * p_1)$ and $(F * p_2)$ convolutions analytically as

$$(F * p_1) = p_1 r\, \text{sgn}\, r \int_0^t \frac{\exp[-r^2/4(t - \tau)]}{4\sqrt{\pi}\,(t - \tau)^{3/2}}\, d\tau = \frac{p_1}{2} (\text{sgn}\, r)\, \text{erf}\, c \left| \frac{r}{2\sqrt{t}} \right|$$

$$(9.23a)$$

where $\text{erf}\, c(z) = 1 - \text{erf}(z)$ and erf is the error function defined by

$$\text{erf}(z) = \frac{2}{\sqrt{\pi}} \int_0^z e^{-\tau^2}\, d\tau$$

If $u_1 =$ constant, say, had been specified in lieu of $p_1$ then convolutions such as $(G * u_1)$ can be evaluated analytically as

$$(G * u_1) = u_1 \int_0^t \frac{\exp\left[-r^2/4(t-\tau)\right]}{2\sqrt{\pi}\,(t-\tau)^{1/2}} = u_1\left[\sqrt{\frac{t}{\pi}}\exp\left(-\frac{r^2}{4t}\right) - \frac{|r|}{2}\left(\mathrm{erf}\,c\left|\frac{r}{2\sqrt{t}}\right|\right)\right]$$

(9-23b)

For convenience consider $\psi = 0$ and $f =$ constant $= \theta$, say, when the final term in (9-22) becomes, using $G(x, t\,|\,\xi, 0)$ from (9-23),

$$\int_0^L (\theta G)\,dx = \int_{-\xi}^{(L-\xi)} \theta \frac{\exp\left[-r^2/4t\right]}{2\sqrt{\pi}\,t^{1/2}}\,dr = 2\theta\left[\mathrm{erf}\left(\frac{L-\xi}{2\sqrt{t}}\right) + \mathrm{erf}\left(\frac{\xi}{2\sqrt{t}}\right)\right]$$

(9-24)

Then, from (9-22), (9-23a), (9-24) we have, taking $\xi$ to the boundaries at $(L^-, 0^+)$ in turn,

$$\begin{Bmatrix} p(L^-, t) \\ p(0, t) \end{Bmatrix} = \begin{Bmatrix} p_1 \\ p_2 \end{Bmatrix} = \begin{bmatrix} -G_1 & G_2 \\ -G_2 & G_1 \end{bmatrix} * \begin{Bmatrix} u_1 \\ u_2 \end{Bmatrix}$$

$$+ \begin{bmatrix} \frac{1}{2} & \frac{1}{2}\mathrm{erf}(0.5L/\sqrt{t}) \\ \mathrm{symm} & \frac{1}{2} \end{bmatrix} \begin{Bmatrix} p_1 \\ p_2 \end{Bmatrix} + 2\theta\begin{Bmatrix} 1 \\ 1 \end{Bmatrix}\mathrm{erf}\left(\frac{L}{2\sqrt{t}}\right)$$

(9-25)

where $G_1 = G(L, t;\, L^-, \tau)$ and $G_2 = G(0, t;\, L^-, \tau)$ which, from the symmetry of $G$, are respectively identical to $G(0, t;\, 0^+, \tau)$ and $G(L, t;\, 0^+, \tau)$.

We can now see the vital point that, whereas the final terms are similar to those in steady state problems, including the $(\frac{1}{2})$ coefficients multiplying $(p_1, p_2)$, the $(G, u)$ products are convolutions. Consequently we have to discretize the time axis, as expected, so that each of the $(G * u)$ terms can be expressed as a matrix product of, say, $n$ discrete boundary element subdivisions of both $u_1$ and $u_2$ within the time interval $(0, t)$ of interest.

At the $N$th time interval we can write the $(G * u)$ terms as

$$\int_0^{t=N\Delta\tau} G(x, t;\, \xi, \tau)\,u(x, \tau)\,d\tau = \int_0^{(N-1)\Delta\tau} G(x, t;\, \xi, \tau)\,u(x, \tau)\,d\tau$$

$$+ \int_{(N-1)\Delta\tau}^{N\Delta\tau} G(x, t;\, \xi, \tau)\,u(x, \tau)\,d\tau \qquad (9\text{-}26)$$

in which the first integral on the right-hand side involves the known information from the solution involving the $(N-1)$th time step. Therefore, if we assume that $(u_1, u_2)$ remain constant over the time step $\Delta\tau$ we can then express (9-25) as

$$\left(\int_{(N-1)\Delta\tau}^{N\Delta\tau} \begin{bmatrix} G_1 & -G_2 \\ G_2 & -G_1 \end{bmatrix} d\tau\right) \begin{Bmatrix} u_1 \\ u_2 \end{Bmatrix} = \int_0^{(N-1)\Delta\tau} \left(\begin{bmatrix} -G_1 & G_2 \\ -G_2 & G_1 \end{bmatrix} \begin{Bmatrix} u_1 \\ u_2 \end{Bmatrix}\right) d\tau$$

$$+ \begin{bmatrix} -\frac{1}{2} & \frac{1}{2}\mathrm{erf}\,c(0.5L/\sqrt{t}) \\ \mathrm{symm} & -\frac{1}{2} \end{bmatrix} \begin{Bmatrix} p_1 \\ p_2 \end{Bmatrix} + 2\theta\begin{Bmatrix} 1 \\ 1 \end{Bmatrix}\mathrm{erf}\left(\frac{L}{2\sqrt{t}}\right) \qquad (9\text{-}27)$$

Equation (9-27) can now be solved for $(u_1, u_2)$ at time $t$ since the right-hand side involves only known quantities. It is important to note also that having obtained the solution for the $N$th time step ($N = 1, 2, ...$), the solution for the $(N + 1)$th step merely requires the right-hand side of (9-27) to be supplemented with new terms involving the convolution product $(F * p)$ and $(G * u)$ over the time intervals $\{N\Delta\tau, (N + 1)\Delta\tau\}$ and $\{(N - 1)\Delta\tau, N\Delta\tau\}$ respectively.

The time-marching process described above is probably the most efficient for use with either direct or indirect BEM and has been described by Shaw[16] and others. However, for some problems, where the diffusion is coupled with elasticity equations such as in transient thermoelasticity and consolidation, the above procedure is not sufficiently general and the method outlined below becomes preferable.

**Method 2** It will be shown in Chapter 12 that non-linearities in the governing differential equation can be handled in a BEM formulation by modifying the value of the source term $Q$ in Eq. (9-11). Thus in a completely general diffusion algorithm, where both the boundary conditions and the internal sources may vary with time and, in addition, are only obtainable by solving a coupled set of differential equations (as in consolidation or thermoelasticity), the following time-marching process is more advantageous.

For simplicity of presentation $p$ and $u$ values will be assumed to remain constant throughout any $\Delta\tau$ time step and $Q(x, \tau)$ will be represented by its average value over each step. The values of $p$ and $u$ over any boundary element can be represented as $p(x, \tau) = \mathbf{N}(x)\mathbf{p}$ and $u(x, \tau) = \mathbf{N}(x)\mathbf{u}$ where $\mathbf{N}(x)$ are isoparametric shape functions and $\mathbf{p}$ and $\mathbf{u}$ are vectors of nodal $(p, u)$ values. Clearly some form of shape function could also be introduced over the $\Delta\tau$ elements along the time axis. This has not yet been explored, as far as we know, but Smith's work,[15] which concerns Padé and Nørsett approximations to exponential functions, could well be very relevant here.[26]

However, we now proceed as before and write, from (9-12), for the $p$th boundary node on a smooth boundary,

$$\tfrac{1}{2}p(\xi^p, t) = \sum_{q=1}^{n} (\mathbf{F}^{pq}\mathbf{p}^q - \mathbf{G}^{pq}\mathbf{u}^q) + \sum_{l=1}^{m} (\mathbf{C}^{pl}\mathbf{\Psi}^l + \mathbf{D}^{pl}\mathbf{f}^l) \tag{9-28}$$

where

$$\mathbf{F}^{pq} = \int_0^{\Delta\tau} \int_{\Delta S_q} F(x^q, t; \xi^p, \tau)\mathbf{N}(x^q)\, dS_q$$

$$\mathbf{G}^{pq} = \int_0^{\Delta\tau} d\tau \int_{\Delta S_q} G(x^q, t; \xi^p, \tau)\mathbf{N}(x^q)\, dS_q$$

$$\mathbf{C}^{pl} = \int_0^{\Delta\tau} d\tau \int_{\Delta V_l} G(x^l, t; \xi^p, \tau)\mathbf{M}(x^l)\, dV_l$$

$$\mathbf{D}^{pl} = \int_{\Delta V_l} G(x^l, t; \xi^p, 0)\mathbf{M}(x^l)\, dV_l$$

$$\tag{9-29}$$

where $\mathbf{M}(x^l)$ is the shape function for the $l$th cell. For the 'constancy over $\Delta\tau$' assumptions used for $(p, u)$ the first two equations of (9-26) can be evaluated using (9-21a, b). If, additionally, $(p, u)$ were assumed constant over $\Delta S$, then the $\mathbf{F}^{pq}$ and $\mathbf{G}^{pq}$ components would be those generated by a line source over $(\Delta\tau, \Delta S)$.[2] Similarly, for a planar body divided into triangular cells, the $\mathbf{C}^{p1}$ and $\mathbf{D}^{p1}$ components can be evaluated analytically.[2] All these results are summarized in Sec. 9-7.

By writing (9-28) at each boundary node for a *particular* instant of time and absorbing the $\alpha p(\xi^p, t)$ coefficients into the diagonal elements of $\mathbf{F}^{pq}$, we can obtain the following matrix equation at $t = \tau$, say:

$$[G]\{u\} - [F]\{p\} = [C]\{\psi\} + [D]\{f\} \tag{9-30}$$

**The indirect method** In the indirect method we assume a linear variation of $\phi$ over the boundary elements, i.e.,

$$\phi(\xi, \tau) = \mathbf{N}(\xi)\boldsymbol{\phi}$$

where $\boldsymbol{\phi}$ are the nodal values of $\phi$, and use linear variations of $Q$ and $f$ as before.

By substituting in Eqs (9-17) and (9-18) for the $p$th boundary element we have

$$p(x^p, t) = \sum_{q=1}^{n} \mathbf{A}^{pq}\boldsymbol{\phi} + \sum_{l=1}^{m} (\mathbf{C}^{pl}\mathbf{Q}^l + \mathbf{D}^{pl}\mathbf{f}^l) \tag{9-31}$$

and

$$u(x^p, t) = \sum_{q=1}^{n} \mathbf{B}^{pq}\phi^q + \sum_{l=1}^{m} (\mathbf{E}^{pl}\mathbf{Q}^l + \mathbf{F}^{pl}\mathbf{f}^l) + \frac{\omega}{4\pi}\phi(x^p, t) \tag{9-32}$$

where

$$\mathbf{A}^{pq} = \int_0^{\Delta\tau} d\tau \int_{\Delta S_q} G(x^p, t; \xi^q, \tau)\,\mathbf{N}(\xi)\,dS_q$$

$$\mathbf{B}^{pq} = \int_0^{\Delta\tau} d\tau \int_{\Delta S_q} F(x^p, t; \xi^q, \tau)\,\mathbf{N}(\xi)\,dS_q$$

$$\mathbf{C}^{pl} = \int_0^{\Delta\tau} d\tau \int_{\Delta V_l} G(x^p, t; \xi^l, \tau)\,\mathbf{M}(\xi)\,dV_l$$

$$\mathbf{D}^{pl} = \int_{\Delta V_l} G(x^p, t; \xi^l, 0)\,\mathbf{M}(\xi)\,dV_l \tag{9-33}$$

$$\mathbf{E}^{pl} = \int_0^{\Delta\tau} d\tau \int_{\Delta V_l} F(x^p, t; \xi^l, \tau)\,\mathbf{M}(\xi)\,dV_l$$

$$\mathbf{F}^{pl} = \int_{\Delta V_l} F(x^p, t; \xi^l, 0)\,\mathbf{M}(\xi)\,dV$$

We can assemble either (9-31), or (9-32), to match the particular boundary conditions specified at each nodal point on the boundary to obtain the following system of equations:

$$\{b\} = [X]\{\phi\} + [Y]\{\bar{Q}\} + [Z]\{f\}$$

or $$[X]\{\phi\} = \{b\} - [Y]\{\bar{Q}\} - [Z]\{f\} \qquad (9\text{-}34)$$

As usual the right-hand side of (9-30) will be known together with half of the $(u, p)$ boundary values <u>at $t = 0$</u>. Solution of this equation will supply the missing boundary values which can then be used in the matrix equivalent of Eq. (9-11) to calculate $p(\xi, \Delta\tau)$ at a sufficient number of nodes to define $f(x, \Delta\tau)$. $\psi(x, \Delta\tau)$ will be known and therefore these $(f, \psi)$ values can be inserted into (9-30) once more to yield (using matrix multiplication operations only) the non-specified boundary information at $t = 2\Delta\tau$. Repeated application of this process (in DBEM or IBEM) will generate a minimum set of $p(\xi, t)$ values up to, say, $t = N\Delta\tau$ and, if required, very detailed information at $t$. Referring to Fig. 9-3, the $(x, t)$ plane is once more 'covered' with solutions.[27]

A major difficulty with all marching processes of this kind is that the solutions obtained for the first time steps are likely to be in error by around 7 to 10 per cent (see below) unless the initial time intervals are extremely small. This arises in part from the 'forward difference' approximation of using, for example, the nominal $(p, u)$ values at $t = 0$ to calculate $f(x, \Delta\tau)$, etc., and partly from the inherent approximation in the diffusion equation as a mathematical description of the response of a system to a step change of input.[7] Apart from Smith's work[15] on one-dimensional problems, cited above, the only other promising improvement in the precision of initial output appears to have been made by Tomlin[2, 6] who introduced the idea of image sources sketched in Fig. 9-4 for use in conjunction with IBEM. In order to satisfy the $t = 0$ boundary conditions over $S$ (Fig. 9-4a) he introduced extra cells ($l'$) external to $S$ which were a mirror image of the adjacent internal cell ($l$). If the source distributions over $l$ and $l'$ are identical then an initial zero flux boundary condition will be achieved (otherwise the initial potential gradient perpendicular to any side of an open triangular cell is infinite!—see Sec. 9-7). If the $l'$ sources are the negative of those over $l$ then a zero boundary potential is being imposed, etc. (He used straight boundary elements and triangular cells with either constant or linearly varying sources distributed over them to solve a variety of problems in two dimensions, including zoned, anisotropic media.) As noted in the following examples in Sec. 9-8, he obtained much improved precision in this way at the early stages of the diffusion process incorporating instantaneous triangular sources [that is, Eq. (9-7) integrated over a triangular cell] to model the distributed internal sources ($Q$) and continuous 'line' sources [that is, Eq. (9-7) integrated over both a line element and time] on his boundary elements. This latter device meant that at each time step his fictitious potentials were merely augmented by $\Delta\phi$ (Fig. 9-4b) rather than impulsively applied from zero each time. Although all of these innovations were justified on purely physical grounds they most certainly worked. Tomlin's IBEM algorithm, using an implicit time-marching procedure, is explained in greater detail in Ref. 6.

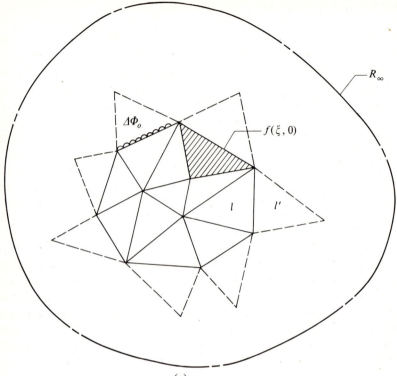

(a)

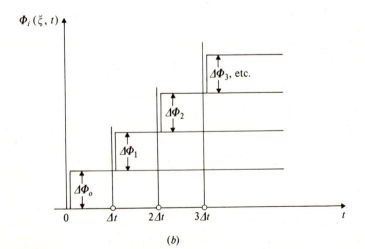

(b)

**Figure 9-4** (a) Discretization scheme for plane diffusion problems. (b) Time sequence of fictitious source increment applications.

## 9-7 EVALUATION OF THE INTEGRALS

In order to construct the matrices for the final system of equations (9-30) and (9-31) it is necessary to evaluate accurately the integrals involved. Due to the introduction of the spatial as well as time integrations it may appear at first glance that the arithmetic calculations involved are prodigious. However, a closer examination of these will reveal that:

1. Sources applied at time $\tau$ can *only* affect events at later times (i.e., discretized events along the time axis are *not* retroactive).
2. The potential and flow velocity at any point are unaffected by a source generated at any other point at the same instant of time. Therefore, if the time increments are kept small, the effect of an element $i$ of a source applied on an element $j$ during the time interval $\Delta T$ is negligible when the elements $i$ and $j$ are at some distance apart.

Nevertheless, it is necessary to evaluate a number of complicated integrals. For the vast majority of these calculations $\xi \neq x$ and $t \neq \tau$; therefore they involve only smooth functions and can be evaluated using ordinary gaussian integration formulae. For singular cases, (i.e., when the load point and the field point coincide over a boundary element or over a volume cell) the integrals need to be evaluated analytically. By using the following results[2] it is possible to calculate all the contributions due to the singular integrals.

The shape functions in the various integrals can be split into a constant part and a variable part containing the integration variable. The constant part can be integrated analytically and the variable part calculated numerically with respect to the spacial variables and analytically with respect to time.

All the following results are developed[2] by integrating the basic two-dimensional free-space Green's function [cf. Eq. (9-7)]. The variables used are not dimensionless, $c = $ diffusivity, and some standard mathematical functions have been used.[2, 18]

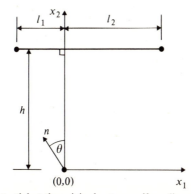

**Figure 9-5** Potential at the origin due to a uniform line segment source.

By constructing a local coordinate system through the field point as shown in Fig. 9-5 the effects of source of strength $q_o$ can be obtained by integrating Eq. (9-7) as

$$G = \int_0^t d\tau \int_{\Delta S} G(0, t; \xi, \tau) q_0 \, dS$$

$$= \frac{q_0}{2\pi \sqrt{ct}} \left\{ L_1 E_1(R_1^2) + L_2 E_1(R_2^2) - 2P\left[ \text{erf} \, xpc\left(\frac{L_1}{P}, P\right) + \text{erf} \, xpc\left(\frac{L_2}{P}, P\right)\right] \right.$$

$$\left. + \sqrt{\pi} \, e^{-P^2}(\text{erf} \, L_1 + \text{erf} \, L_2) \right\} \tag{9-35}$$

where

$$P = \frac{h}{2\sqrt{ct}}$$

$$L_1 = \frac{l_1}{2\sqrt{ct}}$$

$$L_2 = \frac{l_2}{2\sqrt{ct}}$$

$$R_1^2 = P^2 + L_1^2$$

$$R_2 = P^2 + L_2^2$$

erf is called the error function and is defined as

$$\text{erf}(x) = \frac{2}{\pi} \int_0^x e^{-\lambda^2} \, d\lambda$$

$$= \frac{2}{\pi}\left( x - \frac{x^3}{1! \, 3} + \frac{x^5}{2! \, 5} - \frac{x^7}{3! \, 7} + \cdots \right)$$

$$\text{erf} \, xpc(a, b) = \tan^{-1} a - \text{erf} \, xp(a, b)$$

$$\text{erf} \, xp(a, b) = \tan^{-1} a - e^{-b^2}\left\{ \tan^{-1} a + \frac{b^2}{1!}(\tan^{-1} a - a) \right.$$

$$\left. + \frac{b^4}{2!}\left[ \tan^{-1} a - \left(a - \frac{a^3}{3}\right)\right] + \cdots \right\} \qquad \text{for } |a| < 1$$

$$\text{erf} \, xp(a, b) = \frac{a}{|a|}\frac{\pi}{4}(\text{erf} \, b)^2 \qquad \qquad \text{for } |a| = 1$$

$$\text{erf} \, xp(a, b) = \frac{\pi}{2}\text{erf}(b)\text{erf}(ab) - \cot^{-1} a + e^{-a^2 b^2}$$

$$\left\{ \cot^{-1} a + \frac{a^2 b^2}{1!}\left(\cot^{-1} a - \frac{1}{a}\right) + \frac{a^4 b^4}{2!}\left[ \cot^{-1} a - \left(\frac{1}{a} - \frac{1}{3a}\right)\right] + \cdots \right\}$$

$$\text{for } |a| > 1$$

and $\qquad E_1(x) = -0.5772 - \ln x + \dfrac{x}{1! \, 1} - \dfrac{x^2}{2! \, 2} + \dfrac{x^3}{3! \, 3} - \cdots \qquad$ for $x > 0$

The corresponding $F = \partial G / \partial n$ can be obtained directly from (9-35) as

$$F = \frac{q_0}{2\pi t} \left\{ \left[ \operatorname{erf} xpc\left( \frac{L_1}{P}, P \right) + \operatorname{erf} xpc\left( \frac{L_2}{P}, P \right) \right] \cos\theta - \tfrac{1}{2}[E_1(R_1^2) - E_1(R_2^2)] \sin\theta \right\}$$

(9-36)

where $\theta$ is the angle between the normal to the boundary element and the direction $n$ as shown in Fig. 9.5.

In order to take account of the initial condition over the volume it is necessary to integrate the fundamental solutions over a triangular cell. If we define an instant triangular source by the vertices $(-l_1, h)$, $(l_2, h)$, and $(0, 0)$, as shown in Fig. 9-6, we can integrate Eq. (9-7) and obtain the potential at the field

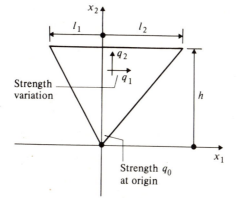

**Figure 9-6** Instant triangular source: vertex at the origin.

point $(0, 0)$ at a time $t$ due to a source varying linearly over the triangle as $q_0 + q_1 \xi_1 + q_2 \xi_2$ (where $q_0$, $q_1$, and $q_2$ are constants) applied at time $\tau = 0$. Thus

$$G(0, t) = \frac{q_0}{2\pi} \left[ \operatorname{erf} xp\left( \frac{L_1}{P}, P \right) + \operatorname{erf} xp\left( \frac{L_2}{P}, P \right) \right]$$

$$+ \frac{1}{2} \sqrt{\frac{ct}{\pi}} \left( \frac{q_1 P + q_2 L_1}{R_1} \operatorname{erf} R_1 - \frac{q_1 P - q_2 L_2}{R_2} \operatorname{erf} R_2 \right)$$

$$- \frac{q_2}{2} \sqrt{\frac{ct}{\pi}} \, e^{-P^2}(\operatorname{erf} L_1 + \operatorname{erf} L_2)$$

(9-37)

For the more general case of a field point lying in the interior of a triangle the potential $G$ can be obtained by summing the effects of the three subtriangles

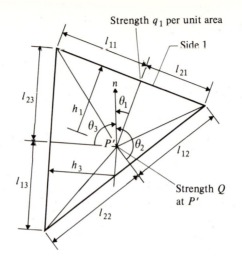

$$L_{1r} = l_{1r}/2\sqrt{ct}$$
$$L_{2r} = l_{2r}/2\sqrt{ct}$$
$$P_r = h_r/2\sqrt{ct} \quad\Bigg\} \quad r = 1, 2, 3$$
$$R_{1r}^2 = L_{1r}^2 + H_r^2$$
$$R_{2r}^2 = L_{2r}^2 + H_r^2$$

**Figure 9-7** Instant triangular source: linearly varying strength.

meeting at the point. By using Eq. (9-37) and the notation of Fig. 9-7 we have

$$G(P', t) = \frac{Q}{2\pi} \sum_{r=1}^{3} \left[ \operatorname{erf} xp\left(\frac{L_{1r}}{P_r}, P_r\right) + \operatorname{erf} xp\left(\frac{L_{2r}}{P_r}, P_r\right) \right]$$

$$- \frac{\sqrt{ct}}{2\sqrt{\pi}} \sum_{r=1}^{3} q_r \exp(-P_r^2)(\operatorname{erf} L_{1r} + \operatorname{erf} L_{2r}) \quad (9\text{-}38)$$

where $L_{1r}, P_r$, etc., denote the quantities $L_1, P$, etc., for the $r$th triangle; $q_r$ denotes the gradients of the source in the direction normal to the outer side of the triangle $r$; and $Q$ denotes the strength of the source at the field point $P'$. The normal velocity at $P'$ due to this triangular source is given by

$$F(P', t) = -\frac{1}{4\sqrt{\pi ct}} \sum_{r=1}^{3} Q_r \exp(-P_r^2)(\operatorname{erf} L_{1r} + \operatorname{erf} L_{2r}) \cos \theta_r$$

$$+ \frac{q_n}{2\pi} \sum_{r=1}^{3} \left[ \operatorname{erf} xp\left(\frac{L_{1r}}{P_r}, P_r\right) + \operatorname{erf} xp\left(\frac{L_{2r}}{P_r}, P_r\right) \right]$$

$$+ \frac{1}{2\pi} \sum_{r=1}^{3} q_r[\exp(-R_{1r}^2) - \exp(-R_{2r}^2)] \sin \theta_r, \quad (9\text{-}39)$$

where $Q_r = Q + q_r h_r$ (i.e., source density at a point represented by the projection of $P'$ on the side $r$) and $q_n$ is the source density gradient in the direction $n$.

For an instant (i.e. applied at time $\tau = 0$) line segment source of linearly varying strength $q = q_0 + q_1 x_1$, the potential at the origin (see Fig. 9-5) is given by

$$G(0, t) = \frac{q_0}{4\sqrt{\pi c t}} e^{-P^2}(\operatorname{erf} L_1 + \operatorname{erf} L_2) + \frac{q_1}{2\pi}[\exp(-R_1^2) + \exp(-R_2^2)] \qquad (9\text{-}40)$$

By replacing the time variable $t$ by $(t-\tau)$ and integrating with respect to $\tau$ expressions can be developed, similar to (9-35) and (9-36), for cases with linear variation of source strength over the elements.

## 9-8 TYPICAL APPLICATIONS

The following four examples are taken from Tomlin.[2] The first (Fig. 9-8) is self-explanatory and illustrates the order of the errors to be expected in a simple, 'one-dimensional' case, with and without the use of exterior image sources in IBEM, when different time increments $\Delta\tau$ are used. The salient points of the solution are that, without image sources, $\Delta\tau = 0.002$ and $0.005$ produce maximum solution errors of around 3 and 5 per cent respectively at both $t = 0.05$ and $0.1$ which have completely disappeared before $t = 0.2$. When 'images' are incorporated even at $t = 0.05$ with $\Delta\tau = 0.025$ the maximum error is about 1 per cent (whereas for this value of $\Delta\tau$ without images the errors are 13 and 10 per cent at $t = 0.05$ and $0.1$).

The second example (Fig. 9-9) shows excellent agreement between numerical and analytical[19] solutions for an axisymmetric diffusion problem. By using two fictitious (inner and outer) ring sources together with an initial, instantaneous disc source over the annulus the problem matrices are only of the order $(2 \times 2)$ although their elements involve Bessel functions (see Ref. 2).

A further test problem is illustrated in Fig. 9-10 which traces the increase in temperature with time throughout a uniform rectangular area heated along one boundary. Agreement between the IBEM results and the analytical solution[30] is again very close, with maximum errors on the boundaries of no more than 2 to 3 per cent rising to around 7 per cent near the singular, lower left-hand corner. External image sources were used but only five and six boundary elements along the shorter and longer rectangle sides, together with two instant triangular sources (formed by the diagonal) within the rectangle ($\Delta\tau = 0.02$). The same problem was also analysed[11] using the DBEM algorithm introduced in this chapter with only four elements per rectangular side but more, four, internal triangular cells. For $t > 0.04$ the results obtained were indistinguishable from the analytical ones.

Finally, Fig. 9-11 is included to illustrate the dissipation of a plausible distribution of excess pore-water pressure under a building foundation on a water-saturated clay. The main point of interest here is that the planar body comprises four contiguous zones of different anisotropic material throughout which the water pressures are diffusing to a 'drained' ground surface. The

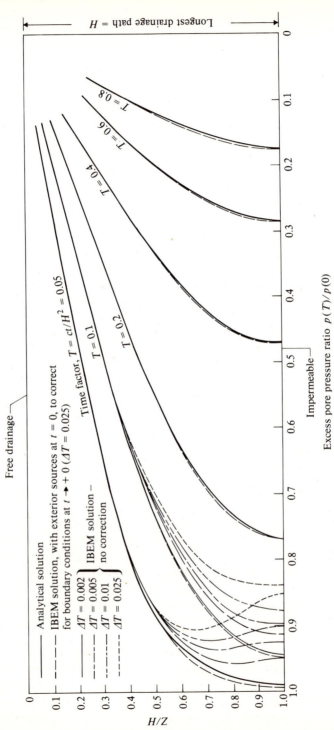

**Figure 9-8** One-dimensional consolidation (diffusion); effect of exterior sources.

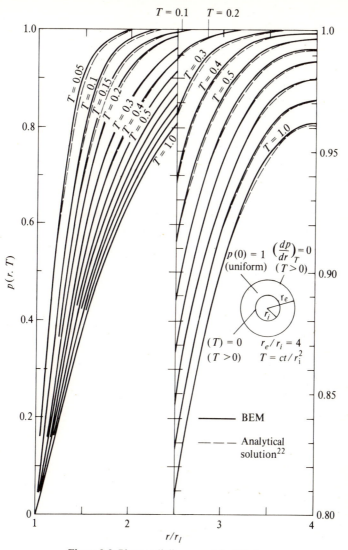

**Figure 9-9** Plane radially symmetric diffusion.

equipotential isochrones have been interpolated from the computer output shown at selected points.

Many engineering problems relate to the melting and solidification of materials. Ice formation and thawing problems arise in environmental engineering and ingot solidification and scrap melting are important metallurgical processes. In these, so-called, moving boundary problems the time-dependent location of the boundary is a major part of the solution. Since BEM are primarily concerned with boundaries they are potentially very efficient tools for solving

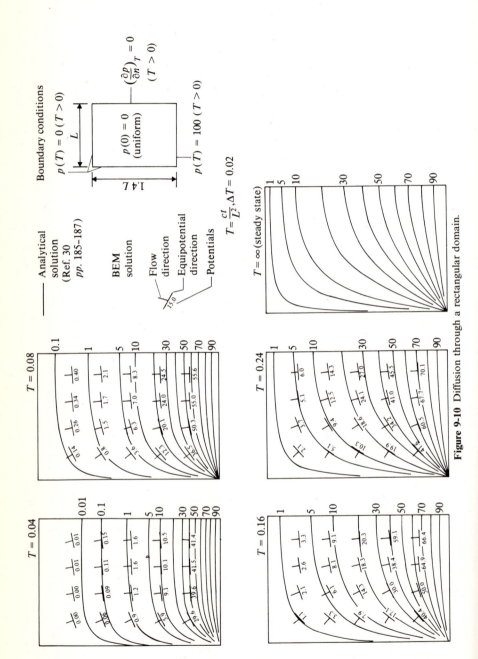

**Figure 9-10** Diffusion through a rectangular domain.

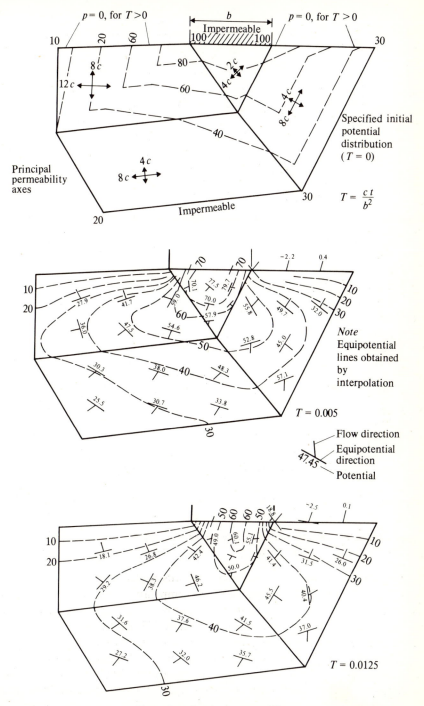

**Figure 9-11** Caption on page 240.

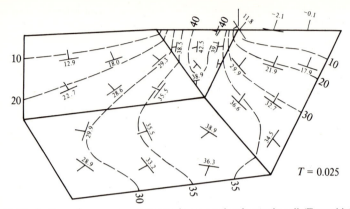

**Figure 9-11** Dissipation of excess pore pressure in a zoned anisotropic soil (Terzaghi–Rendulic theory).

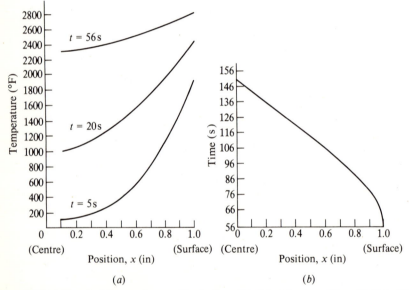

**Figure 9-12** (*a*) Transient temperature profiles in the solid slab, prior to melting. (*b*) Plot of the position of the melt line against time.

such problems. Figure 9-12[20] shows some results related to the melting of a steel slab. The slab, initially ($t = 0$) at a uniform temperature was plunged into well-agitated pure molten iron at a temperature above its melting point. Figure 9-12*a* shows the temperature distribution in the bar at different times prior to melting and Fig. 9-12*b* the time-dependent position of the melt line.

The first application of BEM in this field was by Chuang and Szekely[20, 21] who obtained the solution described above using a DBEM algorithm which they extended to cover the problem in which the melting temperature was mass transfer dependent (carbon diffusion).

Details of BEM algorithms applied to diffusion and other related problems will be found in Refs 2 to 6, 9 to 11, 16, 23 to 25 and 31.

## 9-9 CONCLUDING REMARKS

In this chapter we have presented new DBEM and IBEM algorithms which provide solutions for classical diffusion problems in any number of space dimensions. One rather satisfying result is that, by using the notation for a Riemann convolution integral, the final equations are only minimally different in form from those for steady state flow problems; compare Eqs (9-11) [and (9-16), (9-17)] with (3-30) [and (3-7), (3-8)].

Transient problems of this kind have had less attention from BEM analysts than many other fields and questions concerning optimum values of time steps and precision at early times in the diffusion process are still to be answered satisfactorily. Nevertheless, it is quite clear that, whereas Laplace transform methods do not look at all promising, successful time-marching algorithms have been developed and tested on very general diffusion problems.

## 9-10 REFERENCES

1. Carslaw, H. S., and Jaeger, J. C. (1959) *Conduction of Heat in Solids*, 2nd ed., Oxford University Press.
2. Tomlin, G. R. (1973) 'Numerical analysis of continuum problems in zoned anisotropic media', Ph.D. thesis, Southampton University.
3. Cruse, T. A. (1973) 'Application of the boundary integral equation method for solid mechanics', *Proc. Variational Meth. in Engng*, Southampton University Press.
4. Chang, Y. P., Kang, C. S., and Chen, D. J. (1973) 'The use of fundamental Green functions for solution of problems of heat conduction in anisotropic media', *Int. J. Heat and Mass Transfer*, **16**, 1905–1918.
5. Rizzo, F. J., and Shippey, D. J. (1971) 'A method of solution for certain problems of transient heat conduction', *AIAA J.*, **8**, 2004.
6. Banerjee, P. K., and Butterfield, R. (1976) 'Boundary element methods in geomechanics', in G. Gudehus (ed.), *Finite Elements in Geomechanics*, Chap. 16, Wiley, London.
7. Morse, P., and Feshbach, H. (1953) *Methods of Theoretical Physics*, Vol. II, McGraw-Hill, New York.
8. Lamb, H. (1932) *Hydrodynamics*, Cambridge University Press.
9. Shippy, D. J. (1975) 'Application of the boundary integral equation method to transient phenomenon in solids', *Proc. ASME Conf. on Boundary Integral Equation Method: Computational Applications in Applied Mechanics*, AMD, Vol. II, ASME, New York.
10. Butterfield, R., and Tomlin, G. R. (1972) 'Integral techniques for solving zoned anisotropic continuum problems', *Proc. Int. Conf. Variational Meth. in Engng*, Southampton University, pp. 9/31–9/51.
11. Banerjee, P. K., Butterfield, R., and Tomlin, G. R. (1981) 'Boundary element methods for two-dimensional problems of transient ground-water flow', *Int. J. Num. Meth. in Geomech.* **5**, 15–31.
12. Fung, Y. C. (1965) *Foundations of Solid Mechanics*, Prentice-Hall, Englewood Cliffs, N.-J.
13. Van der Pol, B., and Bremmer, H. (1959) *Operational Calculus Based on the Two Sided Laplace Integral*, Cambridge University Press.

14. Crandall, S. H. (1956) *Engineering Analysis*, McGraw-Hill, New York.
15. Smith, I. M. (1977) 'Some time-dependent soil-structure interaction problems', in G. Gudehus (ed.), *Finite Elements in Geomechanics*, Chap. 8, Wiley, London.
16. Shaw, R. P. (1974) 'An integral equation approach to diffusion', *Int. J. Heat and Mass Transfer*, 17, 693–699.
17. Hwang, C. T., Morgenstern, N. R., et al. (1971) 'On solutions of plane strain consolidation problems by finite element methods', *Can. Geotech. J.*, 8, 109–118.
18. Abramowicz, M., and Stegun, I. A. (1965) *Handbook of Mathematical Functions*, Dover, New York.
19. Johnston, I. W. (1972) 'Electro-osmosis and pore pressures; their effect on the stresses acting on driven piles', Ph.D. thesis, University of Southampton.
20. Chuang, Y. K., and Szekely, J. (1971) 'On the use of Green's function for solving melting or solidification problems', *Int. J. Heat and Mass Transfer*, 14, 1285–1294.
21. Chuang, U. K. (1971) 'The melting and dissolution of a solid in a liquid with a strong exothermic head of solution', Ph.D. thesis, State University of New York, Buffalo.
22. Chuang, Y. K., and Szekely, J. (1972) 'The use of Green's functions for solving melting or solidification problems in cylindrical coordinates system', *Int. J. Heat and Mass Transfer*, 15, 1171–1174.
23. DeMey, G. (1977) 'Numerical solution of a drift–diffusion problem', *Comp. Physics Comm.*, 13, 81–88.
24. DeMey, G. (1978) 'An integral equation method to calculate the transient behaviour of a photovoltaic solar cell', *Solid-State Electronics*, 21, 595–596.
25. DeMey, G. (1976) 'An integral equation approach to a.c. diffusion', *Int. J. Heat and Mass Transfer*, 19, 702–704.
26. Zienkiewicz, O. C. (1971) *The Finite Element Method in Engineering Science*, McGraw-Hill, London.
27. Mitchell, A. R. (1971) *Computational Methods in Partial Differential Equations*, Wiley, New York.
28. Desai, C. S., and Johnson, L. D. (1973) 'Evaluation of some numerical schemes for consolidation', *Int. J. Num. Meth. in Engng*, 7, 243–254.
29. Gurtin, M. E., and Steinberg, E. (1962) 'On the linear theory of visco-elasticity', *Arch. Rational Mech. Anal.*, II(4), 291–356.
30. Sneddon, I. N. (1951) *Fourier Transforms*, McGraw-Hill, London.
31. Chuang, Y. K., and Ehrich, O. (1974) 'On the integral technique for spherical growth problems', *Int. J. Heat and Mass Transfer*, 17, 945–953.

# CHAPTER TEN

# TRANSIENT PROBLEMS IN ELASTICITY

## 10-1 INTRODUCTION

Perhaps unexpectedly, the use of integral formulation for the analysis of transient phenomena in solids and fluids has a long history. In a great many of these problems a part of the boundary is at infinity and, since they are particularly attractive for such cases, boundary element methods have been used quite extensively. References 1 to 12 provide a good account of the classical work in elastodynamics and related topics. Although the basic integral formulations for elastodynamics and wave propagation have been known for well over a hundred years, their adaptation to construct numerical algorithms for the solutions of boundary-value problems is relatively new and the first computer-based developments only appeared in the early 'sixties (e.g. Shaw and Friedman,[13,14] Banaugh and Goldsmith,[15] Chen and Schweikert[16]), followed by others.[17-38] The related problems of quasi-static viscoelasticity have been investigated by Rizzo and Shippy[20,39] and others[40,41] using direct BEM.

## 10-2 VISCOELASTICITY

### 10-2-1 Governing Equations

The formulation presented here is restricted to linear homogeneous viscoelastic materials. With the exception of the stress–strain history relationships all the other field equations follow directly from the linear theory of elasticity with proper recognition of the time-dependent nature of all variables. Thus the displacement form of the equations of equilibrium is

$$\mu(t)\frac{\partial^2 u_j}{\partial x_i \partial x_i} + [\lambda(t) + \mu(t)]\frac{\partial^2 u_i}{\partial x_i \partial x_j} = 0 \tag{10-1}$$

where $u_i(x, t)$ is the displacement vector and $\mu(t)$ and $\lambda(t)$ are analogous to the Lamé constants in elasticity, both of them time dependent.

Only quasi-static theory will be considered. The initial conditions appropriate to an initially stress-free state and the boundary conditions will be taken as

$$u_i(x, t) = 0 \qquad t \leqslant 0$$

$$u_i(x, t) = f_i(x, t) \qquad \text{on } S_1 \tag{10-2}$$

$$t_i(x, t) = \sigma_{ij}(x, t)n_j(x) = g_i(x, t) \qquad \text{on } S_2$$

where $S_1 + S_2 = S$ and $n_i$ are the components of the unit outward normal to the boundary enveloping the body, which is assumed to remain unchanged with time.

The stress tensor $\sigma_{ij}$ is now related to the displacement gradient history, again in a manner closely related to the corresponding elasticity equation (Chapter 4):

$$\sigma_{ij}(x, t) = \int_0^t \mu(t - \tau)\frac{\partial \varepsilon_{ij}}{\partial \tau}(x, \tau)\,d\tau + \delta_{ij}\int_0^t \lambda(t - \tau)\frac{\partial \varepsilon_{\kappa\kappa}}{\partial \tau}(x, \tau)\,d\tau \tag{10-3}$$

where

$$\varepsilon_{ij}(x, t) = \frac{1}{2}\left[\frac{\partial u_i(x, t)}{\partial x_j} + \frac{\partial u_j(x, t)}{\partial x_i}\right]$$

## 10-2-2 Basic Integral Formulation

Once more we can develop the integral equation representation of the above problem by using a reciprocal theorem—this time that for an isotropic viscoelastic body. For such a body, when subjected to two different states of loading, $(u_i, t_i, \text{ and } \psi_i)$ and $(u_i^*, t_i^*, \text{ and } \psi_i^*)$, we can write[40,41]

$$\int_s\int_0^t t_i(x, t - \tau)\frac{\partial u_i^*(x, \tau)}{\partial \tau}\,d\tau\,ds + \int_v\int_0^t \psi_i(x, t - \tau)\frac{\partial u_i^*(x, \tau)}{\partial \tau}\,d\tau\,dv$$

$$= \int_s\int_0^t u_i(x, t - \tau)\frac{\partial t_i^*(x, \tau)}{\partial \tau}\,d\tau\,ds + \int_v\int_0^t u_i(x, t - \tau)\frac{\partial \psi_i^*(x, \tau)}{\partial \tau}\,d\tau\,dv \tag{10-4}$$

Note that the symbol $t$ refers to time, whereas the vector quantity $t_i$ always signifies surface traction components.

If the virtual (*) system is chosen such that it is generated by a unit body force (i.e., Kelvin's solution) of the form

$$\psi_i^* = \delta_{ij}\,\delta(x-\xi)\,\Delta(t-\tau)\,e_j$$

where $\delta$ is the Dirac delta function and $\Delta(t-\tau)$ is the unit (Heaviside) step function, which therefore, in combination, represent a unit force $e_j$ applied at $x = \xi$ at time $t = \tau$ Eq. (10-4) then provides (for $x$ in the interior of $v$)

$$u_j(\xi, t) = \int_s \int_0^t \left[ t_i(x, \tau) \frac{\partial G_{ij}(x, \xi; t-\tau)}{\partial \tau} - u_i(x, \tau) \frac{\partial F_{ij}(x, \xi; t-\tau)}{\partial \tau} \right] d\tau\, ds$$

$$+ \int_v \int_0^t \psi_i(x, \tau) \frac{\partial G_{ij}(x, \xi; t-\tau)}{\partial \tau} d\tau\, dv \qquad (10\text{-}5)$$

By taking $\xi$ in Eq. (10-5) to the boundary as before we arrive at the required boundary constraint equation. An equivalent indirect BEM formulation can be developed by using the principles outlined in Chapters 3 to 6.

### 10-2-3 Numerical Solution

It is possible to solve the boundary equation corresponding to (10-5) by using either a transform method[20, 39] or a direct solution technique.[42, 43]

**A direct method** The displacements and tractions at time $t$ can be approximated by the following implicit time discretization procedure:

$$u_i(x, t) = \frac{1}{\Delta t}\left[(t_r - t)\,u_i(x, t_{r-1}) + (t - t_{r-1})\,u_i(x, t_r)\right]$$

$$\qquad\qquad\qquad\qquad\qquad\qquad\qquad\qquad (10\text{-}6)$$

$$t_i(x, t) = \frac{1}{\Delta t}\left[(t_r - t)\,t_i(x, t_{r-1}) + (t - t_{r-1})\,t_i(x, t_r)\right]$$

where $\Delta t$ is a simple time increment, $\Delta t = t_r - t_{r-1}$ with $r$ and $r-1$ designating successive time steps.

We can now write Eq. (10-5), with $\psi_i(x, \tau) = 0$, for a boundary point $\xi$ at time $t = t_m$ as

$$\beta_{ij}\,u_i(\xi, t_m) = \int_s [G_{ij}^m(x, \xi)\,t_i(x, t_m) - F_{ij}^m(x, \xi)\,u_i(x, t_m)]\,ds + R_j(\xi, t_m) \qquad (10\text{-}7)$$

where

$$G_{ij}^m(x, \xi) = \frac{1}{\Delta t}\int_{t_{m-1}}^{t_m} (\tau - t_{m-1}) \frac{\partial G_{ij}(x, \xi; t_m - \tau)}{\partial \tau}\, d\tau$$

$$F_{ij}^m(x, \xi) = \frac{1}{\Delta t}\int_{t_{m-1}}^{t_m} (\tau - t_{m-1}) \frac{\partial F_{ij}(x, \xi; t_m - \tau)}{\partial \tau}\, d\tau$$

$$R_j(\xi, t_m) = \frac{1}{\Delta t} \int_S \left[ t_i(x, t_{m-1}) \int_{t_{m-1}}^{t_m} (t_m - \tau) \frac{\partial G_{ij}(x, \xi; t_m - \tau)}{\partial \tau} d\tau \right.$$

$$- u_i(x, t_{m-1}) \int_{t_{m-1}}^{t_m} (t_m - \tau) \frac{\partial F_{ij}(x, \xi; t_m - \tau)}{\partial \tau} d\tau \bigg] ds$$

$$+ \sum_{r=1}^{m-1} \int_S \int_{t_{r-1}}^{t_r} \left[ \frac{\partial G_{ij}(x, \xi; t_m - \tau)}{\partial \tau} t_i(x, \tau) \right.$$

$$\left. - \frac{\partial F_{ij}(x, \xi; t_m - \tau)}{\partial \tau} u_i(x, \tau) \right] d\tau \, ds$$

Equation (10-7) can be reduced to an algebraic set of equations in the usual manner. In effect, at time $t = t_m$, an ordinary elastic problem is being solved which incorporates the entire solution history from $t = 0$. The viscoelastic nature of the system is very evident in the $\sum$ term in $R_j(\xi, t_m)$ which appears on the right-hand side of Eq. (10-7) and clearly involves the entire past history up to the time $t = t_m$. Therefore a very large amount of computation relating to all the previous time steps is necessary (see Chapter 9). It is, however, possible to take advantage of the exponential nature of the functions[41] $\lambda(t)$ and $\mu(t)$ which characterize the material properties and use only the more recent history to approximate the forcing term $R_j(\xi, t_m)$. The method outlined above has been used by a number of analysts to solve quite complicated viscoelasticity problems.[40, 42]

**Transform methods** It is often possible to use an integral transform technique to reduce the governing equations and boundary conditions to a time-independent form in the transformed space. The problem can then be solved in the transformed space for a sequence of values of the transform parameter followed by some form of numerical inversion back into a time variable. Examples of such solutions can be found in papers by Rizzo and Shippy,[20, 39] which we follow here.

The constitutive relations for a linear isotropic viscoelastic material may be rewritten in the form

$$\sigma'_{ij} = 2G_1 * de'_{ij}$$
$$\sigma_{\kappa\kappa} = 3G_2 * de_{\kappa\kappa}$$

(10-8)

where $\sigma'_{ij}$ and $e'_{ij}$ are the deviatoric components and $\sigma_{\kappa\kappa}$ and $e_{\kappa\kappa}$ the hydrostatic (volumetric) components of the total stress and strain tensors $\sigma_{ij}$ and $\varepsilon_{ij}$ respectively (that is, $\sigma_{ij} = \sigma'_{ij} + \sigma_{\kappa\kappa} \delta_{ij}$). $G_1$ and $G_2$ are relaxation functions in shear and isotropic compression respectively. In Eq. (10-8) we have used the notation for a convolution (this time that of Stieltjes[44]):

$$g * dh = \int_0^t g(x, t - \tau) \frac{\partial h(x, \tau)}{\partial \tau} d\tau + g(x, t) h(x, 0)$$

(10-9)

where $h(x, 0)$ is the limiting value of $h(x, t)$ as $t \to 0$ from positive time values.

If we assume that the boundary conditions are not time dependent we can take the Laplace transform of the boundary values of $u_i(x, t)$ and $t_i(x, t)$ denoted by $u_i^*(x, s)$ and $t_i^*(x, s)$ respectively. The Laplace transform of a scalar, vector, or any tensor function of space and time is defined as

$$\mathscr{L} f(x, t) = f^*(x, s) = \int_0^\infty f(x, t) e^{-st} dt \qquad (10\text{-}10)$$

where $s$ is the transform parameter.

Using this transformation, Eqs (10-8) become

$$\sigma_{ij}'^* = 2sG_1^*(s) e_{ij}'^*$$

$$\sigma_{\kappa\kappa}^* = 3sG_2^*(s) e_{\kappa\kappa}^* \qquad (10\text{-}11)$$

and the equilibrium and strain-displacement relations respectively simplify to

$$\frac{\partial \sigma_{ij}^*}{\partial x_j} = 0 \quad \text{and} \quad 2\varepsilon_{ij}^* = \frac{\partial u_i^*}{\partial x_j} + \frac{\partial u_j^*}{\partial x_i} \qquad (10\text{-}12)$$

Combinations of (10-11) and (10-12) provide

$$(\lambda^* + \mu^*) \frac{\partial^2 u_j^*}{\partial x_j \partial x_i} + \mu^* \frac{\partial^2 u_i^*}{\partial x_j \partial x_j} = 0 \qquad (10\text{-}13)$$

where $\lambda^* = -(2s/3) G_1^* + sG_2^*$ and $\mu^* = sG_1^*$, equations identical to those governing elastostatic systems but written in terms of the transformed variables. Thus, if we replace $\lambda$ and $\mu$ by $\lambda^*$ and $\mu^*$ and also $u_i$, $t_i$, and $\sigma_{ij}$ by the Laplace transforms of the corresponding viscoelastic variables ($u_i^*$, $t_i^*$, and $\sigma_{ij}^*$), we can use a standard elastostatic algorithm to solve any problem in the transformed space at discrete values of the transform parameter $s$. The direct boundary integral for such a case can be written as

$$\beta_{ij} u_i^*(\xi, s) = \int_s [G_{ij}^*(x, \xi; s) t_i^*(x, s) - F_{ij}^*(x, \xi; s) u_i^*(x, s)] ds \qquad (10\text{-}14)$$

where the functions $G$ and $F$ are identical to those of ordinary elastostatics except that the parameters $\lambda^*$ and $\mu^*$ are now dependent on the transform parameter $s$. For a chosen value of $s$, and hence $\lambda^*$ and $\mu^*$, we can solve Eq. (10-14).

Having obtained $t_i^*$ and $u_i^*$ on $S$, the transformed stresses $\sigma_{ij}^*$ and displacements $u_i^*$ within the region may be calculated in the usual manner. The integrals involved must be interpreted, of course, as belonging to the transformed space (i.e., the kernel functions use $\lambda^*$ and $\mu^*$). The corresponding results (that is, $u_i$, $t_i$, $\sigma_{ij}$, etc.) as functions of time can then be obtained, in principle, by numerical inversions of the Laplace transform.

The method of inversion outlined here is due to Schapery[45] and has been used successfully by Rizzo and Shippy.[20,39,44] We assume that the function

$f(x, t)$, in real space, can be represented by

$$f(x, t) = A + Bt + \sum_{\alpha=1}^{m} a_\alpha e^{-b_\alpha t} \tag{10-15}$$

where $A, B, a_\alpha, b_\alpha$ are constant in time and $m$ selects an arbitrary number of terms in the exponential series.

By taking the Laplace transform of Eq. (10-15) and multiplying both sides by the transform parameter $s$ we get

$$sf^*(x, s) = A + \frac{B}{s} + \sum_{\alpha=1}^{m} \frac{a_\alpha}{1 + b_\alpha/s} \tag{10-16}$$

A value of $m$ and a sequence of values of $s$ are selected (somewhat arbitrarily):

$$s = s_1, s_2, s_3, ..., s_{m-1}, s_M \tag{10-17}$$

where $M = m + 2$. We now choose the $m$ values of $b_\alpha$ to be equal to the first $m$ values of $s_\alpha$, i.e.,

$$b_\alpha = s_\alpha \qquad \alpha = 1, ..., m$$

For each choice of the discrete values of $s$, say $s_\beta$, we have, from Eq. (10-16),

$$s_\beta f^*(x, s_\beta) = A + \frac{B}{s_\beta} + \sum_{\alpha=1}^{m} \frac{a_\alpha}{1 + s_\alpha/s_\beta} \qquad \beta = 1, ..., M \tag{10-18}$$

where $f^*(x, s_\beta)$ is evaluated at a discrete point $x$ on either the boundary or in the interior of the domain for a specific value of the transform parameter $s_\beta$.

Equation (10-18) provides $M$ equations for determining the values of $A$, $B$, and $a_\alpha$ ($\alpha = 1, m$). It has been found that six values of $s$ provide sufficiently accurate results for practical purposes. For optimum results it is important to select a 'proper' sequence of $s$ values. Unfortunately, the sequence appears to change with the type of problem and the behaviour of the transient part of the solution. So far it has not been possible to formulate any general rule.

The inversion method is really a curve-fitting process in which the coefficients are determined to best fit the curve of a selected function. The technique described above selects the coefficients so that the curve passes through a number of points equal to the number of coefficients. Obviously more sophisticated techniques such as least-squares fit could be used, but such elaborations would be useless unless the chosen form of the function (10-15) does represent the physical behaviour of the variables involved.

## 10-2-4 Examples

The only example available is due to Rizzo and Shippy[20, 39] who considered the problem of a thick hollow cylinder enclosed within a thin elastic ring as shown in

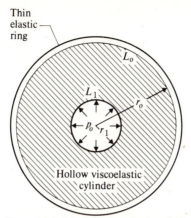

**Figure 10-1** A hollow viscoelastic cylinder.

Fig. 10-1. The viscoelastic material was assumed to behave elastically with respect to spherical stresses (bulk modulus $K$) and as a linear viscoelastic solid in shear, i.e.,

$$G_1(t) = G[\alpha + (1-\alpha)e^{-\lambda t}]$$

$$G_2(t) = K$$

where $\alpha, \lambda, G$, and $K$ are constants. The transformed relaxation functions are then

$$G_1^* = \frac{(G/s)(s+\alpha\lambda)}{s+\lambda}$$

$$G_2^* = \frac{K}{s}$$

They used values of $G/K = 0.6$ and $r_1/r_o = 0.3$ (Fig. 10-1) and characterized the ring by the quantity

$$C = \frac{Eb}{(1-v^2)r_o}$$

where $E$ is Young's modulus, $v$ Poisson's ratio, and $b$ its thickness. The value of $C/K$ was taken to be unity.

Initially there was either a zero traction or a gap between the cylinder and the ring; then the inner boundary $L_1$ was subjected to a uniform pressure $p_o$ applied at time $t = 0$. The mixed boundary conditions are therefore

$$t_r^*(x, s) = \frac{Cu_r^*(x, s)}{r_o} \qquad x \in L_o$$

$$t_r^*(x, s) = \frac{p_o}{s} \qquad x \in L_1$$

$$t_\theta^*(x, s) = 0 \qquad x \in L_o, L_1$$

with $t_r^*$ and $t_\theta^*$ the radial and tangential tractions and $u_r$ the radial displacement on the boundary.

Figure 10-2$a$, $b$ shows the variation of the radial and circumferential stresses with time $\lambda t$ compared with the analytical solution for the same problem (full lines).

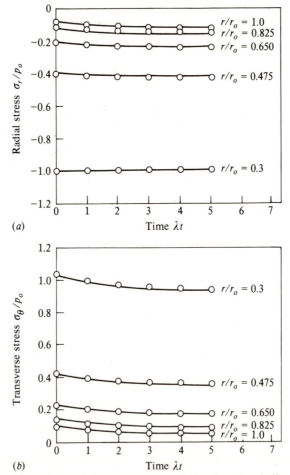

**Figure 10-2** Variations of stresses $\sigma_r$ and $\sigma_\theta$ with time in a viscoelastic cylinder.

## 10-3 THERMOELASTICITY AND CONSOLIDATION

### 10-3-1 Governing Equations

The governing equations for the transient response of a porous elastic solid (consolidation) and those of thermoelasticity are very similar;[46,47] therefore in this section we shall only deal with the former.

As in Chapters 3 and 5 the velocity of pore fluid is given by

$$v_i = -\kappa \frac{\partial h}{\partial x_i} \tag{10-19}$$

where $\kappa$ is the isotropic permeability and $h$ the total head.

The rate of flow of fluid out of unit volume of the porous body is given by $\partial v_i/\partial x_i$ and, if the pores are filled with an incompressible fluid, this must be equal to the rate of decrease of volume which is, in turn, equal but of opposite sign to the volumetric strain rate. Therefore we have

$$\frac{\partial \varepsilon_v}{\partial t} = \kappa \frac{\partial^2 h}{\partial x_i \, \partial x_i} \qquad \varepsilon_v = \varepsilon_{ii}, \text{ the volumetric strain} \tag{10-20}$$

The total head $h$ is composed of the elevation head $h_e$ and the pressure head $h_p = -p/\gamma_w$ ($p$ is the fluid pressure and $\gamma_w$ the unit weight of water).

Since

$$\frac{\partial^2 h_e}{\partial x_i \, \partial x_i} = 0$$

we have

$$\frac{\partial \varepsilon_v}{\partial t} = -\frac{\kappa}{\gamma_w} \frac{\partial^2 p}{\partial x_i \, \partial x_i} \tag{10-21}$$

By using the principle of effective stress ($\sigma_{ij} = \sigma'_{ij} + \delta_{ij} p$) we can easily show that

$$\frac{\partial \varepsilon_v}{\partial t} = \frac{1}{K'} \frac{\partial \sigma'_o}{\partial t} = \frac{1}{K'} \left( \frac{\partial \sigma_o}{\partial t} - \frac{\partial p}{\partial t} \right) \tag{10-22}$$

where 
$K' =$ skeleton bulk modulus
$\sigma'_o =$ mean effective stress $= \frac{1}{3}\sigma'_{ii}$
$\sigma_o =$ mean total stress $= \frac{1}{3}\sigma_{ii}$

By combining Eqs (10-21) and (10-22) we obtain

$$\left( \frac{\kappa K'}{\gamma_w} \right) \frac{\partial^2 p}{\partial x_i \, \partial x_i} = \frac{\partial p}{\partial t} - \frac{\partial \sigma_o}{\partial t} \tag{10-23}$$

which governs the flow of fluid in the porous medium.

Equation (10-23) can be solved for any combination of boundary conditions in terms of $p$ or $\partial p/\partial n$, together with initial conditions for $p$ at $t = 0$, as discussed in Chapter 9, provided that we know 'a priori' the magnitude of $\partial \sigma_o/\partial t$ (a function of time) from a solution of the equations governing the response of the solid skeleton.

Equation (10-23) can be written as

$$C \frac{\partial^2 p(x, t)}{\partial x_i \, \partial x_i} = \frac{\partial p(x, t)}{\partial t} + q(x, t) \tag{10-24}$$

In order to determine the response of the solid elastic skeleton we can convert the equilibrium equations in terms of total stress, i.e.,

$$\frac{\partial \sigma_{ij}}{\partial x_j} = 0$$

to a form involving the effective stresses and gradients of pore-water pressure, i.e.,

$$\frac{\partial \sigma'_{ij}}{\partial x_j} + \delta_{ij}\frac{\partial p}{\partial x_j} = 0 \tag{10-25}$$

The stress–strain relations for the porous solid are, as before,

$$\sigma'_{ij} = \frac{2\mu'v'}{1-2v'}\delta_{ij}\frac{\partial u_m}{\partial x_m} + \mu'\left(\frac{\partial u_i}{\partial x_j} + \frac{\partial u_j}{\partial x_i}\right) \tag{10-26}$$

where $\mu'$ and $v'$ are the shear modulus and Poisson's ratio respectively of the skeleton.

By using Eq. (10-26) and the strain-displacement relations we obtain the governing differential equation for the behaviour of the solid, i.e.,

$$\frac{\mu'}{1-2v'}\frac{\partial^2 u_i(x,t)}{\partial x_i \partial x_j} + \mu'\frac{\partial^2 u_i(x,t)}{\partial x_j \partial x_j} + \delta_{ij}\frac{\partial p(x,t)}{\partial x_j} = 0 \tag{10-27}$$

where $u_i$, $p_i$ are functions of both space and time.

Equation (10-27) can be solved for specified boundary conditions $u_i$ and $t_i = \sigma_{ij}n_j$ (i.e., the total traction) if we know the magnitude of $\partial p/\partial x_j$ from a solution of Eq. (10-24).

Clearly the system of equations (10-24) and (10-27) form a coupled set which, in general, have to be solved simultaneously. In simple consolidation theory the equations are uncoupled by taking $\partial \sigma_o/\partial t = 0$ (real or assumed) and solving (10-23) as explained in Chapter 9.

From Chapter 9 we can write the direct integral representation of Eq. (10-24) as

$$\alpha p(\xi, t) = \int_S [v_n * G - F * p]\, ds + \int_V [q * G + fG]\, dv \tag{10-28}$$

where $\quad v_n$ = normal velocity on the boundary $= \dfrac{\partial p}{\partial n}$

$\qquad\quad F, G$ = kernel functions
$\qquad\qquad f$ = initial distribution of $p(x,t)$ at $t = 0$
and $\qquad\quad *$ denote convolution integrals.

Following Eq. (6-19) for the steady state thermoelastic case the integral representation for Eq. (10-22) becomes

$$\beta_{ij} u_i(\xi) = \int_S [t_i(x) G_{ij}(x, \xi) - F_{ij}(x, \xi) u_i(x)]\, ds - \int_V p(x)\frac{\partial G_{ij}(x, \xi)}{\partial x_i}\, dv \tag{10-29}$$

at any time $t > 0$.

The corresponding indirect BEM equations are

$$p(x, t) = \int_S G * \phi \, ds + \int_V [q * G + fG] \, dv \qquad (10\text{-}30a)$$

$$v_n(x, t) = \alpha' \phi + \int_S F * \phi \, ds + \int_V [q * F + fF] \, dv \qquad (10\text{-}30b)$$

for the fluid system and

$$u_i(x) = \int_S G_{ij}(x, \xi) \, \phi_j(\xi) \, ds - \int_V \frac{\partial G_{ij}(x, \xi)}{\partial \xi_j} p(\xi) \, dv \qquad (10\text{-}31a)$$

$$t_i(x) = \beta'_{ij} \, \phi_j(x) + \int_S F_{ij}(x, \xi) \, \phi_j(\xi) \, ds - \int_V \frac{\partial F_{ij}(x, \xi)}{\partial \xi_j} p(\xi) \, dv + \delta_{ij} \, p(x) \, n_j \qquad (10\text{-}31b)$$

for the skeleton.

The additional term $\delta_{ij} p(x) n_j$ stems from the fact that we can only calculate the effective stresses from Eq. (10-31a) which has therefore to be re-cast in terms of the total stresses and total tractions. Boundary conditions are, of course, always specified in terms of total tractions.

One of the many attractions of BEM is that they can deal with bodies which are completely incompressible at time $t = 0$. By noting that $v = \frac{1}{2}$ for an incompressible solid and that the shear modulus for the bulk solid–fluid system must be identical to that of the skeleton (since the fluid is assumed to have no resistance to shear), we can write Eq. (10-28) for $t = 0$ as

$$\alpha u_j(\xi) = \int_S [t_i(x) \, G'_{ij}(x, \xi) - F'_{ij}(x, \xi) \, u_i(x)] \, ds \qquad (10\text{-}32)$$

where $G'_{ij}$ and $F'_{ij}$ are evaluated with $v = \frac{1}{2}$ and $\mu = \mu'$.

Solution of Eq. (10-32) will now enable us to calculate the total stresses $\sigma_{ij}$ at any interior point at time $t = 0$. The effective stresses $\sigma'_{ij}$ at $t = 0$ can be then obtained from Eq. (10-26) and the excess pore-water pressure $p$ at $t = 0$:

$$\delta_{ij} p = \sigma_{ij} - \sigma'_{ij} \qquad (10\text{-}33)$$

which is easily shown to be equal to $\sigma_{ii}/3$.

A simple numerical solution scheme can be developed by assuming that the quantities $v_n$, $p$, $t_i$, and $u_i$ remain constant over a finite time interval $\Delta t$, as follows:

1. Solve Eq. (10-32) and obtain the distribution of the pore-water pressure $p$ at time $t = 0$.
2. Assume $q = 0$ and the initial condition $f = p$ and solve Eq. (10-28), thus determining the distribution of $p(x, \Delta t)$. This can be done by using the algorithm described in Chapter 9.
3. Use the values of $p(x, \Delta t)$ to solve Eq. (10-29), noting that functions $F_{ij}$ and $G_{ij}$ must now be evaluated using the skeleton moduli. Determine $u_i$, $\sigma'_{ij}$, and the quantity $q(x, t) = 1/\Delta t(\sigma_o | t = \Delta t - \sigma_o | t = 0)$ for the interval.

This completes the solution for the time step $t = 0$ to $t = \Delta t$. For the next time step (that is, $t = \Delta t$ to $t = 2\Delta t$) we use $p(x, \Delta t)$ and $q(x, t)$ from step 3 as the initial condition $f$ and the source function $q$ within the volume respectively, and solve Eq. (10-28) to provide values of $p(x, 2\Delta t)$. We then return to Eq. (10-29) to determine $u_i$, $\sigma'_{ij}$, and the quantity $q(x, t)$ once more, this time for $t = 2\Delta t$. The procedure can be repeated for each successive time interval.

### 10-3-2 Examples

A typical example of a solved problem, which relates to the consolidation (time-dependent settlement) of a strip foundation (plane strain) on a porous, water-saturated, relatively compressible elastic half space, is shown in Fig. 10-3 together with the discretization scheme used. The problem is symmetrical about the centreline and the interior cell integration has been curtailed as shown at about $4B$ from the footing.

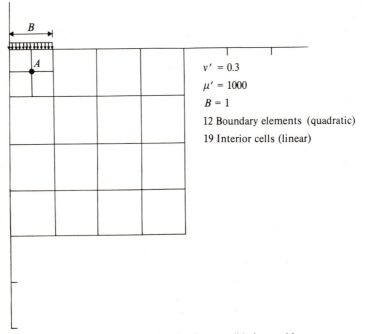

$v' = 0.3$

$\mu' = 1000$

$B = 1$

12 Boundary elements (quadratic)

19 Interior cells (linear)

**Figure 10-3** The discretization for the consolidation problem.

Figure 10-4 shows the calculated degree of consolidation ($U$) under the footing plotted against the non-dimensional time $T = (Ct/4B^2)$ and Fig. 10-5 the variation of vertical stress at the point $A$ beneath the footing.

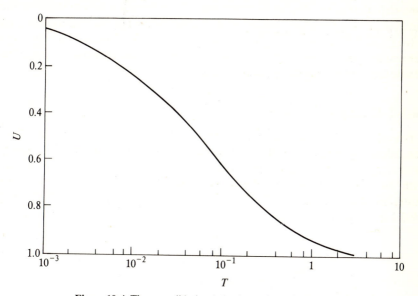

**Figure 10-4** The consolidation behaviour of a strip footing.

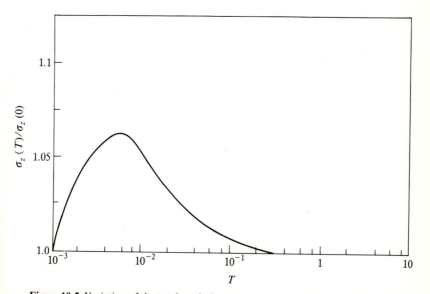

**Figure 10-5** Variation of the total vertical stress at point $A$ under the strip footing.

## 10-4 APPLICATIONS TO ELASTODYNAMICS

### 10-4-1 Governing Equations

The elastodynamic, small-displacement field $u_i(x, t)$ in an isotropic homogeneous elastic body is governed by Navier's equation:

$$(\lambda + \mu) \frac{\partial^2 u_i}{\partial x_i \, \partial x_j} + \mu \frac{\partial^2 u_j}{\partial x_i \, \partial x_i} + \rho(\psi_j - \ddot{u}_j) = 0 \tag{10-34}$$

where   $\lambda, \mu$ = elastic constants
   $\rho$ = mass density of the deformed body
   $\psi_j$ = body forces per unit mass
and   $\ddot{u}_j = \partial^2 u_j / \partial t^2$ are the acceleration.

A domain $V$, bounded by a surface $S$, is considered subject to the following initial conditions:

$$u_i(x, t) = u_i'(x) \qquad\qquad \text{at } t = 0$$

$$\frac{\partial u_i(x, t)}{\partial t} = \dot{u}_i(x, t) = V_i'(x) \qquad \text{at } t = 0 \tag{10-35}$$

and the boundary conditions on $S$ are, in general,

$$u_i(x, t) = f_i(x, t) \qquad\qquad\qquad \text{on } S_1$$

$$t_i(x, t) = \sigma_{ij}(x, t) n_j(x) = g_i(x, t) \qquad \text{on } S_2 \tag{10-36}$$

where $S_1 + S_2 = S$ and the stresses $\sigma_{ij}$ are given by

$$\sigma_{ij} = \rho(C_p - 2C_s)\delta_{ij} \frac{\partial u_m}{\partial x_m} + \rho C_s^2 \left( \frac{\partial u_i}{\partial x_j} + \frac{\partial u_j}{\partial x_i} \right) \tag{10-37}$$

when

$$C_p = \left( \frac{\lambda + 2\mu}{\rho} \right)^{1/2}$$

$$C_s = \left( \frac{\mu}{\rho} \right)^{1/2}$$

Therefore,

$$\left( \frac{C_s}{C_p} \right)^2 = \frac{1 - 2v}{2(1 - v)}$$

The constants $C_p$ and $C_s$, the irrotational and equivoluminal wave velocities respectively, are also called the dilatational and shear wave velocities (or simply the $P$-wave and $S$-wave velocities).

## 10-4-2 The singular solution of Stokes

The fundamental point force solution for the dynamic problem was first obtained by Stokes in 1849 although certain features of the solution were anticipated by Poisson in 1829 (see Refs 1, 3, and 48 for detailed discussions).

In order to demonstrate various features of this solution most simply it is convenient to rewrite Eq. (10-34) in vector notation as

$$\mu \nabla \cdot \nabla \mathbf{u} + (\lambda + \mu) \nabla \nabla \cdot \mathbf{u} + \rho \psi = \rho \frac{\partial^2 \mathbf{u}}{\partial t^2} \tag{10-38}$$

By using the Stokes–Helmholtz resolution theorem,[1] which states that a sufficiently smooth vector field may be decomposed into irrotational and solenoidal parts, $\psi(x, t)$ can be written as

$$\psi = \nabla f + \nabla \times \mathbf{F} \tag{10-39}$$

Similarly, for the displacement vector we have

$$\mathbf{u} = \nabla V + \nabla \times \mathbf{V} \tag{10-40}$$

where $f(x, t)$ and $V(x, t)$ are scalars whereas $\mathbf{F}(x, t)$ and $\mathbf{V}(x, t)$ are vector functions of position and time.

Using Eqs (10-39) and (10-40) and writing $\nabla^2 = \nabla \cdot \nabla$ we can re-write Eq. (10-38) as

$$\nabla \left( C_p^2 \nabla^2 V + f - \frac{\partial^2 V}{\partial t^2} \right) + \nabla \times \left( C_s^2 \nabla^2 \mathbf{V} + \mathbf{F} - \frac{\partial^2 \mathbf{V}}{\partial t^2} \right) = 0 \tag{10-41}$$

This equation is satisfied identically if we choose $V$ and $\mathbf{V}$ to be solutions of the non-homogeneous wave equations

$$C_p^2 \nabla^2 V + f = \frac{\partial^2 V}{\partial t^2} \tag{10-42a}$$

$$C_s^2 \nabla^2 \mathbf{V} + \mathbf{F} = \frac{\partial^2 \mathbf{V}}{\partial t^2} \tag{10-42b}$$

It is now possible to solve Eqs (10-42a, b) and construct the total displacement field $\mathbf{u}$, using Eq. (10-40), as

$$\mathbf{u} = \mathbf{u}^1 + \mathbf{u}^2 \tag{10-43}$$

where $\mathbf{u}^1$ is the solution of Eq. (10-42a) and $\mathbf{u}^2$ that of (10.42b). Clearly the first part, $\mathbf{u}^1$ is curl free whereas the second part, $\mathbf{u}^2$, is divergence free, which means that

$$\nabla \times \mathbf{u}^1 = 0 \quad \text{and} \quad \nabla \cdot \mathbf{u}^2 = 0$$

The solution we require, the fundamental singular solution of the equations of elastodynamics, is that in an infinite region due to a concentrated body force acting at a point $\xi$ in a fixed direction $e_j$ but of time-dependent magnitude $f(t)$, i.e., we need specifically the solution of the elastodynamic equations for a body force

$$\rho \psi_i(x, t) = f(t)\,\delta(x - \xi)\,e_i \tag{10-44}$$

The displacement at a point $x$ at time $t$ due to a force vector of magnitude $f(t)$ acting at $\xi$ in the direction of the $x_i$ axis is given by[3]

$$u_i(x, t) = G_{ij}(x, t; \xi, f)\,e_j(\xi) \tag{10-45}$$

where

$$G_{ij} = G_{ij}^1 + G_{ij}^2$$

with

$$G_{ij}^1 = \frac{1}{4\pi\rho r}\left[-\left(\frac{3y_i y_j}{r^2} - \delta_{ij}\right)\int_0^{C_p^{-1}} \lambda f(t - \lambda r)\,d\lambda + \frac{y_i y_j}{r^2 C_p^2} f\left(t - \frac{r}{C_p}\right)\right]$$

$$G_{ij}^2 = \frac{1}{4\pi\rho r}\left[\left(\frac{3y_i y_j}{r^2} - \delta_{ij}\right)\int_0^{C_s^{-1}} \lambda f(t - \lambda r)\,d\lambda - \frac{y_i y_j}{r^2 C_s^2} f\left(t - \frac{r}{C_s}\right) + \frac{\delta_{ij}}{C_s^2} f\left(t - \frac{r}{C_s}\right)\right]$$

$G_{ij}^1$ and $G_{ij}^2$ are respectively known as the irrotational and equivoluminal parts of the Stokes tensor $G_{ij}$. The functions $f$ vanish for negative arguments reflecting that the longitudinal and shear parts of the wave due to a force $f(t)$ acting at $\xi$ could only be felt at $x$ *after* times $t = r/C_p$ and $r/C_s$ respectively.

The stresses at $(x, t)$ are given by[3]

$$\sigma_{ij} = \rho[(C_p^2 - 2C_s^2)\,G_{mk,m}\,\delta_{ij} + C_s^2(G_{ik,j} + G_{jk,i})]\,e_k$$

$$= T_{ijk}(x, t; \xi, f)\,e_k(\xi) \tag{10-46}$$

where

$$T_{ijk} = T_{ijk}^1 + T_{ijk}^2 \qquad \text{and} \qquad G_{ik,j} \text{ denotes } \frac{\partial G_{ik}}{\partial x_j}, \text{ etc.}$$

with

$$T_{ijk}^1 = \frac{1}{4\pi r^2}\left\{6C_s^2\left(\frac{5y_i y_j y_k}{r^3} - \frac{\delta_{ij} y_k + \delta_{ik} y_j + \delta_{jk} y_i}{r}\right)\int_0^{C_p^{-1}} \lambda f(t - \lambda r)\,d\lambda\right.$$

$$- \frac{1 - 2v}{1 - v}\left(\frac{6y_i y_j y_k}{r^3} - \frac{\delta_{ij} y_k + \delta_{ik} y_j + \delta_{jk} y_i}{r}\right) f\left(t - \frac{r}{C_p}\right)$$

$$- \left(\frac{v}{1 - v}\right)\frac{y_k}{r}\,\delta_{ij}\left[f\left(t - \frac{r}{C_p}\right) + \frac{r}{C_p}\,\dot{f}\left(t - \frac{r}{C_p}\right)\right]$$

$$\left. - \frac{1}{C_p}\frac{y_i y_j y_k}{r^2}\frac{1 - 2v}{1 - v}\,\dot{f}\left(t - \frac{r}{C_p}\right)\right\}$$

$$T_{ijk}^2 = \frac{1}{4\pi r^2}\left\{-6C_s^2\left(\frac{5y_i y_j y_k}{r^3} - \frac{\delta_{ij} y_k + \delta_{ik} y_j + \delta_{jk} y_i}{r}\right)\int_0^{C_s^{-1}} \lambda f(t - \lambda r)\,d\lambda\right.$$

$$+ 2\left(\frac{6y_i y_j y_k}{r^3} - \frac{\delta_{ij} y_k + \delta_{ik} y_j + \delta_{jk} y_i}{r}\right) f\left(t - \frac{r}{C_s}\right)$$

$$\left. + \frac{2y_i y_j y_k}{r^2 C_s}\,\dot{f}\left(t - \frac{r}{C_s}\right) - \frac{\delta_{ik} y_j + \delta_{jk} y_i}{r}\left[f\left(t - \frac{r}{C_s}\right) + \frac{r}{C_s}\,\dot{f}\left(t - \frac{r}{C_s}\right)\right]\right\}$$

The tractions at a point $(x, t)$ can be obtained from

$$t_i(x, t) = F_{ik}(x, \xi, f) e_k(\xi) \tag{10-47}$$

with $$F_{ik} = T_{ijk} n_j(x) = T^1_{ijk} n_j(x) + T^2_{ijk} n_j(x)$$

The above equations, (10-45) to (10-47), are also valid for an impulsive force $\delta(t - \tau)$ applied at time $\tau$, i.e.,

$$f(t) = \delta(t - \tau) \tag{10-48}$$

We can therefore obtain the solution for a unit impulse applied at a point $\xi$ at time $\tau$ by replacing $f$ by $\delta$ and $t$ by $(t - \tau)$. For example, the displacements for this case are given by

$$u_i(x, t) = G_{ij}(x, t; \xi, \tau) e_j(\xi) \tag{10-49}$$

and the tractions $t_i$ are calculable from

$$t_i(x, t) = F_{ik}(x, t; \xi, \tau) e_k(\xi) \tag{10-50}$$

By using (10-48) it is possible to cast the functions $G_{ij}$, $T_{ijk}$, and $F_{ij}$ in more convenient forms (see Eringen and Suhubi[3] for details). It is also possible to derive equivalent two-dimensional solutions. Unfortunately this does not result in any worthwhile simplification and therefore two-dimensional problems of transient elastodynamics may just as well be left as a special class of the general three-dimensional case.

## 10-4-3 Dynamic Reciprocal Theorem

Just as in the elastostatic case, discussed in Chapters 4 and 6, the dynamic reciprocal theorem provides a basis from which the boundary integral equations for elastodynamics can be constructed. The dynamic reciprocal theorem[3, 48 – 51] is in fact a direct extension of Betti's classical theorem in elastostatics and can be simply stated as follows:

If there exist two unrelated elastodynamic states of body forces, boundary tractions, boundary displacements, initial displacements, and initial velocities such as $(\psi_i, t_i, u_i, u_i', V_i')$ and $(\psi_i^*, t_i^*, u_i^*, u_i^{*\prime}, V_i^{*\prime})$ defined in the same region $V$ bounded by a surface $S$ then, for every $t \geqslant 0$,

$$\int_S [t_i * u_i^*](x, t) \, ds + \int_V \rho \left\{ [\psi_i * u_i^*](x, t) + V_i'(x) u_i^*(x, t) + u_i'(x) \frac{\partial u_i^*(x, t)}{\partial t} \right\} dv$$

$$= \int_S [t_i^* * u_i](x, t) \, ds + \int_V \rho \left\{ [\psi_i^* * u_i](x, t) + V_i^{*\prime}(x) u_i(x, t) + u_i^{*\prime}(x) \frac{\partial u_i(x, t)}{\partial t} \right\} dv$$

$$\tag{10-51}$$

where (see Chapter 9)

$$[f * g](x, t) = \begin{cases} \displaystyle\int_0^t f(x, t-\tau)\, g(x, \tau)\, d\tau & \text{for } t \geq 0 \\ 0 & \text{for } t < 0 \end{cases}$$

## 10-4-4 Direct and Indirect Formulations

We may choose the elastodynamic state ($\psi_i$, $t_i$, $u_i$, $u'_i$, and $V'_i$) to be that of the physical problem and the second state that generated by a unit body force $\delta(t-\tau)\,\delta(x-\xi)\,\delta_{ij}\,e_i$ within an infinite solid. By assuming, without any loss of generality, the initial displacements $u_i^{*\prime}$ and velocities $V_i^{*\prime}$ to be zero we can substitute the components of the fundamental solution (10-49) and (10-50) in (10-51) and obtain, for an interior point $\xi$,

$$u_j(\xi, t) = \int_S [G_{ij} * t_i - F_{ij} * u_i]\, ds(x) + \int_V \rho[G_{ij} * \psi_i]\, dv(x)$$
$$+ \int_V \rho\left[ V'_i(x)\, G_{ij}(x, t; \xi, 0) + u'_i(x)\, \frac{\partial G_{ij}(x, t; \xi, 0)}{\partial t} \right] dv(x) \qquad (10\text{-}52)$$

We can now use this direct integral identity to derive the necessary equation for a boundary point, which in turn leads, exactly as before, to the solution algorithm for the complete boundary initial value problem of elastodynamics.

Thus for a general point $\xi_o$ on the boundary we have

$$[\delta_{ij} - C_{ij}]\, u_i(\xi_0, t) = \int_S [G_{ij} * t_i - F_{ij} * u_i]\, ds(x) + \int_V \rho[G_{ij} * \psi_i]\, dv(x)$$
$$+ \int_V \rho\left[ V'_i(x)\, G_{ij}(x, t; \xi_o, 0) + u'_i(x)\, \frac{\partial G_{ij}(x, t; \xi_o, 0)}{\partial t} \right] dv(x)$$
$$(10\text{-}53)$$

where the term $C_{ij}$ is the principal value arising from the treatment of the improper surface integral involving $F_{ij}$. If the boundary at $\xi_o$ is smooth (i.e., has a unique tangent plane) then $C_{ij} = \frac{1}{2}\delta_{ij}$.

By utilizing the development in Chapter 3 which demonstrated the formal equivalence of the direct and indirect formulations the indirect statement for $u_i$ at an interior point $\xi$ is easily shown to be

$$u_j(\xi, t) = \int_S [G_{ij} * \phi_i]\, ds(x) + \int_V \rho[G_{ij} * \psi_i]\, dv(x)$$
$$+ \int_V \rho\left[ V'_i(x)\, G_{ij}(x, t; \xi, 0) + u'_i(x)\, \frac{\partial G_{ij}(x, t; \xi, 0)}{\partial t} \right] dv(x) \qquad (10\text{-}54)$$

in which the indices $i$ and $j$ and the arguments $x$ and $\xi$ can be interchanged because $G_{ij}$ is symmetric in both. The corresponding surface traction $t_i(x, t)$ on a surface through the interior point $x$ with a normal $n_i(x)$ is given by

$$t_i(x, t) = \int_S [F_{ij} * \phi_j] \, ds(\xi) + \int_V \rho [F_{ij} * \psi_j] \, dv(\xi)$$

$$+ \int_V \rho \left[ V'_j(\xi) F_{ij}(x, t; \xi, 0) + u'_j(x) \frac{\partial F_{ij}(x, t; \xi, 0)}{\partial t} \right] dv(\xi) \qquad (10\text{-}55)$$

Equations (10-54) and (10-55) can now be used to derive the discrete indirect boundary element equations in the usual manner.

Whilst the integral representations described above provide very elegant statements of the solution to any transient elastodynamic problem, the computational effort necessary for the complete solution of such boundary-value–initial-value problems is formidable although the methods of discretization, in both space and time, are essentially similar to those explained for transient potential flow problems in Chapter 9.

It is often possible to reduce the computational effort by observing certain features of the propagation of a disturbance. The disturbance from a point is propagated as two uncoupled spherical waves expanding with constant speeds $C_p$ and $C_s$. At any point $\xi$, when reached by the wave moving with speed $C_p$, the disturbance at time $t$ is determined by the sources which emitted a wave at time $(t - r/C_p)$. Similarly, for a wave travelling at speed $C_s$ the disturbance at time $t$ is determined by sources which emitted such a wave at time $(t - r/C_s)$. We shall explore these features of the solution in greater detail later in relation to some special classes of wave-propagation problems.

It is also possible to show[3, 52] that the surface integrals in the boundary integral equations developed above can be separated into two parts. One is associated with the speed $C_p$ (P-waves) while the other is associated with the speed $C_s$ (S-wave). In an unbounded material a P- or an S-wave continue to propagate as P- or S-waves whereas at the surface of a discontinuity such as the interface between two materials mode conversion occurs (i.e., conversion of P-waves into S-waves, and vice versa). For some practical applications it is possible to utilize such simplified separable solutions although in general the separation of P- and S-waves is not a useful procedure.

## 10-4-5 Steady-State Elastodynamics

If we choose our instant of observation of the motion to be a sufficiently long time after the initiating disturbances we can assume that the physical components of the problem are harmonic in time with an angular frequency $\omega$ (i.e., we are then dealing with a steady state elastodynamic problem). The analysis is then greatly simplified since the time variable is thereby eliminated from the governing differential equations and the initial-value–boundary-value problem reduces to a boundary-value problem only.

Thus, if we assume that

$$\psi_i(x, t) = \psi_i(x, \omega)e^{-i\omega t}$$

$$u_i(x, t) = u_i(x, \omega)e^{-i\omega t} \qquad (10\text{-}56)$$

$$t_i(x, t) = t_i(x, \omega)e^{-i\omega t}$$

where $\psi_i(x, \omega)$, $u_i(x, \omega)$, and $t_i(x, \omega)$ are complex amplitudes of body forces, displacements, and tractions respectively.

We can substitute Eq. (10-56) in Eq. (10-38) to obtain

$$[\mu\nabla^2 + (\lambda + \mu)\,\nabla\nabla\cdot]\,\mathbf{u}(x, \omega) + \rho\omega^2\,\mathbf{u}(x, \omega) = -\rho\psi(x, \omega) \qquad (10\text{-}57)$$

which is the governing differential equation of steady state elastodynamics. The fundamental solution of (10-57) for a unit harmonic body force, i.e., the solution of

$$[\mu\nabla^2 + (\lambda + \mu)\,\nabla\nabla\cdot]\,\mathbf{G}(x, \xi, \omega) + \rho\omega^2\,\mathbf{G}(x, \xi, \omega) = -\delta(x, \xi)$$

is given by[3, 12, 52]

$$\mathbf{G}(x, \xi, \omega) \equiv G_{mn}(x, \xi, \omega)$$

$$= \begin{cases} \dfrac{1}{4\pi\rho\omega^2}\left[\delta_{mn}k_s\dfrac{e^{ik_s r}}{r} - \dfrac{\partial^2}{\partial x_m\,\partial x_n}\left(\dfrac{e^{ik_p r}}{r} - \dfrac{e^{ik_s r}}{r}\right)\right] \\ \qquad \text{for three-dimensional problems} \qquad (10\text{-}58a) \\[2ex] \dfrac{i}{4\rho}\left[\dfrac{1}{C_s^2}\delta_{mn}H_o(k_s r) - \dfrac{1}{\omega^2}\dfrac{\partial^2}{\partial x_m\,\partial x_n}\{H_o(k_p r) - H_o(k_s r)\}\right] \\ \qquad \text{for two-dimensional problems} \qquad (10\text{-}58b) \end{cases}$$

where $k_p = \omega/C_p$, $k_s = \omega/C_s$, and $H_0(\eta)$ is the zero-order Hankel function of the first kind with argument $\eta$.

Equation (10-57), which is known as the Helmholtz equation, can be used to derive the reciprocal identity for steady state elastodynamics. If $u_i(x, \omega)$, $u_i^*(x, \omega)$ are two solutions of the reduced elastodynamic equations then

$$\int_S t_i(x, \omega)\,u_i^*(x, \omega)\,ds + \int_V \rho\psi_i(x, \omega)\,u_i^*(x, \omega)\,dv$$

$$= \int_S t_i^*(x, \omega)\,u_i(x, \omega)\,ds + \int_V \rho\psi_i^*(x, \omega)\,u_i(x, \omega)\,dv \qquad (10\text{-}59)$$

By using Eqs (10-58) for the $u_i^*$ system we obtain immediately the following direct formulation for an interior point $\xi$:

$$u_j(\xi, \omega) = \int_S [t_i(x, \omega) \, G_{ij}(x, \xi, \omega) - F_{ij}(x, \xi, \omega) \, u_i(x, \omega)] \, ds$$

$$+ \int_V \rho \psi_i(x, \omega) \, G_{ij}(x, \xi, \omega) \, dv \qquad (10\text{-}60)$$

where $F_{ij}$ is the surface traction at $x$ due to unit body force.

The nature of singularities as $x \to \xi$ is identical to those discussed for the static case.

Equation (10-60) can now be used as a basis for developing both the direct and indirect boundary element algorithms in the usual manner. In effect a series of static problems is being solved, one for each value of the frequency parameter $\omega$.

Such solutions of Eq. (10-60) are unique for any interior region $V$ bounded by the surface $S$ providing that $\omega^2$ is not equal to one of the eigenvalues of the homogeneous part of the differential equation (10-57) under the boundary conditions of the original problem. The exterior region, of course, has no such eigenvalues and therefore one would expect solutions to be attainable for all values of $\omega^2$, provided the radiation and the regularity condition at infinity have been satisfied. Unfortunately this is not always the case, as discussed in Refs 5, 10, and 22 to 24.

Whereas above we have reduced the transient problem to a steady state one by considering the time of our observation to be sufficiently far from the initial disturbance it is also possible to proceed 'in reverse' and to construct transient solutions from the steady state ones by using a superposition technique which depends on the linearity of the governing differential equation.

Consider the integral

$$u_i(x, t) = \int_{-\infty}^{\infty} u_i(x, \omega) \, e^{-i\omega t} \, d\omega \qquad (10\text{-}61)$$

for all the possible frequencies $\omega$.

This equation is a solution of the elastodynamic equation (10-38) for a forcing function

$$\psi_i(x, t) = \int_{-\infty}^{\infty} \psi_i(x, \omega) \, e^{-i\omega t} \, d\omega \qquad (10\text{-}62)$$

for a similarly specified boundary-value problem. It is easy to see that $2\pi\psi_i(x, \omega)$ is the Fourier transform of a body force $\psi_i(x, t)$ with respect to time and hence that $2\pi u_i(x, \omega)$ is the solution of the Fourier-transformed elastodynamic equations. We can therefore construct a general solution of the elastodynamic equations via $u_i(x, \omega)$ over the complete spectrum of frequencies.

It is also possible to arrive at the solution $u_i(x, t)$ via the Laplace transform of the field equations. As before, the Laplace transform of a function $\psi_j(x, t)$ is defined as

$$\bar{\psi}_j(x, s) = \mathscr{L}[\psi_j(x, t)] = \int_0^\infty \psi_j(x, t) e^{-st} dt \qquad (10\text{-}63)$$

We can transform the governing differential equation (10-34) to a form which is similar to Eq. (10-57) with $\omega$ replaced by $is$ and the body force replaced by $[\psi_j(x, s) + V'(x) + su_j'(x)]$. The problem is then reduced to the solution of a static problem for any particular value of the Laplace transform parameter $s$. The major difficulty is then, of course, that of inverting back efficiently to the real space by some numerical means.[64] Cruse and Rizzo,[25, 26] De Hoop,[53] and Doyle[54] have made major contributions in this area.

## 10-4-6 Wave Propagation

**Introduction** In a great many problems of acoustics, electromagnetic field theory, and hydrodynamics, the governing differential equations for the wave-propagation problem are very similar to those of the elastodynamic case discussed above. However, because of the reduced dimensionality of the parameters involved in such problems the analytical complexities of the kernel functions are not so severe.

Let us consider the propagation of small-amplitude acoustic waves in a gas of negligible viscosity. If the particle velocity and the excess pressure are $u_1$ and $p$ respectively then, by defining a potential $\psi$ such that $u_i = \partial\psi/\partial x_i$ and $p = -\rho_0 \partial\psi/\partial t$ ($\rho_0$ being the density of the gas at rest), the governing equations reduce to

$$\frac{\partial^2 \psi(x, t)}{\partial x_i \partial x_i} - \frac{1}{C^2} \frac{\partial^2 \psi}{\partial t^2} = 0 \qquad (10\text{-}64)$$

where $C^2$ is a constant.

In electromagnetic field theory Maxwell's equations for a linear, homogeneous, isotropic, and non-conducting medium of electric conductivity $\varepsilon$ and magnetic permeability $\mu$ are

$$\mathbf{V} \times \mathbf{H} = \varepsilon \frac{\partial \mathbf{H}}{\partial t} \qquad (10\text{-}65a)$$

$$\mathbf{V} \times \mathbf{E} = -\mu \frac{\partial \mathbf{H}}{\partial t} \qquad (10\text{-}65b)$$

$$\mathbf{V} \cdot \mathbf{E} = \mathbf{V} \cdot \mathbf{H} = 0 \qquad (10\text{-}65c)$$

where $\mathbf{E}$ and $\mathbf{H}$ are electric and magnetic vectors at time $t$.

By taking the curl of (10-65a) and using (10-65b) we have

$$\mathbf{V} \times \mathbf{V} \times \mathbf{H} = -\mu\varepsilon \frac{\partial^2 \mathbf{H}}{\partial t^2}$$

which can be reduced, by using the identity

$$\mathbf{V} \times \mathbf{V} \times \mathbf{H} = \mathbf{V}(\mathbf{V} \cdot \mathbf{H}) - \nabla^2 \mathbf{H}$$

and Eq. (10-65c), to the form

$$\nabla^2 \mathbf{H} - \mu \varepsilon \frac{\partial^2 \mathbf{H}}{\partial t^2} = 0 \tag{10-65d}$$

which is similar in character to (10-64).

The governing differential equation for small-amplitude wave propagation in inviscid hydrodynamic problems is also similar to Eq. (10-64) and the differences between the various problems are seen to be physical rather than mathematical. Therefore in what follows we shall describe the systematic solution of the equation

$$\frac{\partial^2 u_i(x, t)}{\partial t^2} = C^2 \frac{\partial^2 u_i(x, t)}{\partial x_j \partial x_j} + \psi_i(x, t) \tag{10-66}$$

and its steady state counterpart. The corresponding analysis of the scalar wave equation (10-64) can then be developed quite easily. Interested readers may also care to study Refs. 3, 4, 5, 8, and 13 to 19, which provide extensive treatment of scalar wave-propagation problems.

**Transient case** The reciprocal identity corresponding to Eq. (10-66) can be written as[3]

$$C^2 \int_S \left[ \frac{\partial u_i}{\partial n} * u_i^* \right] (x, t) \, ds + \int_V \left\{ [\psi_i * u_i^*] (x, t) + V_i'(x) u_i^*(x, t) + u_i'(x) \frac{\partial u_i^*(x, t)}{\partial t} \right\} dv$$

$$= C^2 \int_S \left[ \frac{\partial u_i^*}{\partial n} * u_i \right] (x, t) \, ds + \int_V \left\{ [\psi_i^* * u_i] (x, t) + V_i^{*\prime}(x) u_i(x, t) + u_i^{*\prime}(x) \frac{\partial u_i(x, t)}{\partial t} \right\} dv$$

$$\tag{10-67}$$

where

$$u_i'(x) = u_i(x, 0) \qquad u_i^{*\prime}(x) = u_i^*(x, 0)$$

$$V_i'(x) = \frac{\partial u_i(x, 0)}{\partial t} \qquad V_i^{*\prime}(x) = \frac{\partial u_i^*(x, 0)}{\partial t}.$$

Equation (10-67), like Eq. (10-51), is a statement of a reciprocal relationship between the two solutions of (10-66), that is, $u_i^*(x, t)$ and $u_i(x, t)$ for a body of volume $V$ bounded by a surface $S$ with outward normal $n_i(x)$.

The fundamental solution of Eq. (10-66) for the vector $u_i^*(x, t)$ due to a unit impulsive force $\psi_i^*(x, t) = \delta(t - \tau) \delta(x, \xi) e_i$ is given by

$$u_i^*(x, t) = G_{ij}(x, t; \xi, \tau) e_j \tag{10-68}$$

where

$$G_{ij} = \frac{1}{4\pi C^2 r}\,\delta_{ij}\,\delta\!\left(t-\frac{r}{C}-\tau\right)$$

Similarly,

$$\frac{\partial u_i^*(x,t)}{\partial n} = \frac{\partial u_i^*(x,t)}{\partial x_\kappa}\,n_\kappa = F_{ij}(x,t;\,\xi,\tau)\,e_j \qquad (10\text{-}69)$$

with

$$F_{ij} = \frac{1}{4\pi C^2}\left[\frac{\partial}{\partial n}\!\left(\frac{1}{r}\right)\delta\!\left(t-\frac{r}{C}-\tau\right) - \frac{1}{Cr}\!\left(\frac{\partial r}{\partial n}\right)\overset{\cdot}{\delta}\!\left(t-\frac{r}{C}-\tau\right)\right]$$

By using Eqs (10-68) and (10-69) in (10-67) we obtain

$$u_j(\xi,t) = C^2 \int_S\left[G_{ij}*\frac{\partial u_i}{\partial n} - F_{ij}*u_i\right]ds(x)$$
$$+ \int_V [G_{ij}*\psi_i]\,dv(x) + \int_V\left[V_i'(x)\,G_{ij}(x,t;\,\xi,0) + u_i'(x)\frac{\partial G_{ij}(x,t;\,\xi,0)}{\partial t}\right]dv(x)$$
$$(10\text{-}70)$$

Assuming $V_i'(x) = u_i'(x) = 0$, for convenience, and noting that[3]

$$G_{ij}*\frac{\partial u_i}{\partial n} = \int_0^t G_{ij}(x,t;\,\xi,\tau)\frac{\partial u_i(x,t)}{\partial n}\,d\tau$$
$$= \frac{1}{4\pi C^2 r}\,\frac{\partial u_j(x,t-r/C)}{\partial n} = \frac{1}{4\pi C^2 r}\left[\frac{\partial u_j}{\partial n}\right]$$

and

$$F_{ij}*u_i = \int_0^t F_{ij}(x,t;\,\xi,\tau)\,u_i(x,\tau)\,d\tau$$
$$= \frac{1}{4\pi C^2}\left[\frac{\partial}{\partial n}\!\left(\frac{1}{r}\right)u_j\!\left(x,t-\frac{r}{C}\right) - \frac{1}{Cr}\!\left(\frac{\partial r}{\partial n}\right)\frac{\partial u_j(x,t-r/C)}{\partial t}\right]$$
$$= \frac{1}{4\pi C^2}\left\{\frac{\partial}{\partial n}\!\left(\frac{1}{r}\right)[u_j] - \frac{1}{Cr}\frac{\partial r}{\partial n}\left[\frac{\partial u_j}{\partial t}\right]\right\}$$

we can recast (10-70) into a more convenient form. $[\partial u_j/\partial n]$, $[u_j]$, and $[\partial u_j/\partial t]$ are known as time-retarded values of these functions relative to the point $\xi$ and the retarded time is $(t-r/C)$. The concept of retarded time arises from the fact that a disturbance at $x$ at a time $t$ is felt at a field point $\xi$ only after a time $(t+r/C)$.

By using the above equations we can write (10-70), with $\psi_i = 0$, as

$$u_j(\xi,t) = \frac{1}{4\pi}\int_S\left\{\frac{1}{r}\left[\frac{\partial u_j}{\partial n}\right] - [u_j]\frac{\partial}{\partial n}\!\left(\frac{1}{r}\right) + \frac{1}{Cr}\frac{\partial r}{\partial n}\left[\frac{\partial u_j}{\partial t}\right]\right\}ds \qquad (10\text{-}71)$$

which is the well-known Kirchhoff 'retarded-time' integral equation for $u_j(\xi,t)$. Unfortunately, for two-dimensional transient problems the representation (10-71) cannot be simplified; therefore two-dimensional problems may as well be left as a special class of the general three-dimensional case.

Equation (10-71) can be used to obtain the direct and indirect BEM identities in the usual way. The numerical solution of Eq. (10-71) has received considerable attention in the published literature and the reader is referred for details to a number of excellent papers.[5, 13, 14, 17, 18]

**Steady state case** If we assume a harmonic time dependence for the functions

$$\psi_i(x, t) = \psi_i(x, \omega) e^{-i\omega t}$$

$$u_i(x, t) = u_i(x, \omega) e^{-i\omega t}$$

as before, we can express Eq. (10-66) as

$$\frac{\partial^2 u_i(x, \omega)}{\partial x_j \partial x_j} + k^2 u_i(x, \omega) = -\frac{1}{C^2} \psi_i(x, \omega) \qquad (10\text{-}72)$$

where $k = \omega/C = 2\pi/\lambda C$, with $\lambda$ the frequency.

The differential equation (10-72), the Helmholtz equation, is that governing time harmonic wave scattering. In scattering problems it is convenient to regard the total wave $\mathbf{u}$ at any point as being made up of two parts: (1) a known incident wave $\mathbf{u}^i$ and (2) a scattered wave $\mathbf{u}^s$, which is to be determined, i.e.,

$$\mathbf{u} = \mathbf{u}^i + \mathbf{u}^s \qquad (10\text{-}73)$$

The incident wave is simply the wave function that would be present in the absence of the scattering surfaces and the scattered part is the wave diverging from the scattering region. It is obviously necessary to ensure that $\mathbf{u}^s$ satisfies the radiation condition at infinity, which guarantees the absence of reflected radiation from there. These restrictions apply equally to the transient problem discussed in the earlier sections. For example, the Kirchhoff retarded-time equation, when applied to an exterior scattering problem, must be written in terms of a dependent variable which decays appropriately at a large distance from the scatterer, a condition clearly met by the scattered (i.e., total minus the incident wave) field. The boundary conditions might therefore be expressed in terms of the scattered field although there are other alternatives, as discussed in a recent review by Shaw.[5]

The fundamental solution for Eq. (10-72) is

$$G_{mn}(x, \xi, \omega) = \begin{cases} \dfrac{1}{4\pi} \left( \dfrac{e^{ikr}}{r} \right) \delta_{mn} & \text{for three-dimensional problems} \\[4mm] \dfrac{i}{4} H_o(kr) \delta_{mn} & \text{for two-dimensional problems} \end{cases} \qquad (10\text{-}74)$$

where $H_o(kr)$ is the Hankel function of the first kind of order zero with argument $(kr)$, and

$$F_{mn}(x, \xi, \omega) = \frac{\partial G_{mn}}{\partial n} \qquad (10\text{-}75)$$

For an interior point $\xi$, the direct integral formulation then becomes

$$u_j(\xi, \omega) - u_j^i(\xi, \omega) = \int \left[ G_{ij}(x, \xi, \omega) \frac{\partial u_i(x, \omega)}{\partial n} - F_{ij}(x, \xi, \omega) u_i(x, \omega) \right] ds$$

$$+ \frac{1}{C^2} \int_V G_{ij}(x, \xi, \omega) \psi_i(x, \omega) dv \qquad (10\text{-}76)$$

For a general boundary point $\xi_o$ (assuming $\psi_i = 0$, for convenience) we have

$$[\delta_{ij} - C_{ij}] u_i(\xi_o, \omega) - u_j^i(\xi_o, \omega) = \int_S \left[ G_{ij}(x, \xi_o, \omega) \frac{\partial u_i(x, \omega)}{\partial n} - F_{ij}(x, \xi_o, \omega) u_i(x, \omega) \right] ds \qquad (10\text{-}77)$$

where $C_{ij}$ is the discontinuity term arising from the treatment of the improper integral involving $F_{ij}$.

The interior, time harmonic wave-propagation problem has a unique solution provided $\omega^2$ is not one of the eigenvalues of the system. There are, however, also related difficulties with the corresponding exterior boundary-value problem as expressed by Eq. (10-77), although it does of course satisfy the usual regularity conditions, as well as the radiation condition at infinity. There are an infinite set of values of $\omega$ for which the equation has a multiplicity of solutions which coincide with the 'resonant' wave numbers (or eigenvalues) for a related interior problem. Thus the solution of the exterior Dirichlet and Neumann problems will fail at wave numbers corresponding to eigenvalues of the interior Neumann and Dirichlet problems respectively. This is not a physical difficulty inherent in the exterior problem since there are no exterior eigenvalues. The difficulty of non-uniqueness stems entirely from the formulation of the problem in terms of a boundary integral. Detailed discussion of these difficulties will be found in Refs 5, 10, 21, 23, 24, and 55 to 57, where the authors have proposed modifications to both the direct and indirect formulations to overcome them.

## 10-5 TYPICAL APPLICATIONS

Due to the inherent advantages of BEM for obtaining numerical solutions to exterior problems there are many examples of their application in the published literature, some of which are described below.

**(a) Steady state stresses around cavities of arbitrary shape due to the passage of longitudinal and transverse waves** Niwa, Kobayashi, and Azuma[38] applied the direct boundary element algorithm to this problem to explore the accuracy of the method by comparison with an alternative solution obtained by Pao.[58] An incident longitudinal wave of the form

$$u = A e^{i(kx_1 - \omega t)} \qquad k = \frac{\omega}{C} = \frac{2\pi}{l} \qquad (10\text{-}78)$$

was assumed, where $A$, $k$, $\omega$, and $l$ represent the amplitude, wave number, circular frequency of the wave, and wavelength, respectively. The numerical results obtained, by using 24 boundary elements with constant tractions and displacements over them, are shown compared with Pao's results in Fig. 10-6. Here $a$, represents the radius of the cavity ($a = 1$), (Poisson's ratio $v = 0.35$), and $\sigma_\theta/\sigma_0$ is the ratio of circumferential stress to the applied stress, respectively

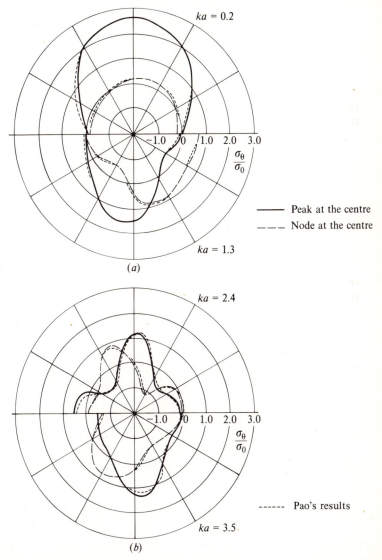

**Figure 10-6** Circumferential stresses on the boundary of the cavity due to a longitudinal sinusoidal wave compared with Pao's results (in the state of plane stress).

As the wavelength increases the stress distribution approaches that of the static case whereas when the wavelength decreases wave scattering reduces the overall stress concentration. Similar results for the case of plane strain (with $v = 0.25$) are shown in Fig. 10-7. The numerical accuracy of the results naturally decreases at shorter wavelengths because of the crude boundary discretization used.

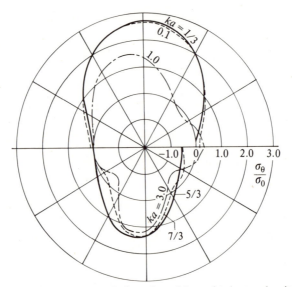

**Figure 10-7** Circumferential stresses on the boundary of the cavity due to a longitudinal sinusoidal wave (in the state of plane strain).

The plane strain stresses on the boundary of the cavity due to a transverse displacement wave of the form (10-78) are shown in Fig. 10-8, where $\tau_o$ is the shear stress carried by the incident wave. The numerical results were obtained by using 48 boundary segments with constant intensities over them ($v = 0.25$).

Once again as the wavelength increases the stress distribution approaches the static case, but is slightly higher, as might be expected. The scattering effect becomes dominant as the wavelength decreases, although when the wavelength becomes short enough to be comparable to the length of the boundary elements the accuracy of the results may be doubtful.

**(b) Transient stresses around cavities of arbitrary shape** Niwa, Kobayashi, and Azuma[38] also considered the problem of transient stresses around cavities of arbitrary shape, due to the passage of travelling waves having an arbitrary time history, by using superposition of the appropriate steady state solutions.[59] The method of solution can be developed in three stages, the first step being to

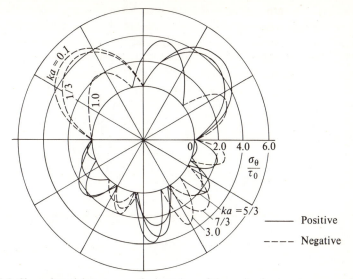

**Figure 10-8** Circumferential stresses on the boundary of the cavity due to a transverse sinusoidal wave.

approximate the transient longitudinal and transverse waves of arbitrary pressure time history by a Fourier expansion

$$F(x) = \frac{1}{2} + \frac{2}{\pi} \sum_{m=1}^{n} \frac{\sigma}{2m-1} \sin(2m-1)x \qquad (10\text{-}79)$$

with Lanczos' factor

$$\sigma = \frac{\sin\{(2m-1)/2n\}\,\pi}{\{(2m-1)/2n\}\,\pi}$$

The second step is then to obtain the steady state solution of the respective sinusoidal waves using the boundary element formulation discussed previously. The third, final, step is the superposition of these solutions so that the original travelling waves are constructed, although it is apparently important to have enough time between pulses to allow the surface energy to be dispersed into the surrounding medium.

Figure 10-9 shows the accuracy of the Fourier series with $n = 10$ in representing the most stringent example of a step-form wave. The resulting solution for the circumferential stress at $\theta = 90°$ (Fig. 10-10) due to such an incident longitudinal wave is shown compared with that of Garnet and Pascal,[59] where the time is taken to be zero at the moment the wavefront arrives at the left boundary of the cavity. The maximum circumferential stress appears to be about $-2.98$ at the non-dimensional time of 3.5 against the value of $-2.67$ for the static case. The corresponding results at $\theta = 0°$ are shown in Fig. 10-11, where the maximum stress concentration is about 0.22 compared to zero for the static

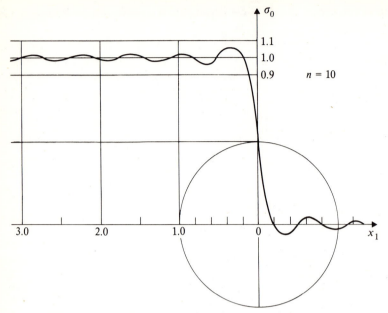

**Figure 10-9** Fourier expansion of step-form pulse with ten terms.

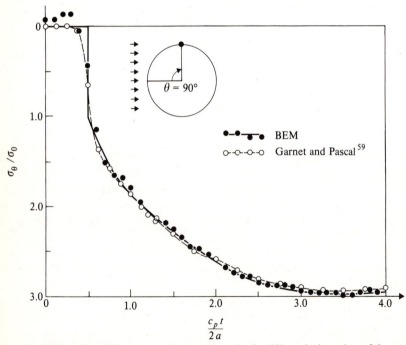

**Figure 10-10** Circumferential stress–time history at a point $\theta = 90°$ on the boundary of the cavity due to longitudinal step-form wave. Time $t$ is taken to be zero at the instant at which the wavefront arrives at the left boundary.

case, although here the BEM results do depart significantly from those of Garnet and Pascal.

Niwa *et al.*[38] also considered the problem of transient stresses on the boundary of a horseshoe-shaped cavity during the passage of a longitudinal step-form pulse travelling in the horizontal direction. Their results are shown in Fig. 10-12 which indicates that high stress concentrations occur at the rounded lower corners having a small radius ($0.2a$), with $a$ the radius of the crown of the cavity.

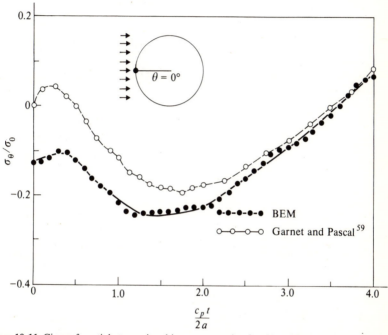

**Figure 10-11** Circumferential stress–time history at a point $\theta = 0°$ on the boundary of the cavity due to a longitudinal step-form wave. Time $t$ is taken to be zero at the instant at which the wave-front arrives at the left boundary.

**(c) Oscillations of harbours of arbitrary shape** Harbour-oscillation problems involve the solution of a scalar Helmholt equation:

$$\frac{\partial^2 \psi}{\partial x_i \partial x_i} + k^2 \phi = 0$$

and therefore naturally fall within the class of problems being discussed in this chapter.

Garrison and Chow,[28] Hwang and Tuck,[29] Lee,[30] and Shaw[31] have all considered the solution of this equation for water wave problems. Figure 10-13 shows a typical solution, by Hwang and Tuck,[29] to the problem of a rectangular

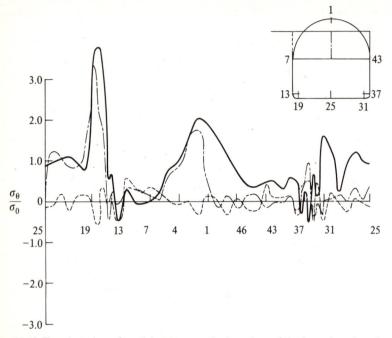

**Figure 10-12** Transient circumferential stresses on the boundary of the horseshoe-shaped cavity due to a travelling longitudinal step-form wave, corresponding to the indicated positions of the wavefront.

harbour connected to open sea, in which the numerical solution is compared with both experimental results and an earlier approximate analytical solution of Ippen and Goda.[60] All results were calculated and measured at the location $A$ shown in the inset figure. In the immediate neighbourhood of the fundamental period, the results obtained using the indirect BEM are slightly larger than those reported by Ippen and Goda, which may well be due to the approximations introduced in their analysis.

**(d) Some applications in acoustics** Most of the features of the integral formulations developed in this chapter have been known to workers in the field of acoustics for many years. However, it is only more recently that such steady state and transient scalar wave integral equations have been treated numerically (see Refs 4 to 6, 10, 13 to 19, and 21 to 24).

Figure 10-14 shows results from Ref. 16 of an analysis of the acoustic radiation from a stiffened cylindrical shell, represented by discrete masses, in water. Chen and Schweikert[16] obtained these results using an indirect BEM formulation of the time harmonic problem.

Mitzner[17] considered the problem of transient scattering from a sphere of radius $a = 1$ centred at the origin due to a pressure pulse of gaussian shape:

$$u_o = \exp\left[-\tfrac{1}{2}(t - Z/c)^2\right]$$

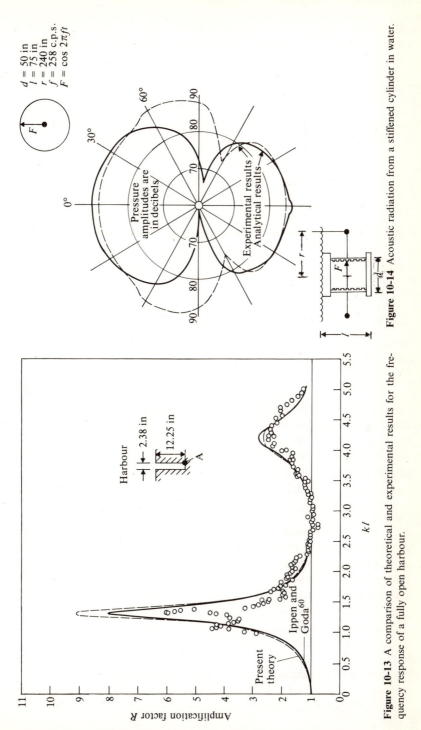

**Figure 10-14** Acoustic radiation from a stiffened cylinder in water.

**Figure 10-13** A comparison of theoretical and experimental results for the frequency response of a fully open harbour.

with velocity $c = 1$. He used a direct BEM analysis (which he described as a retarded potential technique) and compared his results with those from an analytical, separation of variables solution, as shown in Fig. 10-15.

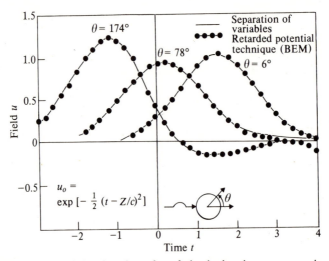

**Figure 10-15** Total pressure induced on the surface of a hard sphere by a pressure pulse of gaussian shape (pulse amplitude, sphere radius, and sound velocity normalized to unity; pulse half width equal to sphere radius).

Other impressive examples of the application of BEM to time harmonic and transient scalar and vector wave-propagation problems can be found in Refs 4, 8, 13 to 15, 18, and 61 to 72. Shippy and Rizzo[63] have now combined their earlier work on viscoelasticity and transient elastodynamics and developed a method of solution for wave propagation through viscoelastic bodies.

## 10-6 CONCLUDING REMARKS

In this chapter, we have established BEM algorithms for time harmonic and transient wave-propagation problems. Although some of the integral formulations are now nearly 150 years old their numerical development in engineering is still in its infancy. It is our hope that we have included enough material here to demonstrate that exploration of the practical applications of such methods could be a very worthwhile activity.

## 10-7 REFERENCES

1. Morse, P. M., and Feshbach, H. (1953) *Methods of Theoretical Physics*, Part 2, McGraw-Hill, New York.
2. Graff, K. F. (1975) *Wave Motion in Elastic Solids*, Clarendon Press, Oxford.

3. Eringen, A. C., and Suhubi, E. S. (1975) *Elasto-dynamics*, Vol. 2, *Linear Theory*, Academic Press, New York.

4. Shaw, R. P. (1970) 'An integral equation approach to acoustic radiation and scattering', *Topics in Ocean Engng*, **II**, 143–163.

5. Shaw, R. P. (1979) 'Boundary integral equation methods applied to wave problems', in P. K. Banerjee and R. Butterfield (eds), *Developments in Boundary Element Methods*, Vol. I, Applied Science Publishers, London.

6. Kanwal, R. P. (1971) *Linear Integral Equations, Theory and Technique*, Academic Press, New York and London.

7. Alawneh, A. D., and Kanwal, R. P. (1972) 'Singularity methods in mathematical physics', *SIAM Rev.*, **19**(3), 437–471.

8. Lean, M. H., Friedman, M., and Wexler, A. (1979) 'Advances in application of the boundary element method in electrical engineering problems', in P. K. Banerjee and R. Butterfield (eds), *Developments in Boundary Element Methods*, Chap. IX, Applied Science Publishers, London.

9. Baker, B. B., and Copson, E. T. (1939) *The Mathematical Theory of Huygen's Principle*, Oxford University Press.

10. Burton, A. J. (1973) 'The solution of Hemholtz equation in exterior domains using integral equations', NPL Report NAC 30, January 1973.

11. Cruse, T. A., and Rizzo, F. J. (eds) (1975) *Proc. ASME Conf. on Boundary Integral Equation Methods*, Report AMD–11, ASME, New York.

12. Kupradze, V. D. (1963) In I. N. Sneddon and R. Hill (eds), *Progress in Solid Mechanics*, Vol. 3, North Holland.

13. Shaw, R. P., and Friedman, M. B. (1962) 'Diffraction of a plane shock wave by a free cylindrical obstacle at a free surface', *Fourth U.S. Natn Congr. on Appl. Mech.*, **2**, 371–379.

14. Friedman, M. B., and Shaw, R. P. (1962) 'Diffraction of a plane shock wave by an arbitrary rigid cylindrical obstacle', *J. Appl. Mech.*, **29**(1), 40–46.

15. Banaugh, R. P., and Goldsmith, W. (1963) 'Diffraction of steady acoustic waves by surfaces of arbitrary shape', *J. Acoust. Soc. Am.*, **35**(10), 1590–1601.

16. Chen, L. H., and Schweikert, J. (1963) 'Sound radiation from an arbitrary body', *J. Acoust. Soc. Am.*, **35**, 1626–1632.

17. Mitzner, K. M. (1967) 'Numerical solution for transient scattering from a hard surface of arbitrary shape—retarded potential technique', *J. Acoust. Soc. Am.*, **42**(2), 391–397.

18. Shaw, R. P. (1966) 'Diffraction of acoustic pulses by obstacles of arbitrary shape with a robin boundary condition—part A', *J. Acoust. Soc. Am.*, **41**(4), 855–859.

19. Shaw, R. P. (1969) 'Diffraction of pulses by obstacles of arbitrary shape with an impendence boundary condition', *J. Acoust. Soc. Am.*, **44**(4), 1962–1968.

20. Shippy, D. J. (1975) 'Application of the boundary integral equation method to transient phenomenon in solids', in T. A. Cruse and F. J. Rizzo (eds), *Boundary Integral Equation Method: Computational Applications in Applied Mechanics*, AMD–11, ASME, New York.

21. Burton, A. J., and Miller, G. F. (1971) 'The application of integral equation methods to the numerical solution of some exterior boundary value problems', *Proc. Roy. Soc., London*, **323**(A), 201–210.

22. Greenspan, D., and Werner, P. (1966) 'A numerical method for the exterior Dirichlet problem for the reduced wave equation', *Arch. Rational Mech. Anal.*, **23**, 288–316.

23. Jones, D. S. (1974) 'Integral equations for the exterior acoustic problems', *Q. J. Mech. Appl. Math.*, **27**(1), 129–142.

24. Schenck, H. A. (1966) 'Improved integral formulation for acoustic radiation problems', *J. Acoust. Soc. Am.*, **44**(1), 41–58.

25. Cruse, T. A., and Rizzo, F. J. (1968) 'A direct formulation and numerical solution of the general transient elasto-dynamic problem—I', *Jour. Math. Anal. Appl.* **22**(1), 244–259.

26. Cruse, T. A., (1968) 'A direct formulation and numerical solution of the general transient elasto-dynamic problem—II', *J. Math. Anal. Appl.*, **22**(2), 341–355.

27. Jain, D. L., and Kanwal, R. P. (1978) 'Scattering of P and S waves by spherical inclusions and cavitities', *J. Sound and Vibration*, **57**(2), 171–202.

28. Garrison, C. J., and Chow, P. Y. (1972) 'Wave forces on submerged bodies', *ASCE*, **98** (WW3), 357–392.
29. Hwang, L. S., and Tuck, E. O. (1970) 'On the oscillations of harbours of arbitrary shape', *J. Fluid Mech.*, **42**, 447–464.
30. Lee, J. J. (1971) 'Wave induced oscillations in harbours of arbitrary shape', *J. Fluid Mech.*, **45**, 375–394.
31. Shaw, R. P. (1975) 'Boundary integral equation methods applied to water waves', in T. A. Cruse and F. J. Rizzo (eds), *Boundary Integral Equation Method: Computational Applications in Applied Mechanics*, AMD-11, ASME, New York.
32. Morita, N. (1978) 'Surface integral representations for electro-magnetic scattering from dielectric cylinders', *IEEE Trans. Antennas and Prop.*, **AP26** (2), 261–266.
33. De-Mey, G. (1976) 'Calculation of eigenvalues of the Helmholtz equation by an integral equation', *Int. J. Num. Meth. in Engng*, **10**, 59–66.
34. Tai, G. R. C., and Shaw, R. P. (1974) 'Helmholtz equation eigenvalues and eigenmodes for arbitrary domains', *J. Acoust. Soc. Am.*, **56** (3), 796–804.
35. Eatock Taylor, R., and Waite, J. B. (1978) 'The dynamics of offshore structures evaluated by boundary integral techniques', *Int. J. Num. Meth. in Engng*, **13**, 73–92.
36. Eatock Taylor, R., and Dolla, J. P. (1978) 'Hydro-dynamic loads on vertical bodies of revolution', Report OEG/78/8, December, Department of Mechanical Engineering, University College, London.
37. Kobayashi, S., Fukui, T., and Azuma, N. (1975) 'An analysis of transient stresses produced around a tunnel by the integral equation method', *Proc. Symp. Earthquake Engng*, Japan, pp. 631–638.
38. Niwa, Y., Kobayashi, S., and Azuma, N. (1975) 'An analysis of transient stresses produced around cavities of arbitrary shape during the passage of travelling waves', Memo, Vol. 36, pp. 1–2, pp. 28–46, Faculty of Engng, Kyoto University, Japan.
39. Rizzo, F. J., and Shippy, D. J. (1971) 'An application of the correspondence principle of linear visco-elasticity theory', *SIAM J. Appl. Math.*, **21** (2), 321–330.
40. Cristensen, R. M. (1971) *Theory of Visco-elasticity*, Academic Press, New York and London.
41. Deak, A. L. (1972) 'Numerical solution of three-dimensional elasticity problems of solid rocket grains based on integral equations', AFSC Report No. AFRPL-TR–71–140–Vol. 1.
42. Noble, B. (1964) 'The numerical solution of non-linear integral equations and related topics', in P. M. Anselone (ed.), *Nonlinear Integral Equations*, University of Wisconsin Press.
43. Buckner, H. (1952) *Die Praktische Behandlung von Integral Gleichungen*, Springer-Verlag, Berlin.
44. Gurtin, M. E., and Sternberg, E. (1962) 'On the linear theory of visco-elasticity', *Arch. Rational Mech. Anal.*, **11** (4), 291–384.
45. Schapery, R. A. (1962) 'Approximate methods of transform inversion for visco-elastic stress analysis', *Proc. Fourth U.S. Natn. Congr. on Appl. Mech.*, **2**, 1065–1085.
46. Biot, M. A. (1941) 'General theory of three-dimensional consolidation', *J. Appl. Physics*, **12**, 155–164.
47. Fung, Y. C. (1965) *Foundations of Solid Mechanics*, Prentice-Hall, Englewood Cliffs, N.-J.
48. Love, A. E. H. (1931) *A Treatise on Mathematical Theory of Elasticity*, Oxford University Press.
49. Gurtin, M. E. (1964) 'Variational principles for linear elasto-dynamics', *Arch. Rational Mech. Anal.*, **16**, 34–50.
50. Gurtin, M. E. (1972) 'The linear theory of elasticity', in C. Truesdell (ed.), *Handbuch der Physik*, Vol. VI a/2, pp. 1–295, Springer-Verlag, Berlin and New York.
51. Wheeler, L. T., and Sternberg, E. (1968) 'Some theories in classical elasto-dynamics', *Arch Rational Mech. Anal.*, **31**, 51–90.
52. Pao, Y. H., and Varatharajulu, V. (1975) 'Huygens' principle, radiation conditions and integral formulae for the scattering of elastic waves', Report No. 2994, Material Science Center, Cornell University, Ithaca, N.Y.
53. De Hoop, A. T. (1958) 'Representation theorems for the displacement in an elastic solid and their application to elasto-dynamic diffraction theory', Doctoral Dissertation, Tech. Hogeschole, Delft.

54. Doyle, J. M. (1966) 'Integration of the Laplace transformed equations of classical elasto-kinetics', *J. Math. Anal. Appl.*, **13**, 118–131.
55. Chertock, G. (1971) 'Integral equation methods in sound radiation and scattering from arbitrary surfaces', NSRDC Rep No. 3538, Washington, D.C.
56. Ursell, F. (1973) 'On exterior problems of acoustics', *Proc. Camb. Phil. Soc.*, **74**, 117–125.
57. Kleinmann, R. E., and Roach, G. F. (1974) 'Boundary integral equations for the three-dimensional Helmholtz equations', *SIAM Rev.*, **16**, 214–236.
58. Pao, Y. H. (1962) 'Dynamic stress concentration in an elastic plate', *J. Appl. Mech.*, **29**, 299–305.
59. Garnet, H., and Pascal, J. C. (1966) 'Transient response of circular cylinder of arbitrary thickness in an elastic medium to a plane dilational wave', *J. Appl. Mech.*, **33**, 521–531.
60. Ippen, A. T., and Goda, Y. (1963) 'Wave induced oscillations in harbours: the solution for a rectangular harbour connected to open sea', Research Report, Hydro. Laboratory, MIT, Cambridge, Mass.
61. Chen, L. H. (1967) 'Projections for the near terms—application of approaches to real submersibles', *Proc. Acoust. of Submerged Structs*, Vol. II, 15–17 Feb., Office of Naval Research, Washington, D.C.
62. Dominguez, J., and Roesset, J. (1978) 'Dynamic stiffness of rectangular foundations', Report on Grant NSF–RANN–ENV–77–18339, Department of Civil Engineering, MIT, Cambridge, Mass.
63. Shippy, D. J., and Rizzo, F. J. (1975) 'Solutions of problems of dynamic visco-elasticity by the boundary integral equation method', Internal report, Department of Engineering Mechanics, University of Kentucky.
64. Papoulis, A. (1957) 'A new method for inversion of Laplace transform', *Q. Appl. Math.*, **14**, 405–414.
65. Banaugh, R. P., and Goldsmith, W. (1963) 'Diffraction of steady elastic waves by surfaces of arbitrary shape', *J. Appl. Mech.*, **30**, 589–597.
66. Cole, D. M., Kosloff, D. D., and Minster, J. B. (1978) 'A numerical boundary integral equation method for elastodynamics I', *Bull. Seism. Soc. Amer.*, **68**, 1331–1357.
67. Manolis, G. D., and Beskos, D. E. (1980) 'Dynamic stress concentration studies by the boundary integral equation method', in R. P. Shaw et al. (eds), *Proc. 2nd Int. Symp. Innov. Num. Analysis in Appl. Engng Sci.*, Virginia, 459–463.
68. Manolis, G. D. (1980) 'Dynamic response of underground structures', Ph.D. Thesis, Univ. of Minn., Minneapolis.
69. Apsel, R. J. (1979) 'Dynamic Green's functions for layered media and applications to boundary-value problems', Ph.D. Thesis, Univ. of Calif., San Diego.
70 Wong, H. L., and Jennings, P. C. (1975) 'Effects of canyon topography on strong ground motion', *Bull. Seism. Soc. Amer.*, **65**, 1239–1257.
71. Wong, H. L., Trifunac, M. D., and Westermo, B. (1977) 'Effects of surface and subsurface irregularities on the amplitudes of monochromatic waves', *Bull. Seism. Soc. Amer.*, **67**, 353–368.
72. Toki, K., and Sato, T. (1977) 'Seismic response analysis of surface layer with irregular boundaries', *Proc. 6th WCEE*, N. Delhi, India, 409–415.

# CHAPTER ELEVEN

## PLATE-BENDING PROBLEMS

## 11-1 INTRODUCTION

The Kirchhoff thin-plate-bending theory, in the absence of membrane forces, is a natural, two-dimensional extension of the simple Bernoulli beam-bending theory discussed in Chapter 2. Both are based on the assumption that 'plane sections remain plane' during bending and that the displacements are small enough for changes of geometry to be negligible and, thereby, small strain theory to be applicable.

Despite the very many applications of plate-bending theory in engineering there appear to be relatively few published papers in which integral-equation-based algorithms have been developed.[1-7] Tottenham[1] mentions using the reciprocal theorem and homogeneous solutions from the early 'sixties; Niwa, Kobayashi, and Fukui[2] published an IBEM algorithm in 1974; Altiero and Sikarskie[3] and Bezine and Gamby,[6] another form of IBEM analysis in 1978; while Segdin and Bricknell[8] discussed a DBEM algorithm for a plate with a re-entrant corner.

The BEM formulations presented below follow the spirit of Tottenham's paper and extended his work on spring-mounted plates to include those with continuous elastic support, all in a rather simpler manner than the majority of the papers cited.

## 11-2 STATEMENT OF THE PROBLEM AND THE GOVERNING DIFFERENTIAL EQUATIONS

An element of our elastic plate, of thickness $h$ and modulus $D = Eh^3/12(1 - v^2)$, is shown in Fig. 11-1$a, b$ together with the sign convention for the edge moments

$(m_{ij})$ and shears $(Q_i)$, all per unit edge length; the plate displacement $(w)$, slopes $(\theta_i)$, all referred to the middle surface of the plate; and the applied surface loads $\psi$ and external moments $m'_i$. The double-arrow sign convention on the moment and slope components indicates their sense as that of a right-handed screw progressing along the arrow.[2]

The thin-plate theory relationships between these quantities are well known to be[9]

$$\theta_i = \frac{\partial w}{\partial x_i} = w_{,i} \tag{11-1a}$$

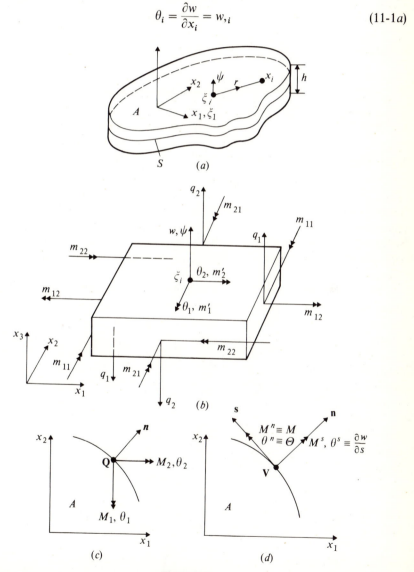

Figure 11-1

with the comma denoting partial differentiation with respect to $x_i$ (Appendix A) and $i$ ranging over $(1, 2)$ only:

$$m_{ij} = m_{ji} = D[(1-v)w_{,ij} + v\delta_{ij} w_{,kk}] \tag{11-1b}$$

(which is somewhat simplified in cases where $v = 0$ can be assumed):

$$q_i = -m_{ij,j} - m'_j = -Dw_{,kki} - m'_i \tag{11-1c}$$

[Note that since the various quantities are per unit length (or area) the dimensions of $m_{ij}$ are force, etc.] Finally, by noting that for vertical equilibrium $q_{i,i} + \psi = 0$ and differentiating (11-1c) again, we arrive at the governing biharmonic equation

$$\psi - m'_{i,i} = Dw_{,iijj} \tag{11-1d}$$

The reader unfamiliar with plate theory will find it instructive to compare Eqs (11-1) with the corresponding set [(2-11), (2-12)] for a simple beam.

As before we shall be very much concerned with the components of $(m, \theta, q)$ 'resolved' along an outward boundary normal $n_i$ at some point on the surface $S$ bounding an arbitrary plan-form plate of area $A$ (Fig. 11-1c): viz.,

$$Q = q_i n_i \quad \text{and} \quad M_i = m_{ij} n_j \tag{11-2}$$

where the sense of the $M_i$ and corresponding $\theta_i$ components is shown in the figure. An immediate, but erroneous, conclusion is that at any point the independent boundary variables will number six $(Q, w, M_i, \theta_i)$. Since Eq. (11-1d) is fourth order we must be able to reduce these to four. This is accomplished by noting that:

1. $M_i$ can be considered more usefully in terms of its boundary normal and tangential components $(M^n, M^s)$ (Fig. 11-1d) with

$$m_{ij} n_j = M_i = -M^n n_i + M^s s_i \tag{11-3}$$

   $s_i$ being a tangential unit boundary vector such that $\mathbf{s} = \mathbf{k} \times \mathbf{n}$ with $\mathbf{k}$ a unit vector directed along the $x_3$ axis (Fig. 11-1d).

2. The edge twisting moment $M^s$ can be combined with $Q$ (Fig. 11-2a) to produce a resultant boundary shear force $V$ as

$$V = Q - \frac{\partial M^s}{\partial s} \tag{11-4}$$

3. The slopes corresponding to $M_i$ will be related to the normal and tangential slope components, $\theta^n$ and $\theta^s$ say (Fig. 11-1d), by an equation equivalent to (11-3): viz.,

$$\theta_i = -\theta^n n_i + \theta^s s_i \tag{11-5a}$$

   whence

$$\theta_i n_i = -\theta^n \quad \text{and} \quad \theta_i s_i = \theta^s = \frac{\partial w}{\partial s} \tag{11-5b}$$

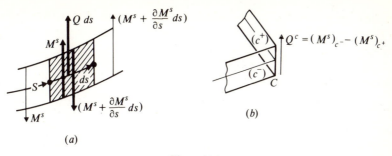

**Figure 11-2**

We have now reduced our boundary variables to four $(V, w, M, \Theta)$ if we write $M \equiv M^n$ and $\Theta \equiv \theta^n$.

However, by incorporating $M^s$ into $V$ in Eq. (11-4) (Fig. 11-2a), we are then obliged to consider the so-called 'corner forces' $Q_c$ (Fig. 11-2b) which arise from $M^s$ at any discontinuous change of boundary direction. From the sign convention of Fig. 11-1d we see that the $M^s$ effects at each side of any corner $(C)$ will be opposed and therefore if $(M^s)_{c-}$ and $(M^s)_{c+}$ relate to the boundary on each side of any corner, we have

$$Q_c = (M^s)_{c-} - (M^s)_{c+} \tag{11-6}$$

These corner forces are analysed more thoroughly in Sec. 11-5 where the DBEM equations are developed.

## 11-3 SINGULAR SOLUTIONS

Our fundamental solution is that for the displacement $w^o(x)$ at any point $(x)$ in a plate of infinite extent produced by a unit load acting at the point $(\xi)$ (Fig. 11-1a), which is

$$w^o(x) = \frac{1}{8\pi D} r^2 \left( \ln \frac{r}{r_0} \right) = G^o(x, \xi) \tag{11-7}$$

where $r = y_i y_i, y_i = (x - \xi)_i$, and $r^o$ locates an arbitrary circle on which the displacement is zero (i.e., once more the displacements are relative to a datum at $r = r^o$ and therefore we shall have to introduce auxiliary terms in the IBEM development to eliminate 'reactions' from the infinite boundary as in Chapter 4). By defining $\rho = r/r_o$ and differentiating, etc., (11-7) in accordance with Eqs (11-1) we obtain the slopes, moments, and shears related to (11-7) as, say,

$$\theta_i^o(x) = \frac{y_i \ln \rho}{4\pi D}$$

whence

$$\Theta^o(x) = -\theta_i^o(x)\, n_i(x) = F^o(x, \xi) \tag{11-8a}$$

$$m_{ij}^o(x) = -\frac{1}{4\pi}\left\{\frac{(1-v)\, y_i\, y_j}{r^2} + \delta_{ij}[(1+v)\ln \rho + v]\right\}$$

whence

$$M_i^o(x) = m_{ij}^o(x)\, n_j(x)$$

and, from (11-3),

$$M^{no} = -M_i^o\, n_i = -m_{ij}^o\, n_i\, n_j = E^o(x, \xi) \tag{11-8b}$$

and

$$M^{so} = M_i^o s_i = C^o(x, \xi)$$

Finally,

$$q_i^o(x) = -\frac{1}{2\pi}\frac{y_i}{r^2}$$

leading to

$$V^o(x) = q_i^o(x)\, n_i(x) - \frac{\partial M^{so}}{\partial s}(x) = D^o(x, \xi) \tag{11-8c}$$

where, in general, $s$ defines any curve of interest through $(x)$.

We shall also require the corresponding results due to a unit moment acting at $(\xi)$, which is most conveniently introduced by considering two 'equal and opposite' forces $(\psi, -\psi)$ acting at $(\xi_i)$ and $(\xi + d\xi)_i$ (Fig. 11-3), with $\mu_i$ a unit vector co-linear with $d\xi_i$.

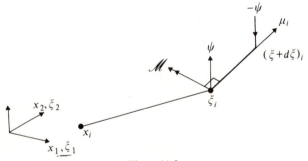

**Figure 11-3**

The applied forces define a couple $(\mathcal{M})$ of magnitude

$$|\mathcal{M}| = \psi\,|d\xi| \qquad \mathcal{M} = \psi\, d\xi \tag{11-9}$$

and if we use a superfix 1, $w^1(x)$, etc., to denote displacements, slopes, etc., due to the 'first-order' moment ($\mathcal{M}$) we have

$$w^1(x) = \psi G^o(x, \xi) - \psi G^o(x, \xi + d\xi)$$

$$= -\psi \frac{\partial G^o}{\partial \xi_i} d\xi_i = -(\psi \,|\, d\xi \,|) \frac{\partial G^o}{\partial \xi_i} \mu_i$$

that is,

$$w^1(x) = \mathcal{M}(\xi) \mu_i(\xi) \frac{\partial G^o}{\partial x_i}(x, \xi) = \mathcal{M} \mu_i G^o,_i \qquad (11\text{-}10a)$$

If, in particular, $\mathcal{M}$ is a unit moment and $\mu_i$ is directed along the normal to $S$ then the solution for a unit 'normal' boundary moment is seen to be

$$w^1(x) = \mu_i G^o,_i = G^1(x, \xi) \qquad (11\text{-}10b)$$

and, for a direction specified by $n_i$ at $x$,

$$\Theta^1(x) = -\theta_j n_j = -w^1,_j n_j = -\mu_i n_j G^o,_{ij} = F^1(x, \xi) \qquad \text{say}$$

By carrying out the appropriate differentiations, etc., a complete set of functions $C^1$, $D^1$, $E^1$, $F^1$ [all of $(x, \xi)$] corresponding to $C^o$ to $F^o$ can be obtained from (11-10b) and (11-1). Tottenham[1] points out that a hierarchy of 'higher' moment unit solutions can be developed using the steps (11-9) to (11-10b). For example, by considering two 'equal and opposite' ($\mathcal{M}, -\mathcal{M}$) couples, in lieu of the ($\psi, -\psi$) forces in Fig. 11-3, and defining a 'second-order' moment ($\mathcal{N}$) such that

$$\mathcal{N} = \mathcal{M} \, d\xi \qquad (11\text{-}11)$$

we obtain, say,

$$w^2(x) = \mathcal{N}(\xi) \mu_j(\xi) \frac{\partial G^1}{\partial x_j}(x, \xi) = \mathcal{N} \mu_j G^1,_j$$

that is,

$$w^2(x) = \mathcal{N} \mu_j \mu_i G^o,_{ij} \qquad (11\text{-}12a)$$

and if, again, $\mu_i$ is normal to $S$ then, due to $|\,\mathcal{N}\,| = 1$,

$$w^2(x) = \mu_i \mu_j G^o,_{ij} = G^2(x, \xi)$$

and

$$\Theta^2(x) = -\mu_i \mu_j \mu_k G^o,_{ijk} = F^2(x, \xi) \qquad (11\text{-}12b)$$

etc., and similarly for higher-order moments generating $w^3(x) = G^3(x, \xi)$, and so on. Tottenham[1] has solved plate problems using different combinations of these (as also have Niwa, Kobayashi, and Fukui[2]) depending upon the form of boundary condition specified. The advantages and limitations related to the use of moments higher than the first do not appear to have been explored systematically, even though they may well have important implications beyond merely plate problems.

## 11-4 INDIRECT BOUNDARY ELEMENT FORMULATION FOR THIN PLATES

By distributing 'fictitious' edge loads, $\phi^0(\xi)$, and normal boundary moments, $\phi^1(\xi)$, say, where $\xi$ is now on the boundary, and following the standard development explained in Chapter 2 (particularly in relation to the beam problem) we arrive at ($M \equiv M^n$ and with $m_i' = 0$)

$$w(x) = \int_S \{\phi^o(\xi)\, G^o(x,\xi) + \phi^1(\xi)\, G^1(x,\xi)\}\, ds(\xi) + \int_A \{\psi(z)\, G^o(x,z)\}\, dA(\xi) + c^o$$

$$\Theta(x) = \int_S \{\phi^o\, F^o + \phi^1\, F^1\}\, ds + \int_A \{\psi F^o\}\, dA + c_i^1\, n_i$$

$$M(x) = \int_S \{\phi^o\, E^o + \phi^1\, E^1\}\, ds + \int_A \{\psi E^o\}\, dA$$

$$V(x) = \int_S \{\phi^o\, D^o + \phi^1\, D^1\}\, ds + \int_A \{\psi D^o\}\, dA$$

$$(11\text{-}13)$$

in which only the first equation has its arguments written out fully.

If the real (as opposed to the fictitious) plate boundary has corners then additional 'corner force' components $\phi_c^o$, one per corner,[1,2,6,7,10] can be included in the discretized $\phi^o$ vector. The normal direction, for $\phi_c^1$, is ambiguous at such a corner but it seems reasonable to use the bisector of the corner angle to define a direction there.[1]

The three constants $(c^o, c_i^1)$ match the three conditions which $\phi^o$ and $\phi^1$ have to satisfy to ensure that the system is in equilibrium without support at any infinitely distant boundary: viz.,

$$\left.\begin{array}{c} \displaystyle\int_S \phi^o\, ds + \int_A \psi\, dA = 0 \\[1.5em] \displaystyle\int_S (\phi^1\, n_i)\, ds = 0 \end{array}\right\}$$

$$(11\text{-}14)$$

The IBEM solution can be developed by taking $(x)$ to the boundary at $(x_o)$ and noting that strong singularities arise in both the shear force boundary integral (due to $\phi^o$) and the moment one (due to $\phi^1$) which have therefore to be interpreted in terms of principal values and Cauchy integrals as before.

Thus the first two equations in (11-13) have only to be modified minimally, replacing $(x)$ by $(x_o)$, whereas the final two become

$$M(x_o) = \frac{\omega}{4\pi}\, \phi^1(x_o) + \int_S \{\phi^o(\xi)\, E^o(x_o,\xi) + \phi^1(\xi)\, E^1(x_o,\xi)\}\, ds(\xi)$$

$$+ \int_A \{\psi(z)\, E^o(x_o,\xi)\}\, dA(z) \qquad (11\text{-}15)$$

and
$$V(x_o) = \frac{\omega}{4\pi} \phi^o(x_o) + \int_S \{\phi^o D^o + \phi^1 D^1\} ds + \int_A \{\psi D^o\} dA$$

with $\omega$ the angle enclosed by the plate boundary at $(x_o)$ (that is, $\omega = 2\pi$ for a smooth boundary).

In a well-posed problem there will always be sufficient boundary information specified to enable equations with known left-hand sides to be selected from the discretized version of (11-13), (11-15) [i.e., known $w$, $\Theta$, $M$, or $V$, at $(x_o)$], which together with (11-14) will provide the distributions of $\phi^o$ and $\phi^1$ around $s$ and the three constants $c^o$, $c_i^1$. Solutions at internal points $(x)$ are then generated, as usual, from (11-13).

## 11-5 THE DIRECT BOUNDARY ELEMENT EQUATIONS

These can be written down immediately from the Maxwell–Betti reciprocal relationships between two distinct equilibrium states in an elastic body. It is quite clear that, considering only normal surface loading $\psi$ initially, one such state can be defined by $(w, \theta_i, M_i, Q, \psi)$ and the second by a corresponding starred (*) set of quantities, whence

$$\int_S (Q^* w + M_i^* \theta_i) ds + \int_A (\psi^* w) dA = \int_S (Qw^* + M_i \theta_i^*) ds + \int_A (\psi w^*) dA \quad (11\text{-}16)$$

However, we really want to work in terms of the resultant boundary shear $V$ [Eq. (11-4)] and the boundary normal moment components $(M^n)$ in order to have numbers of unknowns consistent with our fourth-order equation. We can achieve this by noting that, from (11-3),

$$\int_S M_i \theta_i^* ds = \int_S (M^n n_i - M^s s_i)(\theta^{n*} n_i - \theta^{s*} s_i) ds$$

$$= \int_S (M^n \theta^{n*} + M^s \theta^{s*}) ds = \int_S \left( M^n \Theta^* + M^s \frac{\partial w^*}{\partial s} \right) ds \quad (11\text{-}17)$$

If the partial derivative $\partial F/\partial s$ of any continuous function $F$ is integrated along a smooth curve $S$ between two points $(a, b)$ then

$$I = \int_S \frac{\partial F}{\partial s} ds = \int_a^b dF = F(b) - F(a)$$

and if $S$ is closed $a \equiv b$ and $I = 0$. However, if a corner is introduced at $C$ (Fig. 11-4), we then have

$$I = [F]_a^c + [F]_c^b = F(c^-) - F(c^+) \quad (11\text{-}18a)$$

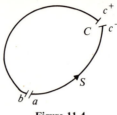

**Figure 11-4**

Thus,[7] along such a boundary $\Delta s$,

$$\int_{\Delta s} \frac{\partial}{\partial s}(M^s w^*)\, ds = \int_{\Delta s}\left(M^s \frac{\partial w^*}{\partial s} + \frac{\partial M^s}{\partial s} w^*\right) ds = (M^s_{(c^-)} - M^s_{(c^+)})\, w^* \quad (11\text{-}18b)$$

If the boundary has $N$ corners a corner force can occur at each of them and the right-hand side of (11-18b) then becomes $Q_c\, w_c^*$, from Eq. 11-6, with $c = 1, 2, ..., N$.

Therefore we can rewrite (11-17), with summation over $c$ implied, as

$$\int_S M_i\, \theta_i^*\, ds = \int_S\left(M^n\,\Theta^* - \frac{\partial M^s}{\partial s} w^*\right) ds + Q_c\, w_c^* \quad (11\text{-}19a)$$

and, similarly,

$$\int_S M_i^*\, \theta_i\, ds = \int_S\left(M^{n*}\,\Theta - \frac{\partial M^{s*}}{\partial s} w\right) ds + Q_c^*\, w_c \quad (11\text{-}19b)$$

Noting that $V = Q - \partial M^s/\partial s$ and $V^* = Q^* - \partial M^{s*}/\partial s$ substitution of Eqs (11-19) into (11-16) produces $(M \equiv M^n)$

$$\int_S (V^* w + M^*\Theta)\, ds + \int_A (\psi^* w)\, dA + Q_c^*\, w_c = \int_S (Vw^* + M\Theta^*)\, ds$$
$$+ \int_A (\psi w^*)\, dA + Q_c\, w_c^* \quad (11\text{-}20)$$

If we take the starred quantities to be the fundamental solution [Eqs (11-7), (11-8)], Eq. (11-20) becomes

$$\alpha w(\xi) - C_c^o\, w_c = \int_S (VG^o + MF^o - \Theta E^o - wD^o)\, ds + \int_A (\psi G^o)\, dA + G_c^o Q_c \quad (11\text{-}21)$$

where $\alpha = \frac{1}{2}$ for $\xi$ on $S$ and $0(1)$ for $\xi$ outside (inside) $S$ and $C_c^o$ is the difference of the $C^o$ values at each side of the corner $c$.

Equation (11-21) has to be complemented by a further one, for $\Theta(\xi) = -[\partial w(\xi)/\partial \xi_i]\, n_i(\xi)$, in order to build up sufficient equations so that, when $(\xi)$ is taken to $(\xi_o)$ on the boundary, the unspecified components of $(V, M, \Theta, w, w_c, Q_c)$ can be determined.

By writing $G' = [\partial G^o(x, \xi)/\partial \xi_i]\, n_i$, $F' = [\partial F^o/\partial \xi_i]\, n_i$, etc., all of which are easily calculated from Eqs (11-7) and (11-8), the additional equation which we need becomes

$$\alpha\Theta(\xi)+C_c^1 w_c = \int_S (VG' + MF' - \Theta E' - wD')\,ds + \int_A (\psi G')\,dA + G_c^1 Q_c \quad (11\text{-}22)$$

In order to clarify the role of the corner forces and matching displacements in Eqs (11-21) and (11-22) it is worth while listing the most common boundary conditions which occur in plate problems:

1. Clamped edges:

$$w = \Theta = 0 \qquad Q_c = w_c = 0$$

2. Simply supported edges:

$$w = M = 0 \qquad w_c = 0, \qquad Q_c \neq 0 \; \bigg\} \; \text{(on } S\text{)}$$

3. Free edges:

$$V = M = 0 \qquad Q_c = 0, \quad w_c \neq 0$$

Had we included external moments $m_i'$ in the loading vector then (11-21) would have been augmented by a further term

$$\int_A -m_i' \frac{\partial G^o(x,\xi)}{\partial x_i}\,dA$$

and (11-22) by the term

$$\int_A -m_i' \frac{\partial G^o(x,\xi)}{\partial x_i\,\xi_j} n_j(\xi)\,dA$$

The discretized form of these equations can now be used, exactly as in all other DBEM solution procedures, to determine the initially unspecified boundary information following which (11-21) and (11-22) will generate $w(\xi)$, $\Theta(\xi)$, $M(\xi)$, $V(\xi)$ at any interior point ($\xi$).

The techniques required for integrating the loading functions ($\psi G$, $\psi G'$, etc.) over the internal cell divisions of a plate are essentially identical to those explained in detail in Chapters 3 and 4; see also Ref. 1.

## 11-6 PLATES AND BEAMS ON ELASTIC SPRING SUPPORT

The simplest practicable form of elastic support to beam and plate structures is that of a series of identical, closely spaced, linear springs without any shear coupling between them—a Winkler foundation. This simplification of truly continuous elastic support is accurately realized in both floating-plate structures and plates supported on quite a large class of orthotropic elastic half spaces in which moduli increase linearly with depth, from zero at the 'ground surface'.[11] It is also used to approximate many other types of structure, particularly in civil engineering.[12]

If the spring support intensity is $k$ per unit of supported area, then the governing equation is a minor extension of (11-1d) to

$$(\psi + m'_{i,i}) = Dw_{,iijj} + kw \tag{11-23}$$

The only modifications required to the IBEM and DBEM algorithms are the use of a different and rather more complicated fundamental solution. Even this is partially mitigated by the fact that such solutions provide absolute values of displacements, are well behaved at infinity, and do not, therefore, require the auxiliary '$c$' parameters in IBEM.

Thus (11-13) and (11-15) without ($c^o$, $c^1_i$) and (11-14) provide the IBEM formulation and (11-21) and (11-22) provide the DBEM formulation again.

Before stating the fundamental solutions for the spring-supported plate it is perhaps useful to revert to another example of a 'one-dimensional' system—an extension of the simple beam of Chapter 2 to its spring-supported counterpart. This is discussed in some detail in Ref. 13 and only the key results are given here.

First, the IBEM equations can be written down immediately, following the pattern of the first equation of (11-13) and Chapter 2, as

$$w(x) = [\phi^o(\xi) G^o(x, \xi) + \phi^1(\xi) G^1(x, \xi)]^{\xi = L}_{\xi = 0}$$
$$+ \int_L [\psi(z) G^o(x, z) + m(z) G^1(x, z)] \, dL \tag{11-24}$$

together with similar equations for $\Theta$, $M$, and $V$, where $L$ is the length of the beam. Four equations corresponding to (11-15) follow immediately, from which the four unknowns $[\phi^o(L^-), \phi^o(0^+), \phi^1(L^-), \phi^1(0^+)]$ can be found. The DBEM equations are equally simply stated and merely involve replacing $\int_s( ) \, ds$ by $[ \ ]^L_o$ and $\int_A( ) \, dA$ by $\int_L( ) \, dL$ in (11-21), (11-22) and removing the corner force terms. The following list of the relevant unit solutions to the governing spring-supported beam equation

$$\psi(x) + \frac{dm'(x)}{dx} = EI \frac{d^4 w}{dx^4} + kw \tag{11-25}$$

should enable the reader to solve all such problems quite readily [$\beta = \sqrt[4]{(k/4EI)}$ and $r = (x - \xi)$]:

For $\psi = 1$:

$$w^o(x) = G^o(x, \xi) = \frac{\beta}{2k} e^{-\beta r} (\cos \beta r + \sin \beta r)$$

$$\theta^o(x) = \Theta^o(x) = F^o(x, \xi) = \frac{\beta^2}{k} e^{-\beta r} (\sin \beta r) \operatorname{sgn}(\xi - x)$$

$$M^o(x) = m^o(x) = E^o(x, \xi) = \frac{e^{-\beta r}}{4\beta} (\cos \beta r - \sin \beta r)$$

$$V^o(x) = Q^o(x) = q^o(x) = D^o(x, \xi) = \frac{e^{-\beta r}}{2} (\cos \beta r) \operatorname{sgn}(\xi - x)$$

For $m' = 1$:

$$w^1(x) = G^1(x, \xi) = F^o(x, \xi)$$

$$\theta^1(x) = F^1(x, \xi) = \frac{-4\beta^4}{k} E^o(x, \xi)$$

$$M'(x) = E^1(x, \xi) = D^o(x, \xi)$$

$$V^1(x) = D^1(x, \xi) = \frac{k}{\beta} G^o(x, \xi)$$

The corresponding solutions for the spring-supported plate were apparently discussed by Hertz (1884) and first presented in Bessel function form by Sleicher (1926) (see Ref. 9). The fundamental solution for unit vertical load on an infinite-extent spring-supported plate is quoted by Tottenham[1] as

$$w^o(x) = G^o(x, \xi) = \frac{1}{4\beta^2 D} H_o(\beta r) \tag{11-26a}$$

and hence from (11-10b) for a unit boundary normal moment

$$w^1(x) = n_i G^o_{,i} = G^1(x, \xi) = \frac{n_i y_i}{4\beta Dr} \frac{\partial H_o}{\partial r} = F^o(x, \xi) \tag{11-26b}$$

where, as before, $r^2 = y_i y_i$, $y_i = (x - \xi)_i$, but $\beta = \sqrt[4]{(k/D)}$ and $H_o(\beta r)$ represents the real part of a Hankel function of the first kind with argument $(\beta\sqrt{i})$.

Utilizing Eqs (11-1) and (11-2) once more the remaining $G^o$ to $D^o$ and $G^1$ to $D^1$ functions can be evaluated, albeit at some analytical complexity. An alternative approach is set out in the following section which is equally applicable to plates with simple spring support.

## 11-7 PLATES SUPPORTED ON AN ELASTIC HALF SPACE

Timoshenko and Woinowsky-Krieger[9] discuss a solution, in the form of an infinite integral, for a loaded plate on quite general elastic support due to Holl (1938) which reduces, for a simple elastic half space, to

$$w^o(x) = \frac{l^3}{2\pi D} \int_o^\alpha \frac{J_o(\alpha r) \, d\alpha}{1 + (\alpha l)^3} = G^o(x, \xi) \tag{11-27}$$

with $J_o(\alpha r)$ a zero-order Bessel function and $l^3 = 2D(1 - v_o^2)/E_o^2$, $(E_o, v_o)$ being Young's modulus and Poisson's ratio for the supporting half space. Whereas one can, in principle, derive all of $G^o$ to $D^o$ and $G^1$ to $D^1$ from (11-27) and use these directly in either the IBEM (11-14) or the DBEM (11-21), (11-22) formulations, the difficulties are formidable and the following alternative procedure is probably much more attractive.

The essential difference is that the plate and the half space are treated separately as components of a 'two-zone' problem with compatibility enforced at the plate-foundation interface.

As an example, consider a DBEM formulation for a plate under vertical loading with the simplest continuity requirement across a smooth interface—that of matching vertical displacements only. (Note that interface couples would be required in addition if local rotations were to be matched; for a perfectly bonded plate, surface tractions, and therefore membrane forces, would arise). The final discretized matrix form of the DBEM equations for the plate alone will be, say,

$$[W]\{w\} + [X]\{\Theta\} + [Y]\{M\} + [Z]\{V\} = [A]\{\psi\} - [B]\{p\} \quad (11\text{-}28)$$

in which all of the coefficient matrices ($W$ to $Z$, $A$, $B$), the loading vector $\psi$, and half of the total number of components in the ($w$, $\Theta$, $M$, $V$) vectors will be known. For, say, $n$ distinct values of each boundary variable the left-hand side of (11-28) will comprise $(2n \times 2n)(2n \times 1)$ matrix and vector pairs, the $[A]\{\psi\}$ terms will be $(2n \times l)(l \times 1)$, for $l$ loaded cells, and the final $[B]\{p\}$ terms will be $(2n \times m)(m \times 1)$ which represent the foundation support reaction vector $\{p\}$ acting over the whole plate surface divided into cells producing $m$ distinct terms.

The half-space loading is now also $\{p\}$ over an identical pattern of surface cells and this will generate an $m \times 1$ displacement vector $\{\bar{w}\}$, say, where

$$\{p\} = [C]\{\bar{w}\} \quad (11\text{-}29)$$

If only a simple spring support foundation is being modelled then $C$ will be a diagonal matrix; otherwise it will be fully populated, with terms calculated using either Boussinesq's solution integrated over the cells for a uniform half space or any other surface-load solutions available for anisotropic or inhomogeneous half spaces.

The components of the $\{\bar{w}\}$ vector can also be calculated from the DBEM equations in terms of ($w$, $\Theta$, $M$, $V$, $\psi$, $p$) in the form, say,

$$\{\bar{w}\} = [\bar{W}]\{w\} + [\bar{X}]\{\Theta\} + [\bar{Y}]\{M\} + [\bar{Z}]\{V\} + [\bar{A}]\{\psi\} - [\bar{B}]\{p\} \quad (11\text{-}30)$$

Substitution of (11-29) into (11-30) provides an equation relating $\{p\}$ to ($w$, $\Theta$, $M$, $V$, $\psi$). $\{p\}$ can then be eliminated between this equation and (11-28) to produce $2n$ equations for the unknown components in ($w$, $\Theta$, $M$, $V$). Thereafter (11-28) generates $\{p\}$ and (11-29) $\{\bar{w}\}$. This is, of course, a well-known procedure for solving foundation–structure interaction problems, but illustrates again how BEM operate very conveniently across interfaces, this time coupling the boundary of one zone to the internal cells of another one.

## 11-8 EXAMPLES

All the published examples which we have located refer solely to plates without elastic support.

**(a) A uniformly loaded square plate with a central square opening** Tottenham[1] analysed this problem by both direct and indirect BEM. The outer edges of the plate were simply supported and the inner edges free. In the IBEM solution, corner forces were ignored and the fictitious moments at the corners applied to one-quarter of the symmetrical system (Fig. 11-5) with nodes at the quarter points along each side. For the direct method solution, using a 'boundary' exterior to $s$, linear variation of parameters was used between the 64 total nodes of which 56 had two nodal unknowns whilst the eight corner ones had three unknowns each.

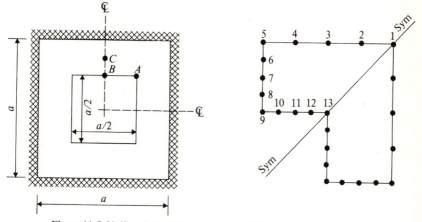

**Figure 11-5** Uniformly loaded square plate with central square opening.

Symmetry reduced the final number of unknowns to 20. Table 11-1 shows a comparison between the displacements calculated at typical points by BEM, finite element, and finite difference analyses, the latter two using mesh sizes of $a/16$ ($v = 0.25$). He also provides other examples, in one of which the BEM equations had been made symmetrical by using a least-squares error technique.

**(b) Clamped plates by an indirect method** Altiero and Sikarskie[3] present solutions to completely clamped 2 to 1 rectangular, equilateral, triangular, and semi-circular plates using a form of IBEM which utilizes the clamped circular

**Table 11-1**

| | Displacement $\times qa^4/100D$ | | | |
|---|---|---|---|---|
| | Boundary element | | | |
| Point | Indirect | Direct | Finite element | Finite difference |
| A | 0.2188 | 0.2188 | 0.2185 | 0.2174 |
| B | 0.3107 | 0.3141 | 0.3156 | 0.3006 |
| C | 0.1558 | 0.1565 | ... | 0.1541 |

plate unit solution to generate the kernels of the various integrals. This form of unit solution is an interesting illustration of an alternative, but algebraically much more cumbersome, formulation to the one which we have used. Table 11-2 shows comparisons between various moments and displacements calculated both by their IBEM technique and analytically.

**Table 11-2** Comparison of selected interior deflections, bending moments with exact solutions

| Plate geometry | Deflection moments | Integral method | Exact[9] | Difference, % |
|---|---|---|---|---|
| Rectangular (point load) | $\omega(0,0)$ | 0.00179 | 0.00180 | −0.56 |
| Rectangular (uniform) | $\omega(0,0)$ | 0.000311 | 0.000317 | −1.89 |
| Rectangular (uniform) | $M_x(0,0)$ | 0.0197 | 0.0206 | −4.37 |
| Rectangular (uniform) | $M_y(0,0)$ | 0.0073 | 0.0079 | −7.59 |
| Triangular | $M_x(0,0)$ | 0.0179 | 0.0196 | −8.67 |
| Triangular | $M_y(0,0)$ | 0.0180 | 0.0191 | −5.76 |
| Semi-circular | $\omega(0,0.406)$ | 0.00128 | 0.00129 | −0.78 |
| Semi-circular | $M_x(0,0525)$ | 0.01219 | 0.01235 | −1.30 |
| Semi-circular | $M_y(0,0.483)$ | 0.0222 | 0.0226 | −1.77 |

**(c) A detailed study of a clamped circular plate with a central load using IBEM and an auxiliary boundary** Niwa, Kobayashi, and Fukui[2] present a study of this problem using an IBEM technique with an auxiliary boundary located at $\delta/a$ outside the circular plate (radius $= a$). They used $N$ straight boundary elements around the periphery of the plate with a uniform distribution of sources (both simple transverse forces and higher-order moments) on each. Table 11-3 shows their results for displacements and radial and tangential moment components compared with analytical solutions over a range of $N$, $\delta/a$, and $r/a$ values. In their paper are also similar comparisons for both simply supported and uniformly loaded circular and rectangular plates. The maximum errors are always in the circumferential moment components ($r/a = 1$) and seem to depend on $N$ rather than $\delta/a$ ($0.05 < \delta/a < 0.2$) for their examples.

Two very recent publications describing the application of DBEM to plate problems are Danson[14] and Stern,[15] both of which contain analyses of simply supported plates. The first of these also provides full details of all the equations needed to evaluate the components of the matrices in equations such as (11-28). Finally, DBEM statements for elastic bending of thin plates have now been extended to included non-linear bending by Marjaria and Mukherjee.[16]

## 11-9 CONCLUDING REMARKS

The plate (and elastically supported beam) analyses presented in this chapter follow the established pattern of IBEM and DBEM formulation and, to that extent, are an advance on the rather ad hoc methods published elsewhere.

**Table 11-3** Deflections and radial and tangential moments of a circular clamped plate subjected to a concentrated load at its centre computed from the boundary element method compared with exact solutions

$w_c(\times Pa^2/8\pi D)$

| $N$ | $\delta/a$ | $r/a$ 0.02 | 0.2 | 0.4 | 0.6 | 0.8 | 1.0 |
|-----|-----------|------------|-----|-----|-----|-----|-----|
| 12 | 0.05 | 0.502 | 0.420 | 0.277 | 0.140 | 0.040 | 0.0 |
|     | 0.2 | 0.490 | 0.407 | 0.267 | 0.132 | 0.035 | 0.0 |
| 24 | 0.05 | 0.495 | 0.412 | 0.271 | 0.134 | 0.036 | 0.0 |
|     | 0.2 | 0.497 | 0.414 | 0.272 | 0.135 | 0.037 | 0.0 |
| 48 | 0.05 | 0.497 | 0.414 | 0.272 | 0.135 | 0.037 | 0.0 |
|     | 0.2 | 0.498 | 0.416 | 0.273 | 0.136 | 0.037 | 0.0 |
| Exact | | 0.498 | 0.416 | 0.273 | 0.136 | 0.037 | 0.0 |

$M_{rr}(\times P/4\pi)$

| $N$ | $\delta/a$ | $r/a$ 0.02 | 0.2 | 0.4 | 0.6 | 0.8 | 1.0 |
|-----|-----------|------------|-----|-----|-----|-----|-----|
| 12 | 0.05 | 4.085 | 1.092 | 0.191 | $-0.332$ | $-0.691$ | $-1.761$ |
|     | 0.2 | 4.071 | 1.077 | 0.176 | $-0.349$ | $-0.711$ | $-1.024$ |
| 24 | 0.05 | 4.080 | 1.087 | 0.186 | $-0.342$ | $-0.713$ | $-1.202$ |
|     | 0.2 | 4.084 | 1.091 | 0.189 | $-0.338$ | $-0.711$ | $-0.990$ |
| 48 | 0.05 | 4.083 | 1.090 | 0.189 | $-0.338$ | $-0.712$ | $-1.010$ |
|     | 0.2 | 4.086 | 1.092 | 0.191 | $-0.336$ | $-0.710$ | $-1.000$ |
| Exact | | 4.086 | 1.092 | 0.191 | $-0.336$ | $-0.710$ | $-1.000$ |

$M_{\theta\theta}(\times P/4\pi)$

| $N$ | $\delta/a$ | $r/a$ 0.02 | 0.2 | 0.4 | 0.6 | 0.8 | 1.0 |
|-----|-----------|------------|-----|-----|-----|-----|-----|
| 12 | 0.05 | 4.785 | 1.792 | 0.891 | 0.358 | 0.092 | 1.412 |
|     | 0.2 | 4.771 | 1·777 | 0.876 | 0.347 | $-0.052$ | $-0.522$ |
| 24 | 0.05 | 4.780 | 1.787 | 0.886 | 0.358 | $-0.021$ | $-0.758$ |
|     | 0.2 | 4.784 | 1.791 | 0.889 | 0.362 | $-0.012$ | $-0.334$ |
| 48 | 0.05 | 4.783 | 1.790 | 0.889 | 0.362 | $-0.012$ | $-0.443$ |
|     | 0.2 | 4.786 | 1.792 | 0.891 | 0.364 | $-0.010$ | $-0.301$ |
| Exact | | 4.786 | 1.792 | 0.891 | 0.364 | $-0.010$ | $-0.300$ |

The thin-plate problem is not only of considerable practical interest but also illustrates how the well-known limitations of the two-dimensional theory, which approximates a three-dimensional problem, can be dealt with by BEM. In addition, the extension of the analysis to include elastically supported plates provides examples of unit solutions of ever-increasing complexity such that, at some point, the attraction of using a uniform algorithm for all of them is outweighed by the intractability of the unit solution itself. The plate and its elastic support are then best separated and treated as a special form of 'two-zone' problem, where the interface links the boundary of one zone with the internal cells of the other. This is seen to be accomplished quite easily by BEM and opens the way to solving a variety of problems involving inhomogeneous slabs on continuous foundations of different types. A further interesting question concerns the possibility of using even simpler unit solutions, e.g., that for the free uniform plate to solve an elastically supported non-uniform one. Some discussion of this problem in relation to elastically supported beams and potential flow will be found in Refs 13 and 17.

The flat, thin, spring-supported plate is closely related to the thin shallow shell. Tottenham[1] and Newton and Tottenham[18] have published analyses of such problems using boundary integral techniques.

## 11-10 REFERENCES

1. Tottenham, H. (1979) 'The boundary element method for plates and shells', in P. K. Banerjee and R. Butterfield (eds), *Developments in Boundary Element Methods*, Applied Science Publishers, London.
2. Niwa, Y., Kobayashi, S., and Fukui, T. (1974) 'An application of the integral equation method to plate bending', Memo, Vol. 36, Pt 2, pp. 140–158, Faculty of Engineering, Kyoto University, Japan.
3. Altiero, N. J., and Sikarskie, D. L. (1978) 'A boundary integral method applied to plates of arbitrary plan form', *Composition and Structs*, **9**, 163–168.
4. Jaswon, M. A., and Maiti, M. (1968) 'An integral formulation of plate bending problems', *J. Eng. Math.*, **2**(1), 83–93.
5. Jaswon, M. A., and Symm, G. T. (1977) *Integral Methods in Potential Theory and Elastostatics*, Academic Press, London.
6. Bezine, G. P., and Gamby, D. A. (1978) 'A new integral equation formulation for plate bending problems', in C. A. Brebbia (ed.), *Recent Advances in Boundary Element Methods*, Pentech Press, London.
7. Bezine, G. P. (1979) 'Boundary integral formulation for plate flexure with arbitrary boundary conditions', *Mech. Res. Commun.* **5**(4),
8. Segdin, C. M., and Bricknell, G. A. (1968) 'Integral equation method for a corner plate,' *J. Struct. Div., ASCE,* **ST.1**, 43–51.
9. Timoshenko, S., and Woinowsky-Krieger, S. (1959) *Theory of Plates and Shells*, McGraw-Hill, New York.
10. Bergman, S., and Schiffer, M. (1953) *Kernel Functions and Elliptic Differential Equations in Mathematical Physics*, New York.
11. Butterfield, R. (1980) 'Simple potential solutions for a class of normally loaded inhomogeneous anisotropic elastic half spaces' (to be published).
12. Hetenyi, M. (1946) *Beams on Elastic Foundation*, University of Michigan Press.
13. Banerjee, P. K., and Butterfield, R. (eds) (1979) *Developments in Boundary Element Methods*, Vol. I, Applied Science Publishers, London.
14. Danson, D. J. (1979) 'Analysis of plate bending problems by the direct boundary element method', M.Sc. dissertation, University of Southampton.
15. Stern, M. (1979) 'A general boundary integral formulation for the numerical solution of plate bending problems', *Int. J. Solid Structs*, **15**, 769–782.

16. Morjaria, M., and Mukherjee, S. (1979) 'Inelastic analysis of transverse deflection of plates by the boundary element method', DOE Rept No: COO-2733-24, Department of Theoretical and Applied Mechanics, Cornell University.
17. Butterfield, R. (1978) 'An application of the boundary element method to potential flow problems in generally inhomogeneous bodies', in C. A. Brebbia (ed.), *Recent Advances in Boundary Element Methods*, Pentech Press, London.
18. Newton, D. A., and Tottenham, H. (1968) 'Boundary value problems in thin shallow shells of arbitrary plan form', *J. Eng. Math.*, **2**(3), 211–224.

# TWELVE

## ELASTOPLASTICITY

## 12-1 INTRODUCTION

In previous chapters we have dealt with problems governed by linear differential equations and explored the essential features of BEM algorithms based on the boundary integral equations developed from them. However, in order to qualify as a completely general problem-solving tool, it is essential that BEM must be demonstrably applicable to non-linear systems. Non-linearities do occur in almost every realistic idealization of practical problems.

Like any other general numerical method, such as finite element or finite difference methods, BEM are fully capable of solving non-linear differential equations by an incremental or iterative procedure—in this case via a volume integral defined over the region within which the non-linearities occur. For the vast majority of such problems the non-linear regions are mainly confined to small subregions of the system and it will be found that, particularly for three-dimensional systems, BEM provides a relatively attractive tool for dealing with non-linearities. Indeed, for a large class of practical problems the method appears likely to be the only reliable means of getting adequately detailed results at reasonable cost. It has already been demonstrated that BEM provide, very efficiently, numerical solutions to linear problems and therefore the introduction of an additional volume integral over part of the body would not be expected to affect their efficiency dramatically. This will be demonstrated in the present chapter.

# 12-2 CONSTITUTIVE RELATIONSHIPS FOR SOLIDS

In order to appreciate the essential features of the solution process it is important to explore first the behaviour of an infinitesimal element of the region. This is described by the constitutive relations for the materials concerned. Many such models have been proposed to cover the wide range of materials that engineers have to deal with and a detailed discussion of these would be out of place in this book. We shall therefore outline below only three, well-established theories which model the major non-linear features of certain classes of solids.

## 12-2-1 Incremental Theory of Plasticity

The necessary ingredients of an incremental stress-strain relationship based on the 'path independence in the small' theory of plasticity (the incremental theory of plasticity) are

1. A yield function defining the limits of the elastic behaviour
2. A flow rule relating the irrecoverable plastic part of the strain rate to the stress state in the material
3. A hardening rule defining the subsequent yield surfaces resulting from continuous plastic deformation

   The nature of the yield surface and the hardening parameters are obviously dependent on the type of material. Therefore, we need to derive stress-strain relationships for quite general cases of (a) isotropic and (b) translational hardening.

**(a) Isotropic hardening** In this theory it is assumed that, during plastic flow, the yield surface expands uniformly about the origin in stress space maintaining its shape, centre, and orientation as shown in Fig. 12-1. For example, the path $OA$ is elastic; at $A$ the material is in a state of incipient yielding and, upon further loading, the yield surface expands to $B$. The path $AB$ represents isotropic strain-hardening behaviour. If we unload along $BC$ the behaviour will be elastic until point $C$ is reached, when the material will start to yield again. If we assume that the continuously changing yield surfaces can be represented by a loading function of type

$$F(\sigma_{ij}, \varepsilon_{ij}^p, h) = 0 \qquad (12-1)$$

where $h$ is a hardening parameter, $\varepsilon_{ij}^p$ the current total plastic strain in the material element, and for plastic states $F = 0$, then for elastic states $F < 0$ and $F > 0$ is meaningless.

Since during plastic deformation the new stress state must lie on a newly developed yield surface characterized by a new strain-hardening parameter $(h + \dot{h})$

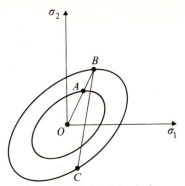

**Figure 12-1** Isotropic hardening.

(where a superior dot will be used to denote an increment, $\dot{h} = dh$, etc.), we have

$$dF = \frac{\partial F}{\partial \sigma_{ij}} \dot{\sigma}_{ij} + \frac{\partial F}{\partial \varepsilon_{ij}^{p}} \dot{\varepsilon}_{ij}^{p} + \frac{\partial F}{\partial h} \dot{h} = 0 \tag{12-2}$$

Using Drucker's postulate for associated plastic flow[1] we obtain

$$\dot{\varepsilon}_{ij}^{p} = \dot{\lambda} \frac{\partial F}{\partial \sigma_{ij}} \tag{12-3}$$

where $\lambda$ is a non-negative scalar variable.

Since the hardening parameter $h$ is a function of plastic strain we can write Eq. (12-2) as

$$\frac{\partial F}{\partial \sigma_{ij}} \dot{\sigma}_{ij} + \frac{\partial F}{\partial \varepsilon_{ij}^{p}} \dot{\varepsilon}_{ij}^{p} + \frac{\partial F}{\partial h} \frac{\partial h}{\partial \varepsilon_{ij}^{p}} \dot{\varepsilon}_{ij}^{p} = 0 \tag{12-4}$$

Substituting (12-3) in (12-4) and solving for $\lambda$ we get

$$\lambda = -\frac{(\partial F/\partial \sigma_{\kappa l}) \dot{\sigma}_{\kappa l}}{[\partial F/\partial \varepsilon_{mn}^{p} + (\partial F/\partial h)(\partial h/\partial \varepsilon_{mn}^{p})] \partial F/\partial \sigma_{mn}} \tag{12-5}$$

Therefore the plastic strain increment can be obtained from[2]

$$\dot{\varepsilon}_{ij}^{p} = G \frac{\partial F}{\partial \sigma_{ij}} \frac{\partial F}{\partial \sigma_{\kappa l}} \dot{\sigma}_{\kappa l} \tag{12-6}$$

where

$$G = -\frac{1}{[\partial F/\partial \varepsilon_{mn}^{p} + (\partial F/\partial h)(\partial h/\partial \varepsilon_{mn}^{p})] \partial F/\partial \sigma_{mn}}$$

Equation (12-6) together with the elastic component of the strain increment ($\dot{\varepsilon}_{ij}^{e}$) may be written as

$$\dot{\varepsilon}_{ij} = \dot{\varepsilon}_{ij}^{e} + \dot{\varepsilon}_{ij}^{p} = C_{ij\kappa l}^{ep} \dot{\sigma}_{\kappa l} \tag{12-7}$$

or

$$\dot{\sigma}_{ij} = D_{ij\kappa l}^{ep} \dot{\varepsilon}_{\kappa l} \tag{12-8}$$

with $D_{ij\kappa l}^{ep}$ a fourth-rank tensor quantity, the incremental elastoplastic material modulus. Equations (12-7) and (12-8) provide the necessary incremental stress-strain relations for a general case of isotropic hardening. Specifically, if we use an isotropic strain-hardening Von Mises material we can write Eq. (12-8) explicitly as

$$\dot{\sigma}_{ij} = 2\mu \left[ \dot{\varepsilon}_{ij} + \frac{v}{1-2v} \delta_{ij} \dot{\varepsilon}_{\kappa\kappa} - \frac{3 S_{ij} S_{\kappa l}}{2\sigma_o^2 (1 + H/3\mu)} \dot{\varepsilon}_{\kappa l} \right]$$ (12-9)

where
$\mu$ = elastic shear modulus

$\sigma_o$ = uniaxial yield stress = $\sqrt{(\tfrac{3}{2} S_{ij} S_{ij})}$

$S_{ij}$ = deviatoric stress = $\sigma_{ij} - \tfrac{1}{3}\delta_{ij}\sigma_{\kappa\kappa}$

$H$ = plastic-hardening modulus, the current slope of the uniaxial plastic stress-strain curve.

With a proper definition of the yield function $F$ and the hardening parameter $h$ in Eq. (12-2) we can derive equations similar to (12-9) for isotropic yielding and hardening of any material.[3-5]

**(b) Kinematic hardening** The kinematic hardening theory of plasticity, originally developed by Prager,[6] assumes that during plastic deformation the loading surface (or the yield surface) translates as a rigid body in stress space. The basic aim of this theory was to model the Bauschinger effect which is apparent during the cyclic loading of metals.

Figure 12-2 shows typical kinematic hardening behaviour. The loading path $OA$ is elastic. At $A$ yielding starts and the loading path $AB$ initiates elastoplastic kinematic hardening. The translation of the yield surface during this loading causes its centre to move from $O$ to $O'$. Any unloading from $B$ along $BC$ results in purely elastic behaviour until the loading path reaches $C$, when the material yields once again—now at a lower yield stress.

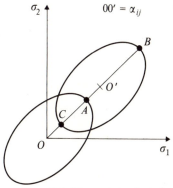

**Figure 12-2** Kinematic hardening.

If we assume that the translation of the centre of the yield surface can be represented by a tensor $\dot{\alpha}_{ij}$, then the loading function may be written as

$$F(\sigma_{ij}, \alpha_{ij}) = 0$$ (12-10)

and the consistency relation during the loading becomes

$$dF = \frac{\partial F}{\partial \sigma_{ij}} \dot{\sigma}_{ij} + \frac{\partial F}{\partial \alpha_{mn}} \frac{\partial \alpha_{mn}}{\partial \varepsilon_{ij}^p} \dot{\varepsilon}_{ij}^p = 0 \tag{12-11}$$

if the translation $\alpha_{mn}$ is assumed to be a function of the plastic strain.

Equation (12-11) represents the kinematic hardening theory of Prager which postulates that the translation increments ($\dot{\alpha}_{ij}$) of the loading surface, in nine-dimensional stress space, occur in the direction of the outward normal to the surface at the current stress state and therefore

$$\dot{\alpha}_{ij} = C\dot{\varepsilon}_{ij}^p \tag{12-12}$$

where $C$ now replaces the hardening parameter $H$ in Eq. (12-9).

By using Eqs (12-3) and (12-12) in Eq. (12-11) we can obtain the plastic flow factor $\dot{\lambda}$ and thus derive an incremental stress-strain relationship in the manner discussed for isotropic hardening.

This was the theory originally proposed by Prager, but when its consequences are investigated in certain subspaces of stress space several inconsistencies arise, as discussed by Shields and Ziegler.[7] To avoid these difficulties Ziegler[8] proposed a modification to Prager's hardening rule. Instead of Eq. (12-12) he proposed that

$$\dot{\alpha}_{ij} = \dot{\mu}_o(\sigma_{ij} - \alpha_{ij}) \tag{12-13}$$

where $\dot{\mu}_o$ is a positive scalar parameter.

Figure 12-3 illustrates the difference between the Prager and the Ziegler postulates. The increment of translation $\dot{\alpha}_{ij}$ in Ziegler's theory is co-linear with the radius vector from the centre of the yield surface $O'$ to the stress point. The

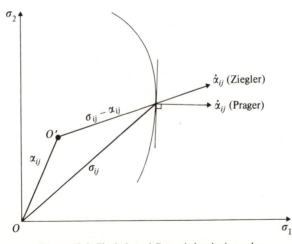

**Figure 12-3** Ziegler's and Prager's hardening rules.

scalar function $\dot{\mu}_o$ can be determined by using the 'consistency condition' that

$$dF(\bar{\sigma}_{ij}) = 0 \qquad \text{where } \bar{\sigma}_{ij} = \sigma_{ij} - \alpha_{ij}$$

that is,

$$\frac{\partial F}{\partial \bar{\sigma}_{ij}} (\dot{\sigma}_{ij} - \dot{\alpha}_{ij}) = 0$$

Substituting Eq. (12-13) in the above gives

$$\dot{\mu}_o = \frac{(\partial F/\partial \sigma_{ij}) \dot{\sigma}_{ij}}{(\sigma_{\kappa l} - \alpha_{\kappa l}) \partial F/\partial \sigma_{\kappa l}} \tag{12-14}$$

In order to determine the plastic strain we have to determine $\dot{\lambda}$ in the flow rule (12-3). Ziegler assumed that the vector $C\dot{\varepsilon}_{ij}^p$ is the projection of $\dot{\sigma}_{ij}$ on the outward normal through the instantaneous stress state as shown in Fig. 12-4. This

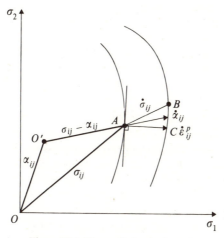

**Figure 12.4** Kinematic hardening rule.

assumption is consistent with the 'path independent in the small' theory of plasticity and we can therefore write

$$(\dot{\sigma}_{ij} - C\dot{\varepsilon}_{ij}^p) \frac{\partial F}{\partial \sigma_{ij}} = 0$$

Substituting Eq. (12-3) in the above we have

$$\dot{\lambda} = \frac{1}{C} \frac{(\partial F/\partial \sigma_{ij}) \dot{\sigma}_{ij}}{(\partial F/\partial \sigma_{\kappa l})(\partial F/\partial \sigma_{\kappa l})}$$

Therefore the plastic strain increments are given by

$$\dot{\varepsilon}_{ij}^p = \frac{1}{C} \frac{(\partial F/\partial \sigma_{ij})(\partial F/\partial \sigma_{mn}) \dot{\sigma}_{mn}}{(\partial F/\partial \sigma_{\kappa l})(\partial F/\partial \sigma_{\kappa l})} \tag{12-15}$$

The elastic and plastic strain components may be added to provide the complete stress-strain relations [Eqs (12-7) or (12-8)]. Specifically, if we choose a Von Mises yield criterion, i.e.,

$$F = \tfrac{1}{2}\bar{S}_{ij}\,\bar{S}_{ij} - \tfrac{1}{3}\bar{\sigma}_o^2 = 0$$

where

$$\bar{S}_{ij} = \bar{\sigma}_{ij} - \tfrac{1}{3}\delta_{ij}\,\bar{\sigma}_{\kappa\kappa}$$

$$\bar{\sigma}_{ij} = \sigma_{ij} - \alpha_{ij}, \quad \text{which may be called a reduced stress state}$$

and $\qquad \bar{\sigma}_o = \sqrt{(\tfrac{3}{2}\bar{S}_{ij}\,\bar{S}_{ij})}, \quad$ the reduced uniaxial yield stress

we can write the explicit stress-strain relations as

$$\dot{\sigma}_{ij} = 2\mu\left[\dot{\varepsilon}_{ij} + \frac{v}{1-2v}\delta_{ij}\dot{\varepsilon}_{\kappa\kappa} - \frac{3\bar{S}_{ij}\,\bar{S}_{\kappa l}}{2\bar{\sigma}_o^2(1+C/2\mu)}\dot{\varepsilon}_{\kappa l}\right] \qquad (12\text{-}16)$$

and $\qquad \dot{\alpha}_{ij} = C\dot{\varepsilon}_{ij}^p = \dfrac{3\bar{S}_{ij}\,\bar{S}_{\kappa l}}{2\bar{\sigma}_o^2}\dot{\sigma}_{\kappa l}$

It is of interest to note that since the normal to the Von Mises yield surface is co-linear with the line joining the centre of the yield surface and the material point, the stress-strain relations derived from Prager's and Ziegler's hardening rules become identical.

By examining the uniaxial loading case it is easy to see that the relationship between $C$ and $H$ is $C = 2H/3$. Equations (12-9) and (12-16) are the stress-strain relationships that we need for carrying out a non-linear stress analysis involving a Von Mises material. Similar stress-strain relations can be derived[9] for any material to describe its mechanical behaviour under monotonic and cyclic loading provided that an appropriate choice of the yield functions $F$ and the hardening parameters are made from physical test data. An interesting alternative kinematic hardening model was proposed by Mroz.[9] Equations (12-9) and (12-16) can be integrated along a prescribed loading path to obtain the current state of both stresses and strains.

## 12-2-2 Viscoplasticity

A viscoplastic model is often used to describe time-dependent inelastic deformation in solids. One of the most attractive features of this model is that steady state viscoplasticity turns out to be an alternative description of elastoplastic behaviour.

In such a model a set of viscoplastic strains $\varepsilon_{ij}^{vp}$ are present in addition to the elastic strains $\varepsilon_{ij}^e$. Whenever a certain threshold value of stress (such as yield) is exceeded a non-zero value of $\dot{\varepsilon}_{ij}^{vp}$ will occur. (A comprehensive account of the present state of development of viscoplasticity can be found in a review by Perzyna[10].) Such a viscoplastic model was used by Zienkiewicz and

Cormeau[11-13] in a very efficient finite element algorithm for a non-linear deformation analysis of solids.

In viscoplasticity the stress state can be outside the yield surface but then creep occurs. When the creep stops the stress state returns to the current yield surface, an elastoplastic solution is established, and it is therefore often possible to use viscoplasticity as a purely fictitious device to obtain elastoplastic solutions. There are, of course, in addition many problems where a viscoplastic idealization is used to represent the time-dependent behaviour.

Viscoplastic strains can be defined if a yield condition is specified as

$$F(\sigma_{ij}, h) = F^o(\sigma_{ij}, h) - Y(h) = 0 \qquad (12\text{-}17)$$

where $\quad Y(h)$ = yield stress which is in turn a function of the magnitude of the hardening ($h$)

$\quad F < 0$ indicates purely elastic behaviour

If we define a plastic potential function $Q(\sigma_{ij})$ we can obtain the viscoplastic strain rate[12] from

$$\frac{\partial \varepsilon_{ij}^{vp}}{\partial t} = \dot{\varepsilon}_{ij}^{vp} = \langle \phi(F) \rangle \frac{\partial Q}{\partial \sigma_{ij}} \qquad (12\text{-}18)$$

where $\langle \phi(F) \rangle$ is a viscoplastic flow factor which may be dependent upon certain state variables such as time, strain invariants, etc.

To ensure that there is no viscoplastic strain in the elastic regime we can require the function $\phi$ to satisfy the following conditions:

$$\langle \phi \rangle = 0 \qquad \text{if } F \leqslant 0, \text{ in the elastic region}$$
$$\langle \phi(F) \rangle = \phi(F) \qquad \text{if } F > 0, \text{ in the viscoplastic region} \qquad (12\text{-}19)$$

A sufficiently general viscoplastic strain rate expression for our purposes is[12]

$$\phi(F) = \gamma F^n \qquad (12\text{-}20)$$

where $\gamma$ is a fluidity parameter which again may be a function of time, strain invariants, etc., and $n$ a suitable material parameter.

If the plastic flow is assumed to be governed by the associated flow rule (i.e., the plastic strain increment vector is normal to the yield surface at the material point) then $Q = F$; on the other hand, if the flow is non-associated the plastic potential function $Q$ may be chosen to be different from $F$.[5, 10, 12]

By defining $Q$ and/or $F$ surfaces for a given material and using Eqs (12-19) and (12-20) we can express Eq. (12-18) as[10, 12]

$$\dot{\varepsilon}_{ij}^{vp} = C_{ij\kappa l}^{vp} \sigma_{\kappa l} \qquad (12\text{-}21)$$

Equation (12-21) differs from that developed for incremental elastoplasticity in one, most important, respect—it specifies the viscoplastic strain (or the non-linear strain) rate as a function of the current stress state. In the incremental theory of plasticity, the fact that the strain rates are functions not only of the current stress level but also of the stress increments is the major source of difficulty in all solution processes (compare Eqs (12-21) and (12-7)).

## 12-2-3 State Variable Theories for Inelastic Deformation of Metals

Whilst the incremental theory of plasticity provides the most general theory for the inelastic deformation of a wide class of materials, it is essentially restricted to strain hardening and elastic, ideally plastic materials, primarily as a result of adopting Drucker's postulate for a stable material. Therefore the behaviour of metals at elevated temperatures and that of some geological materials which may violate this postulate cannot be described entirely satisfactorily by such a theory.

In recent years considerable effort has been directed towards developing so-called 'state variable theories'. Many of these state variable models have a mathematical structure similar to the viscoplastic model outlined above (i.e., the inelastic strain rate at any time is a function of stress and state variables but not of the rates of stress).

Since no specific yield surfaces or special loading criteria in the sense of classical plasticity appear in such theories, inclusion of strain softening does not present any problem.

State variable theories assume that elastic and inelastic behaviour are present at every stage of loading, unloading, and reloading and include within them features such as strain rate sensitivity, work hardening, Bauschinger's effect, history dependence, creep recovery, strain softening, etc., that may be present in the deformation of metals at elevated temperature, etc.

According to these models the total strain rate tensor $\dot{\varepsilon}_{ij}$, at any time $t$, is the sum of the elastic part $\dot{\varepsilon}_{ij}^e$, the inelastic part $\dot{\varepsilon}_{ij}^n$, and the thermal strain rate $\dot{\varepsilon}_{ij}^T$, i.e.,

$$\dot{\varepsilon}_{ij} = \dot{\varepsilon}_{ij}^e + \dot{\varepsilon}_{ij}^n + \dot{\varepsilon}_{ij}^T \tag{12-22}$$

where the dot now denotes differentiation with respect to the real time variable $t$ (as opposed to any time-like monotonically increasing parameter in elastoplasticity). The elastic strain is related to the stress tensor $\sigma_{ij}$ by Hooke's law and $\dot{\varepsilon}_{ij}^T = \delta_{ij} \alpha \dot{T}$, where $\alpha$ is the coefficient of thermal expansion. The inelastic strain rate tensor $\dot{\varepsilon}_{ij}^n$ is time and history dependent in nature and usually consists of a recoverable part and a permanent part. The inelastic strain rates $\dot{\varepsilon}_{ij}^n$ are given by

$$\dot{\varepsilon}_{ij}^n = M_{ij}(\sigma_{ij}, q_{ij}, T) \tag{12-23}$$

where $q_{ij}$ denotes the state variables of the model

with $$\dot{q}_{ij} = N_{ij}(\sigma_{ij}, q_{ij}, T) \tag{12-24}$$

and $$\dot{\varepsilon}_{\kappa\kappa}^n = 0 \tag{12-25}$$

(i.e., the inelastic strain is a deviatoric strain and therefore the material behaves elastically under isotropic compression).

The state variables $q_{ij}$ are assumed to completely characterize the present deformation state of the material. They vary along the deformation path according to certain laws, and their values at some time $t$, depend upon the deformation history up to that time. Moreover, the history dependence of the rate of inelastic strain $\dot{\varepsilon}_{ij}^n$ up to time $t$ is taken into account completely by $q_{ij}$ at $t$ and no further knowledge of the prior deformation history is required. The number of state variables $q_{ij}$ varies in different models and they can be either scalar or tensor quantities. For example:

1. Miller[14, 15] uses a scalar 'drag' stress $D$ and a 'back' stress tensor $R_{ij}$.
2. Hart and his coworkers[16, 17] use a strain tensor $\varepsilon_{ij}^a$ and a scalar hardness $\sigma^*$.
3. Lagneborg[18] and Robinson[19] use an internal flow stress $S_{ij}$.
4. Bodner and Pantom[20] use the plastic work $W^p$.

Mukherjee and his coworkers[21-24] used Hart's constitutive model and developed a DBEM algorithm to describe the time-dependent inelastic deformation of metals.

## 12-3 GOVERNING DIFFERENTIAL EQUATIONS FOR ELASTOPLASTICITY

An elastoplastic material must obey the equilibrium equations for the stress rate (increments) which are, in the absence of time-dependent body forces,

$$\frac{\partial \dot{\sigma}_{ij}}{\partial x_j} = 0 \tag{12-26}$$

where the dot indicates an increment.

The strain-displacement relations may be written as

$$\dot{\varepsilon}_{ij} = \frac{1}{2}\left(\frac{\partial \dot{u}_i}{\partial x_j} + \frac{\partial \dot{u}_j}{\partial x_i}\right) \tag{12-27}$$

The total strain rate $\dot{\varepsilon}_{ij}$ may be decomposed into its elastic and plastic parts:

$$\dot{\varepsilon}_{ij} = \dot{\varepsilon}_{ij}^e + \dot{\varepsilon}_{ij}^p \tag{12-28}$$

The stresses are related to elastic strains via

$$\dot{\sigma}_{ij} = \lambda \delta_{ij} \dot{\varepsilon}_{\kappa\kappa}^e + 2\mu \dot{\varepsilon}_{ij}^e \tag{12-29}$$

where $\lambda$, $\mu$ are elastic constants.

This equation may be written by using Eq. (12-28) as

$$\dot{\sigma}_{ij} = \lambda \delta_{ij}(\dot{\varepsilon}_{\kappa\kappa} - \dot{\varepsilon}_{\kappa\kappa}^p) + 2\mu(\dot{\varepsilon}_{ij} - \dot{\varepsilon}_{ij}^p) \tag{12-30}$$

By substituting Eq. (12-30) in (12-26) and using the strain-displacement relations (12-27) we obtain the governing differential equations for elastoplastic flow (see Lin[25, 26] and Swedlow and Cruse[27]):

$$\mu \frac{\partial^2 \dot{u}_i}{\partial x_j^2} + (\lambda + \mu) \frac{\partial^2 \dot{u}_j}{\partial x_i \partial x_j} - \left(\lambda \delta_{ij} \frac{\partial \dot{\varepsilon}^p_{\kappa\kappa}}{\partial x_j} + 2\mu \frac{\partial \dot{\varepsilon}^p_{ij}}{\partial x_j}\right) = 0 \qquad (12\text{-}31)$$

In the absence of any plastic strain the second pair of terms in parentheses disappear and Eq. (12-31) becomes Navier's equation for elasticity in incremental form. An elastoplastic problem may therefore be posed as the solution of an elastic problem governed by the differential equation

$$\mu \frac{\partial^2 \dot{u}_i}{\partial x_j^2} + (\lambda + \mu) \frac{\partial^2 \dot{u}_j}{\partial x_i \partial x_j} + \dot{f}_i = 0 \qquad (12\text{-}32)$$

over the volume and subject to the boundary condition

$$\dot{t}'_i = \dot{\sigma}_{ij} n_j - \dot{t}^o_i \qquad (12\text{-}33)$$

over the surface, where

$$\dot{f}_i = -\left(\lambda \delta_{ij} \frac{\partial \dot{\varepsilon}^p_{\kappa\kappa}}{\partial x_j} + 2\mu \frac{\partial \dot{\varepsilon}^p_{ij}}{\partial x_j}\right)$$

$$\dot{t}^o_i = -(\lambda \delta_{ij} \dot{\varepsilon}^p_{\kappa\kappa} + 2\mu \dot{\varepsilon}^p_{ij}) n_j$$

with $n_j$ the outward normal to the boundary.

We see at once that Eq. (12-32) is identical to the incremental Navier equation with suitably modified body forces $\dot{f}_i$ generated by the irrecoverable component of the deformation field. Moreover, the non-linear terms occur only in the body force term of the inhomogeneous differential equation. Hence Eq. (12-32) is quasi-linear in character.

We may assume an *imaginary* elastic body whose body forces and the traction boundary conditions have been modified according to Eqs (12-32) and (12-33). The displacement field obtained from the solution of Eq. (12-32) would thus be correct for the real elastoplastic body. The stresses corresponding to this displacement field would then be given by the elastic stress-strain relations in the elastic regions and elastoplastic stress-strain relations in the elastoplastic regions.

This procedure for solving an elastoplastic problem by means of a modified elastic one is not new. Reisner[28] suggested an essentially identical scheme using intuitive reasoning, and called it an initial stress approach (*eigenspannungen*). Zienkiewicz and his coworkers[29, 30] also developed an initial stress algorithm for the solution of elastoplastic problems by the finite element method.

The governing differential equations for the viscoplastic and the state variable models are similar in character to those developed for the elastoplastic case outlined above. However, the variations of quantities such as displacements, stresses, etc., with time give rise to stiff systems of differential equations such as[13, 24]

$$\frac{d\sigma_{ij}}{dt} = f(\sigma_{ij}, t) \qquad (12\text{-}34)$$

It is therefore important to choose the correct automatic time step in a time-integration scheme. This integration scheme is discussed in Sec. 12-7.

## 12-4 DIRECT AND INDIRECT FORMULATIONS FOR MATERIAL NON-LINEARITIES

It is evident, from the discussion in the preceding section, that the general direct and indirect BEM formulations developed in Chapter 6 for dealing with an interior distribution of body forces, initial strains, and initial stresses will be directly applicable to the present case. Thus, depending on the type of formulation used, the BEM algorithms for dealing with material non-linearities may be classified as (a) modified body forces and modified surface tractions, (b) initial stress, and (c) initial strain.

**(a) Modified body forces and surface tractions** For the modified body force and surface traction approach, we can develop, at a general boundary point $\xi_o$, the following integral representations based on Eqs (12-32) and (12-33):

$$[\delta_{ij} - C_{ij}] u_i(\xi_o) = \int_S [\dot{t}'_i(x) G_{ij}(x, \xi_o) - F_{ij}(x, \xi_o) \dot{u}_i(x)] \, ds$$

$$+ \int_V G_{ij}(x, \xi_o) \dot{f}_i(x) \, dv \qquad (12\text{-}35)$$

for the DBEM. For a boundary point $x_o$ in IBEM we have

$$\dot{u}_i(x_o) = \int_S G_{ij}(x_o, \xi) \phi_j(\xi) \, ds + \int_V G_{ij}(x_o, \xi) \dot{f}_j(\xi) \, dv \qquad (12\text{-}36)$$

$$\dot{t}'_i(x_o) = D_{ij} \phi_j(x_o) + \int_S F_{ij}(x_o, \xi) \phi_j(\xi) \, ds + \int_V F_{ij}(x_o, \xi) \dot{f}_j(\xi) \, dv \qquad (12\text{-}37)$$

where

$$\dot{t}'_i = \text{modified surface traction}$$
$$= \dot{\sigma}_{ij} n_j - \dot{t}^o_i = \dot{t}_i - \dot{t}^o_i$$
$$\dot{t}^o_i = -(\lambda \delta_{ij} \dot{\varepsilon}^p_{\kappa\kappa} + 2\mu \dot{\varepsilon}^p_{ij}) n_j$$
$$\dot{f}_i = \text{modified body force}$$
$$= -\left(\lambda \delta_{ij} \frac{\partial \dot{\varepsilon}_{\kappa\kappa}}{\partial x_j} + 2\mu \frac{\partial \dot{\varepsilon}^p_{ij}}{\partial x_j}\right)$$

An algorithm based on such a formulation was developed by Banerjee and Mustoe,[31] although the convergence and the accuracy of the solution were not entirely satisfactory. One of the main attractions of this method is that the integral equations (12-35) to (12-37) lead to simpler expressions for interior stresses than do the alternatives.

**(b) Initial stress approach** In initial stress formulations the initial stress rates are defined as

$$\dot{\sigma}_{ij}^o = \dot{\sigma}_{ij}^e - \dot{\sigma}_{ij}^{ep} \tag{12-38}$$

where

$$\dot{\sigma}_{ij}^e = D_{ijkl}^e \dot{\varepsilon}_{kl} \qquad \dot{\sigma}_{ij}^{ep} = D_{ijkl}^{ep} \dot{\varepsilon}_{kl}$$

and $D_{ijkl}^e$, $D_{ijkl}^{ep}$ are the elastic and elastoplastic constitutive tensors respectively. The equilibrium equation then becomes

$$\frac{\partial \dot{\sigma}_{ij}^{ep}}{\partial x_j} = 0 \qquad \text{or} \qquad \frac{\partial \dot{\sigma}_{ij}^e}{\partial x_j} - \frac{\partial \dot{\sigma}_{ij}^o}{\partial x_j} = 0 \tag{12-39}$$

It is immediately evident that the term $(-\partial \dot{\sigma}_{ij}^o / \partial x_j)$ is equivalent to a body force $\dot{f}_i$ and the boundary traction is given by

$$\dot{t}_i = \dot{\sigma}_{ij}^{ep} n_j = \dot{\sigma}_{ij}^e n_j - \dot{\sigma}_{ij}^o n_j$$

$$= \dot{t}_i' - \dot{t}_i^o \tag{12-40}$$

Thus a system of boundary integral equations similar to Eqs (12-35) to (12-37) can be obtained, the only difference between them being in the volume integrals involving the body force term. For example, in Eq. (12-35) the body force integral would become

$$\int_V G_{ij}(x, \xi_o)\, \dot{f}_i(x)\, dv = -\int_V G_{ij}(x, \xi_o) \frac{\partial \dot{\sigma}_{ik}^o}{\partial x_\kappa}\, dv \tag{12-41}$$

By using the divergence theorem the volume integral on the right-hand side of this equation can be written as

$$\int_V G_{ij}(x, \xi_o) \frac{\partial \dot{\sigma}_{ik}^o}{\partial x_\kappa}\, dv = -\int_V \frac{\partial G_{ij}(x, \xi_o)}{\partial x_\kappa}\, \dot{\sigma}_{ik}^o(x)\, dv$$

$$+ \int_S G_{ij}(x, \xi_o)\, \dot{\sigma}_{ik}^o(x)\, n_\kappa(x)\, ds$$

$$= -\int_V \frac{\partial G_{ij}(x, \xi_o)}{\partial x_\kappa}\, \dot{\sigma}_{ik}^o(x)\, dv + \int_S G_{ij}(x, \xi_o)\, \dot{t}_i^o(x)\, ds \tag{12-42}$$

Thus, by substituting Eqs (12-40) to (12-42) in Eq. (12-35), we obtain the DBEM statement for an initial stress formulation:

$$[\delta_{ij} - C_{ij}]\, \dot{u}_i(\xi_o) = \int_S [\dot{t}_i(x)\, G_{ij}(x, \xi_o) - F_{ij}(x, \xi_o)\, \dot{u}_i(x)]\, ds$$

$$+ \int_V B_{ikj}(x, \xi_o)\, \dot{\sigma}_{ik}^o(x)\, dv \tag{12-43}$$

where the tractions on the boundary are the real tractions and

$$B_{i\kappa j}(x, \xi_o) = \frac{\partial G_{ij}(x, \xi_o)}{\partial x_\kappa}$$

The corresponding indirect formulation can be developed from Eqs (12-36) and (12-37) by replacing the body forces $\hat{f}_j(\xi)$ by $-\partial \dot{\sigma}^o_{j\kappa}(\xi)/\partial \xi_\kappa$.

$$\dot{u}_i(x_o) = \int_S G_{ij}(x_o, \xi)\, \phi_j(\xi)\, ds - \int_V G_{ij}(x_o, \xi)\, \frac{\partial \dot{\sigma}^o_{j\kappa}(\xi)}{\partial \xi_\kappa}\, dv$$

$$= \int_S G_{ij}(x_o, \xi)\, \phi_j(\xi)\, ds - \int_S G_{ij}(x_o, \xi)\, \dot{\sigma}^o_{j\kappa}(\xi)\, n_\kappa(\xi)\, ds$$

$$+ \int_V \frac{\partial G_{ij}(x_o, \xi)}{\partial \xi_\kappa}\, \dot{\sigma}^o_{j\kappa}(\xi)\, dv$$

$$= \int_S G_{ij}(x_o, \xi)\, [\phi_j(\xi) - \dot{\sigma}^o_{j\kappa}(\xi)\, n_\kappa(\xi)]\, ds + \int_V \frac{\partial G_{ij}}{\partial \xi_\kappa}\, \dot{\sigma}_{j\kappa}(\xi)\, dv \quad (12\text{-}44)$$

Noting that $\partial G_{ji}/\partial \xi_\kappa = \partial G_{ij}/\partial \xi_\kappa = -\partial G_{ij}/\partial x_\kappa$ and that $[\phi_j - \dot{\sigma}^o_{j\kappa} n_\kappa]$ can be replaced by another fictitious traction $\phi'_j$, Eq. (12-44) can be written as

$$\dot{u}_i(x_o) = \int_S G_{ij}(x_o, \xi)\, \phi_j(\xi)\, ds - \int_V B_{j\kappa i}(x_o, \xi)\, \dot{\sigma}^o_{j\kappa}(\xi)\, dv \quad (12\text{-}45)$$

where, for convenience, we have dropped the accent on $\phi$. The reader should note that Eqs (6-39) and (12-45) are identical since $B_{j\kappa i}(\xi, x) = -B_{j\kappa i}(x, \xi)$.

The corresponding equation for the surface traction $t_i(x_o)$ can be developed either directly from Eq. (12-45), as shown in Chapter 6, or from Eq. (12-37) in the manner outlined above. This leads to

$$\dot{t}_i(x_o) = D_{ij}\, \phi_j(x_o) + \int_S F_{ij}(x_o, \xi)\, \phi_j(\xi)\, ds - \int_V M_{j\kappa i}(x_o, \xi)\, \dot{\sigma}^o_{j\kappa}(\xi)\, dv - \dot{\sigma}^o_{ij}(x_o)\, n_j(x_o)$$

$$(12\text{-}46)$$

Equations (12-45) and (12-46) can now be used to solve any well-posed boundary-value problem.

This initial stress formulation has been developed and implemented by a number of workers[32-37] who applied it to a variety of two- and three-dimensional problems.

**(c) Initial strain approach** In an initial strain formulation the initial strain rates are defined as

$$\dot{\varepsilon}^o_{ij} = \dot{\varepsilon}_{ij} - \dot{\varepsilon}^e_{ij} \quad (12\text{-}47)$$

where        $\dot{\varepsilon}_{ij}$ = total strain rate

$\dot{\varepsilon}^e_{ij}$ = elastic strain rate = $C^e_{ij\kappa l}\, \dot{\sigma}_{\kappa l}$

$C^e_{ij\kappa l}$ = elastic compliance tensor

$\dot{\sigma}_{\kappa l}$ = stress increment

For elastoplasticity the initial strain is simply the plastic strain increment but for viscoplasticity and state variable theories the initial strain rate takes on the role of the viscoplastic strain rate or the inelastic strain rate respectively. The governing equations for the initial strain problem are then exactly those given by Eqs (12-32) and (12-33). Therefore, following the procedure outlined above, we obtain the following DBEM statement for a boundary point $\xi_o$:

$$[\delta_{ij} - C_{ij}]\, \dot{u}_i(\xi_o) = \int_S [\dot{t}_i(x)\, G_{ij}(x, \xi) - F_{ij}(x, \xi)\, \dot{u}_i(x)]\, ds$$

$$+ \int_V T_{i\kappa j}(x, \xi_o)\, \dot{\varepsilon}_{i\kappa}^o(x)\, dv \qquad (12\text{-}48)$$

where $T_{i\kappa j}(x, \xi)$ is the stress at a point $x$ due to a unit force vector $e_j$ acting at $\xi$.

Although Eqs (12-43) and (12-48) look very similar, their implementation in an incremental solution process involves the development of quite distinct algorithms, as explained later. The equivalent IBEM formulation of an initial strain algorithm can be developed similarly. Equation (12-48) was first derived by Swedlow and Cruse.[27] Riccardella,[38] Mendelson,[39, 40] and Mukherjee[21-24] and coworkers have developed numerical algorithms based on this formulation.

## 12-5 INCREMENTAL COMPUTATIONS FOR ELASTOPLASTICITY

If the initial stresses and strains are known the equations developed above provide an exact formulation of the problem for any general, well-posed, set of boundary conditions and any numerical errors in the solution arise solely due to subsequent discretization of these equations. Such errors can be minimized by implementing an advanced numerical-solution scheme such as that outlined in Chapter 8.

For our present purposes we represent boundaries by straight line segments for two-dimensional problems and by triangular or quadrilateral elements for three-dimensional ones. The interior region, which is expected to yield as a result of the loading, is then divided into a suitable number of triangular or quadrilateral cells for two-dimensional problems and tetrahedra or brick-like cells in three dimensions. Although such a discretization has the appearance of that used in the finite element method, the cells are used here simply to evaluate the various volume integrals as a piecewise summation. The formation of the discretized system equations is therefore essentially identical to that described in Chapters 3 to 8. Thus, for example, for Eq. (12-43), we can write

$$\mathbf{G}\dot{\mathbf{t}} - \mathbf{F}\dot{\mathbf{u}} = \mathbf{B}\dot{\sigma}^o \qquad (12\text{-}49)$$

which can be solved to obtain the unknown boundary data, provided that the initial stress increments are known. The determination of the initial stresses is discussed in the next section.

Having obtained the unknown boundary displacement components and tractions the displacements and stresses at interior points can be calculated by using the interior point versions of the various equations. However, it is often more efficient to evaluate the displacements at a sufficient number of interior nodes and to calculate the strains from them using either a finite element or a finite difference expression, say,

$$\dot{\varepsilon} = \mathbf{M}\dot{\mathbf{u}}_n \qquad (12\text{-}50)$$

where $\dot{\mathbf{u}}_n$ are the nodal values of displacements. The matrix $\mathbf{M}$ depends on the order of variation of the displacements (i.e., linear, quadratic, or cubic) between the nodes.

The stresses corresponding to these strains may be calculated from

$$\dot{\boldsymbol{\sigma}}^e = \mathbf{D}^e \dot{\boldsymbol{\varepsilon}} \qquad \text{in the elastic region}$$

or $\qquad \dot{\boldsymbol{\sigma}}^{ep} = \mathbf{D}^{ep} \dot{\boldsymbol{\varepsilon}} \qquad \text{in the elastoplastic region} \qquad (12\text{-}51)$

where $\qquad \mathbf{D}^e$ = elasticity matrix

$\mathbf{D}^{ep}$ = elastoplastic matrix calculated by using a suitably updated stress history

The above analysis requires complete knowledge of the initial stress distribution $\dot{\sigma}^o_{ik}$ within the yielded region, which is not known a priori for a particular loading increment. This must be obtained from an interactive process as described below. For this purpose it is convenient to write Eq. (12-49) as

$$\mathbf{A}\dot{\mathbf{x}} = \dot{\mathbf{b}} + \mathbf{B}\dot{\boldsymbol{\sigma}} \qquad (12\text{-}52)$$

where

$\dot{\mathbf{x}}$ = vector of unknown boundary tractions and boundary displacements

$\dot{\mathbf{b}}$ = matrix formed by multiplying the appropriate columns of $\mathbf{G}$ and $\mathbf{F}$ and the given increments of the boundary tractions and boundary displacements

The steps of the incremental solution may then be described as follows:

1. Apply a load increment $\mathbf{b}$, assuming $\dot{\boldsymbol{\sigma}}^o$ to be zero. Calculate $\dot{\mathbf{x}}$, the strains $\dot{\boldsymbol{\varepsilon}}$, and the elastic stress increments $\dot{\boldsymbol{\sigma}}$. Scale the solution so that the most highly stressed cell is at the point of yielding. Store the current values of stresses $\boldsymbol{\sigma}_1$.
2. Apply a small-load increment. Calculate the stress increments in all the cells by using the elastic stress-strain relations, that is, $\dot{\boldsymbol{\sigma}}^e = \mathbf{D}^e \dot{\boldsymbol{\varepsilon}}$. Calculate the value of $\sigma_o$, the equivalent stresses, by using $\boldsymbol{\sigma}_2 = \boldsymbol{\sigma}_1 + \dot{\boldsymbol{\sigma}}^e$ as the stress history and compile a list of yielded cells. Calculate the correct stresses in the elastoplastic cells by using the elastoplastic stress-strain relations $\dot{\boldsymbol{\sigma}}^{ep} = \mathbf{D}^{ep} \dot{\boldsymbol{\varepsilon}}$, using the elastic strain increments as a first approximation. The initial stresses generated are $\dot{\boldsymbol{\sigma}}^o = \dot{\boldsymbol{\sigma}}^e - \dot{\boldsymbol{\sigma}}^{ep}$ as a first approximation. Modify the stress history for the yielded cells to $\boldsymbol{\sigma}_2 = \boldsymbol{\sigma}_1 + \dot{\boldsymbol{\sigma}}^{ep}$ and make $\boldsymbol{\sigma}_1 = \boldsymbol{\sigma}_2$.

3. Assume $\dot{\mathbf{b}} = 0$ and, with the generated initial stresses $\dot{\boldsymbol{\sigma}}^o$, calculate a new $\dot{\mathbf{x}}$ vector by using Eq. (12-52) and also the nodal displacement increments $\dot{\mathbf{u}}_n$, the strain increment $\dot{\boldsymbol{\varepsilon}}$, and the stress increments by using the elastic stress-strain relations $\dot{\boldsymbol{\sigma}}^e = \mathbf{D}^e \dot{\boldsymbol{\varepsilon}}$. Calculate the equivalent stresses by using the history $\boldsymbol{\sigma}_2 = \boldsymbol{\sigma}_1 + \dot{\boldsymbol{\sigma}}^e$ and compile a list of yielded cells. For the elastoplastic cells calculate the correct stresses $\dot{\boldsymbol{\sigma}}^{ep} = \mathbf{D}^{ep} \dot{\boldsymbol{\varepsilon}}$. The initial stresses generated are $\dot{\boldsymbol{\sigma}}^o = \dot{\boldsymbol{\sigma}}^e - \dot{\boldsymbol{\sigma}}^{ep}$. Modify the stress history for the yielded cells to $\boldsymbol{\sigma}_2 = \boldsymbol{\sigma}_1 + \dot{\boldsymbol{\sigma}}^{ep}$ and make $\boldsymbol{\sigma}_1 = \boldsymbol{\sigma}_2$.
4. Check if the initial stresses $\dot{\boldsymbol{\sigma}}^o$ are less than an acceptable norm; if so, go to step 2, if not, go back to step 3. If the number of iterations exceeds, say, 50 then it is reasonable to assume that 'collapse' has occurred.

Within the framework of this algorithm several improvements are possible. For example, a certain degree of forward extrapolation may result in quicker convergence. The accuracy can be improved if a check is carried out on the sign of the dissipated plastic work rate to ensure that all the plastic cells are, in fact, plastic. The incremental algorithm outlined above is essentially similar to that described by Zienkiewicz.[30]

An algorithm based on an initial strain formulation would be somewhat different, and for explicit details the reader is referred to Mendelson and Albers[39] although for conventional elastoplasticity, the initial stress method which we have described appears to provide the most generally applicable procedure.

## 12-6 INCREMENTAL COMPUTATIONS FOR VISCOPLASTICITY

The viscoplastic algorithm[11, 12] is essentially an initial strain approach and we can write the final system of equations for (12-48) in a form similar to (12-52), i.e.,

$$\mathbf{A}\Delta\mathbf{X} = \Delta\mathbf{b} + \mathbf{T} \cdot \Delta\boldsymbol{\varepsilon}^{vp} \tag{12-53}$$

where we have denoted the incremental quantities by $\Delta$ since in viscoplasticity the superior dots indicate the true time rate and the loading path has been divided into an arbitrary number of increments $\Delta\mathbf{b}$. Assuming that we have complete knowledge of the stresses $\boldsymbol{\sigma}^n$, displacements $\mathbf{u}^n$, viscoplastic strain $(\Delta\boldsymbol{\varepsilon}^{vp})^n$, and $(\Delta\mathbf{b})^n$ at the $n$th state at time $t$, the simplest algorithm for obtaining the $(n+1)$ state, at time $t + \Delta t$, can be described as follows:

1. Starting with the known values of the state variables at the $n$th state calculate the viscoplastic strain rate $\dot{\boldsymbol{\varepsilon}}^{vp}$ from Eq. (12-21), i.e.,

$$(\dot{\boldsymbol{\varepsilon}}^{vp})^n = \mathbf{C}^{vp} \boldsymbol{\sigma}^n \tag{12-54}$$

2. Determine the change of the viscoplastic strain $\boldsymbol{\varepsilon}^{vp}$ as

$$(\delta\boldsymbol{\varepsilon}^{vp})^n \simeq (\dot{\boldsymbol{\varepsilon}}^{vp})^n \Delta t \tag{12-55}$$

3. Using the values of the known $(\Delta \mathbf{b})^{n+1}$ (i.e., time-dependent boundary loading) and $(\Delta \boldsymbol{\varepsilon}^{vp})^{n+1} = (\Delta \boldsymbol{\varepsilon}^{vp})^{n} + (\delta \boldsymbol{\varepsilon}^{vp})^{n}$ compute the new right-hand side of Eq. (12-53), solve for the new $(\Delta \mathbf{X})^{n+1}$, and then calculate $(\Delta \mathbf{u})^{n+1}$, $(\Delta \boldsymbol{\varepsilon})^{n+1}$, and $(\Delta \boldsymbol{\sigma})^{n+1}$ from the equation $(\Delta \boldsymbol{\sigma})^{n+1} = \mathbf{D}^{e}(\Delta \boldsymbol{\varepsilon} - \Delta \boldsymbol{\varepsilon}^{vp})^{n+1}$.
4. Update the stresses and start a new time step $(n+2)$.

It should be noted that Eq. (12-53) need not be written in terms of $\Delta \mathbf{X}$, $\Delta \mathbf{b}$, and $\Delta \boldsymbol{\varepsilon}^{vp}$. Since the theory allows for the path-dependent features that are present in elastoplasticity through Eq. (12-18), we could write (12-53) in terms of the current values of $\mathbf{X}$, $\mathbf{b}$, and $\boldsymbol{\varepsilon}^{vp}$ at time $t$. If, however, we wish to solve elastoplasticity problems using a viscoplastic approach, then we are obliged to introduce incremental quantities into the equation. In such an algorithm the exact form of the viscoplastic flow parameters $n$ (often $n = 1$) and $\gamma$ in Eq. (12-20) is immaterial and time plays the role of a fictitious variable.[12] For each increment of loading $\Delta \mathbf{b}$ an elastic solution is followed by a relaxation process (in time) back to a steady state equilibrium position. The time stepping is stopped when the stresses reach a point sufficiently close to the yield surface (i.e., when the steady state viscoplastic solution has been reached). Figures 12-5 and 12-6 illustrate the difference between the viscoplastic and initial stress methods for the analysis of elastoplastic solids.

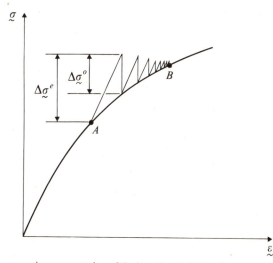

**Figure 12-5** Diagrammatic representation of the iterative algorithm from $A$ to $B$ in the initial stress method for elastoplasticity.

The major problem associated with this approach is that the magnitude of the time step $(\Delta t)$ which can be used is severely limited by stability and convergence requirements.[12, 13] These are discussed below in relation to the state variable theory where analogous problems arise. Further details are available in Refs 13 and 24.

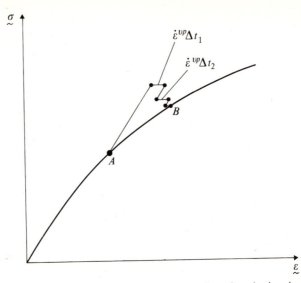

**Figure 12-6** Diagrammatic representation of the iterative viscoplastic algorithm used in elastoplasticity.

## 12-7 A NUMERICAL ALGORITHM FOR THE TIME-DEPENDENT INELASTIC DEFORMATION OF METALS

The numerical algorithm for calculating the inelastic deformation of metals using a state variable model is also based on the initial strain approach. The final system equation can therefore be written as

$$\mathbf{A}\dot{\mathbf{X}}(t) = \dot{\mathbf{b}}(t) + T\dot{\boldsymbol{\varepsilon}}^n(t) \tag{12-56}$$

where the superior dots indicate actual rates of change of variables at time $t$. Starting with the known values of stresses, strains, etc., at $t = 0$ (which are obtainable from a linear solution with prescribed initial conditions) the algorithm for obtaining values of the variables at time $t = \Delta t$ may be developed as follows:[23, 24]

1. Using the information at $t = 0$ obtain $\dot{\boldsymbol{\varepsilon}}^n$ from Eq. (12-23) which can be used in Eq. (12-56) to obtain the boundary values of $\dot{\mathbf{t}}$ and $\dot{\mathbf{u}}$ and hence the interior values of $\dot{\mathbf{u}}$, $\dot{\boldsymbol{\sigma}}$, and $\dot{\boldsymbol{\varepsilon}}$.
2. The displacements, stresses, and strains at time $t = \Delta t$ are then given by $\mathbf{u} = \mathbf{u}\big|_{t=0} + \dot{\mathbf{u}}\big|_{t=0} \Delta t$, $\boldsymbol{\sigma} = \boldsymbol{\sigma}\big|_{t=0} + \dot{\boldsymbol{\sigma}}\big|_{t=0} \Delta t$, etc.
3. Update the stresses $\boldsymbol{\sigma}$ and the state variables $\mathbf{q}$ so that these are now the current values at time $t = \Delta t$.

Step 2 above is essentially an Euler style integration procedure with respect to time. It is important therefore to adopt an efficient integration scheme with automatic time step control to ensure both stability and convergence of the solution. A number of such methods are available in the published literature[13] and we shall demonstrate a simple, recently developed technique[24] applied to a differential equation of the type (12-34), i.e.,

$$\frac{dy}{dt} = f(y, t) \tag{12-57}$$

The value of $y(t + \Delta t)$ in terms of $y(t)$ is

$$y(t + \Delta t) = y(t) + f(y, t)\Delta t \tag{12-58}$$

and the error in this step may be defined as

$$E = \frac{\Delta t |\nabla f|}{|y(t)|} \tag{12-59}$$

where $\nabla f = f(y, t) - f(y, t - \Delta t)$ is the backward difference of $f$.

Two error bounds $E_{max}$ and $E_{min}$ are initially prescribed and the algorithm can then proceed as follows:

$E > E_{max}$: replace $\Delta t$ by $0.5\Delta t$ and recompute $E$

$E \leqslant E_{max}$: accept $\Delta t$ and calculate $y(t + \Delta t)$ via (12-58)

The time step for the next step $\Delta t_{next}$ is decided according to the following:

$E_{max} \geqslant E > E_{min}$: $\Delta t_{next} = \Delta t$

$E_{min} \geqslant E$: $\Delta t_{next} = 2\Delta t$

In an actual problem the quantity $y$ will of course be the stress components at boundary and interior points. Therefore the error for the $i$th variable is defined as

$$E_i = \frac{\Delta t \sum (\nabla f)^\kappa}{\sum |y^\kappa(t)|} \tag{12-60}$$

where the summation extends over the values of the $i$th variable over all nodes and the error $E$ becomes

$$E = \max |E_i| \tag{12-61}$$

## 12-8 APPLICATIONS TO OTHER RELATED SYSTEMS

There are many problems[41-46] in engineering science where the non-linearities involved in the governing differential equation are similar to those explored in the preceding sections. These include problems of non-Darcy flow through porous media,[41,42] compressible flow of fluids,[44] non-linear viscous flow, magnetic saturation problems,[43,45,46] etc., where in addition to the boundary integrals a volume integral over the non-linear region may be introduced in the manner discussed in this chapter. Some of these applications are discussed in a recent article by Banerjee.[46]

Boundary element methods applied to such problems make extensive use of the fact that the region over which non-linearities exist is usually quite small and therefore very efficient numerical solution techniques for non-linear problems can be developed using BEM.

## 12-9 EXAMPLES

**(a) Expansion of a circular disc**[24] Numerical results for a circular disc with a circular cut-out, under internal pressure increasing at a constant rate, are shown in Fig. 12-7. The problem was solved by Morjaria and Mukherjee using Hart's state variable model. Figure 12-8 shows the boundary and internal cell subdivisions, in which they used linear variations over boundary elements together with uniformly distributed inelastic strain rates over the interior cells.

**(b) Expansion of a circular cylinder**[32] Two solutions to this problem were obtained, using coarse and fine boundary and cell discretizations. The coarse

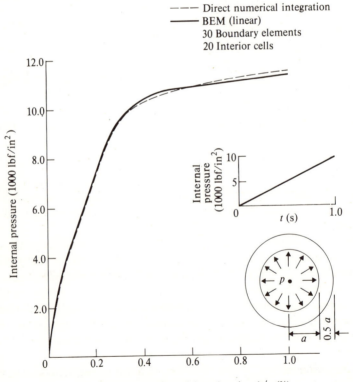

**Figure 12-7** Comparison of direct and BEM solutions for circular plate with concentric circular cut-out under internal pressure increasing at a constant rate (304 SS at 200 °C).

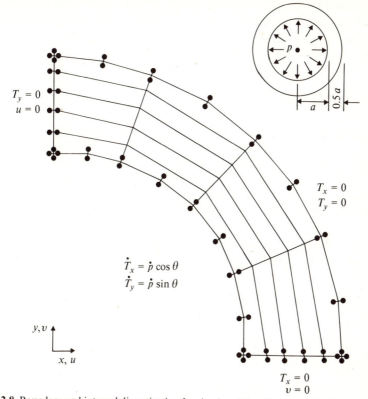

**Figure 12-8** Boundary and internal discretization for circular plate with concentric circular cut-out under internal pressure (30 boundary elements and 20 internal cells).

discretization was similar to that shown in Fig. 12-8, the fine one had twice the number of boundary segments and interior cells. The material was assumed to be an elastic ideally plastic Von Mises one and an initial stress algorithm was used in the solution.

The computed results for the expansion of the outer surface of the cylinder with a gradually increasing internal pressure together with the analytical results are shown in Fig. 12-9. The small differences between the two numerical solutions at first yield are to be expected since these are strongly influenced by the size of the cells used near the inner boundary. However, the comparison between analytical and computed values is still very good and the results from the two different discretization schemes establish the stability and convergence of the solution method.

Typical circumferential stresses at first yield from the analysis using the finer discretization at an instant when the elastoplastic boundary is at $r = 1.6a$ are shown in Fig. 12-10 and again the computed results and the analytical solution are in reasonably good agreement. Solutions obtained by using load increments

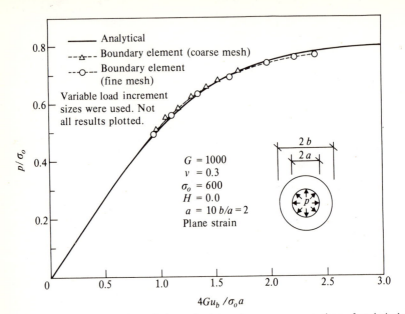

**Figure 12-9** Expansion of a thick cylinder under internal pressure—comparison of analytical and numerical results.

of 5 and 7.5 per cent of the load at first yield were essentially identical at a given load level.

**(c) A perforated strip in tension**[33] Theocaris and Marketos[30] studied perforated aluminium strips loaded in tension experimentally. The boundary and cell discretizations used to analyse this problem are shown in Fig. 12-11. It is of interest to note that the cell discretization is deliberately confined to the region which is likely to become elastoplastic as the load increases. Experienced engineers can usually define such areas adequately by considering the problem geometry and the loading and thereby reduce the computing and data-preparation costs substantially.

The experimental and computed load-deflection characteristics of the strip are presented non-dimensionally in Fig. 12-12, along with the results of an initial stress finite element analysis. Agreement between the solutions is quite satisfactory.

The longitudinal stresses at the root of the plate are compared with the measured values in Fig. 12-13 at a load level just below that which would cause failure.

**(d) Square plate with elliptic cut-out**[24] A square plate with elliptic cut-out was analysed recently by Morjaria and Mukherjee using Hart's constitutive model. Figure 12-14 shows the boundary and cell discretization scheme chosen. The

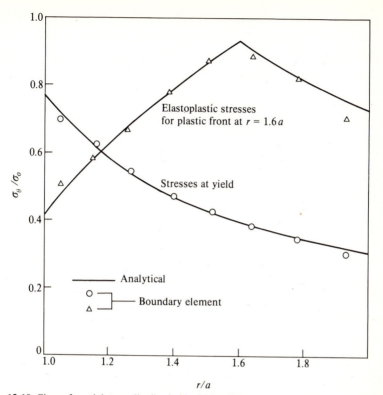

**Figure 12-10** Circumferential stress distribution in thick cylinder—analytical and boundary element results.

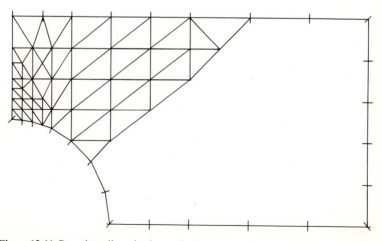

**Figure 12-11** Boundary discretization and cell geometry for perforated plate solution.

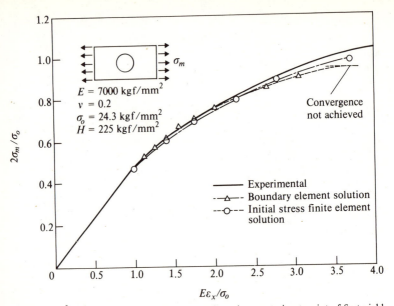

**Figure 12-12** Perforated plate—development of maximum strain at point of first yield.

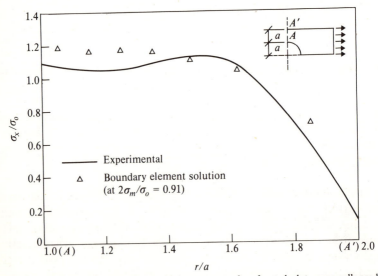

**Figure 12-13** Measured and computed stresses at root of perforated plate near collapse load.

region $ABCD$ was subdivided further into a much finer cell network. It is of interest to note the arbitrariness of the interior discretization which is possible with BEM (i.e., it does not have to match that adopted on the boundary of the region).

Numerical results for the stress concentration at different times are shown in Fig. 12-15. The elliptic cut-out had an axis ratio of 4 which leads to an elastic

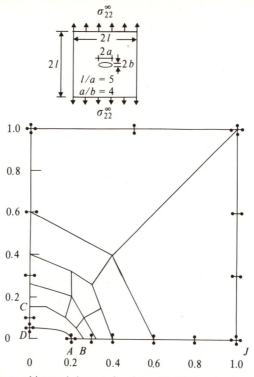

**Figure 12-14** Boundary and internal elements for plate with elliptic cut-out under uniaxial loading (38 boundary elements and 30 internal cells).

stress concentration factor of about 10 at point $A$. Thus there is a severe spatial stress gradient initially and plastic flow causes very rapid stress relaxation near the cut-out. The elastic solution is within 4 per cent of the correct solution near the stress concentration.

**(e) A flexible footing on an elastoplastic half space**[32, 33] Whilst, for the examples shown earlier, the computational cost using a first-generation computer programme was about 50 per cent higher than the corresponding solutions obtained using the finite element method, BEM is particularly attractive for problems where the elastic region extends to infinity and the yielded region is confined to the vicinity of the loaded area. The infinitely distant elastic boundary is automatically included without discretization, a feature unique to BEM.

In order to solve the problem of an elastoplastic half space loaded by a gradually increasing surface pressure (plane strain problem), symmetry about the centreline was exploited. This quarter space was then represented by 40 boundary elements and 64 triangular cells confined to the vicinity of the loaded strip.

The problem was solved for strain hardening ($H > 0$), elastic–ideally plastic ($H = 0$), and strain softening ($H < 0$, restricted) materials. The load-displacement

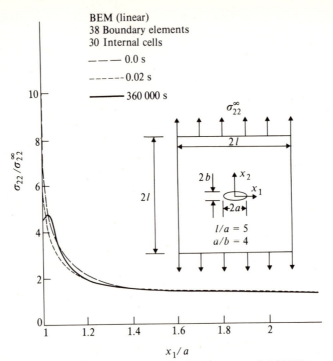

**Figure 12-15** Stress redistribution on line $AJ$ of Fig. 12-14 in an annealed 304 SS square plate with elliptic cut-out under uniaxial tension at 400 °C ($\sigma_{22}^{\infty} = 4000\,\text{lbf/in}^2$).

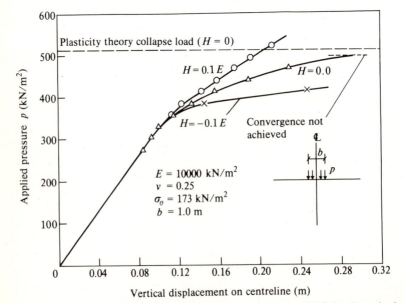

**Figure 12-16** Load-deflection characteristics of flexible footing on semi-infinite elastoplastic half space.

plots are presented in Fig. 12-16 for the three cases. It is of interest to note that, for the elastic–ideally plastic case, the lowest load at which convergence could not be obtained is about 3 per cent below the exact analytical collapse load of 514. The solution for the strain softening case cannot of course be valid for an unrestricted choice of the strain softening parameter ($H < 0$) since during the incremental process the possibility of a non-unique solution may arise. Furthermore, the loading and unloading criteria of the incremental theory of plasticity do not allow for strain softening.

**(f) Displacements due to the presence of plastic strains in a cube embedded in a half space**[32, 35] In order to test the accuracy of the three-dimensional elastoplastic analysis, the recently solved problem (Chiu[47]) of the displacement field due to plastic strains in a cube embedded in the interior of an elastic half space was analysed. Chiu presented a very precise solution to this problem, obtained by an integral transform method, and plotted the vertical displacement of a surface point vertically above the cube against the depth of the cube below the surface for three different plastic strain distributions in the cube. These were

(i) the vertical plastic strain rate $\dot{\varepsilon}^p_{11}$ only,
(ii) the horizontal plastic strain rate $\dot{\varepsilon}^p_{22}$ only, and
(iii) the uniform plastic strain rate in all directions:

$$\dot{\varepsilon}^p_{11} = \dot{\varepsilon}^p_{22} = \dot{\varepsilon}^p_{33}$$

In order to reproduce Chiu's results, the traction-free surface was discretized by using 34 square boundary elements. The symmetry of the problem was utilized by using the unknowns for one quadrant only and assuming identical element patterns in the other quadrants. Numerical results for the vertical displacements of the point on the free surface vertically above the cube due to the initial stresses corresponding to the three values of plastic strain are plotted against the depth of the cube below the surface (Fig. 12-17). It can be seen that these are in excellent agreement with Chiu's results except for the case when the cube is very close to the surface. This is due to the difficulties involved in evaluating the weakly singular volume integral numerically. Nevertheless, the maximum error is only about 4 per cent.

**(g) The problem of an indentation of a half space by a rigid square punch**[32, 35] Figure 12-18 shows the load-displacement curve for a square footing resting on the surface of a semi-infinite elastic–ideally plastic solid. The exact solution to this problem is not known, but the collapse load for a rigid circular footing resting on the surface of a semi-infinite solid is given by $P = 6\pi r^2 C_u$ where $r$ is the radius of the footing and $C_u = \sigma_o/\sqrt{3}$. The collapse load for a circular footing having the same contact area as the square punch is shown in Fig. 12-18 for comparison. The result of the present analysis gives the collapse load

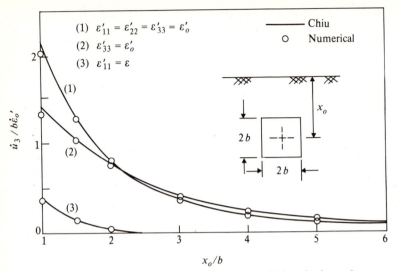

**Figure 12-17** Vertical surface displacement due to initial strains in a cube.

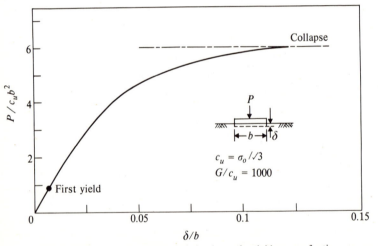

**Figure 12-18** Load-displacement behaviour of a rigid square footing.

slightly higher than $6c_u b^2$, which supports an engineering expectation that it would be so in reality.

It is worth noting that first yield under the corner of the footing occurs at a very early stage of loading. The surface discretization used for this problem is identical to that used in the previous example; in addition 60 cubical cells were used to calculate the contributions of the generated initial stresses. The computation time was about 6 min by using a CDC 7600 computer.

**(h) The problem of a laterally loaded pile subjected to cyclic loading**[32, 35] Whilst all the previous examples were solved using the general method of analysis

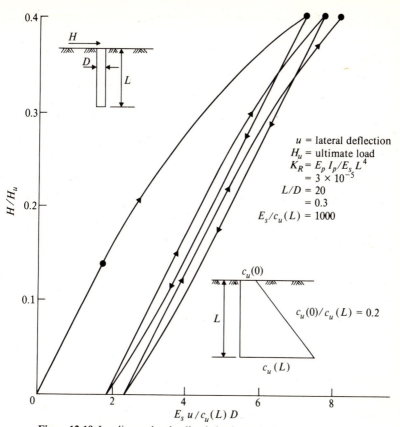

**Figure 12-19** Loading and unloading behaviour of a laterally loaded pile.

$u$ = lateral deflection
$H_u$ = ultimate load
$K_R = E_p I_p / E_s L^4$
$= 3 \times 10^{-5}$
$L/D = 20$
$= 0.3$
$E_s/c_u(L) = 1000$

$c_u(0)/c_u(L) = 0.2$

$\frac{E_s u}{c_u(L) D}$

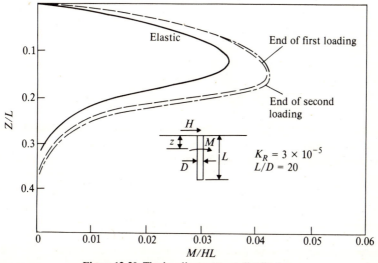

Elastic

End of first loading

End of second loading

$K_R = 3 \times 10^{-5}$
$L/D = 20$

$M/HL$

**Figure 12-20** The bending moment distribution.

incorporating (i) Kelvin's kernel functions and (ii) an isotropic strain hardening model, this particular problem was analysed by using Mindlin's solution and a kinematic hardening model. The traction-free surface of the half space is automatically satisfied by Mindlin's solution; therefore only discretization of the pile-soil interface is necessary. The pile was represented as a linear structural element (see Banerjee and Driscoll,[48] Banerjee,[49] and Banerjee and Davies[50]).

Although the material was assumed to be elastically homogeneous the yield stress $c_u = \sigma_o \sqrt{3}$ varied linearly with depth. The hardening parameter for the kinematic hardening model was assumed to be $C = 0.001E_s$, where $E_s$ = 'Young's modulus' for the soil.

Figure 12-19 shows the loading, unloading, and reloading curves for the pile and Fig. 12-20 the initial elastic bending moment distribution and the subsequent ones at the end of each reloading. An analysis of this type is clearly useful to investigate any possible lack of shakedown load for cyclically loaded structures.

## 12-10 CONCLUDING REMARKS

We have attempted to demonstrate in this chapter that BEM are a useful problem-solving tool for non-linear problems in solid mechanics and, with imagination, the principles and methodologies outlined here could be applied to many other areas of engineering science.

The numerical results presented were obtained from 'first-generation' BEM programmes using either piecewise constant or linear variations of functions over the boundaries and piecewise constant approximations of non-linear components over the volume. There is therefore considerable scope for improving both the efficiency and the accuracy of all these programmes.

## 12-11 REFERENCES

1. Drucker, D. C. (1951) 'A more fundamental approach to plastic stress-strain solutions', *Proc. First U.S. Nat. Congr. in Appl. Mech.*, pp. 487–491.
2. Fung, Y. C. (1965) *Foundations of Solid Mechanics*, Prentice-Hall, Englewood Cliffs, N.J.
3. Chen, W. F. (1975) *Limit Analysis and Soil Plasticity*, Elsevier, New York.
4. Zienkiewicz, O. C., and Naylor, D. J. (1971) 'The adoption of critical state soil mechanics for use in finite element', *Proc. Roscoe Meml Symp.*, pp. 537–547, Cambridge University, Foulis.
5. Banerjee, P. K., and Stipho, A. S. (1978) 'Associated and non associated constitutive relations for undrained behaviour of isotropic soft clay', *Int. J. Num. Anal. Meth. in Geomech.*, **2**(1), 35–56.
6. Prager, W. (1955) 'Theory of plasticity—a survey of recent achievement' (James Clayton Lecture), *Proc. Inst. Mech. Engrs*, **169**, 41–57.
7. Shields, R. T., and Ziegler, H. (1958) 'On Prager's hardening rule', *ZAMP*, **9a**, 260–276.
8. Ziegler, H. (1959) 'A modification of Prager's hardening rule', *Q. Appl. Math.*, **17**(1), 55–65.
9. Mroz, Z. (1969) 'An attempt to describe the behaviour of metals under cyclic loads using a more general work hardening model', *Act. Mech.*, **7**, 199–212.
10. Perzyna, P. (1966) 'Fundamental problems in visco-plasticity', *Adv. Mech.*, **9**, 243–277.

11. Zienkiewicz, O. C., and Cormeau, I. C. (1972) 'Visco-plasticity solution by finite element process', *Arch. Mech.*, **24**, 5–6, 873–888.
12. Zienkiewicz, O. C., and Cormeau, I. C. (1974) 'Visco-plasticity, plasticity and creep in elastic solids—a unified numerical solution approach', *Int. J. Num. Meth. in Engng*, **8**, 821–845.
13. Cormeau, I. C. (1975) 'Numerical stability in quasi-static elasto-visco-plasticity', *Int. J. Num. Meth. in Engng*, **9**, 109–128.
14. Miller, A. (1976) 'An inelastic constitutive model for monotonic, cyclic and creep deformation: Part 1—equations development and analytical procedures', *J. Engrs Mat. Tech.*, *ASME*, **98**, 97–105.
15. Miller, A. (1976) 'An inelastic constitutive model for monotonic, cyclic and creep deformation: Part 2—applications to type 304 stainless steel', *J. Engrs Mat. Tech.*, *ASME*, **98**, 106–113.
16. Hart, E. W., Li, C. Y., and Yamada, H. (1976) 'Phenomenological theory—a guide to constitutive relations and fundamental deformation properties', in A. S. Argon (ed.), *Constitutive Equations in Plasticity*, pp. 149–197, MIT Press, Cambridge, Mass.
17. Hart, E. W. (1976) 'Constitutive relations for nonelastic deformation of metals', *J. Engrs Mat. Tech.*, *ASME*, **98**(3), 193–202.
18. Lagneborg, R. (1972) 'A modified recovery-creep model and its evaluation', *Metal Sci. J.*, **6**, 127–133.
19. Robinson, D. N. (1975) 'A candidate creep-recovery model of $2\frac{1}{2}$ Cr-1 MO steel and its experimental implementation', Report No. ORNL–TM–5110, Oak Ridge National Laboratory.
20. Bodner, S. R., and Partom, Y. (1975) 'Constitutive equations for elastic visco-plastic strain hardening materials', *J. Appl. Mech.*, *ASME*, **42**(2), 385–389.
21. Kumar, V., and Mukherjee, S. (1977) 'A boundary-integral equation formulation for time-dependent inelastic deformation in metals', *Int. J. Mech. Sci.*, **19**(12), 713–724.
22. Mukherjee, S. (1977) 'Corrected boundary integral equations in planar thermoelastoplasticity', *Int. J. Solids and Structs*, **13**(4), 331–336.
23. Mukherjee, S., and Kumar, V. (1978) 'Numerical analysis of time dependent inelastic deformation in metallic media using boundary integral equation method', *J. Appl. Mech.*, *ASME*, **45**(4), 785–790.
24. Marjaria, M., and Mukherjee, S. (1979) 'Improved boundary integral equation method for time-dependent inelastic deformation in metals', *Int. J. Num. Meth. in Engng* **15**, 97–111.
25. Lin, T. Y. (1967) 'Reciprocal theorem for displacements in inelastic bodies', *J. Comp. Matter*, **1**, 144–151.
26. Lin, T. Y. (1969) *Theory of Inelastic Structures*, Wiley, Chichester.
27. Swedlow, J. L., and Cruse, T. A. (1971) 'Formulation of boundary integral equations for three-dimensional elasto-plastic flow', *Int. J. Solids and Structs*, **7**, 144–151.
28. Reisner, H. (1931) 'Initial stresses and sources of initial stresses' (in German), *ZAMM*, **11**, 1–8.
29. Zienkiewicz, O. C., Valliappan, S., and King, I. P. (1969) 'Elasto-plastic solution of engineering problems by initial stress, finite element approach', *Int. J. Num. Meth. in Engng*, **1**, 75–100.
30. Zienkiewicz, O. C. (1971) *Finite Element Method in Engineering Science*, McGraw-Hill.
31. Banerjee, P. K., and Mustoe, G. G. W. (1978) 'The boundary element method for two-dimensional problems of elasto-plasticity', *Proc. Int. Conf. Rec. Adv. in Boundary Element Meth.* pp. 283–300, Pentech Press, London.
32. Banerjee, P. K., Cathie, D. N., and Davies, T. G. (1979) 'Two and three-dimensional problems of elasto-plasticity', in P. K. Banerjee and R. Butterfield (eds), *Developments in Boundary Element Methods*, Applied Science Publishers, London.
33. Banerjee, P. K., and Cathie, D. N. (1979) 'A direct formulation and numerical implementation of the boundary element method for two dimensional problems of elasto-plasticity', *Int. J. Mech. Sci.* **22**, 233–245.
34. Banerjee, P. K., and Davies, T. G. (1979) 'Analysis of some case histories of laterally loaded pile groups', *Proc. Int. Conf. Num. Meth. in Offshore Piling*, Institution of Civil Engineers, London.
35. Davies, T. G. (1979) 'Linear and nonlinear analyses of pile groups', Ph.D. thesis, University of Wales, University College, Cardiff.

36. Chaudonneret, M. (1977) 'Boundary integral equation method for visco-plasticity analysis' (in French), *J. de Méc. Appliq.*, **1**(2), 113–131.
37. Chaudonneret, M. (1978) 'Calcul des concentrations de contrainte en élasto-viscoplasticité' (in French), Ph.D. thesis and ONERA Publication No. 1978-1.
38. Riccardella, P. (1973) 'An implementation of the boundary integral technique for plane problems of elasticity and elasto-plasticity', Ph.D. thesis, Carnegie Mellon University, Pittsburg.
39. Mendelson, A., and Albers, L. U. (1975) 'Application of boundary integral equation method to elasto-plastic problems', in T. A. Cruse and F. J. Rizzo (eds), *Proc. ASME Conf. on Boundary Integral Equation Meth.*, AMD-11, ASME, New York.
40. Rzasnicki, W., and Mendelson, A. (1975) 'Application of boundary integral equation method to elasto-plastic analysis of V-notched beams', NASA Technical Report TMX-71472.
41. Volker, R. E. (1969) 'Nonlinear flow in porous media by finite elements', *Proc. Am. Soc. C.E.*, **95** (H76), 2093–2114.
42. Ahmed, H., and Suneda, D. K. (1969) 'Nonlinear flow in porous media', *Proc. Am. Soc. C.E.*, **95**(H76), 1847–1859.
43. Winslow, A. M. (1967) 'Numerical solution of quasi-linear Poissons equation in a non-uniform triangle mesh', *J. Comp. Physics*, **1**, 149–172.
44. Luu, T. S., and Coulmy, G. (1977) 'Method of calculating the compressible flow round an aerofoil or a cascade up to the shockfree transonic range', *Computers and Fluids*, **5**, 261–275.
45. Karmaker, H. C., and Robertson, S. D. T. (1978) 'An integral equation formulation for electromagnetic field analysis in electrical apparatus', Presented at the IEEE PES Summer meeting, Los Angeles, Calif., 16–21 July.
46. Banerjee, P. K. (1979) 'Nonlinear problems of potential flow', in P. K. Banerjee and R. Butterfield (eds), *Developments in Boundary Element Methods*, Chap. II, Applied Science Publishers, London.
47. Chiu, Y. (1978) 'On the stress field and surface deformation in a half space with a cuboidal zone in which initial strains are uniform', *J. Appl. Mech.*, **45**, 302–306.
48. Banerjee, P. K., and Driscoll, R. M. C. (1976) 'Three-dimensional analysis of raked pile groups', *Proc. Inst. Civ. Engrs*, **61**, 653–670.
49. Banerjee, P. K. (1978) 'Analysis of axially and laterally loaded pile groups', in C. R. Scott (ed.), *Developments in Soil Mechanics*, Chap. 9, pp. 317–340, Applied Science Publishers, London.
50. Banerjee, P. K., and Davies, T. G. (1978) 'Behaviour of axially and laterally loaded single piles embedded in non-homogenous soils', *Géotechnq.*, **28**(3), 309–326.

# THIRTEEN

## EXAMPLES IN FLUID MECHANICS

## 13-1 INTRODUCTION

Most workers in fluid mechanics are fully aware of the similarities between the governing equations for solid mechanics problems and those of fluid mechanics.[1,2] Just as developments in the application of the finite element method to complex problems in solid mechanics inspired a parallel development in fluid mechanics so, we hope, might the extensive treatment of linear, non-linear, steady state, and transient solid mechanics problems outlined in previous chapters convince the reader that the application of BEM to problems in fluid mechanics is equally well worth while.

The great majority of fluid mechanics problems involve fluid regions that are extensive and very often those that extend to infinity. Although the governing differential equations are usually highly non-linear, they can be reformulated in such a way that the non-linear terms are only present over a localized part of the entire region. Examples of such an algorithm have already been described in Chapter 12, in which no interior discretization was needed within the dominant linear regions. BEM are, to this extent, unique among numerical methods since they can accommodate infinitely distant boundaries without any discretization whatsoever.

Following the pioneering work of Hess and Smith (see Chapter 5) remarkable progress has been made, particularly during the past five years, in developing the potential of BEM in fluid mechanics. Most of this work is summarized in the present chapter.

## 13-2 GOVERNING EQUATIONS AND THEIR INTEGRAL FORMULATIONS

There are a number of excellent books and monographs[3-6] which develop the governing equations of fluid mechanics, both comprehensively and elegantly, and familiarity with one or other of these texts will be assumed in this section. We shall simply present the basic differential equations for each class of problem and discuss the integral representations which can be developed to deal with them.

### 13-2-1 Navier–Stokes Equations for the Motion of Compressible and Incompressible Viscous Fluids

In an eulerian orthogonal cartesian coordinate system $x_i$ the basic governing equations for the motion of a compressible fluid are

$$\rho\left(\frac{\partial u_i}{\partial t}+u_j\frac{\partial u_i}{\partial x_j}\right)=-\frac{\partial p}{\partial x_i}+\mu\frac{\partial^2 u_i}{\partial x_j\partial x_j}+(\lambda+\mu)\frac{\partial^2 u_j}{\partial x_i\partial x_j}+F_i \qquad (13\text{-}1)$$

where 

$\rho$ = mass density
$t$ = time
$u_i$ = velocity in the $x_i$ direction
$p$ = pressure
$\mu$ = viscosity
$\lambda$ = a coefficient of viscosity ($\lambda = 2\mu/3$ for a monatomic gas)
$F_i$ = body force density

If we compare Eq. (13-1) for steady state flow (that is, $\partial u_i/\partial t = 0$) with the corresponding solid mechanics equation [Eq. (4-4)] we see that the major difference between them is the convective term ($u_j\,\partial u_i/\partial x_j$). By assuming that this term can be represented by a pseudo-body force density $\chi_i$ an integral formulation of (13-1) can be developed.

For an incompressible fluid (13-1) simplifies to

$$\rho\left(\frac{\partial u_i}{\partial t}+u_j\frac{\partial u_i}{\partial x_j}\right)=-\frac{\partial p}{\partial x_i}+\mu\frac{\partial^2 u_i}{\partial x_i\partial x_i}+F_i \qquad (13\text{-}2)$$

whereas for steady ($\partial u_i/\partial t = 0$), slow ($u_j\,\partial u_j/\partial x_j \simeq 0$) motion of a fluid the left-hand side of Eq. (13-2) vanishes and the equation becomes identical to that for the displacement field in an incompressible solid. The various solutions for solid bodies developed in earlier chapters are then directly applicable.

By introducing the concept of vorticity it is possible to recast Eq. (13-2) in a more convenient form and it is then possible to develop very useful BEM formulations of (13-2). In order to demonstrate this most simply it is convenient to rewrite (13-2) in vector notation:

$$\rho\left[\frac{\partial \mathbf{u}}{\partial t}+(\mathbf{u}\cdot\mathbf{V})\mathbf{u}\right]=-\nabla p+\mu\nabla^2\mathbf{u}+\mathbf{F} \qquad (13\text{-}3)$$

## 13-2-2 Equations of Motion in Terms of Vorticity

Vorticity is defined as

$$\mathbf{w} = \nabla \times \mathbf{u} \tag{13-4}$$

By taking curl of both sides of Eq. (13-3) and using Eq. (13-4) and the continuity equation

$$\nabla \cdot \mathbf{u} = 0 \tag{13-5}$$

we can obtain the vorticity transport equation

$$\frac{\partial \mathbf{w}}{\partial t} = \nabla \times (\mathbf{u} \times \mathbf{w}) + \nu \nabla^2 \omega \tag{13-6}$$

where $\nu$ is the kinematic viscosity.

Thus the set of equations (13-4) to (13-6) with $\mathbf{u}$ and $\mathbf{w}$ as variables can replace (13-3) and (13-5) in which $\mathbf{u}$ and $p$ are the variables.

By taking the curl of (13-4) and using (13-5) we get

$$\nabla^2 \mathbf{u} = -\nabla \times \mathbf{w} \tag{13-7}$$

which is a vector form of Poisson's equation for $\mathbf{u}$.

BEM formulations for the solution of any time-dependent incompressible viscous flow problem can now be developed from Eqs (13-6) and (13-7). This has in fact been done by Wu and his coworkers[7-14] as described below.

Since the velocity vector $\mathbf{u}$ is solenoidal, a vector potential can be defined (see Chapter 10) such that

$$\mathbf{u} = \nabla \times \psi \tag{13-8}$$

Substituting (13-8) in (13-4) we get

$$\nabla \times \nabla \times \psi = \mathbf{w} \tag{13-9}$$

Now, if $\mathbf{A}$ and $\mathbf{B}$ are two vectors that are single valued and have continuous second derivatives then, by Green's theorem in vector form[1,7] (Appendix B), we have

$$\int_V [\mathbf{A} \cdot (\nabla \times \nabla \times \mathbf{B}) - \mathbf{B} \cdot (\nabla \times \nabla \times \mathbf{A})] \, dV = \int_S (\mathbf{B} \times \nabla \times \mathbf{A} - \mathbf{A} \times \nabla \times \mathbf{B}) \cdot \mathbf{n} \, dS \tag{13-10}$$

where $\mathbf{n}$ is the normal to $S$.

We can identify the vector $\mathbf{B}$ with $\psi$ in (13-9) and the vector $\mathbf{A}$ as the fundamental three-dimensional solution of the vector equation

$$\nabla \times \nabla \times \mathbf{A} = 0$$

that is,

$$\mathbf{A} = \nabla \left( \frac{1}{4\pi r'} \right) \times \mathbf{e} = \nabla \times \left( \frac{\mathbf{e}}{4\pi r'} \right) \tag{13-11}$$

where **e** is a unit vector

$$r' = |\mathbf{r}_o - \mathbf{r}| \qquad (r')^2 = y_i y_i \qquad y_i = (\xi - x)_i \qquad (13\text{-}12)$$

Substituting $\psi$ for **B** and the fundamental solution (13-11) for **A** in (13-10) and using Eqs (13-8) and (13-9) we arrive at

$$\int_V (\mathbf{G} \times \mathbf{e}) \cdot \mathbf{w}\, dV = \int_S [\psi + \nabla(\mathbf{e} \cdot \mathbf{G}) \cdot \mathbf{n} - (\mathbf{G} \times \mathbf{e}) \times \mathbf{u} \cdot \mathbf{n}]\, dS \qquad (13\text{-}13)$$

where

$$\mathbf{G} = \nabla \frac{1}{4\pi r'} \qquad (13\text{-}14)$$

which can be rewritten as

$$\int_V \mathbf{e} \cdot (\mathbf{w} \times \mathbf{G})\, dV = \int_S \{(\mathbf{e} \cdot \mathbf{G})\mathbf{u} \cdot \mathbf{n} - \mathbf{e} \cdot [(\mathbf{u} \times \mathbf{n}) \times \mathbf{G}]\}\, dS \qquad (13\text{-}15)$$

The surface integral in Eq. (13-15) is improper since **G** is infinite when $r' = |\mathbf{r}_o - \mathbf{r}|$ tends to zero (i.e., the load point and the field point coincide). By considering a small region of exclusion around the singularity, letting it shrink to zero, and eliminating the arbitrary unit vector **e** from (13-15) we obtain

$$\beta \cdot \mathbf{u}(r_o) = \int_V \mathbf{w} \times \mathbf{G}\, dV - \int_S [(\mathbf{u} \cdot \mathbf{n})\mathbf{G} - (\mathbf{u} \times \mathbf{n}) \times \mathbf{G}]\, ds \qquad (13\text{-}16)$$

where

$$\beta = \begin{cases} 1 & \text{if } r_o \text{ is in the flow region} \\ \frac{1}{2} & \text{if } r_o \text{ is located on a smooth boundary} \end{cases}$$

Equation (13-16) is also valid for any two-dimensional problem if the following fundamental solution is used for (13-4).

viz

$$\mathbf{G} = \nabla \left[ \frac{1}{2\pi} \ln \left( \frac{1}{r'} \right) \right] \qquad (13\text{-}17)$$

where once again $r' = |\mathbf{r}_o - \mathbf{r}|$.

Equation (13-16) can also be written to include the free-stream velocity as[7]

$$\beta \mathbf{u}(r_o) = \frac{1}{m} \int_V \frac{\mathbf{w} \times (\mathbf{r}_o - \mathbf{r})}{|\mathbf{r}_o - \mathbf{r}|^d}\, dV - \int_S \left[ \frac{(\mathbf{u} \cdot \mathbf{n})(\mathbf{r}_o - \mathbf{r})}{|\mathbf{r}_o - \mathbf{r}|^d} - \frac{(\mathbf{u} \times \mathbf{n}) \times (\mathbf{r}_o - \mathbf{r})}{|\mathbf{r}_o - \mathbf{r}|^d} \right] dS + \mathbf{u}^\infty \qquad (13\text{-}18)$$

where  $\mathbf{u}^\infty$ = free-stream velocity (see Fig. 13-1)

$m = 4\pi$ and $d = 3$ for three-dimensional problems

$m = 2\pi$ and $d = 2$ for two-dimensional problems

Equation (13-18) is now in a form which can be used for the solution of well-posed boundary-value problems.

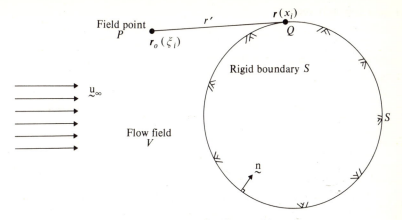

**Figure 13-1** A typical flow problem.

It might be noted that it is not really necessary to have $\mathbf{V} \cdot \mathbf{u} = 0$ in Eq. (13-15); in fact, if we had $\mathbf{V} \cdot \mathbf{u} = g$ instead we would end up with[12]

$$\beta \mathbf{u}(\mathbf{r}_o) = \int_V (g + \mathbf{w} \times) \, \mathbf{G} \, dV - \int_S [\mathbf{u} \cdot \mathbf{n} - (\mathbf{u} \times \mathbf{n}) \times ] \, \mathbf{G} \, dS \qquad (13\text{-}19)$$

It is important to note that the volume integral is evaluated with respect to $\mathbf{r}$ in space and the surface integral with respect to $\mathbf{r}$ on the surface.

For the vorticity transport equation we note that, in Chapter 9, we have already dealt with equations of the form [see (9-1)]

$$c \nabla^2 \phi = \frac{\partial \phi}{\partial t} - q \qquad (13\text{-}20)$$

with the direct boundary integral for this equation given by Eq. (9-11), i.e.,

$$\alpha \phi(P, t) = \int_S \left( G * \frac{\partial \phi}{\partial n} - F * \phi \right) dS + \int_V (G * q + fG) \, dV \qquad (13\text{-}21)$$

Functions $G$ and $F$ are defined in (9-7) and (9-8), $f$ denotes the distribution of the function $\phi$ in the region $V$ at time $t = 0$, and $*$ between the quantities denotes a convolution integral [see (9-10)].

By noting the similarity between Eqs (13-6) and (13-20) and then using (13-21) we can obtain an integral formulation of the vorticity transport equation.

By imposing a no-slip condition on a non-rotating surface $S$ for two-dimensional flow the required integral representations simplify to[13]

$$\beta \mathbf{u}(\mathbf{r}_o, t) = \frac{1}{2\pi} \int_V \frac{[w]_{\tau=t} \, \mathbf{e} \times (\mathbf{r}_o - \mathbf{r})}{|\mathbf{r}_o - \mathbf{r}|^2} \, dV + \mathbf{u}^\infty \qquad (13\text{-}22)$$

and for vorticity transport

$$\beta w(\mathbf{r}_o, t) = \int_V [Gw]_{\tau=0} \, dV + \int_0^t d\tau \int_V w[\mathbf{u} \cdot \nabla G] \, dV$$

$$+ v \int_0^t d\tau \int_S [G \nabla w - w \nabla G] \cdot \mathbf{n} \, dS \qquad (13\text{-}23)$$

Equation (13-22) is readily interpreted as a statement of the Biot–Savart law[15–17] for distributed vortex lines. The volume integrals in (13-23) show that the vorticity distribution changes as a result of the convective process which continuously affects the subsequent distribution of vorticity in the fluid. The surface integral in (13-23) represents the effect of continuous generation (or depletion) of vorticity on the solid boundary $S$. Because the velocity $\mathbf{u}(\mathbf{r}_o, t)$ on $S$ is zero (the no-slip condition) the vorticity generated can only leave $S$ through diffusion.

The numerical solution[13] of the coupled pair of equations (13-22) and (13-23) would obviously follow a time-marching scheme for which the discretization would be identical to one of those discussed in earlier chapters. If the initial time level in (13-23) is taken to be the interval $(t - \Delta t)$, known values of the velocity and vorticity at this time would be used to calculate a new set of vorticity values at time $t$. These values of $\mathbf{w}(\mathbf{r}, t)$ in the vorticity volume cells are then used to calculate the velocity $\mathbf{u}(\mathbf{r}_0, t)$ at any point within the field using Eq. (13-22). Numerical solution of the problem thus progresses with repeated execution of such loops which, in effect, simulates the physical processes of vorticity diffusion, convection, and generation. This algorithm, developed by Wu, offers time-dependent solutions together with steady state or periodic (vortex-shedding) solutions at later times.

It is important to appreciate that the vorticity cells within the volume need only be introduced in the specific regions where the vorticity is non-zero. For the problem shown in Fig. 13-1 these would be confined to the immediate vicinity of the surface $S$. Such an algorithm is therefore likely to be more efficient than those which can be developed using finite element or finite difference procedures.

### 13-2-3 Stream Functions and Velocity Potential

In two dimensions the flow may be described by a stream function $\psi$. Physically, the stream function is constant along a streamline and the flow rate between any two streamlines is proportional to the numerical difference between the stream functions on them.

For two-dimensional incompressible flow the stream function is defined by

$$u_1 = -\frac{\partial \psi}{\partial x_2} \qquad u_2 = \frac{\partial \psi}{\partial x_1} \qquad (13\text{-}24)$$

where $u_1$ and $u_2$ are velocities in the $x_1$ and $x_2$ directions respectively. For two-dimensional compressible flow[6]

$$u_1 = -\frac{\rho_o}{\rho}\frac{\partial \psi}{\partial x_2} \qquad u_2 = \frac{\rho_o}{\rho}\frac{\partial \psi}{\partial x_1} \tag{13-25}$$

where $\rho_o$ and $\rho$ are an arbitrary reference density and the current density respectively.

A velocity potential $\phi$ can be defined if the velocity field is irrotational (that is, $\nabla \times \mathbf{u} = 0$) and therefore, in an orthogonal cartesian system, we would have, alternatively,

$$u_i = -\frac{\partial \phi}{\partial x_i} \tag{13-26}$$

### 13-2-4 Equations of Motion for Low Reynolds Number in Terms of a Stream Function

For low Reynolds numbers the convective terms in Eqs (13-1), (13-2), etc., become negligible and the governing equations become linear. Thus in two-dimensional steady motion the stream function $\psi$ satisfies the equation

$$\nabla^4 \psi = 0 \tag{13-27}$$

In a more general situation, at intermediate values of Reynolds number, the governing equation for steady, incompressible viscous flow may be written as[18]

$$\nabla^4 \psi = q(\psi, \nabla^2 \psi) \tag{13-28}$$

Equation (13-28) is clearly non-linear since the right-hand side depends upon both $\psi$ and $\nabla^2 \psi$, neither of which will be known initially.

It is immediately evident (see Chapter 11) that we can develop an integral representation of (13-28) by the well-known Rayleigh–Green[1] identity for two biharmonic functions $\psi$ and $\chi$:

$$\int_V (\psi\nabla^4\chi - \chi\nabla^4\psi)\,dV = \int_S \left[\psi\frac{\partial}{\partial n}(\nabla^2\chi) - \nabla^2\chi\frac{\partial\psi}{\partial n} + \nabla^2\psi\frac{\partial\chi}{\partial n} - \chi\frac{\partial}{\partial n}(\nabla^2\psi)\right]dS \tag{13-28a}$$

defined within a region $V$ bounded by its boundary $S$ with an outward normal $\mathbf{n}$.

If we choose $G$ to be a solution of the biharmonic equation

$$\nabla^4 G = \delta(x, \xi)$$

and substitute this fundamental solution for the variable $\chi$ in Eq. (13-28a) (see Chapter 11) we can develop an integral representation

$$\beta\psi(\xi) = \int_S \left[\psi\frac{\partial}{\partial n}(\nabla^2 G) - \nabla^2 G\frac{\partial\psi}{\partial n} + \nabla^2\psi\frac{\partial G}{\partial n} - G\frac{\partial}{\partial n}(\nabla^2\psi)\right]dS + \int_V qG\,dV \tag{13-29}$$

where
$$\beta = \begin{cases} 1 & \xi \text{ within the flow region} \\ 0 & \xi \text{ outside the flow region} \\ \frac{1}{2} & \xi \text{ on the boundary} \end{cases}$$

The integrations in (13-29) are with respect to $x$ and the normal is evaluated at $x$ also. For the linear case (13-27) the volume integral in (13-29) vanishes and the problem is then one of boundary discretization only. However, for the solution of problems governed by (13-28a) volume cells would have to be introduced as explained in Chapter 12.

### 13-2-5 Inviscid, Irrotational, and Incompressible Flow

For steady state inviscid, irrotational, incompressible flow the governing equations can be written in terms of the velocity potential $\phi$ as

$$\nabla^2 \phi = q \tag{13-30}$$

Problems governed by this equation have been discussed at length in Chapters 3 and 5 where this equation was used as the main vehicle for introducing BEM initially.

We can, however, introduce here an additional class of related problem, one involving water waves.[19] The two-dimensional motion of water of variable depth can be described by

$$\nabla^2 \phi = 0 \qquad u_1 = \frac{\partial \phi}{\partial x_1} \qquad u_2 = \frac{\partial \phi}{\partial x_2} \tag{13-31}$$

throughout the fluid.

If the surface is parametrically represented by the arc length $s$ taken counterclockwise from the upper right-hand corner of Fig. 13-2, the coordinates

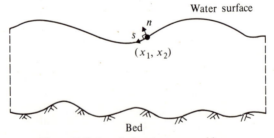

**Figure 13-2** A typical water-wave problem.

of the surface points $x_1(t, s)$, $x_2(t, s)$ are governed by

$$\frac{\partial x_1}{\partial t} - \frac{\partial x_2}{\partial s} \frac{\partial \phi}{\partial n} = 0$$

$$\frac{\partial x_2}{\partial t} + \frac{\partial x_1}{\partial s} \frac{\partial \phi}{\partial n} = 0 \tag{13-32}$$

where $ds^2 = dx_1^2 + dx_2^2$ and $t$ is the time, and

$$\frac{\partial}{\partial n} = \frac{\partial x_2}{\partial s}\frac{\partial}{\partial x_1} - \frac{\partial x_1}{\partial s}\frac{\partial}{\partial x_2}$$

$$\frac{\partial}{\partial s} = \frac{\partial x_1}{\partial s}\frac{\partial}{\partial x_1} + \frac{\partial x_2}{\partial s}\frac{\partial}{\partial x_2}$$

(13-33)

These equations state merely that the point $(x)$ remains on the free surface at all times. For constant atmospheric pressure Bernoulli's equation is

$$\frac{\partial \phi}{\partial t} + \frac{1}{2}\left[\left(\frac{\partial \phi}{\partial n}\right)^2 + \left(\frac{\partial \phi}{\partial s}\right)^2\right] + gx_2 = f(t)$$

(13-34)

with $g$ the acceleration due to gravity; $f(t)$ does not depend on the spatial variables and is usually taken as constant.

The problem is then one of solving (13-31) with boundary conditions (13-32) and (13-34) starting from an initially prescribed boundary geometry. Since only the shape of the boundary at different times is of interest BEM have an immediate appeal and the relevant algorithm has been developed recently by Marder[19] who describes a number of solutions to problems involving the propagation of a surface disturbance as well as impulsively initiated flow over an obstacle.

Generally speaking BEM provide a very efficient numerical technique for a large class of moving boundary problems, a feature which has already been exploited by a number of people.[20-25]

## 13-2-6 Inviscid, Irrotational, and Compressible Flow

Steady state inviscid, irrotational, compressible flow may be described by[26-28] the equation

$$\nabla^2 \phi = M^2 \frac{\partial u}{\partial s} = q(M, u)$$

(13-35)

where $s$ represents the tangential direction of the streamline, $u$ the velocity, and $M$ the local Mach number.

Clearly Eq. (13-35) may be used to develop BEM formulations similar to those outlined in Chapters 3 and 12 with a numerical-solution algorithm involving a boundary discretization and a cell discretization[26-30] scheme in regions of non-zero $q$ values.

## 13-2-7 Transient and Steady State Wave Equations for Fluids

These are closely analogous to the propagation of dilatational and shear waves through elastic solids for which BEM formulations have been developed and described in Chapter 10.

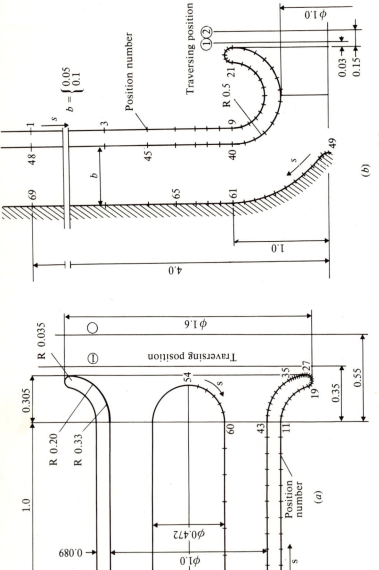

**Figure 13-3** (*a*) Schematic view of axial entrance. (*b*) Schematic view of radial entrance.

## 13-3 EXAMPLES

**(a) Potential flow through an axisymmetric entry to a turbomachine**[31] Figure 13-3a, b shows schematic views of the entry and the boundary discretization used and Fig. 13-4a, b the experimental and calculated surface velocity distribution for the axial entrances with and without a hub. The agreement is generally satisfactory although the experimental values are slightly higher than the calculated ones near the transition between the entrance and the straight section for the case without a hub, possibly due to separation. Calculated momentum

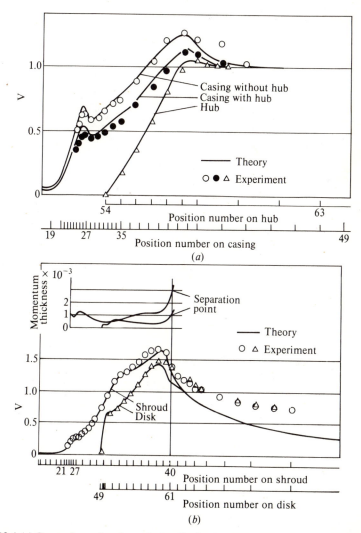

**Figure 13-4** (a) Comparison of surface velocity distribution for axial entrance. (b) Comparison of surface velocity distribution for radial entrance.

thickness of the surface boundary layer and predicted separation point are shown in Fig. 13-4b.

The calculated surface velocity distribution agrees with experimental data up to the separation point on the shroud surface and up to the maximum velocity position in the disc. Figure 13-5a, b show the velocity profiles and flow directions at the traversing stations (1) and (2) marked in Fig. 13-3 in front of the axial and radial entrances respectively. It is of considerable interest to note the good agreement between the experimental and the calculated data in front of the radial entrance in spite of the fully developed separation of the inner flow.

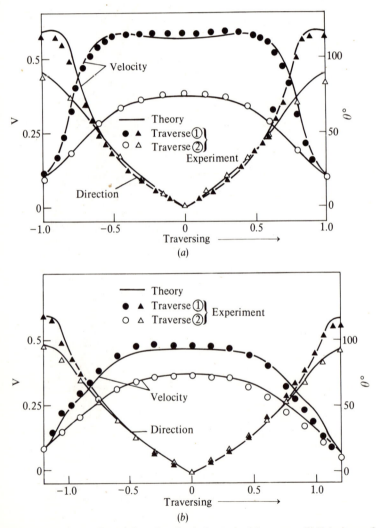

**Figure 13-5** (a) Velocity profile and flow direction in front of axial entrance. (b) Velocity profile and flow direction in front of radial entrance.

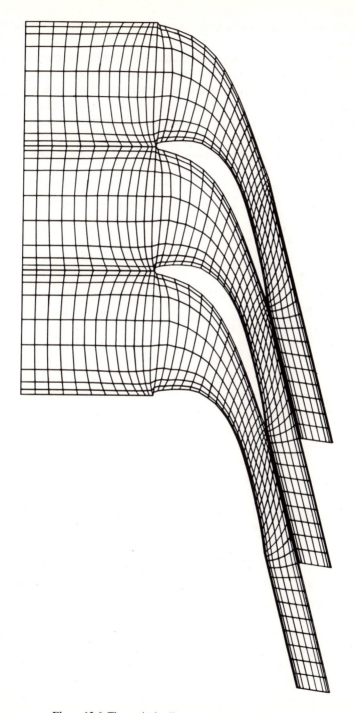

**Figure 13-6** The typical cell pattern around a cascade.

**(b) Flow through a cascade up to the transonic shock-free range**[26,27] Solutions for compressible potential flow can be used to investigate the pressure distribution on the boundary up to the transonic, shock-free range. Figure 13-6 shows a typical cell discretization pattern around a turbine cascade and Fig. 13-7 the pressure distribution over each profile given by different values of free-stream Mach number. Any further increase of the upstream Mach number leads to supercritical flow with shock, in which case the compressible potential flow solution is no longer relevant.

The problem is thus very similar to collapse points in solid mechanics (see Chapter 12-9, examples **e**, **g**).

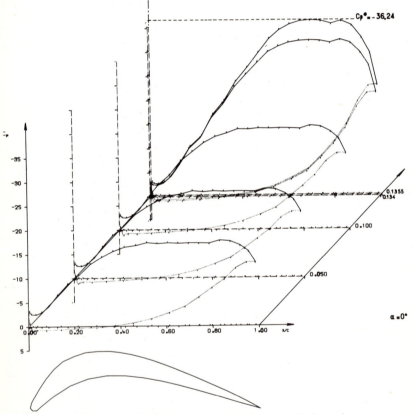

**Figure 13-7** The pressure distribution for various free-stream Mach numbers.

**(c) Slow viscous flow within a circle**[18] A typical solution for the impinging jet problem is shown in Fig. 13-8. No circulation occurs at zero Reynolds number (Fig. 13-8a) and the eddies first appear at a value of Reynolds number between 3 and 3.1. These solutions were obtained by Mills[18] using BEM in which he adopted a fundamental solution which satisfies the boundary conditions a priori. This is obviously not necessary although in some problems it may well result in a

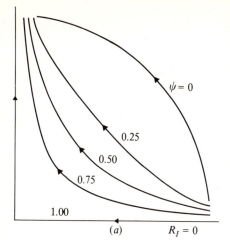

$\psi = 0$

0.25

0.50

0.75

1.00

*(a)*  $R_I = 0$

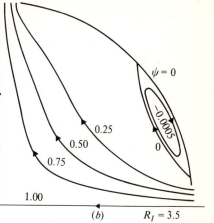

$\psi = 0$

−0.0005

0

0.25

0.50

0.75

1.00

*(b)*  $R_I = 3.5$

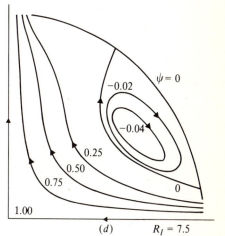

$\psi = 0$

−0.005
−0.01

0.25

0.50

0.75

0

1.00

*(c)*  $R_I = 5$

$\psi = 0$

−0.02

−0.04

0.25

0.50

0.75

0

1.00

*(d)*  $R_I = 7.5$

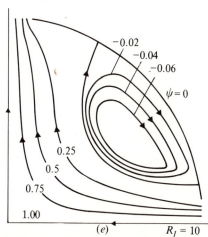

−0.02

−0.04

−0.06

$\psi = 0$

0.25

0.5

0.75

1.00

*(e)*  $R_I = 10$

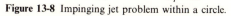

**Figure 13-8** Impinging jet problem within a circle.

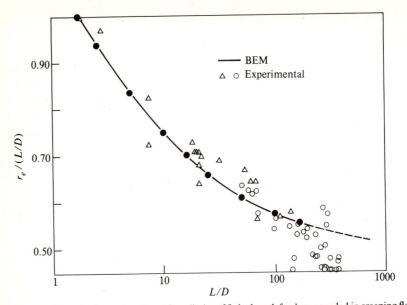

**Figure 13-9** The equivalent axis ratio $r_e$ of a cylinder of finite length freely suspended in creeping flow.

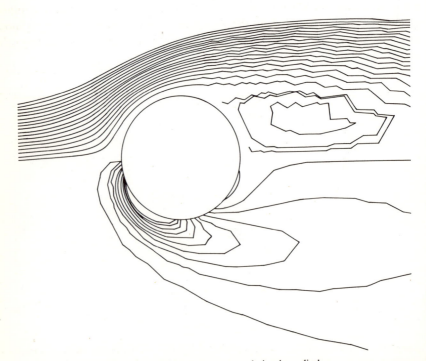

**Figure 13-10** Flow pattern around circular cylinder.

substantial computer cost saving. He noted that his iterative algorithm became unstable at a Reynolds number ($R_I$) of about 15.

**(d) Stokes flow past an arbitrary obstacle**[32] The problem of creeping flow past surfaces of arbitrary shape was considered by Youngren and Acrivos,[32] in particular three-dimensional flow around a cylinder of finite length. The problem has been studied experimentally many times and their solution is therefore of considerable practical interest. The angular velocity of a cylinder freely suspended in a general linear shear flow is related to a single, unknown, scalar parameter—the equivalent axis ratio $r_e$, defined as the axis ratio of that spheroid which would, when freely suspended in the same flow field at infinity, experience the same periodic motion as the cylinder (see Fig. 13-9).

They solved this problem by using direct BEM and compared the results with experimental data obtained by others. It can be seen that agreement with the experimental data is very good for $L/D < 100$, but for $L/D > 100$ the experimental results differ significantly from the calculated values although the experimental scatter is also very high in this region.

**(e) Viscous flow past a cylinder**[13] The computed streamlines and equivorticity contours are shown in Fig. 13-10 for flow past a cylinder with a Reynolds number of 40 based on the free-stream velocity and the cylinder diameter. The flow patterns shown are for the asymptotically steady state time. The computed pressure coefficient divided by the free-stream kinetic energy on the cylinder surface is shown in Fig. 13-11$a$ where the angle $\theta$ has been measured from the stagnation point in front of the cylinder.

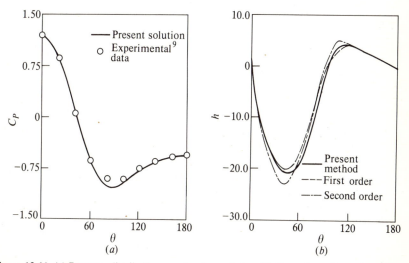

**Figure 13-11** ($a$) Pressure distribution on circular cylinder. ($b$) Normal gradient of vorticity on circular cylinder.

Figure 13-11*b* shows the distribution of normal vorticity gradient around the cylinder boundary, a result which is automatically provided by BEM but has to be calculated by using one-sided difference formulae of different order in most other methods.

Other examples of BEM applied to problems in fluid mechanics will be found in Refs 7 to 14, 19 to 27, 29, and 31 to 52 and in Chapters 3 and 5.

## 13-4 CONCLUDING REMARKS

We hope that this chapter has demonstrated adequately the very considerable potential of BEM in the field of fluid mechanics, some of which has been exploited already. We firmly believe that BEM will very soon have established their superiority over other numerical methods throughout the vast range of fluid mechanics problems which are waiting to be explored.

Although we have only described DBEM formulations in this chapter a series of equivalent IBEM statements can also be established for all of these problems.

## 13-5 REFERENCES

1. Morse, P. M., and Feshback, H. (1953) *Methods of Theoretical Physics*, Vol. I, McGraw-Hill, New York.
2. Zienkiewicz, O. C. (1977) *The Finite Element Method*, 3rd ed., McGraw-Hill, London.
3. Lamb, H. (1932) *Hydrodynamics*, 6th ed., Dover, New York.
4. White, F. M. (1974) *Viscous Fluid Flow*, McGraw-Hill, New York.
5. Yih, C. S. (1969) *Fluid Mechanics*, McGraw-Hill, New York.
6. Hughs, W. F., and Gaylord, E. W. (1964) *Basic Equations of Engineering Science*, Schaum, New York.
7. Wu, J. C., and Thompson, J. F. (1973) 'Numerical solution of time dependent incompressible Navier–Stokes equations using an integro differential formulation', *J. Comp. Fluids*, **1**(2), 197–215.
8. Wu, J. C., Sanker, N. L., and Sampath, S. (1978) 'A numerical study of unsteady viscous flows around airfoil', *AGARD Conf. Proc.*, **227**, 24–1–18.
9. Wu, J. C., Spring, A. H., and Sankar, N. L. (1974) 'A flow field segmentation method for numerical solution of viscous flow problems', *Lecture Notes in Physics*, Vol. 35, pp. 452–457, Springer-Verlag, Berlin and New York.
10. Wu, J. C., and Wahbah, M. M. (1976) 'Numerical solutions of viscous flow equations using integral representations', *Lecture Notes in Physics*, Vol. 59, pp. 448–453, Springer-Verlag, Berlin and New York.
11. Wahbah, M. M. (1978) 'Computation of internal flows with arbitrary boundaries using integral representation methods', Report, School of Aerospace Engineering, Georgia Institute of Technology.
12. Wu, J. C. (1976) 'Finite element solution of flow problems using integral representations', *Proc. Second Int. Symp. on Finite Element Meth. in Engng*, International Centre for Computer Aided Design, Conf. Series No. 2/76, pp. 205–216.
13. Wu, J. C., and Rizk, Y. M. (1976) 'Integral representation approach to time-dependent viscous flows', Report Grant No. DAAG 29–75–G–0147, US Army Research Office, Department of Aerospace Engineering, Georgia Institute of Technology.

14. Sankar, N. L., and Wu, J. C. (1978) 'Viscous flow around oscillating airfoil—a numerical study', *AIAA Eleventh Fluid and Plasma Dynamics Conf.*, Seattle, Washington.
15. Karmakar, H. C., and Robertson, S. D. T. (1978) 'An integral equation formulation for electromagnetic field analysis in electrical apparatus', IEEE PES Summer Meeting, Los Angeles, Calif., 16–21 July.
16. Zaky, S. G., and Robertson, S. D. T. (1973) 'Integral equation formulations for the solution of magnetic field problems, Part I and II', *IEEE Trans.*, **2**(PAS-92), 808–823.
17. Lean, M. H., Friedman, M., and Wexler, A. (1979) 'Advances in application in boundary element method in electrical engineering problems', in P. K. Banerjee and R. Butterfield (eds), *Developments in Boundary Element Methods*, Chap. IX, Applied Science Publishers, London.
18. Mills, R. D. (1977) 'Computing internal viscous flow problems for the circle by integral methods', *J. Fluid Mech.*, **79**(3), 609–624.
19. Marder, B. (1979) 'Computing water waves without solving Laplace's equation', Report Sandia Laboratories, Division 2642, Albuquerque, N. Mexico.
20. Banerjee, P. K. (1979) 'Nonlinear problems of potential flow', in P. K. Banerjee and R. Butterfield (eds), *Developments in Boundary Element Methods*, Chapter II, Applied Science Publishers, London.
21. Liggett, J. (1977) 'Location of free surface in porous media', *J. Hydrauls Div.*, ASCE, **HY4**, 353–365.
22. Liu, P. L. F., and Liggett, J. (1977) 'Boundary integral solutions to ground water problems', *Proc. Int. Conf. Appl. Num. Modelling*, Southampton University, pp. 559–569.
23. Liggett, J. A., and Liu, P. L. F. (1979) 'Boundary solutions to two problems in porous media', *J. Hydrals Div.*, ASCE, **HY3**, 171–183.
24. Liu, P. L. F., and Liggett, J. A. (1978) 'An efficient numerical method of two-dimensional steady ground water problems', *Water Resources Res.*, **14**(3), 385–390.
25. Niwa, Y., Kobayashi, S., and Fukui, T. (1974) 'An application of the integral equation method to seepage problems', *Proc. Twenty-fourth Nat. Congr. on Appl. Mech.*, pp. 479–486.
26. Coulmy, G. (1976) 'The calculation of flow field by means of the panel methods', Euromech Colloquium 75, Braunschweig/Rhode, 10–13 May.
27. Luu, T. S., and Coulmy, G. (1977) 'Method of calculating the compressible flow round an aerofoil or a cascade up to the shockfree transonic range', *Computers and Fluids*, **5**, 261–275.
28. Ogana, W., and Sprieter, J. R. (1977) 'Derivation of an integral equation for transonic flows', *AIAA J.*, **15**, 281–283.
29. Ogana, W. (1977) 'Numerical solutions for subcritical flows by a transonic integral equation method', *AIAA J.*, **15**, 444–446.
30. Ogana, W. (1979) 'Derivation of an integral equation for three-dimensional transonic flows', *AIAA J.*, **17**(3), 305–307.
31. Inoue, M., Kuroumaru, M., and Yamaguchi, S. (1979) 'A solution of Fredholm integral equation by means of spline fit approximation', *Computers and Fluids*, **7**, 33–46.
32. Youngren, G. K., and Acrivos, A. (1975) 'Stokes flow past a particle of arbitrary shape: a numerical method of solution', *J. Fluid Mech.*, **69**(2), 377–403.
33. Banerjee, P. K., and Bate, C. J. (1978) 'Laminar flow measurements in square ducts using laser doppler anemometry', *Proc. Symp. on Laminar and Turbulent Flow*, University College, Swansea, pp. 191–199.
34. Zaroodny, S. J., and Greenberg, M. D. (1973) 'On vortex sheet approach to the numerical calculation of water waves', *J. Comp. Physics*, **11**,(3), 440–446.
35. Coulmy, G., Luu, T. S., and Malavard, L. (1974) 'Aerodynamique des systems portante et propulsifs en regime de fonctionnement periodique optimisation de leurs performances', *AAAF Eleventh Colloque D'Aerodyn. Appliq.*, 6–8 Nov., Faculté des Sciences de Bordeaux, Talence.
36. Luu, T. S., and Coulmy, G. (1975) 'Methode des singularites repartiou discretisee dans le domaine de l'hydro et de l'aerodynamique', Report Laboratoire d'Informatique pour la Méchanique et les Sciences, Orsay.
37. Hunt, B. (1977) 'Relationships between volume, surface and line distributions of vorticity source and doublicity', BAC (MAD) Report Ae/384, November.

38. Semple, W. G. (1977) 'A note on the relationship of the influence of sources and vortices in incompressible and linearised compressible flow', BAC (MAD) Report Ae/A/541, October.
39. Morino, L., and Kuo, C. C. (1974) 'Subsonic potential aerodynamics for complex configurations: a general theory', *AIAA J.*, **12**(2), 191–197, February.
40. Johnson, F. T., Ehlers, F. E., and Rubbert, P. E. (1976) 'A higher order panel method for general analysis and design applications in subsonic flow', *Proc. Fifth Int. Conf. Num. Meth. in Fluid Dynamics*, Twente University, Entschede, Holland; also *Lecture Notes in Physics*, Vol. 59, Springer-Verlag, Berlin and New York.
41. Rubbert, P. E., Saaris, G. R., et al. (1967) 'A general method for determining the aerodynamic characteristics of fan-in-wing configurations', Vol. 1, Theory and Application USAAVLABS Technical Report 67–61A.
42. Labrujere, Th. E., Loeve, W., and Slooff, J. W. (1970) 'An approximate method for the calculation of the pressure distribution on wing-body combinations at subcritical speeds', *AGARD Conf. Proc.*, No. 71, September, NLR MP 70014U.
43. Roberts, A., and Rundle, K., (1972) 'Computation of incompressible flow about bodies and thick wings using the spline mode system', BAC (CAD) Report Aero MA19.
44. Johnson, F. T., and Rubbert, P. E. (1975) 'Advanced panel-type influence coefficient methods applied to subsonic flows', AIAA Paper No. 75–50, January.
45. Hess, J. L. (1962) 'Calculation of potential flow about arbitrary 3-D lifting bodies: final technical report', McDonnell Douglas, Report No. NDC J5679–01, October.
46. Labrujere, Th. E. (1972) 'A survey of current collocation methods in inviscid subsonic lifting surface theory, Part 1: numerical aspects', VKI Lecture Series 44, February.
47. Hess, J. L. (1974) 'The problem of 3-D lifting potential flow and its solution by means of surface singularity distributions', *Computer Meth. in Appl. Mech. Engng.* **4**, 283–319.
48. Hunt, B., and Semple, W. G. (1976) 'Economic improvements to the mathematical model in a plane/constant strength panel method', Paper presented at Euromech Colloq. No. 75, at Rhade (Braunschweig, W. Germany), May.
49. Kraus, W. (1976) 'Panel methods in aerodynamics', Paper given at VKI Lecture Series 87, 15–19 March.
50. Smith, P. D. (1974) 'An integral prediction method for three-dimensional compressible turbulent boundary layers', ARC R & M 3739.
51. Hunt, B. (1977) 'The prediction of external store characteristics by means of the panel method', BAC Report Ae/372, January.
52. Bristow, D. R. (1976) 'A new surface singularity method for multi-element airfoil analysis and design', AIAA Paper 76-20, January.

# FOURTEEN

## COMBINATION OF BOUNDARY ELEMENT METHODS WITH OTHER NUMERICAL METHODS

## 14-1 INTRODUCTION

So far in the book we have, we believe, demonstrated that BEM are very efficient numerical tools for solving problems involving bulky two- and three-dimensional regions. On the other hand, finite element or finite difference methods are attractive in both finite regions and in regions where there are high degrees of either geometrical or material non-linearity. Therefore for some problems it may be advantageous to use a combination of more than one method in what is often called a 'hybrid solution'.

In fact a very large group of problems in solid and fluid mechanics fall into the category where the use of either domain discretization (finite element or finite differences) or boundary discretization alone is impractical. Problems such as:

1. the interaction between a finite size structure and an extensive solid supporting material (e.g., soil–structure interaction problems),
2. problems in fracture mechanics,
3. interaction between a structure and the fluid medium in which it is immersed, etc.,

have attracted the attention of many workers who have developed algorithms combining differential and integral methods of analysis. For example, Cruse,

Shaw, Silvester, Wexler, Zienkiewicz, and others[1-13] developed solutions based on a combination of finite elements and BEM, while Banerjee, Butterfield, Wu, and others[14-24] have combined finite difference methods with BEM.

The system matrices which arise in finite element methods are usually symmetric whereas the BEM algorithms described in earlier chapters lead to non-symmetric system matrices. Therefore if a small BEM system needs to be incorporated into a large finite element system, it is necessary for an efficient solution to modify the BEM formulation so that the matrices it generates are symmetric. The contrary case leads to no loss of efficiency since any large non-symmetric BEM system can accommodate a symmetric finite element one. In either case the two systems can be assembled together if care is taken to satisfy the interface compatibility conditions.

The BEM system equations can be made symmetric by using an energy minimization scheme to establish them rather than the point-matching scheme used in earlier chapters. One byproduct of such an operation is that, in the indirect method at least, it leads to smaller errors at edges and corners and usually enables a better estimate of errors to be made. However, such solutions are rather more expensive than those using point matching and therefore as far as possible attempts should be made to achieve a non-symmetrical coupling in terms of the BEM boundary variables.

## 14-2 BOUNDARY ELEMENT METHOD SOLUTIONS DERIVED BY AN ENERGY METHOD

### 14-2-1 Introduction

There is a considerable quantity of published literature[1-8,24-34] in which boundary integral statements are developed from energy considerations; e.g., the method of moments, the Galerkin method, and the Rayleigh–Ritz method. Although the use of any of these methods can lead to symmetric systems of matrices the weighted residual approach is probably preferable from an engineer's point of view since the physical process involved can be more readily understood.

### 14-2-2 The General Theory of Weighted Residuals

To illustrate the weighted residual procedure we shall consider the determination of a function $(u)$, which may be either a scalar or a vector quantity within a region $V$ bounded by $S$, defined by the general equation

$$L(u) = 0 \quad \text{in } V \tag{14-1}$$

subject to the boundary conditions

$$A(u) - f = 0 \quad \text{on } S_1$$

$$B(u) - g = 0 \quad \text{on } S_2 \tag{14-2}$$

where $f$ and $g$ are prescribed on $S = S_1 + S_2$.

The operators $L$, $A$, and $B$ may be either differential or integral operators and also either linear or non-linear in nature.

If $u^o$ is some approximation to $u$ then Eqs (14-1) and (14-2) will not be satisfied exactly. Let us assume that errors involved are

$$L(u^o) = E_1$$

$$A(u^o) - f = E_2 \tag{14-3}$$

and
$$B(u^o) - g = E_3$$

where $E_1$, $E_2$, and $E_3$ are residual error functions.

To determine the approximate solution $u^o$ some weighted integral of the errors, defined in (14-3), is set to zero, so that[5-7]

$$\int_V W^k L(u^o)\,dV + \int_{S_1} \bar{W}^k [A(u^o) - f]\,dS + \int_{S_2} \bar{\bar{W}}^k [B(u^o) - g]\,dS = 0 \tag{14-4}$$

where $W^k$, $\bar{W}^k$, and $\bar{\bar{W}}^k$ are a set $(k = 1, 2, 3, ..., m)$ of independent weighting functions.

Equation (14-4) is then the weighted residual statement for our problem. For a detailed account of this procedure see Zienkiewicz and others.[5]

## 14-2-3 The Indirect Boundary Element Method as a Special Class of Weighted Residual Method

Consider the potential flow problem

$$\frac{\partial^2 p}{\partial x_i \partial x_i} = 0 \quad \text{in } V \tag{14-5}$$

and
$$p = f \text{ on } S_1 \quad \text{and} \quad \frac{\partial p}{\partial x_i} n_i = \frac{\partial p}{\partial n} = g \text{ on } S_2$$

The general weighted residual statement for (14-5) can be written as

$$\int_V W^k \left( \frac{\partial^2 p}{\partial x_i \partial x_i} \right) dV + \int_{S_1} \bar{W}^k (p - f)\,dS + \int_{S_2} \bar{\bar{W}}^k \left( \frac{\partial p}{\partial n} - g \right) dS = 0 \tag{14-6}$$

The volume integral can be transformed by noting that

$$W^k \frac{\partial^2 p}{\partial x_i \partial x_i} = \frac{\partial}{\partial x_i} \left( W^k \frac{\partial p}{\partial x_i} \right) - \frac{\partial W^k}{\partial x_i} \frac{\partial p}{\partial x_i} \tag{14-7}$$

Using the divergence theorem we can express (14-6) as[7]

$$-\int_V \frac{\partial W^k}{\partial x_i} \frac{\partial p}{\partial x_i}\,dV + \int_S W^k \frac{\partial p}{\partial n}\,dS + \int_{S_1} \bar{W}^k (p - f)\,dS$$

$$+ \int_{S_2} \bar{\bar{W}}^k \left( \frac{\partial p}{\partial n} - g \right) dS = 0 \tag{14-8}$$

in which we can, once again, use Eq. (14-7) with $W^k$ and $p$ interchanged and apply the divergence theorem to obtain

$$\int_V \frac{\partial^2 W^k}{\partial x_i \partial x_i} p \, dV - \int_S p \frac{\partial W^k}{\partial n} \, dS + \int_S W^k \frac{\partial p}{\partial n} \, dS$$

$$+ \int_{S_1} \bar{W}^k (p - f) \, dS + \int_{S_2} \bar{\bar{W}}^k \left( \frac{\partial p}{\partial n} - g \right) dS = 0 \qquad (14\text{-}9)$$

It should be noted that all integrations in these equations are carried out with respect to $x_i$ and all variables are functions of $x_i$ only ($x_i \in V$, $x_i \in S$).

We can describe the distribution of $p(x)$ in terms of a surface source distribution $\phi(\zeta)$ via

$$p = \int_S G\phi \, dS = N^r \phi^r \qquad (14\text{-}10)$$

where $\phi^r$ are the nodal values $\phi$ on the surface and the integration variable here is $\xi_i$. Similarly, $\partial p / \partial n$ is obtainable from, say,

$$\frac{\partial p}{\partial n} = \int_S F\phi \, dS = \frac{\partial N^r}{\partial n} \phi^r = M^r \phi^r \qquad (14\text{-}11)$$

If we choose the weighting functions $W^k$, $\bar{W}^k$, and $\bar{\bar{W}}^k$ such that[7,8]

$$\begin{aligned}
W^k &= -\tfrac{1}{2} N^k & x_i \in V \text{ and } S \\
\bar{W}^k &= -M^k & x_i \in S_1 \qquad\qquad (14\text{-}12)\\
\bar{\bar{W}}^k &= N^k & x_i \in S_2
\end{aligned}$$

We note immediately that the volume integral in (14-9) vanishes since $W^k$ (that is, $N^k$) satisfies the governing differential equation. Thus, substituting (14-10) to (14-12) in (14-9) we obtain[7,8]

$$K^{kr} \phi^r + F^k = 0$$

where

$$K^{kr} = \frac{1}{2} \left[ \int_{S_2} (M^k N^r + N^k M^r) \, dS - \int_{S_1} (M^k N^r + N^k M^r) \, dS \right] \qquad (14\text{-}13)$$

$$F^k = \int_{S_1} M^k f \, dS - \int_{S_2} N^k g \, dS$$

It is important to appreciate that there is a double integration process involved in generating the system of equations (14-13); the first one (the inner integral) is with respect to $\xi_i$ [see Eqs (14-10) and (14-11)] and the second one (the outer integral) is with respect to $x_i$. The system matrix **K** in (14-13) is symmetric and can be combined with any symmetric finite element system matrix. This specific way of deriving a symmetric indirect formulation is due to Mustoe[7] who developed it for two-dimensional elastostatics which generates an identical final system of equations.

Although such a symmetric IBEM algorithm appears to be a little more expensive computationally than the standard non-symmetric form it does lead to more accurate solutions, particularly near geometrical discontinuities.[6-8]

## 14-2-4 Symmetric Direct Boundary Element Formulation for Elasticity

DBEM formulations clearly do not follow the philosophy of trial function procedures described in the earlier section and therefore weighted residual procedures cannot be used to find a symmetric set of equations in this case. However, for an elastic system (see Chapter 4), we can consider the total energy functional as

$$\Pi = \frac{1}{2} \int_V \sigma_{ij} \varepsilon_{ij} - \int_V u_i \psi_i \, dV - \int_{S_2} u_i g_i \, dS \qquad (14\text{-}14)$$

The volume integral can be recast in terms of displacements by using the strain-displacement equations as

$$\frac{1}{2} \int_V \sigma_{ij} \varepsilon_{ij} \, dV = \frac{1}{4} \int_V (\sigma_{ij} u_{i,j} + \sigma_{ij} u_{j,i}) \, dV$$

$$= \frac{1}{4} \int_V [(\sigma_{ij} u_i)_{,j} + (\sigma_{ij} u_j)_{,i} - \sigma_{ij,j} u_i - \sigma_{ij,i} u_j] \, dV \qquad (14\text{-}15)$$

By applying the divergence theorem and the stress equilibrium equation to (14-15) we obtain

$$\frac{1}{2} \int_V \sigma_{ij} \varepsilon_{ij} \, dV = \frac{1}{2} \int_S u_i t_i \, dS - \frac{1}{2} \int_V u_i \psi_i \, dV \qquad (14\text{-}16)$$

Substituting (14-16) in (14-14) and assuming $\psi_i = 0$ we get the total energy functional as [7]

$$\Pi = \frac{1}{2} \int_S u_i t_i \, dS - \int_S u_i g_i \, dS \qquad (14\text{-}17)$$

or, in matrix notation,

$$\Pi = \frac{1}{2} \int_S \mathbf{u}^T \mathbf{t} \, dS - \int_S \mathbf{u}^T \mathbf{g} \, dS \qquad (14\text{-}18)$$

By assuming a suitable variation of $u$ and $t$ in the form

$$\mathbf{u} = \mathbf{N} \mathbf{u}^n$$

$$\mathbf{t} = \mathbf{M} \mathbf{t}^n \qquad (14\text{-}19)$$

where $\mathbf{N}$ and $\mathbf{M}$ are shape functions, we can write (14-18) as

$$\Pi = \frac{1}{2}(\mathbf{u}^n)^T \left( \int_S \mathbf{N}^T \mathbf{M} \, dS \right) \mathbf{t}^n - (\mathbf{u}^n)^T \int_S \mathbf{N}^T \mathbf{g} \, dS \tag{14-20}$$

It should be noted that the shape functions $\mathbf{N}$ and $\mathbf{M}$ need not, in general, be the same.

Now consider the direct boundary integral equation

$$\beta_{ij} u_i(\xi) = \int_S [t_i(x) G_{ij}(x, \xi) - F_{ij}(x, \xi) u_i(x)] \, dS \tag{14-21}$$

which can be written, following the usual BEM procedure, as

$$\mathbf{A}\mathbf{u}^n = \mathbf{B}\mathbf{t}^n \tag{14-22}$$

where the variations of $\mathbf{u}$ and $\mathbf{t}$ over the boundary elements are assumed to be identical to those in Eq. (14-19) but with $\mathbf{M} = \mathbf{N}$ to ensure that both $\mathbf{A}$ and $\mathbf{B}$ are square matrices. If we rewrite (14-22) as

$$\mathbf{t}^n = \mathbf{B}^{-1} \mathbf{A}\mathbf{u}^n \tag{14-23}$$

this equation can be used to eliminate $\mathbf{t}^n$ from (14-20). However, if this relationship is used the resulting formulation may not, due to rounding errors, satisfy equilibrium exactly.[7] This problem may be overcome by introducing an auxiliary equilibrium equation

$$\int_S \mathbf{t} \, dS = 0 \tag{14-24}$$

the discretized form of which is

$$\left( \int_S \mathbf{M} \, dS \right) \mathbf{t}^n = \mathbf{Q}\mathbf{t}^n = 0 \tag{14-25}$$

The nodal tractions $\mathbf{t}^n$ are found by linking together a modified form of Eq. (14-22), i.e.,

$$\mathbf{A}\mathbf{u}^n = \mathbf{B}\mathbf{t}^n + \mathbf{Q}^T \boldsymbol{\lambda} \tag{14-26}$$

where $\boldsymbol{\lambda}$ (for two-dimensional problems) is a two-component vector of pseudo-Lagrange multipliers. Here the role of such multipliers is to introduce a controlled perturbation in each of Eqs (14-22) so that (14-25) can be satisfied exactly.

Equations (14-25) and (14-26) can now be written as

$$\begin{bmatrix} \mathbf{A} \\ \mathbf{0} \end{bmatrix} \{\mathbf{u}^n\} = \begin{bmatrix} \mathbf{B} & \mathbf{Q}^T \\ \mathbf{Q} & 0 \end{bmatrix} \begin{Bmatrix} \mathbf{t}^n \\ \boldsymbol{\lambda} \end{Bmatrix} \tag{14-27}$$

from which the nodal tractions $\mathbf{t}^n$ can be determined by simple matrix algebra:

$$\mathbf{t}^n = \mathbf{E}\mathbf{u}^n \tag{14-28}$$

If we substitute (14-28) in (14-20) the functional expression becomes

$$\Pi = (\mathbf{u}^n)^T \mathbf{K} \mathbf{u}^n + (\mathbf{u}^n)^T \mathbf{F} \tag{14-29}$$

where
$$\mathbf{K} = \frac{1}{2}\left( \int_S \mathbf{N}^T \mathbf{M} \, dS \right) \mathbf{E} \quad \text{and} \quad \mathbf{F} = -\int_{S_2} \mathbf{N}^T \mathbf{g} \, dS$$

The functional $\Pi$ may be minimized to obtain

$$\delta\Pi = \left( \frac{\partial\Pi}{\partial\mathbf{u}^n} \right)^T \delta\mathbf{u}^n = 0$$

and since this is valid for arbitrary variations $\delta\mathbf{u}^n$ we must have

$$\frac{\partial\Pi}{\partial\mathbf{u}^n} = 0 \tag{14-30}$$

which leads to the final system of equations

$$\mathbf{K}^o \mathbf{u}^n + \mathbf{F} = 0 \tag{14-31}$$

where

$$\mathbf{K}^o = \frac{1}{2}\left\{ \left( \int_S \mathbf{N}^T \mathbf{M} \, dS \right) \mathbf{E} + \left[ \left( \int_S \mathbf{N}^T \mathbf{M} \, dS \right) \mathbf{E} \right]^T \right\}$$

Equation (14-31) is now a symmetric form of DBEM statement.[7,8]

It is interesting to note that an alternative symmetric DBEM formulation can be obtained by using a very simple method. The nodal forces $\mathbf{F}$ corresponding to a surface traction $\mathbf{t}$ on the boundary are given by

$$\mathbf{F} = \int_S \mathbf{N}^T \mathbf{t} \, dS \tag{14-32}$$

where $\mathbf{N}$ is the shape function matrix for displacements. Substituting the boundary shape functions (14-19) and the relationship between the nodal tractions and nodal displacements of (14-28) into (14-32) produces the force displacement relationship

$$\mathbf{F} = \left( \int_S \mathbf{N}^T \cdot \mathbf{M} \, dS \right) \mathbf{E} \cdot \mathbf{u}^n = \mathbf{K}' \mathbf{u}^n \tag{14-33}$$

In general the matrix $\mathbf{K}'$ is not symmetric and therefore the Maxwell–Betti reciprocal theorem will not be satisfied. In order to make the equations symmetric $\mathbf{K}'$ can be replaced by

$$\mathbf{K}^o = \tfrac{1}{2}[\mathbf{K}' + \mathbf{K}'^T] \tag{14-34}$$

which leads to a formulation superficially identical to (14-31). However, (14-34) is an entirely intuitive modification of (14-33), its only justification being that it produces the same form of equation as that deduced from the energy method.[7]

## 14-2-5 An Alternative Energy Approach to Symmetric Boundary Element Method Algorithms

Consider the integral equation

$$K\phi = g \tag{14-35}$$

where $K$ is an integral operator, $\phi$ an unknown function, and $g$ are the known boundary values.

For a self-adjoint $K$ we can write a functional $\Pi$ as

$$\Pi = \langle K\phi, \phi \rangle - 2\langle \phi, g \rangle \tag{14-36}$$

Where $\langle \ \rangle$ denotes an inner product defined by

$$\langle u, v \rangle = \int_S uv \, dS \tag{14-37}$$

If we adopt a distribution of $\phi$ such that

$$\phi = \mathbf{N}^T \boldsymbol{\phi}_n \tag{14-38}$$

where $\boldsymbol{\phi}_n$ are the nodal values of $\phi$, we can rewrite (14-36) as

$$\Pi = \boldsymbol{\phi}_n^T \langle K\mathbf{N}, \mathbf{N}^T \rangle \boldsymbol{\phi}_n - 2\boldsymbol{\phi}_n^T \langle \mathbf{N}, g \rangle$$

By taking the variation of the functional with respect to $\boldsymbol{\phi}_n$ and equating $\partial \Pi / \partial \boldsymbol{\phi}_n$ to zero we get

$$\langle K\mathbf{N}, \mathbf{N}^T \rangle \boldsymbol{\phi}_n = \langle \mathbf{N}, g \rangle \tag{14-39}$$

or

$$\mathbf{A}\boldsymbol{\phi} = \mathbf{b}$$

Note that the integral form, say of (14-39) written out explicitly, is

$$\left[ \int_{S(x)} \mathbf{N}(x) \int_{S(\xi)} K(x, \xi) \mathbf{N}^T(\xi) \, dS(\xi) \, dS(x) \right] \boldsymbol{\phi}_n = \int_{S(x)} \mathbf{N}(x) g(x) \, dS(x) \tag{14-40}$$

which is a Galerkin formulation of the original boundary integral equation leading to a symmetric system matrix $\mathbf{A}$, a formulation which has been developed and used by a number of workers, principally in electrical engineering.[3,4,17,26,33,34] A somewhat different approach for obtaining BEM system matrices by using the least-square error on the boundary was described by Tottenham.[35]

## 14-3 EXAMPLES OF SOLVED PROBLEMS USING AN ENERGY APPROACH

(a) **Potential flow in an L-shaped domain**[8] Figure 14-1 shows an L-shaped domain divided into three regions and analysed by the symmetric DBEM and IBEM procedures outlined above. The problem is defined by the equation

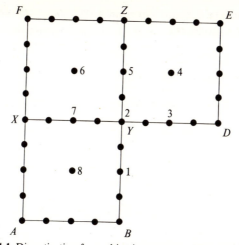

**Figure 14-1** Discretization for multiregion symmetric BEM procedures.

$\nabla^2 p = 0$ and boundary conditions along $ABYD$ of $\partial p/\partial n = 0$, on $DE$ of $p = 0$, along $EZF$ of $\partial p/\partial n = 0$, and along $AF$ of the potential $p = 1$. Table 14-1 shows that both the symmetric BEM formulations gave results which were in very good agreement with an accurate solution obtained by Jaswan and Symm,[36] whereas a non-symmetric IBEM analysis would lead to much greater errors near the geometric discontinuities. Table 14-2 shows the nodal values of the interface $\partial p/\partial n$ values along section $XY$ and $YZ$. It can now be seen that equilibrium at the interfaces is being achieved in the nodal force sense rather than by point matching. It is also worth noting that the symmetric IBEM solution produces $\partial p/\partial n$ values for the two regions which are exactly equal and opposite along the interfaces, whereas the potential values in them are not exactly equal (Table 14-3).

**Table 14-1  A comparison of symmetric IBEM and DBEM results for values of $p$**

| Node | Accurate solution | Indirect boundary integral symmetric three-region | Direct boundary integral symmetric three-region |
|---|---|---|---|
| 1 | 0.8640 | 0.8644 | 0.8587 |
| 2 | 0.6667 | 0.6733 | 0.6667 |
| 3 | 0.2972 | 0.2997 | 0.3023 |
| 4 | 0.2881 | 0.2883 | 0.2875 |
| 5 | 0.5680 | 0.5692 | 0.5706 |
| 6 | 0.8081 | 0.8039 | 0.8040 |
| 7 | 0.8514 | 0.8478 | 0.8488 |
| 8 | 0.9109 | 0.9063 | 0.9085 |
| Maximum error, % | | 0.85 | 1.6 |

**Table 14-2 Interface $\partial\phi/\partial n$ values on the L-shaped domain problem obtained by the symmetric BEM multiregion discretizations**

| Node | Symmetric DBEM (quadratic $\phi$, $\partial\phi/\partial n$ discontinuous) | | Symmetric IBEM | |
|---|---|---|---|---|
| | $\dfrac{\partial\phi}{\partial n_1}$ | $\dfrac{\partial\phi}{\partial n_2}$ | $\dfrac{\partial\phi}{\partial n_1}$ | $\dfrac{\partial\phi}{\partial n_2}$ |
| X | −0.0030 | 0.0043 | −0.0052 | 0.0052 |
| | −0.0111 | 0.0119 | | |
| | −0.0237 | 0.0170 | −0.0180 | 0.0180 |
| | −0.0280 | 0.0362 | | |
| | −0.0452 | 0.0555 | −0.0296 | 0.0296 |
| Y | −0.0990 | 0.0692 | −0.1111 | 0.1111 |
| | −0.1095 | 0.1480 | | |
| | −0.1198 | 0.1075 | −0.1616 | 0.1616 |
| | −0.1115 | 0.1012 | | |
| | −0.0994 | 0.1077 | −0.0984 | 0.0984 |
| | −0.1045 | 0.1022 | −0.0978 | 0.9078 |
| Z | −0.1030 | 0.1015 | −0.1088 | 0.1088 |

**Table 14-3 Interface potential values given by the symmetric IBEM multiregion discretization**

| Node | Symmetric IBEM | |
|---|---|---|
| | $\phi_1$ | $\phi_2$ |
| X | 0.9752 | 0.9700 |
| | 0.8888 | 0.8850 |
| | 0.8104 | 0.8073 |
| Y | 0.7439 | 0.7456 |
| | 0.6247 | 0.6234 |
| | 0.5792 | 0.5795 |
| | 0.5588 | 0.5590 |
| Z | 0.5617 | 0.5562 |

**(b) A cylinder of elliptic cross section in a uniform flow**[26] Figure 14-2 shows the convergence of a numerical solution of the problem using the Galerkin IBEM formulation outlined in Sec. 14-2-5 and a quadratic representation for the functions and geometry all due to Fried. Hess, and Smith (see Chapter 5) solved this problem very precisely using 180 nodes in a point collocation scheme. Fried's solution actually converged to three significant figures using only 16 nodes.

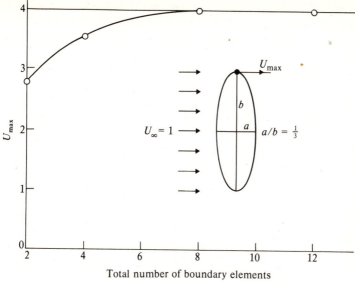

**Figure 14-2** Potential flow past an elliptical obstruction.

## 14-4 COMBINATION OF THE FINITE ELEMENT AND BOUNDARY ELEMENT METHODS

### 14-4-1 Finite Element Formulations via Weighted Residuals

For all elasticity problems the weighted residual statement (14-4) becomes

$$\int_V W_i^k(\sigma_{ij,j}+\psi_i)\,dV+\int_{S_1}\bar{W}_i^k(u_i-f_i)\,dS+\int_{S_2}\bar{\bar{W}}_i^k(t_i-g_i)\,dS=0 \quad (14\text{-}41)$$

We can apply the divergence theorem again to the volume integral to obtain

$$\int_V W_i^k\sigma_{ij,j}\,dV=-\int_V W_{i,j}^k\sigma_{ij}\,dV+\int_{S_1}W_i^k t_i\,dS \quad (14\text{-}42)$$

Using (14-42) in (14-41) and assuming $\psi_i=0$ for convenience, we obtain the 'weak formulation'

$$-\int_V W_{i,j}^k\sigma_{ij}\,dV+\int_S W_i^k t_i\,dS+\int_{S_1}\bar{W}_i^k(u_i-f_i)\,dS$$

$$+\int_{S_2}\bar{\bar{W}}_i^k(t_i-g_i)\,dS=0 \quad (14\text{-}43)$$

If we utilize the symmetry of the stress tensor ($\sigma_{ij}$) and use the stress–strain and strain–displacement relations, we can express the volume integral in (14-43) as

$$\int W_{i,j}^k \sigma_{ij} \, dV = \frac{1}{2} \int (W_{i,j}^k + W_{j,i}^k) \sigma_{ij} \, dV$$

$$= \int \mu (W_{i,j}^k + W_{j,i}^k) \left[ \alpha \delta_{ij} u_{r,r} + \tfrac{1}{2}(u_{i,j} + u_{j,i}) \right] dV \qquad (14\text{-}44)$$

where $\mu$ is the shear modulus and $\alpha = \nu/(1 - 2\nu)$.

By interpolating displacements in terms of the nodal values $u_n^k$ for the $k$th node,

$$u_i = N_i^k u_n^k \qquad (14\text{-}45)$$

we can rewrite (14-44) as [7]

$$\frac{1}{2} \int (W_{i,j}^k + W_{j,i}^k) \sigma_{ij} \, dV = \left\{ \int \mu (W_{i,j}^k + W_{j,i}^k) \right.$$

$$\left. \times \left[ \alpha \delta_{ij} N_{r,r}^k + \tfrac{1}{2}(N_{i,j}^k + N_{j,i}^k) \right] dV \right\} u_n^k \qquad (14\text{-}46)$$

It can be seen immediately that if $W_i^k$ is selected so that $W_i^k = N_i^k$ Eq. (14-46) is symmetric with respect to the indices $i$ and $j$. Therefore if the weighting functions are chosen such that[7,37]

$$W_i^k = \begin{cases} -N_i^k & \text{in } V \\ 0 & \text{on } S_1 \end{cases}$$

$$\bar{W}_i^k = 0 \qquad \text{on } S_1 \qquad\qquad (14\text{-}47)$$

$$\bar{W}_i^k = N_i^k \qquad \text{on } S_2$$

we can use (14-46) via (14-44) in (14-43) to obtain the symmetric set of linear equations

$$\left\{ \int_V \mu (N_{i,j}^k + N_{j,i}^k) \left[ \alpha \delta_{ij} N_{r,r}^k + \tfrac{1}{2}(N_{i,j}^k + N_{j,i}^k) \right] dV \right\}$$

$$\times \mathbf{u}_n^k - \int_{S_2} N_i^k g_i \, dS = 0 \qquad (14\text{-}48a)$$

or, simply,

$$\mathbf{K} \mathbf{u}_n + \mathbf{F} = 0 \qquad (14\text{-}48b)$$

which is the displacement finite element formulation.[37]

## 14-4-2 Symmetric Coupling of Direct Boundary Element and Finite Element Methods

If the finite element region is designated as $V_A$ and the BEM region $V_B$ the system equation for the two regions can be written (in elasticity, for example) as

$$\mathbf{K}_A \mathbf{u}_A + \mathbf{F}_A = 0 \qquad (14\text{-}49a)$$

$$\mathbf{K}_B \mathbf{u}_B + \mathbf{F}_B = 0 \qquad (14\text{-}49b)$$

where $\mathbf{u}_A$ and $\mathbf{F}_A$ are vectors of nodal displacements and forces within the domain $V_A$ including the boundary nodes. $\mathbf{u}_B$ and $\mathbf{F}_B$ are the vectors of nodal displacements and forces on the boundary of the region $V_B$.

Clearly, Eq. (14-49$b$) is of the same form as any finite element system contribution from a new element. Therefore Eqs (14-49$a, b$) can be coupled by satisfying interface–displacement compatibility in the usual manner. Interface compatibility can only be ensured if the boundary shape function for displacements is identical to the variation of displacements in the adjoining finite elements. Equilibrium is satisfied in a nodal sense—simply that the sum of the nodal forces at every node has to be equal to the resultant external force at that node.

### 14-4-3 Symmetric Coupling of Indirect Boundary Element and Finite Element Methods

To apply symmetric IBEM to the region $V_B$ and FEM to the region $V_A$, it is necessary to regard the common interface boundary as a part of the boundary $S_1$ where the potential $p$ (in the potential flow problem, for example) is specified. The equivalent system equations for the region $V_B$ given by Eq. (14-13) can be modified to

$$\begin{bmatrix} \mathbf{K} & \bar{\mathbf{K}} \\ \bar{\mathbf{K}}^T & \mathbf{0} \end{bmatrix} \begin{Bmatrix} \boldsymbol{\phi} \\ \mathbf{f} \end{Bmatrix} = \begin{Bmatrix} \mathbf{F} \\ \mathbf{0} \end{Bmatrix} \tag{14-50}$$

where $\mathbf{K}$, $\bar{\mathbf{K}}$, and $\bar{\mathbf{K}}^T$ are defined by Eq. (14-13) in an obvious way. The second matrix equation in (14-50) simply states that the integral of $\boldsymbol{\phi}$ over the boundary of $V_B$ is zero (i.e., the auxiliary uniqueness condition discussed in Chapters 2 to 4).

Equation (14-50) can now be incorporated into a finite element system for $V_A$ as a new element $V_B$ with nodeless variables $\boldsymbol{\phi}$ and nodal variables $\mathbf{f}$. Further details of this procedure are given by Kelly, Mustoe, and Zienkiewicz[6, 8] and Mustoe.[7]

### 14-4-4 Examples

**(a) The machine-component problem**[8] Figure 14-3 shows the geometry of a machine-component which has been analysed by Kelly, Mustoe, and Zienkiewicz[8] using (i) a finite element analysis with 95 eight-noded isoparametric elements, generating a total of 366 nodes of which 140 were boundary nodes, (ii) coupled DBEM and FEM, and (iii) symmetric DBEM. In the DBEM analysis the boundary was modelled using quadratic variations of the boundary potentials and the region subdivided into six subregions to facilitate the construction of a banded system matrix. It should be noted that in problems of this type BEM is not a very satisfactory method since the surface-to-volume ratio is very high.

Figure 14-4 shows the distribution of the potential along the boundaries given by the three analyses. The finite element solutions were obtained in about one-third of the computing time required for the symmetric DBEM analysis. This is

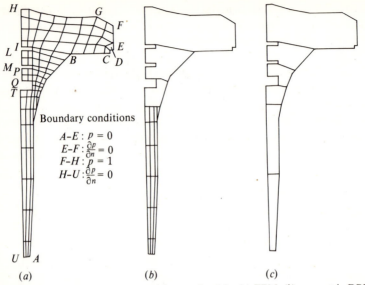

**Figure 14-3** The machine-component problem analysed by (*a*) FEM, (*b*) symmetric DBEM and FEM, (*c*) symmetric DBEM.

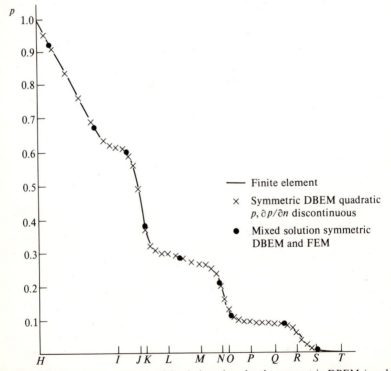

**Figure 14-4** Machine-component potential solution given by the symmetric DBEM (quadratic) procedure and the finite element method, along the boundary *H* to *T*.

principally due to two reasons: (i) the symmetric analysis is about twice as expensive as the non-symmetric analysis and (ii) for regions of this shape domain discretization schemes are shown to advantage.

**(b) Waves incident upon a cylinder with a porous protecting wall**[8]  The discretization for this problem is shown in Fig. 14-5 where the region between the cylinder and a finite region outside the porous protective wall was modelled by FEM, beyond which symmetric DBEM modelling was used. The porous wall itself was simulated by a six-noded finite element to reproduce the flow condition through the wall; thus $\partial p/\partial n = K(\phi_2 - \phi_1)$ where $\phi_1$ and $\phi_2$ are the velocity potentials

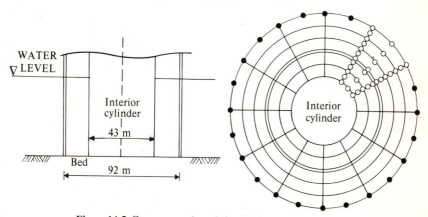

**Figure 14-5** Geometry and mesh for Ekofisk-type structure.

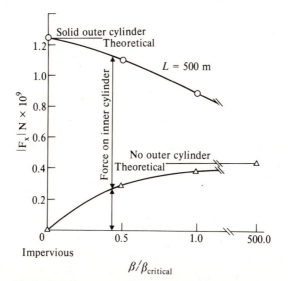

**Figure 14-6** Forces on cylinder with protecting wall.

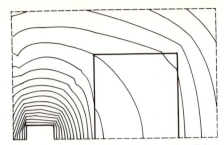

**Figure 14-7** Equipotential lines near square conductor placed parallel to dielectric slab of relative permittivity 6.

interior and exterior to the porous surface respectively. The wave forces on the cylinder and the porous wall are shown in Fig. 14-6 for different values of the permeability ratio $(\beta/\beta_{critical})$.

**(c) A square conductor flanked by a dielectric slab**[1] Silvester and Hseih described the coupling of IBEM and FEM to model exterior field problems. Figure 14-7 shows the equipotential lines when a charged square metallic conductor is placed parallel to and near a rectangular dielectric slab of relative permeability 6. The region bounded by the dotted lines was represented by FEM and the exterior region by a Galerkin formulation of IBEM.

## 14-5 EXAMPLES OF PROBLEMS SOLVED USING A COMBINATION OF FINITE DIFFERENCE AND BOUNDARY ELEMENT METHODS

**(a) The problem of a structural framework embedded in a non-homogeneous elastic soil mass**[16,17] In the mid-'sixties the authors became interested in the possibility that soil–structure interaction problems might be solved using a combination of a domain discretization method with IBEM. Boundary element methods are not very attractive for modelling thin structural elements in bending, whereas, conversely, domain discretization methods are unattractive for massive three-dimensional solids. A combination of the two techniques would therefore be the most logical choice for problems of the following kind.

Figure 14-8 shows the solutions for a number of three-dimensional structures (pile groups with vertical or inclined piles) embedded in a three-dimensional solid with a modulus of elasticity increasing linearly with the depth, compared with those obtained from a series of full-scale field tests. In these the pile structure was modelled by a finite difference scheme and the extensive solid (soil) medium by IBEM. An approximate point force solution[17] for the non-homogeneous solid was constructed so that it satisfied the ground-surface boundary conditions a priori and thus the only discretization needed was surface elements on the pile–soil interfaces.

| Test | Schematic diagram | | Lateral load for ¼ in deflection | | Lateral deflection due to vertical load of 20 $T$/pile | |
|---|---|---|---|---|---|---|
| | Plan | Elevation | Test | Theory | Test | Theory |
| 1 | 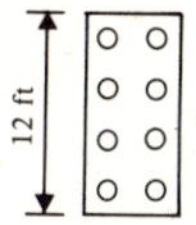 | 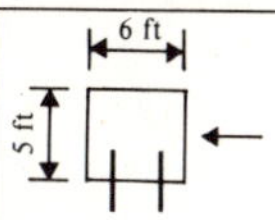 | 4.8 | 4.8 | 0.0 | 0.0 |
| 2 | 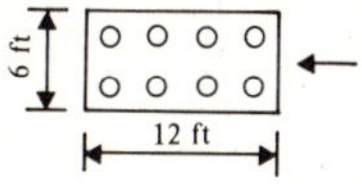 | 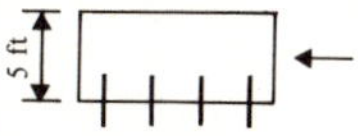 | 5.8 | 5.3 | 0.0 | 0.0 |
| 3 | 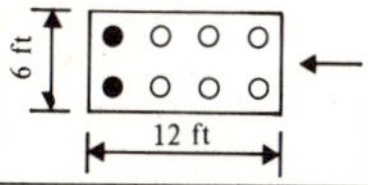 | 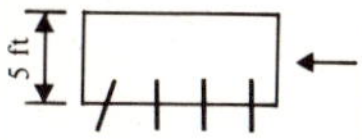 | 7.0 | 7.3 | 0.04 | 0.06 |
| 4 | 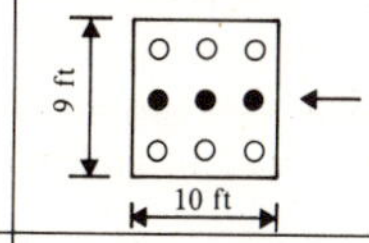 | 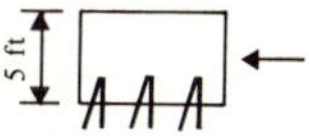 | 7.1 | 6.8 | 0.06 | 0.08 |
| 5 | 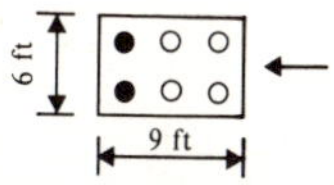 | 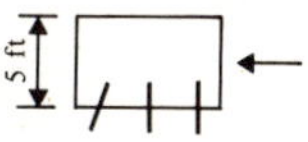 | 7.3 | 8.1 | 0.05 | 0.07 |
| 6 | 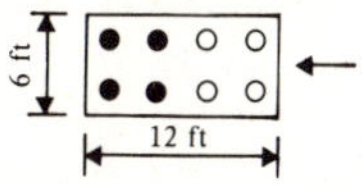 | 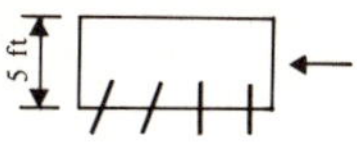 | 9.0 | 8.4 | 0.07 | 0.11 |
| 7 | 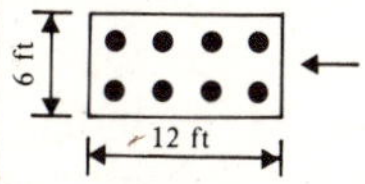 |  | 9.0 | 8.2 | 0.21 | 0.27 |
| 8 | 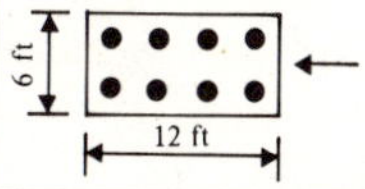 | 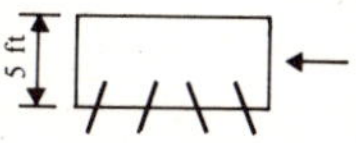 | 15.8 | 11.7 | 0.0 | 0.0 |

**Figure 14-8** Comparisons with full-scale tests.

**(b) Oscillations in a bay**[23] Olsen and Hwang[23] studied the tidal oscillations in a bay using a combined finite difference–BEM scheme. The variable depth region of the bay was modelled using finite differences and the region beyond that (to infinity) by the BEM scheme. Figure 14-9 shows the bay which they studied and Figs 14-10 and 14-11 their result for the wave-amplification field and the wave spectra.

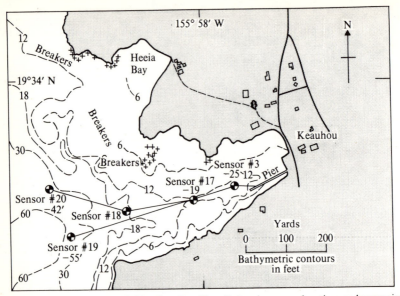

**Figure 14-9** Wave-sensor array in Keauhou Bay, Hawaii, showing sensor locations and approximate cable runs (all depths are in feet).

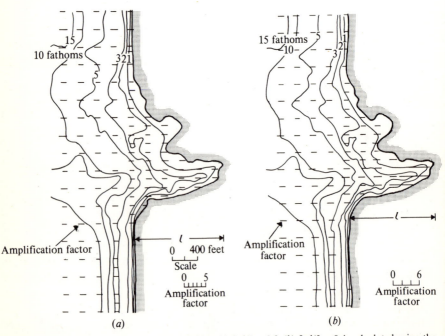

**Figure 14-10** Wave-amplification factor field at (*a*) $2\pi l/L = 0.2$, (*b*) $2\pi l/L = 0.4$, calculated using the combined algorithm.

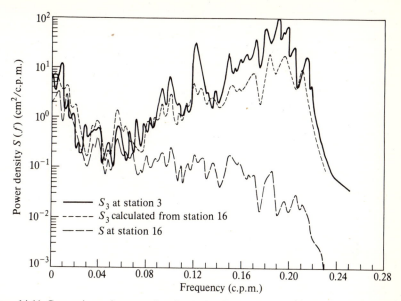

**Figure 14-11** Comparison of measured and calculated wave spectra at station 3. The calculated spectrum is based on the measured (input spectrum) at station 16 which is in deep water 3.5 km away from the harbour.

**(c) A problem involving the elastohydrodynamics of lubrication**[38-40] Solutions to problems of lubrication elastohydrodynamics involve the solution of the Reynolds equation

$$\frac{\partial}{\partial x}\left(\beta\frac{\partial p}{\partial x}\right)+\frac{\partial}{\partial y}\left(\beta\frac{\partial p}{\partial y}\right)=12u\frac{\partial}{\partial x}(\rho h) \qquad (14\text{-}51)$$

where $\quad \beta = \rho h^3/\eta$

$\eta$ = viscosity

$h$ = lubrication film thickness

$\rho$ = density

$p$ = pressure

$u$ = hydrodynamic velocity

The pressure $p$ in Eq. (14-51) affects the film thickness $h$ and the viscosity $\eta$. Equation (14-51) is clearly highly non-linear; moreover, the elastic deformation of the contact region, which can be solved by BEM, affects the lubrication film thickness $h$.

Snidle and others[37-40] have solved this problem using a finite difference approximation for (14-51) coupled to a BEM formulation for calculating the elastic deformation of the solid under the initially unknown generated pressures.

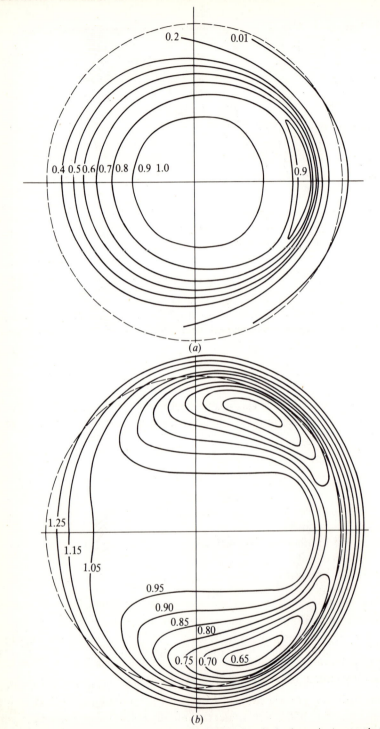

**Figure 14-12** A typical solution to a problem in lubrication elastohydrodynamics (contact between a sphere and a half space).

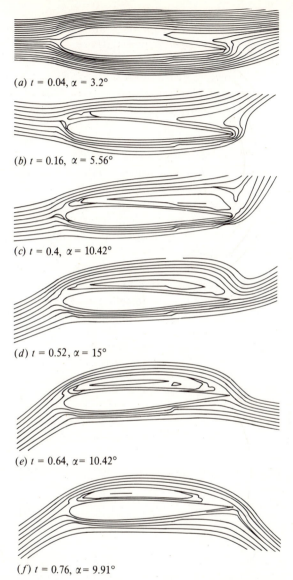

(a) $t = 0.04$, $\alpha = 3.2°$

(b) $t = 0.16$, $\alpha = 5.56°$

(c) $t = 0.4$, $\alpha = 10.42°$

(d) $t = 0.52$, $\alpha = 15°$

(e) $t = 0.64$, $\alpha = 10.42°$

(f) $t = 0.76$, $\alpha = 9.91°$

**Figure 14-13** Streamlines in the rotating reference frame for an oscillating Joukowski 12 per cent airfoil at selected time levels.

The final solution was obtained by iterating between the BEM and the finite difference equations. They found it convenient to transform Eq. (14-51) in terms of a state variable $q$, where $q = (1/\alpha)(1 - e^{-\alpha p})$, to give

$$\nabla^2(\rho h^3 q) = q\nabla^2(\rho h^3) - \rho h^3 \nabla^2 q + 24\eta_o u\frac{\partial(\rho h)}{\partial x} \tag{14-52}$$

Figure 14-12$a$ shows the pressure distribution over the contact area between a sphere and a half space. The relevant data are the hydrodynamic velocity $u = 0.5$ m/s, the load on the sphere $W = 120$ N, $\alpha = 10^{-8}$ m²/N, the radius of the sphere $R = 0.0254$ m, Young's modulus $E = 10.80 \times 10^{10}$ N/m², and $\eta_o = 0.52$ N s/m². The pressure distribution is given as non-dimensional values $p/p_o$, where $p_o$ is the corresponding maximum hertzian pressure for a dry contact and the broken circle represents the hertzian dry contact area. Figure 14-12$b$ shows film thickness contours $h/h_o$ with $h_o$ the central film thickness.

**(d) Viscous flow around an oscillating airfoil**[21] Sankar and Wu[21] described the numerical solution of this problem using a finite difference scheme for the vorticity transport equation and a DBEM formulation for the kinematic vorticity equation [see Chapter 13, where it is also explained why it is not really necessary to use a finite difference scheme for the vorticity transport equation since such initial value problems, governed by a parabolic (diffusion) equation, can be tackled equally well by BEM].

Figure 14-13 shows the solution of the viscous flow problem around an oscillating Joukowski 12 per cent airfoil for a large amplitude–high frequency case $\alpha = 9° - 6° \cos(6t)$, where $\alpha$ is the instantaneous angle of attack.

Other examples of hybrid solutions using finite differences with BEM can be found in Refs 15 to 24 and those using combinations of the finite element and boundary element methods in Refs 1 to 13.

## 14-6 CONCLUDING REMARKS

We hope that we have shown in this chapter how, by using the most advantageous features of different numerical methods, it is often possible to achieve the most satisfactory algorithm for solving many realistic engineering problems. Therefore the skilful analyst will utilize the relevant properties of each rather than slavishly extol the virtues of any single one of them.

## 14-7 REFERENCES

1. Silvester, P. P., and Hsieh, M. S. (1971) 'Finite element solution of two-dimensional exterior field problem', *Proc. IEE*, **118**(12), 1743–1747.
2. Silvester, P. P., Carpenter, C. J., and Wyatt, E. A. (1977) 'Exterior finite elements for two-dimensional field problems, with open boundaries', *Proc. IEE*, **124**(12), 1267–1270.
3. Lean, M. H., Friedman, M., and Wexler, A. (1979). 'Advances in application of the boundary element method in electrical engineering', in P. K. Banerjee and R. Butterfield (eds), *Developments in Boundary Element Methods*, Chap. IX, Applied Science Publishers, London.
4. McDonald, B. H., and Wexler, A. (1972) 'Finite element solution of unbounded field problems', *IEEE Trans. on Microwave Theory and Technqs*, **MTT-20**(12), 841–847.
5. Zienkiewicz, O. C., Kelly, D. W., and Bettess, P. (1977) 'The coupling of the finite element method and boundary solution procedure', *Int. J. Num. Meth. in Engng*, **11**(12), 355–375.

6. Kelly, D. W., Mustoe, G. G. W., and Zienkiewicz, O. C. (1978) 'On hierarchical order for trial functions based on the satisfaction of the governing equations', *Proc. Int. Symp. on Rec. Adv. in Boundary Element Methods*, Southampton University.
7. Mustoe, G. G. W. (1979) 'Coupling of boundary solution procedures and finite elements for continuum problems', Ph.D. thesis, University of Wales, University College, Swansea.
8. Kelly, D. W., Mustoe, G. G. W., and Zienkiewicz, O. C. (1979) 'Coupling of boundary element methods with other numerical methods', in P. K. Banerjee and R. Butterfield (eds), *Developments in Boundary Element Methods*, Chap. X, Applied Science Publishers, London.
9. Shaw, R. P., and Falby, W. (1978) 'FE-BIE combination of the finite element and boundary integral method', *J. Computers and Fluids*, **6**, 153–160.
10. Cruse, T. A., and Wilson, R. B. (1978) 'Advanced application of boundary integral equation methods', *Nucl. Engng Des.*, **46**, 223–234.
11. Cruse, T. A., Osias, J. R., and Wilson, R. B. (1976) 'Boundary integral equation method for elastic fracture mechanics analysis', Report AFOSR–TR–76–0878, Pratt and Whitney Aircraft, Connecticut.
12. Wilton, D. R. (1978) 'Acoustic radiation and scattering from elastic structures', *Int. J. Num. Meth. in Engng*, **13**, 123–138.
13. Patel, J. S. (1978) 'Radiation and scattering from an arbitrary elastic structure using consistent fluid structure formulation', *Computers and Structures*, **9**, 287–291.
14. Banerjee, P. K., and Butterfield, R. (1971) 'The problem of pile-cap group interaction', *Géotechnq.*, **21**(3), 135–142.
15. Banerjee, P. K., and Driscoll, R. M. C. (1976) 'Three-dimensional analysis of raked pile groups', *Proc. Inst. of Civ. Engrs, Res. and Theory*, **61**, 653–671.
16. Banerjee, P. K., and Davies, T. G. (1979) 'Analysis of some reported case histories of laterally loaded pile groups', *Int. Conf. Num. Meth. in Offshore Piling*, Institute of Civil Engineers, London.
17. Davis, T. G. (1979) 'Linear and nonlinear analysis of pile groups', Ph.D. thesis, University of Wales, University College, Cardiff.
18. Wu, J. C., and Thomson, J. F. (1973) 'Numerical solutions of time-dependent incompressible Navier–Stokes equations using an integro-differential formulation', *Computers and Fluids*, **1**, 197–215.
19. Wu, J. C. (1976) 'Numerical boundary conditions for viscous flow problems', *AIAA J.*, **14**(8), 1042–1049.
20. Wu, J. C., and Sugavanam, A. (1978) 'Method of numerical solution of turbulent flow problems', *AIAA J.*, **16**, 948–955.
21. Sankar, N. L., and Wu, J. C. (1978) 'Viscous flow around oscillating airfoil—a numerical study', *AIAA Eleventh Fluid and Plasma Dynamics Conf.*, Seattle, Washington.
22. Coulmy, G., and Luu, T. S. (1977) 'Solution of the Navier–Stokes equation for an incompressible flow by an integro-differential method' (in French), *Proc. Inst. Symp. on Innovative Num. Analysis in Appl. Engng Sci.*, Versailles, France.
23. Olsen, K., and Hwang, L. S. (1971) 'Oscillations in a bay of arbitrary shape and variable depth', *J. Geophys. Res.*, **76**(21), 5048–5064.
24. Cermak, I. A., and Silvester, P. P. (1968) 'Solution of two-dimensional field problems by boundary relaxation', *Proc. IEEE*, **155**(9), 1341–1348.
25. Poggio, A. J., and Miller, E. K. (1973) 'Integral equation solutions of three-dimensional scattering problems', in R. Mittra (ed.), *Computers in Electromagnetics*, Chap. III, Pergamon Press, Oxford.
26. Fried, I. (1968) 'Finite element analysis of problems formulated by an integral equation, application to potential flow', Report, University of Stuttgart, October.
27. Collatz, L. (1966) *The Numerical Treatment of Differential Equations*, 3rd ed., Springer-Verlag, Berlin, Heidelberg, and New York.
28. Mikhlin, S. G. (1964) *Variational Methods in Mathematical Physics*, Pergamon Press, Oxford.
29. Ames, W. F. (1963) *Nonlinear Partial Differential Equations in Engineering*, Academic Press, London.

30. Silvester, P., and Hsieh, M. S. (1971) 'Projective solution of integral equations arising in electric and magnetic field problems', *J. Comp. Physics*, **8**, 73–82.

31. Hassan, M. A., and Silvester, P. (1977) 'Radiation and scattering by wire antenna structures near a rectangular plate reflector', *Proc. IEE*, **124**(5), 429–435.

32. Gopinath, A., and Silvester, P. (1973) 'Calculation of inductance of finite length strips and its variation with frequency', *IEEE Trans. on Microwave Theory and Technqs*, **MIT-21**(6), 380–386.

33. Jeng, G., and Wexler, A. (1977) 'Isoparametric, finite element variational solution of integral equations for three-dimensional fields', *Int. J. Num. Meth. in Engng*, **11**, 455–71.

34. Jeng, G., and Wexler, A. (1978) 'Self-adjoint variational formulation of problems having a non-self adjoint operator', *IEEE Trans. on Microwave Theory and Technqs*, **MTT-26**, 91–94.

35. Tottenham, H. (1979) 'The boundary element method for plates and shells', in P. K. Banerjee and R. Butterfield (eds), *Developments in Boundary Element Methods*, Chap. 8, Applied Science Publishers, London.

36. Jaswan, M. A., and Symm, G. T. (1977) *Integral Equation Methods in Potential Theory and Elastostatics*, Academic Press, London.

37. Zienkiewicz, O. C. (1977) *The Finite Element Method*, McGraw-Hill, London.

38. Biswas, S., and Snidle, R. W. (1977) 'Calculation of surface deformation in point contact elasto-hydrodynamics', *J. Lubrication Technol.*, **99**, 313.

39. Evans, H. P., and Snidle, R. W. (1978) 'Towards a refined solution of the isothermal point contact elasto-hydrodynamics problem', Report No. 409, Mechanical Engineering Department, University College, Cardiff, University of Wales; also presented at the *Int. Conf. of Fundamentals of Tribology*, MIT, Cambridge, Mass.

40. Snidle, R. W. (1979) Private communication.

## FIFTEEN

## COMPUTER IMPLEMENTATION OF BOUNDARY ELEMENT METHODS

## 15-1 INTRODUCTION

No matter how powerful or elegant a numerical method may be its full potential can only be realized when it has been programmed efficiently. In this respect BEM probably require greater effort from the programmer and less from a prospective user than do finite element methods. The first generation of BEM programmes were not very efficient since they were mainly developed by workers as exercises in the process of investigating the method. This situation has altered somewhat in recent years and so-called second-generation computer programmes such as BASQUE,[1-4] BINTEQ,[5] PESTIE,[6] PGROUP,[7-10] etc., have followed from the original research efforts. No doubt further developments of BEM during the next decade will lead to programmes comparable to the major finite element systems such as NASTRAN, ASKA, MARC, etc.

In this chapter we shall briefly outline the basic principles involved in developing simple BEM programmes; having mastered these, a reader with established programming skill may then proceed to develop his own system.

## 15-2 THE STRUCTURE OF A BOUNDARY ELEMENT PROGRAMME

The following logical sequence of steps is basic to all BEM analyses:

1. Generation of input data defining the geometry of the surface elements.
2. Integration of the kernel-shape function products to generate the system matrices.

3. Assembly of equations for each subregion.
4. Solution of the system of equations to generate the unknown boundary data.
5. Backsubstitution of boundary data into the integrals at interior points to obtain the interior information.

The task of developing a BEM programme is facilitated by adopting a modular approach based on these primary subdivisions.

## 15-3 SPECIFICATION AND GENERATION OF INPUT DATA

The boundaries may be represented by line elements in two dimensions and surface elements in three dimensions, defined by their nodal coordinates and some specified variation of the surface geometry. It is also necessary to have a global numbering system for elements and nodes so that a directory for each surface element and its connectivity (through the nodes) with adjoining elements may be specified. The node numbers for each element should be specified in either clockwise or anticlockwise order when viewed in the direction of the outward normal. Thus for a flat, triangular element defined by three nodes (Fig. 15-1a):

| Element number | Node numbers |
|---|---|
| 2 | 2 4 3 |
| 3 | 3 4 5 |
| etc. | |

This entry shows that the element numbers 2 and 3 are connected via common nodes 3 and 4. Therefore, when the integrations have been performed on these elements, coefficients associated with the values of the function at these nodes are summed during the formation of the system matrices. This means that only one set of unknowns is being allowed for each node (i.e., there is a continuous distribution of functions through the nodes). Discontinuity of either the geometry or the functions can be allowed for by assigning different numbers to coincident nodes. For example, a node numbering of 5, 6, and 7 for the element 3 would imply that functions at nodes 3 and 4 are different from those in 5 and 6 respectively, even though the global coordinates $x_i$ of nodes 3 and 4 are identical to those of 5 and 6 respectively (see Fig. 15-1b).

It is desirable to keep the maximum difference in node numbers for an element to a minimum. This improves the stability of the system matrices since the larger coefficients are then closer to the leading diagonal.

A global boundary condition matrix needs to be defined for each surface node of the system so that a directory of surface boundary conditions can be constructed. A similar matrix is also necessary to identify the interface nodes between two regions.

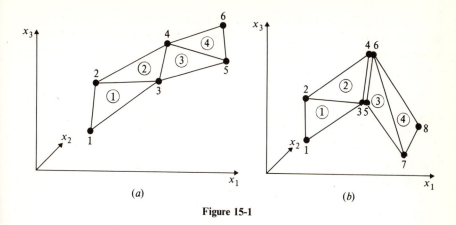

**Figure 15-1**

A good computer programme will of course have a comprehensive input data-checking facility so that only a correctly specified data set is passed for execution.

## 15-4 INTEGRATION OF THE KERNEL-SHAPE FUNCTION PRODUCTS

### 15-4-1 Introduction

In order to form the system matrices it is necessary to evaluate the integrals in the discretized boundary integral equations (here body forces are assumed to be zero for convenience); i.e., for DBEM,

$$c_{ij}(\xi_o^p)\,u_i(\xi_o^p) = \sum_{q=1}^{N}\left(\int_{\Delta S_q}\left\{G_{ij}[x(\eta),\ \xi_o^p]\sum_{k=1}^{n} N^k(\eta)\,J(\eta)\,ds(\eta)\right\}t_i^k\right.$$

$$\left. -\int_{\Delta S_q}\left\{F_{ij}[x(\eta),\ \xi_o^p]\sum_{k=1}^{n} N^k(\eta)\,J(\eta)\,ds(\eta)\right\}u_i^k\right) \quad (15\text{-}1)$$

where $N$ is the total number of boundary elements and $n$ is the total number of nodes on the $q$th boundary element. For the indirect method we have similarly the discretized system:

$$u_i(x_o^p) = \sum_{q=1}^{N}\left(\int_{\Delta S_q}\left\{G_{ij}[x_o^p,\ \xi(\eta)]\sum_{k=1}^{n} N^k(\eta)\,J(\eta)\,ds(\eta)\right\}\phi_j^k\right) \quad (15\text{-}2a)$$

$$t_i(x_o^p) = \beta_{ij}(x_o^p)\,\phi_j(x_o^p) + \sum_{q=1}^{N}\left(\int_{\Delta S_q}\left\{F_{ij}[x_o^p,\ \xi(\eta)]\sum_{k=1}^{n} N^k(\eta)\,J(\eta)\,ds(\eta)\right\}\phi_j^k\right)$$

$$(15\text{-}2b)$$

It is implied in Eq. (15-1) that the cartesian coordinates $x_i$ [$\xi_i$ in Eqs (15-2) for the indirect method] of an arbitrary point on the $q$th boundary element are defined in terms of the nodal cartesian coordinates $x_i^k$ ($k = 1, 2, ..., n$, for example) and shape functions $N^k(\eta)$:

$$x_i(\eta) = N^k(\eta) x_i^k \tag{15-3}$$

where the shape functions $N^k(\eta)$ can be linear (flat elements), quadratic, cubic, etc. (see Chapter 8).

By differentiating (15-3) with respect to $\eta$ (the local axis directed along the line boundary element) we obtain a vector defining the tangent to the element at the point $\eta$:

$$s_i(\eta) = \frac{dx_i}{d\eta} = \frac{dN^k(\eta)}{d\eta} x_i^k \tag{15-4}$$

The jacobian $J(\eta)$ is equal to $\sqrt{[(dx_i/d\eta)(dx_i/d\eta)]}$ and the normals are given by either $(s_2, -s_1)$ or $(-s_2, s_1)$ depending on which side of the element the elastic body lies.

For a surface boundary element in a three-dimensional analysis we can similarly construct a local axes system $\eta_j$ ($j = 1, 2$), and differentiating (15-3) with respect to $\eta_j$ vectors, tangent to the coordinate lines, are obtained:

$$s_{ij}(\eta) = \frac{\partial x_i}{\partial \eta_j} = \frac{\partial N^k(\eta)}{\partial \eta_j} x_i^k \tag{15-5}$$

A vector normal to the element at $\eta$ can be obtained by taking the cross-product of the two vectors defined by (15-5), and the jacobian is the modulus of this cross-product. Having thus defined the jacobian $J(\eta)$ and the normal $n_i(\eta)$ we can now attempt to integrate the terms within the braces { } in Eqs (15-1) and (15-2).

## 15-4-2 Evaluation of the Non-singular Integrals

When the field point $\xi_o^p$ in Eq. (15-1) is not on the node $k$ the kernel, shape function, and jacobian product remain bounded and therefore can be integrated numerically by ordinary gaussian integration formulae with weight function unity:

$$\int_{-1}^{1} f(\xi) \, d\xi = \sum_{i=1}^{m} A_i \, f(\xi_i) \tag{15-6a}$$

and $$\int_{-1}^{1} f(\xi_1, \xi_2) \, d\xi_1 \, d\xi_2 = \sum_{i=1}^{m_1} \sum_{j=1}^{m_2} A_i^{m_1} B_j^{m_2} \, f(\xi_i, \xi_j) \tag{15-6b}$$

for the line and surface area integrals respectively. Explicit details of such integration formulae can be found in Appendix C. It is clearly necessary to map our boundary elements before these formulae, which refer to unit intervals, can be used to evaluate the kernel-shape function products.

Since the evaluation of the non-singular integrals occupies a significant amount of the computational time it is necessary to optimize this whenever possible. This may be accomplished (e.g., in Lachat and Watson's[1-4] three-dimensional elasticity programme) by specifying a maximum upper bound to the error in the numerical integrations. In simple terms this means that the order of the integrations should be varied depending on the ratio of the distance between the 'loaded' boundary element and the field point to a characteristic dimension of the 'loaded' element and the strength of the singularity. A comprehensive discussion of approximate formulae for determining the order of the integration formulae can be found in Lachat and Watson[1-4, 11] and Mustoe.[12] Watson[11] discusses additionally an integration scheme over infinite boundary elements (see also Chapter 8).

## 15-4-3 Evaluation of the Singular Integrals

When the field point $\xi_o^p$ [or $x_o^p$ in Eq. (15-2)] coincide with the node $k$ the integration scheme outlined above cannot be used.

For a two-dimensional analysis the integral involving $G_{ij}$ can be split into a singular part and a non-singular part, i.e.,

$$\int_{\Delta S} [G_{ij} N^k(\eta) J(\eta)] \, ds \, t_i^k = A\delta_{ij} \int_{\Delta S} [\ln r \, N^k(\eta) J(\eta)] \, ds \, t_i^k$$

$$+ \int_{\Delta S} [G_{ij}^o N^k(\eta) J(\eta)] \, ds \, t_i^k \qquad (15\text{-}7)$$

where $A$ is a constant and the second integral containing $G_{ij}^o$ is non-singular and can therefore be evaluated as before. The first integral, however, has a logarithmic singularity and needs some special attention.

This particular singular integral in (15-7) can be calculated using a quadrature formula due to Stroud and Secrest[13] (see Appendix C):

$$\int_0^1 \log \left( \frac{1}{\xi} \right) f(\xi) \, d\xi = \sum_{i=1}^m w_i f(\xi_i) \qquad (15\text{-}8)$$

which integrates (15-8) exactly when $f(\xi)$ is any polynomial whose order is not greater than $2m - 1$.

To use this formula, each boundary element is divided into several sub-elements so that the singular point is always at the origin $O$ of (15-8).

Unfortunately, singular integrals involving the function $F_{ij}$ cannot be evaluated in this way. However, by noting that the shape function is unity at the singular node (say at node 1), we can write, for (15-2b),

$$\int_{\Delta S} \{F_{ij} N^1 J\} \, ds \, \phi_j^1 = \int_{\Delta S} \{F_{ij}(1 - N^1) J\} \, ds \, \phi_j^1 - \int_{\Delta S} (F_{ij} J) \, ds \, \phi_j^1 \qquad (15\text{-}9)$$

The first integral, which is non-singular since $(1 - N^1)$ approaches zero at the singular point and the integrand remains reasonably well behaved, can be evaluated numerically but the second integral must be calculated analytically. For flat boundary elements, these integrals are easy to evaluate. For quadratic and cubic boundary elements (i.e., curved boundary elements), however, it is necessary to split this second integral into a part involving a flat tangent plane through the singular point which can be calculated analytically and a part involving the curvature of the boundary element which can be evaluated numerically. The discontinuity term $\beta_{ij}$ is then added to these coefficients to give the dominant diagonal block of coefficients. It is of course possible to use the above-mentioned procedure for integrals involving $G_{ij}$.

While the quite elaborate technique described above can be used also in the direct method, the most elegant way of evaluating the contributions from these integrals and the free term $C_{ij}$ is to consider the rigid-body displacement modes discussed in Chapters 3 and 4 and also in Refs 1 to 5, 11, and 12. Watson[11] has given a particularly lucid account of this idea applied to the calculation of the singular contributions over infinite boundary elements (see Chapter 8).

For the three-dimensional case, integrals involving $F_{ij}$ can be evaluated by any of the methods outlined above, i.e., via either (15-9) or utilizing rigid-body displacements. As far as integrals involving $G_{ij}$ are concerned, when the singular point is on a corner of a triangular (degenerate quadrilateral) element (see Fig. 15-2a) the integrations can be carried out with respect to the coordinates $\bar{\eta}_1$ and $\bar{\eta}_2$.

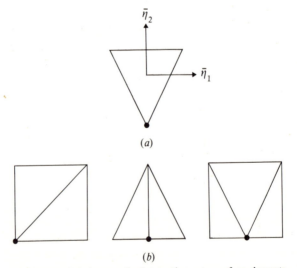

(a)

(b)

**Figure 15-2** Subelements for integration over surface elements.

Note that, since the lines $\bar{\eta}_1 = +1$ and $\bar{\eta}_2 = -1$ converge to the field point, the product $(G_{ij} NJ)$ tends to a finite limit as the field point is approached and therefore the formula (15-6b) may be used. When the field point is on a corner node of a rectangle or a mid-side node (see Fig. 15-2b), the element must be subdivided into triangular subelements and integration over each subelement is carried out with respect to the axes shown in Fig. 15-2a. Further discussions on these integration schemes can be found in Refs 1, 11, 12, 14, and 15.

## 15-5 ASSEMBLY OF EQUATIONS

Since the boundary nodes are shared by a number of elements it is necessary to compile an element connectivity matrix which contains a directory of these nodal connections so that after integration over individual elements the coefficients relating to the common nodes can be summed and put in the appropriate location of the system matrices for each region. For example, with reference to Fig. 15-1a, the displacements at node 1 in an indirect BEM analysis may be written as

$$\mathbf{u}^1 = (\mathbf{a}^1 \, \boldsymbol{\phi}^1 + \mathbf{b}^1 \, \boldsymbol{\phi}^2 + \mathbf{c}^1 \, \boldsymbol{\phi}^3) + (\mathbf{a}^2 \, \boldsymbol{\phi}^2 + \mathbf{b}^2 \, \boldsymbol{\phi}^3 + \mathbf{c}^2 \, \boldsymbol{\phi}^4)$$
$$+ (\mathbf{a}^3 \, \boldsymbol{\phi}^3 + \mathbf{b}^3 \, \boldsymbol{\phi}^4 + \mathbf{c}^3 \, \boldsymbol{\phi}^5) + (\mathbf{a}^4 \, \boldsymbol{\phi}^4 + \mathbf{b}^4 \, \boldsymbol{\phi}^5 + \mathbf{c}^4 \, \boldsymbol{\phi}^6) \tag{15-10}$$

where $\boldsymbol{\phi}^i$ denotes the unknown $\boldsymbol{\phi}$ vector at node $i$ and $\mathbf{a}^j$, $\mathbf{b}^j$, and $\mathbf{c}^j$ are submatrices of coefficients obtained from the integration over the $j$th element of the system. Because of the common nodes between the elements we can look up a directory as discussed in Sec. 15-3 and instead of (15-10) write the following equation directly:

$$\mathbf{u}^1 = \mathbf{a}^1 \, \boldsymbol{\phi}^1 + (\mathbf{b}^1 + \mathbf{a}^2) \, \boldsymbol{\phi}^2 + (\mathbf{c}^1 + \mathbf{b}^2 + \mathbf{a}^3) \, \boldsymbol{\phi}^3 + (\mathbf{c}^2 + \mathbf{b}^3 + \mathbf{a}^4) \, \boldsymbol{\phi}^4$$
$$+ (\mathbf{c}^3 + \mathbf{b}^4) \, \boldsymbol{\phi}^5 + \mathbf{c}^4 \, \boldsymbol{\phi}^6 \tag{15-11}$$

Such an assembly process clearly implies a continuity of $\boldsymbol{\phi}$ across elements via nodes. This is the same well-established procedure as that used in the finite element methods.

Having assembled the system equations for each subregion the global system matrices for multiple regions can be constructed in the manner discussed in Chapter 3 and Refs 1, 3, 4, 16, 17. From the point of view of efficient programming it is desirable to combine such an assembly process for multiple regions with the solution algorithm for the block-banded system of equations it generates as discussed by Tomlin.[16] This obviously requires considerable programming skill.

## 15-6 SOLUTION OF EQUATIONS

Unless some special approach is used in the discretization procedure (see Chapter 14), a BEM analysis generates a non-symmetric fully populated matrix for a single region and a non-symmetric block-banded system matrix for multiple regions. Only rarely does the time required for the solution of such a system of equations exceed that required for the formation of the system matrices.

The numerical formulation of an $n$-zone problem will be represented by a system of equations, conveniently expressed in block form as (see Chapter 3) where each block row of the matrix corresponds to an individual zone. This

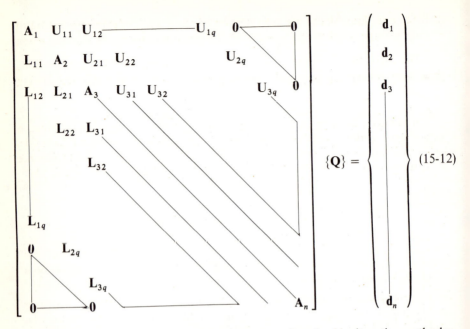

$$\{Q\} = \qquad (15\text{-}12)$$

matrix is more densely populated than that normally solved by iterative methods; furthermore, although it is diagonally strong it is not necessarily diagonally dominant and iterative convergence cannot be ensured. Such a system of equations is therefore best solved by a direct reduction.

For a given number of equations, the smaller the bandwidth, the less the amount of computational effort required whether the equations are block banded or not. The minimum block bandwidth of the coefficient matrix in Eq. (15-12) corresponds to an ordering of the zones wherein the maximum difference between any two adjacent zones is minimized. This is again analogous to the scheme used in finite element analyses for element ordering. In the matrix of Eq. (15-12) the maximum difference between adjacent zones is $q$.

An efficient procedure for solving the block-banded system of Eq. (15-12) was developed by Tomlin[16] who described IBEM solutions of multizone problems using a procedure based on gaussian elimination analogous to established routines for solving banded systems by triangular decomposition, with the principal exception that block (matrix) multiplication and division operations are performed instead of single-entry multiplication and division. Comparing the procedure with routines for solving banded systems:

1. Advantage is taken of the block form of the coefficient matrix, thereby avoiding wasted storage space which would result from the use of a routine designed for single-entry banded matrices.
2. Housekeeping is easier, because the identification of the position of each entry within a block is retained, instead of being converted to identify the position relative to the complete matrix.

3. Since only four blocks are operated on at any stage, considerable saving of core storage is achieved by reserving the remaining blocks in backing store.

The algorithm is listed in Fig. 15-3 as a sequence of matrix operations (inversion, multiplication, etc.).[16] As is common in other routines for solving

In each of the following statements, the matrix operation on the right-hand side of each 'equals' sign is performed and the result is overwritten on the matrix on the left-hand side, e.g., in the first statement the inverse of $A_n$ is overwritten on $A_n$ and any subsequent reference to $A_n$ furnishes the inverse of the original $A_n$. Note the penultimate para. of Sec. 15-6.

Loop to 30 for $n' = n, n-1, ..., 1$

$$A_{n'} = A_{n'}^{-1}$$

Loop to 30 for $q' = 1, 2, ..., q; \; n' - q' \geq 1$

$$U_{n'-q', q'} = U_{n'-q', q'} A_{n'}$$

If $q' \geq 2$, loop to 10 for $q'' = 1, 2, ..., q'-1$

10. $U_{n'-q', q'-q''} = U_{n'-q', q'-q''} - U_{n'-q', q'} L_{n'-q', q''}$

If $q' \leq q-1$, loop to 20 for $q'' = q'+1, q'+2, ..., q; \; n'-q'' \geq 1$

20. $L_{n'-q'', q''-q'} = L_{n'-q'', q''-q'} - U_{n'-q', q'} L_{n'-q', q''}$

For cases where there is only one right-hand side vector,

$$d_{n'-q'} = d_{n'-q'} - U_{n'-q', q'} d_{n'}$$

30. $A_{n'-q'} = A_{n'-q'} - U_{n'-q', q'} L_{n'-q', q'}$

Triangular decomposition is now complete. Back substitution starts. If there is more than one right-hand side vector, loop to 80 for each new right-hand side vector.

If there is only one right-hand side vector, jump to 50 and proceed only once to 80.

Loop to 40 for $n' = n, n-1, ..., 1$

Loop to 40 for $q' = 1, 2, ..., q; \; n'+q' \leq n$

40. $d_{n'} = d_{n'} - U_{n', q'} d_{n'+q'}$

50. Loop to 70 for $n' = 1, 2, ..., n$

Loop to 60 for $q' = q, q-1, ..., 1; \; n'-q' \geq 1$

60. $d_{n'} = d_{n'} - L_{n'-q', q'} d_{n'-q'}$

70. $d_{n'} = A_{n'} d_{n'}$

80. The array $\{d_1, d_2, ..., d_n\}^T$ now contains the solution vector $\{Q\}$.

**Figure 15-3** Algorithm for solving linear block-banded system.

linear systems by triangular decomposition, a facility is provided where, if the solution is required for several right-hand-side vectors and the same coefficient matrix, the decomposition is performed only once. The BEM solutions of diffusion and elastoplasticity (see Chapters 9 and 12) have been obtained using this procedure.

The algorithm of Fig. 15-3 can be applied to any non-singular linear block-banded system where all in-band blocks are fully occupied. For the coefficient matrices considered herein, the blocks **L** and **U** are sparsely occupied or, if the zones represented by the matrices **A** in the same row and column are not adjacent, are zero at the outset of elimination. Non-zero entires appear as complete rows. The occupancy changes as elimination proceeds, but considerable economy of computation can be made by recognizing zero blocks and rows so as to avoid operating on them.

Similar routines for dealing with a block-banded system of equations in BEM have been developed by Lachat,[1] Lachat and Watson,[3] Mustoe,[12] Das,[18] and Davies.[19]

## 15-7 CALCULATIONS AT INTERIOR POINTS

The stresses and displacements may be obtained at selected interior points by expressing the interior integral identities in the form (15-1) and (15-2), which are then integrated numerically. The cost of evaluating such integrals for interior points is high. For example, the cost of calculating the system equation at a boundary node using Eq. (15-1) is almost identical to that at an interior point.

In DBEM it is sometimes more efficient to divide a homogeneous region into a number of subregions so that the initial solution of the system equations provides displacements and tractions at a number of interface nodes which, if carefully chosen, are then useful interior results. Lachat and Watson[2] used this idea to advantage and described solutions for a number of three-dimensional problems of considerable geometrical complexity. The stresses at a surface point can be calculated, using the following relationships:[1, 20]

1. For two-dimensional problems:

$$\bar{\sigma}_{11} = \frac{1}{1-v'} \left[ \frac{E'}{1+v'} \bar{\varepsilon}_{11} + v' \bar{t}_2 \right]$$

$$\bar{\sigma}_{22} = \bar{t}_{22}$$

$$\bar{\sigma}_{12} = \bar{\sigma}_{21} = \bar{t}_1$$

(15-13)

where $\bar{\varepsilon}_{11}$ and $\bar{t}_1$ are tangential strain and traction respectively, $\bar{t}_2$ is the traction in the normal direction at the boundary point, $E' = E$ and $v' = v$ for plane strain, and $E' = E(1-v'^2)$ and $v' = v/(1+v)$ for plane stress.

2. For three-dimensional problems:

$$\bar{\sigma}_{11} = \frac{v}{1-v}t_3 + \frac{Ev}{1-v^2}(\bar{\varepsilon}_{11} + \bar{\varepsilon}_{22}) + \frac{E}{1+v}\bar{\varepsilon}_{11}$$

$$\bar{\sigma}_{12} = \bar{\sigma}_{21} = \frac{E}{2(1+v)}\bar{\varepsilon}_{12}$$

$$\bar{\sigma}_{22} = \frac{v}{1-v}t_3 + \frac{Ev}{1-v^2}(\bar{\varepsilon}_{11} + \bar{\varepsilon}_{22}) + \frac{E}{1+v}\bar{\varepsilon}_{22} \qquad (15\text{-}14)$$

$$\bar{\sigma}_{33} = t_3$$

$$\bar{\sigma}_{32} = \bar{\sigma}_{23} = t_2$$

$$\bar{\sigma}_{31} = \bar{\sigma}_{13} = t_1$$

It should be noted that the quantities in Eqs (15-13) and (15-14) are referred to the local axes through the point at which the stresses are required (See Fig. 15-4). The strains and tractions in the tangential and the normal directions used in the above equations are easily calculated from the nodal tractions and displacements.

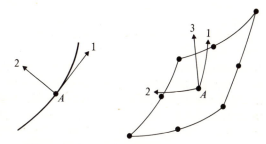

**Figure 15-4** Local axes through a surface point $A$.

In the indirect methods it is possible to use Eq. (15-2b) twice to calculate the stress tensor at an interior point $x_i$.[12] First, the kernel $F_{ij}$ is calculated with the outward normal $n_i(x)$ as the unit vector in the direction of the global axes 1 (that is, $n_i(x) = \delta_{i1}$; Eq. (15-2b) then gives

$$\sigma_{11}(x) = t_1(x) \qquad \sigma_{12}(x) = t_2(x) \qquad (15\text{-}15)$$

Then, choosing $n_i(x) = \delta_{i2}$ (i.e., the unit vector in the direction of the global axes 2) we get

$$\sigma_{21}(x) = t_1(x) \qquad \sigma_{22}(x) = t_2(x) \qquad (15\text{-}16)$$

Obviously $\sigma_{21} = \sigma_{12}$.

It is often preferable to subdivide a region bounded by a complex surface into a number of subregions. This not only improves the accuracy (particularly with IBEM) but also is computationally efficient, as mentioned earlier, if interior results are required at a large number of points.

## 15-8 A DIRECT BOUNDARY ELEMENT PROGRAMME FOR TWO-DIMENSIONAL ELASTOSTATICS

A DBEM programme for two-dimensional elastostatics is listed below. The programme uses quadratic variations for both the geometry and functions $u$ and $t$ over the boundary elements. The boundary displacements $u$ are continuous but the tractions $t$ are discontinuous at a corner node.

### 15-8-1 The Programme Listing

The main segment of the programme which is named DBEM calls subroutines GAUSS, INPUT, BELMAT, BISOLV, and BIPTS. The computer programme has been written in Standard FORTRAN language and should be operable on any computer system. The only machine-dependent feature is the use of two temporary disc files with designated logical unit numbers NDISKA $-$ 17 and NDISKB $=$ 18. These two cards must be altered to suit the user's computer system.

The input data is read in subprogrammes DBEM, INPUT, and BIPTS. The segment DBEM reads data relating to the number of problems, title of the job, and an index card for the variable JBUG, which provides considerable amounts of intermediate output. This may help a new user to understand the workings of the programme. Subroutine INPUT reads the data relating to the geometry, boundary conditions, and material properties. It calls subroutine NODEXY to interpolate the coordinates of the mid-side nodes of various boundary elements if they were omitted by the user. The coordinates of the interior points at which stresses and displacements need to be calculated are read in subroutine BIPTS.

Subroutine BELMAT sets up the coordinates of the boundary nodes for point collocation and calls subroutine INTAB to calculate the **F** and **G** matrices for each boundary collocation point. These are then transferred to the two sequential access files NDISKA and NDISKB. At corners or at a known loading discontinuity the tractions **t** are assumed to be discontinuous. Subroutines INTAB sets up the local axes system on the boundary elements, calculates the order of integration formulae, the jacobian, and integrates the kernel-shape function products to generate coefficients of the matrices **F** and **G** for a specified boundary or interior field point.

Subroutine BISOLV brings **F** and **G** matrices (variables AMT and BMT) back in core and assembles the final system of equations for the specified boundary conditions. The system equations are then rearranged (if necessary) to give a well-conditioned matrix, which is then reduced for a given right-hand side in subroutine SIME. The values of the boundary tractions and boundary displacements are calculated and printed out.

Subroutine BIPTS calculates the displacements at interior points by multiplying **F** and **G** matrices (generated by INTAB) with **u** and **t**. The variables JSING and XSI are zero if the field point is an interior point. For a boundary point JSING must set to the boundary segment number and XSI is the local coordinate of the field point. Thus if the field point is the first node (for an anticlockwise numbering system) of boundary element 3 then JSING = 3 and XSI = 0, for the middle node of element 3 we have JSING = 3, XSI = 0.5, etc.

The direct BEM programme listed here does not have the facility to calculate the stresses at an interior point. In order to include such a facility it would be necessary to write a new subroutine INTABS (say) which will contain the coding relevant to XSI = 0 and JSING = 0 from INTAB with displacement and traction kernel functions replaced by the stress kernel functions $D_{ijk}$ and $S_{ijk}$. The subroutine INTABS could then be called in BIPTS to calculate the stresses at interior points.

**Input specification**

(a) TITLE (18A4)

Title of the problem

(b) JBUG(I), I = 1, (80I1)

An integer array which controls the intermediate output. It is set to zero or 1 depending on whether the intermediate output is not required or required.

(c) NVFIX, NOTRBC, NTYPE, NOEQ, (4I5)

NVFIX = total number of boundary nodes having prescribed displace-ment boundary conditions

NOTBRC = total number of boundary elements over which traction boundary conditions have been specified

NTYPE = 1 for plane strain, = 0 for plane stress

NOEQ = 1 for automatic choice of gaussian integration formulae, = 0 for a specified order of integration formula

(d) NBDATA(1, 3) (3I5)

NBDATA(1, 2) = number of boundary nodes used to define the geometry

NBDATA(1, 2) = 0 for interior problem
            = 1 for exterior problem

NBDATA(1, 3) = nominal order of gaussian integration formula to be used throughout (2, 4, 8, or 12)

If NOEQ = 1 the programme optimizes the integration rule.

(e) IPOIN, COORD (IPOIN, IDIME) (I5, 2F10.5)

The node number, x coordinate and y coordinate

The first node must be at the bottom left-hand corner and the nodes must be numbered in anticlockwise manner in ascending order.

(f) NOFIX(IVFIX), IFPRE(IVFIX, IDOFN), PRESC(IVFIX, IDOFN) (1X, I4, 3X, 2I1, 2F10.6)

NOFIX(IVFIX) = node number of the node for which displacement boundary conditions are specified

IFPRE(IVFIX, 1) = 1 if the displacement in the x direction is specified; otherwise = 0

IFPRE(IVFIX, 2) = 1 if the displacement in the $y$ direction
is specified; otherwise = 0

PRESC(IVFIX, 1) = the prescribed value of the displacement in the
$x$ direction; if specified otherwise leave blank

PRESC(IVFIX, 2) = the prescribed value of the displacement in the
$y$ direction; if specified otherwise leave blank

(g) NUMAT, PROPS(NUMAT, IPROP) (I5, 2F10.5)

NUMAT = material number = 1

PROPS(1, 1) = Young's modulus

PROPS(1, 2) = Poisson's ratio

(h) NSEG, NTRKBC(1, NSEG), (TRX(I), TRY(I), I = 1, 3) (I5, 4X, I1, 6F10.4)

NSEG = boundary element number

NTRKBC(1, NSEG) = 1 if $t_x$ only specified

= 2 if $t_y$ only specified

= 3 if both $t_x$ and $t_y$ specified

TRX(I), TRY(I) = values of $t_x$ and $t_y$ respectively at the local node I
over the boundary element (I = 1 for left node,
I = 2 for middle node, and I = 3 for the right node
for an anticlockwise numbering system)

(i) NOINPT (I5)

Number of interior results required

(j) JSING, XIN, YIN, XSI (I5, 3F10.3)

JSING = 0 for an interior point

= the boundary element number for a boundary point

XSI = 0 for an interior point

= the local coordinate for a boundary point (e.g., for left-hand
endpoint XSI = 0, for the middle point XSI = 0.5, and for
the right-hand endpoint XSI = 1.0)

## Size limitations

(a) One boundary enveloping one region

(b) A maximum of 30 boundary elements

(c) A maximum of 120 unknown boundary displacements and tractions (that
is, 60 boundary nodes)

(d) Displacement boundary conditions only cannot be specified on two
boundary elements meeting at a corner (see Chapter 7)

## 15-8-2 A Sample Problem, Input Data, and Output

The cantilever problem shown in Fig. 15-5 was analysed using this programme.
Figure 15-6 shows the nodes at which the displacements and the boundary
tractions have been printed out in the sample output.

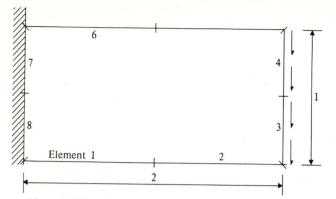

**Figure 15-5** Boundary element discretization for the cantilever.

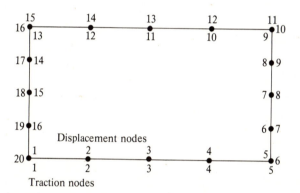

**Figure 15-6** Node numbering for boundary points for displacements and tractions in the sample output.

## 15-9 AN INDIRECT BOUNDARY ELEMENT PROGRAMME FOR TWO-DIMENSIONAL ELASTOSTATICS

A simple IBEM programme for two-dimensional elastostatics is also listed together with the input specification.

The programme uses quadratic variations of both geometry and fictitious tractions ($\phi_j$) over the boundary elements. In general $\phi_j$ are assumed to be continuous across the element boundaries except at a boundary discontinuity such as a corner, an edge crack, or a discontinuity in the applied loading.

### 15-9-1 The Programme Listing

The main segment of this programme is called IBEM. The overall structure as well as the names and functions of the various subroutines of this programme are identical to those of DBEM described in Sec. 15-8-2. Since the FORTRAN codes of parts or all of subprogrammes IBEM, NODEXY, INTAB, SIME, and GAUSS are identical to the DBEM programme described earlier, these parts have been omitted from the listing and the user has to duplicate the approrpiate sections of the programme DBEM and insert them as indicated.

**Input specification**
Card types (a) and (b) are identical to those of DBEM.
(c)  NVFIX, NTYPE, NOEQ, (3I5)

> NVFIX = total number of boundary nodes on which boundary conditions have been specified
>
> NTYPE = 1 for plane strain
> = 0 for plane stress
>
> NOEQ = 1 for automatic choice of integration formulae,
> = 0 for a specified order of integration (that is, 2, 4, 8, or 12)

Card type (d) is identical to that of DBEM.
(e)  INTYPE (1, ISEG) (16I5)

> For each boundary element (ISEG = 1 to NSEG where NSEG = the total number of boundary elements) INTYPE is set to either 1 or 0 depending on whether or not the end of the boundary element has a corner or a loading discontinuity. The elements must be numbered in anticlockwise manner in ascending order with the first element being at the bottom left-hand corner.

Card type (f) is identical to card type (e) of DBEM.
(g)  NOFIX(IVFIX), IFPRE(IVFIX, IDOFN), PRESC(IVFIX, IDOFN) (1X, I4, 3X, 2I1, 2F.10.6)

> For each boundary node at which boundary conditions are specified (IVFIX = 1 to NVFIX): NOFIX(IVFIX) = the node number, IFPRE(IVFIX, 1) and IFPRE (IVFIX, 2) = the indices for boundary conditions in the directions of $x$ and $y$ axes. The index in IFPRE is equal to 1 if displacement boundary conditions are specified and is equal to 2 if traction boundary conditions are specified. PRESC(IVFIX, 1) and PRESC(IVFIX, 2) are the prescribed values of boundary conditions in $x$ and $y$ directions respectively.

Card type (h) is identical to the card type (g) of DBEM.
Card (i) is identical to that of DBEM.
(j)  JSING, XIN, YIN, XNP, YNP, XSI (I5, 5F 10.3)

> See explanations for the same named variables in card type (j) in DBEM.
> XIN, YIN are the $x$ and $y$ coordinates of the point
> XNP, YNP = 1 and 0 respectively if $\sigma_{11}$ and $\sigma_{12}$ are required
> = 0 and 1 respectively if $\sigma_{22}$ and $\sigma_{21}$ are required

**Size limitations**
(a)  One boundary enveloping a region
(b)  Approximately a maximum of 25 boundary elements
(c)  A maximum of 60 boundary nodes at which boundary conditions can be specified

```
00000100            PROGRAM DBEM
00000200            DIMENSION TITLE(12)
00000300            COMMON/DBUG/JBUG(80)
00000400            COMMON/GSPWT/GSP(3),GSPW(3)
00000500            COMMON/SOLV/RLHS(120,121),UFIX(120),TFIX(120),NUFIX(120),
00000600           .NTFIX(120),NUPOS(120),NTPOS(120),RHS(120),BCBAR(120)
00000700            COMMON/CONTRO/NPOIN,NNODE,NDOFN,NTYPE,NGAUS,NPROP,NMATS,NVFIX
00000800           .,NPROB,NTCOMP,NCALCD
00000900            COMMON/LGDATA/COORD(60,2),PROPS(1,5),NOFIX(60),IFPRE(60,2),
00001000           .MATNO(1),PRESC(60,2)
00001100            COMMON/BMATS/NBDATA(1,4),LNODBI(1,60),NBIREG,TRXBC(1,90),TRYBC
00001200           .(1,90),XCRDBI(61),YCRDBI(61),BIDIS(120),TRACT(180),
00001300           .NTRKBC(1,30),CNRTOL,INTYP(1,61)
00001400            COMMON/MATCON/RNU,CONS1,CONS2,CONS4,DMAX,NGP,NGGP,GP(12,12)
00001500           .  ,GPW(12,12),ISUPS,NBB,NBBM,CONS3,MAXSZ,NBB2,NBB2P
00001600           .  ,NCOMP,NCLPT,RMAXL,NUNIT,NOEQ,GLP(12,12),GLPW(12,12)
00001700            COMMON /LDISK/ NDISKA,NDISKB
00001800            NDISKA=17
00001900            NDISKB=18
00002000            MAXSZ=120
00002100            CALL GAUSS
00002200            NPROB=1
00002300            WRITE(6,905) NPROB
00002400            DO 20 IPROB=1,NPROB
00002500            REWIND NDISKA
00002600            REWIND NDISKB
00002700            READ(5,910) TITLE
00002800            WRITE(6,915) IPROB,TITLE
00002900            READ(5,901)(JBUG(I),I=1,80)
00003000            WRITE(6,901)(JBUG(I),I=1,80)
00003100            CALL INPUT
00003200            CALL BELMAT(1,1)
00003300            CALL BISOLV
00003400            CALL BIPTS
00003500         20 CONTINUE
00003600        900 FORMAT(16I5)
00003700        905 FORMAT(1H0,5X,23HTOTAL NO. OF PROBLEMS =,I5)
00003800        910 FORMAT(18A4)
00003900        915 FORMAT(//////,6X,12HPROBLEM NO. ,I3,10X,18A4)
00004000        901 FORMAT(80I1)
00004100            STOP
00004200            END
00004300            SUBROUTINE INPUT
00004400            DIMENSION TRX(3),STR(3),TRY(3)
00004500 C   ***INCLUDE COMMON BLOCKS ******
00004600 C   *** MATCON CONTRO LGDATA BMATS ****
00005500            READ(5,900)  NVFIX,NOTRBC,NTYPE,NOEQ
00005600            NUMBI=1
00005700            NELEM=1
00005800            NDOFN=2
00005900            NMAT=1
00006000            NPROP=2
00006100            NDIME=2
00006200            WRITE(6,905)  NVFIX,NOTRBC,NTYPE,NOEQ
00006300            READ(5,900) (NBDATA(NUMBI,J),J=1,3)
00006400            WRITE(6,900)(NBDATA(NUMBI,J),J=1,3)
00006500            NBB=NBDATA(NUMBI,1)
00006600            NPOIN=NBB
00006700            NBBM=NBB
00006800            CNRTOL=0.0
00006900 C NO OF QUADRATIC SEGS..............
00007000            NBBQ=NBBM/2
00007100            MATNO(1)=1
00007200            DO 10 I=1,NBB
00007300         10 LNODBI(NUMBI,I)=I
00007400 CSET UP DEFAULT PARAMETERS ..........
00007500            IF((CNRTOL-1.0E-06).LT.0.0) CNRTOL=30.0
00007600            NREG=NUMBI
00007700            NODPT=NPOIN
```

```
00007800           DO 20 IPOIN=1,NPOIN
00007900           DO 20 IDIME=1,NDIME
00008000        20 COORD(IPOIN,IDIME)=0.0
00008100  C*** READ SOME NODAL COORDINATES, FINISHING WITH THE LAST NODE OF ALL
00008200           WRITE(6,920)
00008300           WRITE(6,925)
00008400        30 READ(5,930) IPOIN,(COORD(IPOIN,IDIME),IDIME=1,NDIME)
00008500           IF(IPOIN.NE.NPOIN) GO TO 30
00008600  C*** INTERPOLATE COORDINATES OF MID-SIDE NODES
00008700           CALL NODEXY
00008800        40 CONTINUE
00008900           DO 50 IPOIN=1,NPOIN
00009000        50 WRITE(6,935) IPOIN,(COORD(IPOIN,IDIME),IDIME=1,NDIME)
00009100  C*** READ THE FIXED VALUES.
00009200           WRITE(6,940)
00009300           WRITE(6,945)
00009400           DO 60 IVFIX=1,NVFIX
00009500           READ(5,950) NOFIX(IVFIX),(IFPRE(IVFIX,IDOFN),IDOFN=1,NDOFN),
00009600         . (PRESC(IVFIX,IDOFN),IDOFN=1,NDOFN)
00009700        60 WRITE(6,950) NOFIX(IVFIX),(IFPRE(IVFIX,IDOFN),IDOFN=1,NDOFN),
00009800         . (PRESC(IVFIX,IDOFN),IDOFN=1,NDOFN)
00009900           WRITE(6,960)
00010000           WRITE(6,965)
00010100           READ(5,930) NUMAT,(PROPS(NUMAT,IPROP),IPROP=1,NPROP)
00010200           WRITE(6,970) NUMAT,(PROPS(NUMAT,IPROP),IPROP=1,NPROP)
00010300  C                     READ TRACTION B.C.
00010400           WRITE(6,1020) NOTRBC
00010500           WRITE(6,1040)
00010600           DO 120 IBC=1,NOTRBC
00010700           READ(5,1030) NSEG,NTRKBC(1,NSEG),(TRX(I),TRY(I),I=1,3)
00010800           DO 110 I=1,3
00010900           IP=3*(NSEG-1)+I
00011000           TRXBC(NREG,IP)=TRX(I)
00011100       110 TRYBC(NREG,IP)=TRY(I)
00011200           WRITE(6,1031) NSEG,NTRKBC(1,NSEG),(TRX(I),TRY(I),I=1,3)
00011300       120 CONTINUE
00011400       903 FORMAT(4I5,F10.4)
00011500       900 FORMAT(6I5)
00011600       905 FORMAT(//8H NVFIX =,I4,4X,8H NOTRBC=,I4,4X,8H NTYPE =,I4,
00011700         .4X,8H  NOEQ =,I4,//)
00011800       904 FORMAT(16I5)
00011900       920 FORMAT(//25H  NODAL POINT COORDINATES)
00012000       925 FORMAT(6H  NODE,7X,1HX,9X,1HY)
00012100       930 FORMAT(I5,2F10.5)
00012200       935 FORMAT(1X,I5,3F10.3)
00012300       940 FORMAT(//17H RESTRAINED NODES)
00012400       945 FORMAT(5H NODE,1X,4HCODE,6X,12HFIXED VALUES)
00012500       950 FORMAT(1X,I4,3X,2I1,2F10.6)
00012600       960 FORMAT(//21H  MATERIAL PROPERTIES)
00012700       965 FORMAT(8H  NUMBER,7X,11H PROPERTIES)
00012800      1020 FORMAT(6X,'NUMBER OF TRACTION B.CS = ',I5)
00012900      1040 FORMAT(6X,' TRACTION BOUNDARY CONDITIONS')
00013000      1030 FORMAT(I5,4X,I1,6F10.4)
00013100      1031 FORMAT(1X,I5,4X,I1,6F10.4)
00013200       970 FORMAT(1X,I5,7X,5E14.6)
00013300           RETURN
00013400           END
00013500           SUBROUTINE NODEXY
00013600  C   *** INCLUDE COMMON BLOCKS *****
00013700  C   *** BMATS LGDATA CONTRO ****
00014300           IELBI=0
00014400           NBIR=1
00014500           DO 30 IELEM=1,NBIR
00014600        22 IELBI=IELBI+1
00014700           NODBI=NBDATA(IELBI,1)
00014800           DO 26 IN=1,NODBI,2
00014900           NST=LNODBI(IELBI,IN)
00015000           INP=IN+2
00015100           IF(INP.GT.NODBI) INP=1
```

```
00015200          NFN=LNODBI(IELBI,INP)
00015300          NMID=LNODBI(IELBI,IN+1)
00015400          TOT=ABS(COORD(NMID,1))+ABS(COORD(NMID,2))
00015500          IF(TOT.GT.0.0) GO TO 25
00015600          COORD(NMID,1)=0.5*(COORD(NST,1)+COORD(NFN,1))
00015700          COORD(NMID,2)=0.5*(COORD(NST,2)+COORD(NFN,2))
00015800       25 CONTINUE
00015900       26 CONTINUE
00016000       30 CONTINUE
00016100          RETURN
00016200          END
00016300          SUBROUTINE BELMAT(M,NBBI)
00016400          DIMENSION AMT(2,120),BMT(2,180),ISING(40),SCOL(3)
00016500 C   **** INCLUDE COMMON BLOCKS ****
00016600 C   **** DBUG MATCON CONTRO LGDATA BMATS LDISK ****
00017700          SHN1(S)= 2.0*(S-0.5)*(S-1.0)
00017800          SHN2(S)=-4.0*S*(S-1.0)
00017900          SHN3(S)= 2.0*S*(S-0.5)
00018000          NBB=NBDATA(NBBI,1)
00018100          NBB2=2*NBB
00018200          NBB3=3*NBB
00018300          CCORN=COS(.01745329*CNRTOL)
00018400          ISUPS=NBDATA(NBBI,2)
00018500          NGP=NBDATA(NBBI,3)
00018600          NBBM=NBB
00018700          MT=MATNO(M)
00018800          PI=3.14159
00018900          IF(NTYPE.EQ.1)  GOTO 1
00019000          RNU=PROPS(MT,2)
00019100          RNU=RNU/(1.0+RNU)
00019200          ES=PROPS(MT,1)
00019300          ES=ES/(1.0-RNU*RNU)
00019400          PROPS(MT,2)=RNU
00019500          PROPS(MT,1)=ES
00019600        1 CONTINUE
00019700          RNU=PROPS(MT,2)
00019800          CONS1=-1.0/(8.0*PI*(1.0-RNU))
00019900          CONS2=3.0-4.0*RNU
00020000          CONS3=-1.0/4.0/PI/(1.0-RNU)
00020100          CONS4=1.0-2.0*RNU
00020200          IF(JBUG(70).NE.0) WRITE(6,1004) CONS1,CONS2,CONS3,CONS4
00020300          J=0
00020400          DO 2 I=1,NBB
00020500          J=J+1
00020600          NUMC=LNODBI(NBBI,I)
00020700          XCRDBI(J)=COORD(NUMC,1)
00020800          YCRDBI(J)=COORD(NUMC,2)
00020900        2 CONTINUE
00021000          NBBP=NBB+1
00021100          XCRDBI(NBBP)=XCRDBI(1)
00021200          YCRDBI(NBBP)=YCRDBI(1)
00021300 C FIND MAX ELEMENT LENGTH  .........
00021400          RMAXL=0.0
00021500          NBBQ=NBB/2
00021600          DO 3 I=1,NBBQ
00021700        3 INTYP(NBBI,I)=0
00021800          NCOMP=NBB
00021900          DO 5 I=1,NBBQ
00022000          I1=2*I-1
00022100          I2=2*I+1
00022200          IP=I1+1
00022300          IM=I1-1
00022400          IF(I.EQ.1) IM=NBB
00022500          IF(I.EQ.NBBQ) I2=1
00022600          RL=SQRT((XCRDBI(I2)-XCRDBI(I1))**2 +
00022700        .          (YCRDBI(I2)-YCRDBI(I1))**2)
00022800          IF(RL.GT.RMAXL) RMAXL=RL
00022900          XA=XCRDBI(I1)-XCRDBI(IM)
00023000          YA=YCRDBI(I1)-YCRDBI(IM)
```

```
00023100          XB=XCRDBI(IP)-XCRDBI(I1)
00023200          YB=YCRDBI(IP)-YCRDBI(I1)
00023300          DOT=XA*XB+YA*YB
00023400          RA=XA*XA+YA*YA
00023500          RA=SQRT(RA)
00023600          RB=SQRT(XB*XB+YB*YB)
00023700          DOT=DOT/RA/RB
00023800          IF(DOT.LT.CCORN) INTYP(NBBI,I)=1
00023900 C   ANGLE L.E 30 DEGREES NO CORNER .............
00024000          NCOMP=NCOMP+INTYP(NBBI,I)
00024100        5 CONTINUE
00024200          IF(JBUG(70).NE.0) WRITE(6,2004)(XCRDBI(I),YCRDBI(I),I=1,NBBP)
00024300          NCOMP=2*NCOMP
00024400          IF(JBUG(70).NE.0) WRITE(6,1010)(INTYP(NBBI,I),I=1,NBBQ)
00024500          NBBIF=NBB
00024600          IF(JBUG(70).NE.0) WRITE(6,1000)(LNODBI(NBBI,I),I=1,NBBIF)
00024700 C PERFORM BOUNDARY COLLOCATION
00024800          NCLPT=0
00024900          NCOL=2
00025000          SCOL(1)=0.5
00025100          SCOL(2)=1.0
00025200          DO 70 ISEG=1,NBBQ
00025300 C   SET UP 3 NODAL COORDS ON ASEG.............
00025400          ISA=2*ISEG-1
00025500          ISB=ISA+1
00025600          ISC=ISB+1
00025700          IF(ISEG.EQ.NBBQ) ISC=1
00025800          XA=XCRDBI(ISA)
00025900          YA=YCRDBI(ISA)
00026000          XB=XCRDBI(ISB)
00026100          YB=YCRDBI(ISB)
00026200          XC=XCRDBI(ISC)
00026300          YC=YCRDBI(ISC)
00026400 C   NULL SINGULARITY FLAG................
00026500          DO 25 I=1,NBBQ
00026600       25 ISING(I)=0
00026700          ISING(ISEG)=1
00026800          DO 60 IC=1,NCOL
00026900 C   FIND A AND B MATRICES...........
00027000          XSI=SCOL(IC)
00027100          XPT=SHN1(XSI)*XA+SHN2(XSI)*XB+SHN3(XSI)*XC
00027200          YPT=SHN1(XSI)*YA+SHN2(XSI)*YB+SHN3(XSI)*YC
00027300          CALL INTAB(XPT,YPT,XSI,ISING,AMT,BMT,NBBI)
00027400 C UPDATE TOTAL NO OF COLLOCATIONS
00027500          NCLPT=NCLPT+1
00027600          DO 50 I=1,2
00027700          WRITE (NDISKA) (AMT(I,J),J=1,NBB2)
00027800          WRITE (NDISKB) (BMT(I,J),J=1,NBB3)
00027900       50 CONTINUE
00028000       60 CONTINUE
00028100          IF(JBUG(70).EQ.0) GO TO 69
00028200          WRITE(6,1011)
00028300          DO 61 I=1,2
00028400       61 WRITE(6,2004)(AMT(I,J),J=1,NBB2)
00028500          DO 62 I=1,2
00028600       62 WRITE(6,2004)(BMT(I,J),J=1,NBB3)
00028700       69 CONTINUE
00028800       70 CONTINUE
00028900       20 RETURN
00029000     1004 FORMAT(1X,'C1,C2,C3,C4     ',4F12.6)
00029100     1000 FORMAT(1X,17HREDUCED B.I NODES,/,1X,16I5)
00029200     1010 FORMAT(1X,16I5)
00029300     1011 FORMAT(6X,' A AND B MATRICES ')
00029400     2004 FORMAT(1X,10E12.4)
00029500          END
00029600          SUBROUTINE INTAB(XPT,YPT,XSI,ISING,AMT,BMT,NBBI)
00029700          DIMENSION AMT(2,120),BMT(2,180),ISING(40),SGP(12)
00029800          DIMENSION ISNGSG(3),AII(3),BII(3),SLGP(12),GLWPJ(3)
00029900          DIMENSION AI(3),BI(3),AIII(3),BIII(3),SHP(3),AAI(3),BBI(3)
```

```
00030000          DIMENSION GWPJ(3)
00030100 C  **** INCLUDE COMMON BLOCKS ****
00030200 C  **** GSPWT DBUG BMATS MATCON ****
00030300 C THIS ROUTINE CALCULATES THE DISPLACEMENT AND TRACTION MATRICES
00030400 C LOCAL FUNCTIONS,SHAPE FUNCTIONS AND DERIVATIVES
00030500          SHN1(S)= 2.0*(S-0.5)*(S-1.0)
00030600          SHN2(S)=-4.0*S*(S-1.0)
00030700          SHN3(S)= 2.0*S*(S-0.5)
00030800          DSH1(S)= 4.0*S -3.0
00030900          DSH2(S)=-8.0*S +4.0
00031000          DSH3(S)= 4.0*S -1.0
00031100          JSING=0
00031200 C    CHANGE DIRECTION OF THE NORMAL EXT PROBLEM
00031300          RFACT=1.0
00031400          IF(NBDATA(NBBI,2).EQ.1) RFACT=-1.0
00031500          DO  5  J=1,3
00031600          AIII(J)=0.0
00031700     5    BIII(J)=0.0
00031800 C SET UP ORDER FOR LOG RULE.....
00031900          T13=1./3.
00032000          T23=2./3.
00032100 C INITIALISE A AND B MATRICES
00032200          NBB2=2*NBB
00032300          NBB3=3*NBB
00032400          RAT=0.0
00032500          SLNG=0.0
00032600          SLNGS=0.0
00032700          DO 21 I=1,2
00032800          DO 20 J=1,NBB2
00032900     20   AMT(I,J)=0.0
00033000          DO 21 J=1,NBB3
00033100     21   BMT(I,J)=0.0
00033200 C QUADRATIC SEGMENT LOOP
00033300          NQSEG= NBB/2
00033400          IF(MOD(NBB,2).NE.0) NQSEG=(NBB-1)/2
00033500 C SET UP TRANS VARIABLES FOR SINGULAR CASE.......
00033600 C    NODAL COLLOCATION.........................
00033700          IF(ABS(XSI-0.5).GT.0.001) GOTO 22
00033800          AI(1)=0.0
00033900          AI(2)=0.5
00034000          BI(1)=0.5
00034100          BI(2)=0.5
00034200          AII(1)=.5
00034300          AII(2)=.5
00034400          BII(1)=-.5
00034500          BII(2)=.5
00034600          ISNGSG(1)=1
00034700          ISNGSG(2)=1
00034800          GOTO 26
00034900     22 CONTINUE
00035000          IF(ABS(XSI).GT.0.001.AND.ABS(XSI-1.0).GT.0.001) GOTO 15
00035100 C   FIELD PT XSI=0.0 OR 1.0............
00035200          AI(1)=0.0
00035300          AI(2)=T13
00035400          AI(3)=T23
00035500          BI(1)=T13
00035600          BI(2)=T13
00035700          BI(3)=T13
00035800          DO 10 I=1,3
00035900          AII(I)=0.0
00036000          BII(I)=0.0
00036100     10 ISNGSG(I)=0
00036200          IF(ABS(XSI).GT.0.001) GOTO 12
00036300 C    XSI=0.0    .....
00036400          ISNGSG(1)=1
00036500          AII(1)=0.0
00036600          BII(1)=T13
00036700 C SET UP NEIGHBOURING SINGULAR INTEGRATION PARAMETERS
00036800          AIII(3)=1.0
```

```
00036900          BIII(3)=-T13
00037000          GOTO 26
00037100       12 CONTINUE
00037200 C    XSI=1.0         .....
00037300          ISNGSG(3)=1
00037400          AII(3)=1.0
00037500          BII(3)=-T13
00037600 C NIEGHBOURING INTEGRATION PARAMETERS
00037700          AIII(1)=0.0
00037800          BIII(1)=T13
00037900          GOTO 26
00038000       15 CONTINUE
00038100          IF(XSI.GT.0.5) GOTO 23
00038200 C BETWEEN NODES 1 AND 2......
00038300          AI(1)=0.0
00038400          AI(2)=XSI
00038500          AI(3)=2.0*XSI
00038600          BI(1)=XSI
00038700          BI(2)=XSI
00038800          BI(3)=1.0-2.0*XSI
00038900          AII(1)=XSI
00039000          AII(2)=XSI
00039100          AII(3)=0.0
00039200          BII(1)=-XSI
00039300          BII(2)=XSI
00039400          BII(3)=0.0
00039500          ISNGSG(1)=1
00039600          ISNGSG(2)=1
00039700          ISNGSG(3)=0
00039800          GOTO 26
00039900       23 CONTINUE
00040000          AI(1)=0.0
00040100          AI(2)=2.0*XSI-1.0
00040200          AI(3)=XSI
00040300          BI(1)=2.0*XSI-1.0
00040400          BI(2)=1.0-XSI
00040500          BI(3)=BI(2)
00040600          AII(1)=0.0
00040700          AII(2)=XSI
00040800          AII(3)=XSI
00040900          BII(1)=0.0
00041000          BII(2)=XSI-1.0
00041100          BII(3)=1.0-XSI
00041200          ISNGSG(1)=0
00041300          ISNGSG(2)=1
00041400          ISNGSG(3)=1
00041500       26 CONTINUE
00041600          MPOS=0
00041700          DO 200 IS=1,NQSEG
00041800 C SET UP USUAL ORDER OF INTEGRATION .........
00041900          NGP=NBDATA(NBBI,3)
00042000          ISM=IS-1
00042100          ISP=IS+1
00042200          IF(IS.EQ.1) ISM=NQSEG
00042300          IF(IS.EQ.NQSEG) ISP=1
00042400          ITP=INTYP(NBBI,IS)
00042500          IF(IS.NE.1) MPOS=MPOS+2*ITP
00042600 C SET INTEGRAL TRANSF VARIABLE.........
00042700          TRANSL=1.0
00042800          IS6=6*(IS-1)
00042900          IS44=4*IS-4
00043000 C SET UP COORDS OF 3 NODES ON SEGMENT (A,B,C)
00043100          ISA=2*IS-1
00043200          ISB=ISA+1
00043300          ISC=ISB+1
00043400          IF(IS.EQ.NQSEG) ISC=1
00043500          XA=XCRDBI(ISA)
00043600          YA=YCRDBI(ISA)
00043700          XB=XCRDBI(ISB)
```

```
00043800          YB=YCRDBI(ISB)
00043900          XC=XCRDBI(ISC)
00044000          YC=YCRDBI(ISC)
00044100 C FIND APPROX LENGTH OF SEGMENT ABC.......
00044200          RR1=SQRT((XB-XA)*(XB-XA)+(YB-YA)*(YB-YA))
00044300          RR2=SQRT((XC-XB)*(XC-XB)+(YC-YB)*(YC-YB))
00044400          RABC=RR1+RR2
00044500 C TEST FOR SINGULAR CASE
00044600          NSUBSG=1
00044700 C NIEGHBOURING SEGMENTSREFINED INTEGRATION............
00044800          IF(ISING(ISM).EQ.1.OR.ISING(ISP).EQ.1) NSUBSG=3
00044900          IF(ABS(XSI).LT.0.0001.AND.ISING(ISM).EQ.1) GOTO 34
00045000          IF(ABS(XSI-1.0).LT.0.0001.AND.ISING(ISP).EQ.1) NSUBSG=1
00045100          IF(ABS(XSI).GT.0.001.AND.ABS(XSI-1.0).GT.0.001) GOTO 29
00045200          IF(ABS(XSI).LT.0.001) GOTO 33
00045300 C....XSI=0.0 ............
00045400          IF(IS.EQ.1.AND.ISING(ISM).EQ.1) GOTO 34
00045500          IF(ISING(IS).NE.1) GOTO 29
00045600 C....XSI=1.0   ISING   =1..........
00045700          GOTO 34
00045800      33 CONTINUE
00045900          IF(ISING(ISP).NE.1) GOTO 29
00046000      34 ISD=2*ISP-1
00046100          IF(ISING(ISM).EQ.1) ISD=2*ISM-1
00046200          ISE=ISD+1
00046300          ISF=ISE+1
00046400          XAA=XCRDBI(ISD)
00046500          YAA=YCRDBI(ISD)
00046600          XBB=XCRDBI(ISE)
00046700          YBB=YCRDBI(ISE)
00046800          XCC=XCRDBI(ISF)
00046900          YCC=YCRDBI(ISF)
00047000 C DEFAULT PARAMS FOR INTGRATION OF NEIGHBOURING SEGS........
00047100          AAI(1)=0.0
00047200          AAI(2)=T13
00047300          AAI(3)=T23
00047400          BBI(1)=T13
00047500          BBI(2)=T13
00047600          BBI(3)=T13
00047700 C FIND LENGTHS OF SEGS ABC AND DEF............
00047800          SLNGS=0.0
00047900          SLNG=0.0
00048000          DO 35 IG=1,3
00048100          S=GSP(IG)
00048200          D1=DSH1(S)
00048300          D2=DSH2(S)
00048400          D3=DSH3(S)
00048500          DXIS=D1*XA+D2*XB+D3*XC
00048600          DYIS=D1*YA+D2*YB+D3*YC
00048700          DXI=D1*XAA+D2*XBB+D3*XCC
00048800          DYI=D1*YAA+D2*YBB+D3*YCC
00048900          RJACS=SQRT(DXIS*DXIS+DYIS*DYIS)
00049000          RJAC=SQRT(DXI*DXI+DYI*DYI)
00049100          SLNGS=SLNGS+GSPW(IG)*RJACS
00049200      35 SLNG=SLNG+GSPW(IG)*RJAC
00049300          IF(ISING(ISM).NE.1) GOTO 31
00049400          TMPL=SLNG
00049500          SLNG=SLNGS
00049600          SLNGS=TMPL
00049700      31 CONTINUE
00049800          IF(SLNGS.LE.SLNG) GOTO 37
00049900 C   LENGTH OF SINGULAR SEG IS LARGER........
00050000 C   ADJUST SINGULAR INTEGRATION
00050100          RAT=SLNG/SLNGS/3.
00050200          IF(ABS(XSI).LE.0.001) GOTO 36
00050300          AI(1)=0.
00050400          AI(2)=(1.-RAT)*0.5
00050500          AI(3)=1.-RAT
00050600          BI(1)=AI(2)
```

```
00050700           BI(2)=AI(2)
00050800           BI(3)=RAT
00050900           AII(1)=0.0
00051000           AII(2)=0.0
00051100           AII(3)=1.0
00051200           BII(1)=0.0
00051300           BII(2)=0.0
00051400           BII(3)=-RAT
00051500  C NIEGHBOURING INTGRATION PARAMETERS
00051600           AIII(1)=0.0
00051700           BIII(1)=T13
00051800           GOTO 29
00051900        36 CONTINUE
00052000  C   SINGULAR SEG IS SMALLER......
00052100  C ADJUST NEIGHBOURING SEG INTEGRATION...........
00052200           AAI(1)=0.0
00052300           AAI(2)=(1.-RAT)*0.5
00052400           AAI(3)=1.-RAT
00052500           BBI(1)=AAI(2)
00052600           BBI(2)=AAI(2)
00052700           BBI(3)=RAT
00052800  C NIEGHBOURING INTEGRATION PARAMS
00052900           AIII(3)=1.0
00053000           BIII(3)=-RAT
00053100           GOTO 29
00053200        37 CONTINUE
00053300           RAT=SLNGS/SLNG/3.
00053400           IF(ABS(XSI).LE.0.001) GOTO 38
00053500  C   XSI=1.0    ...........
00053600           AAI(1)=0.0
00053700           AAI(2)=RAT
00053800           AAI(3)=(1.+RAT)*0.5
00053900           BBI(1)=RAT
00054000           BBI(2)=(1.-RAT)*0.5
00054100           BBI(3)=BBI(2)
00054200  C NIEGHBOURING INTEGRATION PARAMS
00054300           AIII(1)=0.0
00054400           BIII(1)=RAT
00054500           GOTO 29
00054600        38 CONTINUE
00054700  C   XSI=0.0...........
00054800           AI(1)=0.0
00054900           AI(2)=RAT
00055000           AI(3)=(1.+RAT)*0.5
00055100           BI(1)=RAT
00055200           BI(2)=(1.-RAT)*0.5
00055300           BI(3)=BI(2)
00055400           AII(1)=0.0
00055500           AII(2)=0.0
00055600           AII(3)=0.0
00055700           BII(1)=RAT
00055800           BII(2)=0.0
00055900           BII(3)=0.0
00056000  C NIEGHBOURING INTAGRATION PARAMS
00056100           AIII(3)=1.0
00056200           BIII(3)=-T13
00056300        29 CONTINUE
00056400           IF(ABS(XSI).LT.0.001.AND.ISING(ISM).EQ.1) GOTO 39
00056500           IF(ABS(XSI-1.).LT.0.001.AND.ISING(ISP).EQ.1) GOTO 39
00056600           IF(ABS(XSI-0.5).LT.0.499) GOTO 39
00056700           GOTO 41
00056800        39 AAI(1)=0.0
00056900           AAI(2)=T13
00057000           AAI(3)=T23
00057100           BBI(1)=T13
00057200           BBI(2)=T13
00057300           BBI(3)=T13
00057400        41 CONTINUE
00057500           IF(ISING(IS).EQ.0) GOTO 32
```

```
00057600            NSUBSG=3
00057700            JSING=IS
00057800            IF(ABS(XSI-0.5).LT.0.0001) NSUBSG=2
00057900         32 CONTINUE
00058000 C SINGULAR SEGMENT INTEGRATED IN 2 PARTS
00058100 C SET UP GUASS PTS
00058200 C AUTOMATIC SELECTION OF INTEGRATION ORDER BY APPROX
00058300 C ERROR CONSIDERATION ..........
00058400            IF(NOEQ.EQ.0) GOTO 54
00058500            IF(NSUBSG.NE.1) GOTO 54
00058600            EROR=.00001
00058700            RMIN1=SQRT((XA-XPT)**2+(YA-YPT)**2)
00058800            RMIN2=SQRT((XB-XPT)**2+(YB-YPT)**2)
00058900            RMIN3=SQRT((XC-XPT)**2+(YC-YPT)**2)
00059000            RMINN=RMIN1
00059100            IF(RMIN2.LT.RMIN1) RMINN=RMIN2
00059200            IF(RMINN.GT.RMIN3) RMINN=RMIN3
00059300            RJACX=RABC
00059400            RJC=0.25*RJACX/RMINN
00059500            IF(ABS(RJC-1.0).LT.1.0E-05) RJC=RJC+.00001
00059600            ALG1=ALOG(EROR*0.125)
00059700            ALG2=ALOG(RJC)
00059800            RORDN=0.5*ABS(ALG1/ALG2-1.0)
00059900            NORD=IFIX(RORDN) +1
00060000            IF(NORD.GT.8) NORD=12
00060100            NGP=NORD
00060200            IF(JBUG(9).EQ.1) WRITE(6,1008) RORDN ,IS,NORD
00060300         54 CONTINUE
00060400 C SET UP GAUSS PTS AND WTS ACCORDINGLY .....
00060500            DO 40 I=1,NGP
00060600            SLGP(I)=GLP(NGP,I)
00060700         40 SGP(I)=GP(NGP,I)
00060800            DO 150 ISB=1,NSUBSG
00060900            IF(NSUBSG.EQ.1) GOTO 70
00061000            TRANSL=BI(ISB)
00061100            IF(ISING(IS).EQ.0) TRANSL=BBI(ISB)
00061200            DO 50 I=1,NGP
00061300            SLGP(I)=AII(ISB)+BII(ISB)*GLP(NGP,I)
00061400            IF(ISING(IS).NE.1) SLGP(I)=AIII(ISB)+BIII(ISB)*GLP(NGP,I)
00061500            SGP(I)=AI(ISB)+BI(ISB)*GP(NGP,I)
00061600            IF(ISING(IS).EQ.0) SGP(I)=AAI(ISB)+BBI(ISB)*GP(NGP,I)
00061700         50 CONTINUE
00061800         70 CONTINUE
00061900 C START GUASSIAN INTEGRATION LOOP
00062000            DO 130 IG=1,NGP
00062100 C FIND X,Y COORDS OF G-PT
00062200            S=SGP(IG)
00062300            SL=SLGP(IG)
00062400            XGP=SHN1(S)*XA +SHN2(S)*XB +SHN3(S)*XC
00062500            YGP=SHN1(S)*YA +SHN2(S)*YB +SHN3(S)*YC
00062600            XLGP=SHN1(SL)*XA+SHN2(SL)*XB+SHN3(SL)*XC
00062700            YLGP=SHN1(SL)*YA+SHN2(SL)*YB+SHN3(SL)*YC
00062800            IF(JBUG(15).EQ.1) WRITE(6,1027)XGP,YGP,XLGP,YLGP
00062900 C FIND 1-D JACOBIAN (SCALE FACTOR )
00063000            DXIS= DSH1(S)*XA +DSH2(S)*XB + DSH3(S)*XC
00063100            DYIS= DSH1(S)*YA +DSH2(S)*YB + DSH3(S)*YC
00063200            DLXIS=DSH1(SL)*XA+DSH2(SL)*XB+DSH3(SL)*XC
00063300            DLYIS=DSH1(SL)*YA+DSH2(SL)*YB+DSH3(SL)*YC
00063400            RLJAC=SQRT(DLXIS*DLXIS+DLYIS*DLYIS)
00063500            RJAC=SQRT(DXIS*DXIS + DYIS*DYIS)
00063600 C COORDS OF VECTOR BETWEEN GUASS AND FIELD PTS
00063700            XPG = XGP -XPT
00063800            YPG = YGP -YPT
00063900            XLPG=XLGP-XPT
00064000            YLPG=YLGP-YPT
00064100 C OUTWARD NORMAL AT GUASS PT
00064200            XNGP = RFACT*DYIS/RJAC
00064300            YNGP =-RFACT*DXIS/RJAC
00064400            RPG =SQRT(XPG*XPG +YPG*YPG)
```

```
00064500          RPG2=1.0/RPG/RPG
00064600          XRP = XPG*XPG*RPG2
00064700          YRP = YPG*YPG*RPG2
00064800          XYRP= XPG*YPG*RPG2
00064900          RNXY= XNGP*XPG + YNGP*YPG
00065000 C INTEGRATION KERNELS
00065100          GLPWJ=GLPW(NGP,IG)*RLJAC*TRANSL
00065200          GLWPJ(1)=SHN1(SL)*GLPWJ
00065300          GLWPJ(2)=SHN2(SL)*GLPWJ
00065400          GLWPJ(3)=SHN3(SL)*GLPWJ
00065500          GPWJ=GPW(NGP,IG)*RJAC*TRANSL
00065600          GWPJ(1)=SHN1(S)*GPWJ
00065700          GWPJ(2)=SHN2(S)*GPWJ
00065800          GWPJ(3)=SHN3(S)*GPWJ
00065900          RTERM = CONS1*CONS2*ALOG(RPG)
00066000          DO 80 K=1,3
00066100          K2=2*K+IS44+MPOS
00066200          IF(K2.GT.NBB2.AND.INTYP(NBBI,1).EQ.0) K2=K2-NBB2
00066300          KM=K2-1
00066400          RTERMG=CONS1*CONS2*GLWPJ(K)
00066500          BMT(1,KM)=BMT(1,KM)-GWPJ(K)*CONS1*XRP
00066600          BMT(2,K2)=BMT(2,K2)-GWPJ(K)*CONS1*YRP
00066700          IF(ISING(IS).EQ.1.AND.ISNGSG(ISB).EQ.1) GOTO 71
00066800          IF(ISING(ISM).EQ.1.AND.ISB.EQ.1.AND.ABS(XSI-1.).LT.0.0001) GOTO 7
00066900          IF(ISING(ISP).EQ.1.AND.ISB.EQ.3.AND.ABS(XSI).LT.0.0001) GOTO 71
00067000          GOTO 72
00067100 C SINGULAR INTEGRATION OF LOG TERMS
00067200       71 CONTINUE
00067300          RBI=BI(ISB)*RABC
00067400          IF(ISING(IS).EQ.1) GOTO 73
00067500          IF(ISB.EQ.1) RBI=ABS(BIII(1))*SLNG
00067600          IF(ISB.EQ.3) RBI=ABS(BIII(3))*SLNGS
00067700       73 CONTINUE
00067800          IF(ABS(RBI).LT.0.00001) WRITE(6,8061) SLNG,SLNGS,BI(ISB),BIII(ISB
00067900         .,ISING(IS),ISB
00068000            STERM=-RTERMG+CONS1*CONS2*ALOG(RBI)*GWPJ(K)
00068100          BMT(1,KM)=BMT(1,KM)+STERM
00068200          BMT(2,K2)=BMT(2,K2)+STERM
00068300          GOTO 76
00068400       72 CONTINUE
00068500            BMT(1,KM)=BMT(1,KM)+GWPJ(K)*RTERM
00068600          BMT(2,K2)=BMT(2,K2)+GWPJ(K)*RTERM
00068700       76 CONTINUE
00068800          BMT(1,K2)=BMT(1,K2)   -GWPJ(K)*CONS1*XYRP
00068900       80 BMT(2,KM)=BMT(1,K2)
00069000 C SET NUMBER OF TRACTION COMPONENTS =NCOMP.........
00069100          NCOMP=K2
00069200          GWPJ(1)= SHN1(S)*GPWJ
00069300          GWPJ(2)= SHN2(S)*GPWJ
00069400          GWPJ(3)= SHN3(S)*GPWJ
00069500          RNYX= YNGP*XPG -XNGP*YPG
00069600          DO 100 K=1,3
00069700          K2=2*K+IS44
00069800          IF(K.EQ.3.AND.IS.EQ.NQSEG) K2=K2-NBB2
00069900          KM=K2-1
00070000          GTERM=GWPJ(K)*CONS3*RPG2
00070100          AMT(1,KM)=AMT(1,KM)+ GTERM*(CONS4 +2.0*XRP)*RNXY
00070200          AMT(2,K2)=AMT(2,K2)+ GTERM*(CONS4 +2.0*YRP)*RNXY
00070300          AMT(2,KM)=AMT(2,KM)+ GTERM*(CONS4*RNYX +2.0*XYRP*RNXY)
00070400      100 AMT(1,K2)=AMT (1,K2)+GTERM*(-CONS4*RNYX+2.*XYRP*RNXY)
00070500      130 CONTINUE
00070600      150 CONTINUE
00070700 C INSERT A AND B TEMP MATRICES INTO GLOBAL MATRICES
00070800      200 CONTINUE
00070900 C   CALCULATION OF DIAGONAL TERMS OF MATRIX AMT
00071000          IF(JBUG(10).EQ.0) GOTO 207
00071100          WRITE(6,1009)
00071200          WRITE(6,1010)((AMT(I,J),J=1,NBB2),I=1,2)
00071300          WRITE(6,1010)((BMT(I,J),J=1,NBB3),I=1,2)
```

```
00071400     207 CONTINUE
00071500         JS4=4*JSING-6
00071600 C   IF PT IS INTERIOR TO REGION SINGULAR TERM IS NOT ADJUSTED
00071700         IF(JSING.EQ.0) GOTO 255
00071800         SUM11=0.0
00071900         SUM22=0.0
00072000         SUM12=0.0
00072100         SUM21=0.0
00072200 C ADJUST SINGULAR TERMS FOR EXTERIOR PROBLEMS ......
00072300         IF(NBDATA(NBBI,2).EQ.0) GOTO 227
00072400         SUM11=-1.0
00072500         SUM22=-1.0
00072600     227 CONTINUE
00072700 C FIND COEFF SUMS ..............
00072800         DO 230 I=1,NBB
00072900         I2=2*I
00073000         IM=I2-1
00073100         SUM11=SUM11+AMT(1,IM)
00073200         SUM12=SUM12+AMT(1,I2)
00073300         SUM21=SUM21+AMT(2,IM)
00073400     230 SUM22=SUM22+AMT(2,I2)
00073500         JS4=4*(JSING-1)
00073600         SHP(1)=SHN1(XSI)
00073700         SHP(2)=SHN2(XSI)
00073800         SHP(3)=SHN3(XSI)
00073900         DO 250 I=1,3
00074000         I2=2*I+JS4
00074100         IM=I2-1
00074200         IF(I2.GT.NBB2) I2=I2-NBB2
00074300         IF(IM.GT.NBB2) IM=IM-NBB2
00074400         AMT(1,IM)=AMT(1,IM)-SUM11*SHP(I)
00074500         AMT(2,I2)=AMT(2,I2)-SUM22*SHP(I)
00074600         AMT(2,IM)=AMT(2,IM)-SUM21*SHP(I)
00074700     250 AMT(1,I2)=AMT(1,I2)-SUM12*SHP(I)
00074800     255 CONTINUE
00074900         IF(JBUG(10).EQ.0.0) GOTO 300
00075000         WRITE(6,1000)
00075100         WRITE(6,1010)((AMT(I,J),J=1,NBB2),I=1,2)
00075200         WRITE(6,1010)((BMT(I,J),J=1,NBB3),I=1,2)
00075300     300 CONTINUE
00075400    1008 FORMAT(1X,'AUTO INTEGRATION ORDER = ',F15.5,'SEG NO =',I5,
00075500     . 'NEW ORDER =',I5 )
00075600    1027 FORMAT(1X,'ORDINARY G PTS',2E12.4,'LOG G PTS',2E12.4)
00075700    8061 FORMAT(1X,'RBI=0',4F10.5,2I5)
00075800    1009 FORMAT(1X,' A AND B AFTER 200 CONTINUE ')
00075900    1000 FORMAT(6X,'A AND MATRICES')
00076000    1010 FORMAT(1X,10E10.3)
00076100         RETURN
00076200         END
00076300         SUBROUTINE SIMF( A,M,N,N1 )
00076400         DIMENSION A(M,121)
00076500         NPN1 = N + N1
00076600 C                            DETM. MAX. DIAGN. TERM FOR SING. CHECK
00076700         RMAX=0.
00076800         DO 70 I=1,N
00076900      70 IF(ABS(A(I,I)).GT.ABS(RMAX)) RMAX=A(I,I)
00077000         DO 25 J=1,N
00077100         Z=A(J,J)
00077200         IF(ABS(Z/RMAX).LT.1.0E-08) GO TO 30
00077300         DO 10 L=J,NPN1
00077400      10 A(J,L) = A(J,L) / Z
00077500         DO 20 I = 1 , N
00077600         IF(I.EQ.J .OR. A(I,J).EQ.0.0) GO TO 19
00077700         Z =  A(I,J)
00077800         DO 15 L=J,NPN1
00077900      15 A(I,L) = A(I,L) - Z * A(J,L)
00078000      19 CONTINUE
00078100      20 CONTINUE
00078200      25 CONTINUE
```

```
00078300          RETURN
00078400      30 WRITE(6, 35) J
00078500      35 FORMAT(1H0,23HSINGULAR MATRIX, COLUMN,I3)
00078600          STOP
00078700          END
00078800          SUBROUTINE BISOLV
00078900          DIMENSION AMT(2,120),BMT(2,180),UD(2),TR(2)
00079000 C    **** INCLUDE COMMON BLOCKS ****
00079100 C    **** DBUG SOLV CONTRO LGDATA BMATS MATCON LDISK ****
00079200          NREG=1
00079300          NBB=NBDATA(NREG,1)
00079400          NBBQ=NBB/2
00079500          NBB2=2*NBB
00079600          NBB3=3*NBB
00079700          GMOD=0.5*PROPS(1,1)/(1.0+PROPS(1,2))
00079800          NCLPT2=2*NCLPT
00079900 C SETUP BC FLAGS
00080000          DO 5 I=1,NBB2
00080100          NUPOS(I)=0
00080200          UFIX(I)=0.0
00080300       5 NUFIX(I)=0
00080400          DO 10 I=1,NBB3
00080500          NTPOS(I)=0
00080600          TFIX(I)=0.0
00080700      10 NTFIX(I)=0
00080800          MPOS=0
00080900          DO 100 IS=1,NBBQ
00081000          IS3=3*(IS-1)
00081100          IS4=4*(IS-1)
00081200          IS6=IS4
00081300          ITP=INTYP(NREG,IS)
00081400          IF(IS.NE.1) MPOS=MPOS+2*ITP
00081500 C TEST IF 1ST 2 NODES ON SEGMENT HAVE BCS
00081600          DO 40 I=1,2
00081700          IPP=2*(I-1)
00081800          IP=2*(IS-1)+I
00081900          LNOD=LNODBI(NREG,IP)
00082000 C SET UP DISP FLAGS IN VECTOR FORM
00082100          DO 30 IF=1,NVFIX
00082200          IF(LNOD.NE.NOFIX(IF)) GO TO 29
00082300 C   LOOP OVER DEG OF FREEDOM
00082400          DO 20 IFD=1,2
00082500          IFP=IFD+IPP+IS4
00082600          IF (IFPRE(IF,IFD).NE.1) GO TO 19
00082700          NUFIX(IFP)=1
00082800          UFIX(IFP)=PRESC(IF,IFD)*GMOD
00082900      19 CONTINUE
00083000      20 CONTINUE
00083100      29 CONTINUE
00083200      30 CONTINUE
00083300      40 CONTINUE
00083400          DO 50 I=1,3
00083500 C 2 OR 3 TRACTION NODES ON A SEGMENT.............
00083600          NTRK=NTRKBC(NREG,IS)
00083700          IF(NTRK.EQ.0) GOTO 90
00083800          IF(NTRK.EQ.1) IP=2*I-1+IS6+MPOS
00083900          IF(NTRK.EQ.2) IP=2*I+IS6+MPOS
00084000          IF(NTRK.EQ.3) GOTO 60
00084100          ISS=IS3+I
00084200          IF(NTRK.EQ.1) TFIX(IP)=TRXBC(NREG,ISS)
00084300          IF(NTRK.EQ.2) TFIX(IP)=TRYBC(NREG,ISS)
00084400          NTFIX(IP)=1
00084500      50 CONTINUE
00084600          GOTO 90
00084700      60 CONTINUE
00084800          I3=3
00084900          I6=6
00085000          IF(IS.NE.NQSEG)GOTO 59
00085100          IF(INTYP(NREG,1).NE.0)GOTO 59
```

```
00085200        I3=2
00085300        I6=4
00085400    59  CONTINUE
00085500        DO 70 I=1,I6
00085600        IP=I+IS6+MPOS
00085700    70  NTFIX(IP)=1
00085800        DO 80 I=1,I3
00085900        I1=2*I-1+IS6+MPOS
00086000        I2=I1+1
00086100        ISS=IS3+I
00086200        TFIX(I1)=TRXBC(NREG,ISS)
00086300    80  TFIX(I2)=TRYBC(NREG,ISS)
00086400    90  CONTINUE
00086500        IF(JBUG(40).EQ.1) WRITE(6,3000)(NTFIX(L),L=1,NBB3)
00086600   100  CONTINUE
00086700        IF(JBUG(40).EQ.0) GOTO 104
00086800 C WRITE FLAGS BCS VECTORS
00086900        WRITE(6,3000) (NUFIX(I),I=1,NBB2)
00087000        WRITE(6,3000)(NTFIX(I),I=1,NBB3)
00087100        WRITE(6,3001)
00087200        WRITE(6,3002)(UFIX(I),I=1,NBB2)
00087300        WRITE(6,3002)(TFIX(I),I=1,NBB3)
00087400   104  CONTINUE
00087500 C LOOP OVER COLLOCATION PTS
00087600 C AND SET UP MATRIX EQNS
00087700        REWIND NDISKA
00087800        REWIND NDISKB
00087900        DO 200 ICP=1,NCLPT
00088000        ICP2=2*(ICP-1)
00088100        DO 110 I=1,2
00088200        READ (NDISKA) (AMT(I,J),J=1,NBB2)
00088300        READ (NDISKB) (BMT(I,J),J=1,NBB3)
00088400   110  CONTINUE
00088500 C  FIND R.H.S VECTOR            ...........
00088600        DO 140 I=1,2
00088700        IP=ICP2+I
00088800        RHS(IP)=0.0
00088900        DO 120 J=1,NBB2
00089000   120  RHS(IP)=RHS(IP)-AMT(I,J)*UFIX(J)
00089100        DO 130 J=1,NBB3
00089200   130  RHS(IP)=RHS(IP)+BMT(I,J)*TFIX(J)
00089300 C RHS IS DUE TO BOUNDARY CONDITIONS ONLY
00089400   140  CONTINUE
00089500        IDST=0
00089600 C  SET UP FINAL EQNS ..............
00089700        MPOS=0
00089800        MSUM=0
00089900        NTCOMP=0
00090000        DO 190 ISEG=1,NBBQ
00090100        NCOMPT=6
00090200        IF(ISEG.EQ.1) GOTO 145
00090300        ITP=INTYP(NREG,ISEG)
00090400        IF(ITP.EQ.0) NCOMPT=4
00090500        MPOS=MPOS+2*ITP
00090600   145  CONTINUE
00090700        NTCOMP=NTCOMP+NCOMPT
00090800        IS4=4*(ISEG-1)
00090900        IS6=4*(ISEG-1)
00091000 C  LOOP OVER 4 CURRENT DISP COMPONENTS
00091100        DO 160 I=1,4
00091200        IP=I+IS4
00091300        IF(NUFIX(IP).EQ.1) GO TO 159
00091400        IDST=IDST+1
00091500        DO 150 IR=1,2
00091600        IROW=IR+ICP2
00091700        NUPOS(IP)=IDST
00091800   150  RLHS(IROW,IDST)=AMT(IR,IP)
00091900   159  CONTINUE
00092000   160  CONTINUE
```

```
00092100 C   LOOP OVER CURRENT 4 OR 6 TRACTION COMPONENTS
00092200         DO 180 I=1,NCOMPT
00092300         IP=I+MSUM
00092400         IF(NTFIX(IP).EQ.1) GO TO 179
00092500         IDST=IDST+1
00092600         DO 170 IR=1,2
00092700         IROW=IR+ICP2
00092800         NTPOS(IP)=IDST
00092900   170 RLHS(IROW,IDST)=-BMT(IR,IP)
00093000   179 CONTINUE
00093100   180 CONTINUF
00093200         MSUM=IP
00093300   190 CONTINUE
00093400         IF(JBUG(41).EQ.0) GOTO 197
00093500         WRITE(6,3000)(NUPOS(I),I=1,NBB2)
00093600         WRITE(6,3000)(NTPOS(I),I=1,NBB3)
00093700   197 CONTINUE
00093800   200 CONTINUE
00093900 C
00094000         JPDST=IDST+1
00094100         DO 210 I=1,IDST
00094200         RLHS(I,JPDST)=RHS(I)
00094300   210 CONTINUE
00094400         IP=JPDST
00094500         IF(JBUG(41).EQ.0) GOTO 215
00094600         WRITE(6,1030)
00094700         DO 211 I=1,NCLPT2
00094800   211 WRITE(6,1020)(RLHS(I,J),J=1,IP)
00094900   215 CONTINUE
00095000         NTCOMP=NTCOMP/2
00095100         IF(IDST.NE.NCLPT2) WRITE(6,1000) IDST,NCLPT2
00095200 C SOLVE LINEAR EQNS
00095300         RMAX=0.0
00095400         DO 240 J=1,IDST
00095500         IROW=J
00095600         Z=0.0
00095700         DO 220 K=J,NCLPT2
00095800         IF(ABS(RLHS(K,J)).LT.Z) GO TO 219
00095900         Z=ABS(RLHS(K,J))
00096000         IROW=K
00096100   219 CONTINUE
00096200   220 CONTINUE
00096300         IF(Z.GT.RMAX) RMAX=Z
00096400         DO 230 K=1,IP
00096500         VAR=RLHS(J,K)
00096600         RLHS(J,K)=RLHS(IROW,K)
00096700   230 RLHS(IROW,K)=VAR
00096800   240 CONTINUE
00096900         IF(JBUG(41).EQ.0) GOTO 253
00097000         WRITE(6,1031)
00097100         DO 250 I=1,NCLPT2
00097200   250 WRITE(6,1020)(RLHS(I,J),J=1,IP)
00097300   253 CONTINUE
00097400         CALL SIME(RLHS,MAXSZ,IDST,1)
00097500         IF(JBUG(41).EQ.0) GOTO 270
00097600         WRITE(6,1010)
00097700         WRITE(6,8885) IDST,JPDST
00097800         WRITE(6,1020)(RLHS(I,JPDST),I=1,IDST)
00097900   270 CONTINUE
00098000         WRITE(6,2000)
00098100 C SET OUT PRINT OUT OF DISPLACEMENTS AND TRACTIONS
00098200         JPOS1=0
00098300         JPOS2=0
00098400         DO 330 IN=1,NTCOMP
00098500         I2=2*(IN-1)
00098600         DO 320 J=1,2
00098700         JP=I2+J
00098800         IF(IN.GT.NBB) GOTO 300
00098900         NUP=NUPOS(JP)
```

```
00099000          IF(NUP.EQ.0) GOTO 290
00099100          UD(J)=RLHS(NUP,JPDST)
00099200          UFIX(JP)=UD(J)
00099300          GOTO 295
00099400      290 UD(J)=UFIX(JP)
00099500      295 BIDIS(JP)=UD(J)
00099600      300 CONTINUE
00099700          NUT=NTPOS(JP)
00099800          IF(NUT.EQ.0) GOTO 310
00099900          TR(J)=RLHS(NUT,JPDST)
00100000          TFIX(JP)=TR(J)
00100100          GO TO 319
00100200      310 TR(J)=TFIX(JP)
00100300      319 CONTINUE
00100400      320 TRACT(JP)=TR(J)
00100500          IF(IN.GT.NBB) GOTO 325
00100600          DO 311 I=1,2
00100700      311 UD(I)=UD(I)/GMOD
00100800          WRITE(6,2010) IN,(UD(I),I=1,2),(TR(J),J=1,2)
00100900          GO TO 329
00101000      325 WRITE(6,2011) IN,(TR(J),J=1,2)
00101100      329 CONTINUE
00101200      330 CONTINUE
00101300          RETURN
00101400     3000 FORMAT(1X,'BC FLAGS ',60I1)
00101500     3001 FORMAT(6X,' VECTOR OF BCS ')
00101600     3002 FORMAT(1X,10E12.4)
00101700     1030 FORMAT(6X,'FINAL SYSTEM OF EQNS',/)
00101800     1000 FORMAT(1X,'***ERROR EQN SET NOT SQUARE***- NO OF COLS
00101900        . =',2I5)
00102000     1031 FORMAT(6X,'REARRANGED MATRIX ')
00102100     1010 FORMAT(1X,'BOUNDARY DISPLACEMENTS AND TRACTIONS')
00102200     8885 FORMAT(1X,'IDST,JPDST =',2I5)
00102300     1020 FORMAT(1X,10E12.4)
00102400     2000 FORMAT(5X,'NODE',3X,'DISPLACEMENTS',15X,'TRACTIONS ')
00102500     2010 FORMAT(3X,I5,2(2E12.4,2X))
00102600     2011 FORMAT(3X,I5,26X,2E12.4)
00102700          END
00102800          SUBROUTINE BIPTS
00102900          DIMENSION DPMAT(2,120),SRMAT(2,120),TRMAT(2,180),UIN(2),TIN(2)
00103000          DIMENSION SPT(3),ISING(40)
00103100 C  **** INCLUDE COMMON BLOCKS ****
00103200 C  **** DBUG CONTRO LGDATA BMATS MATCON ****
00103300 C THIS ROUTINE FINDS DISPS AND TRACTIONS IN AND ON
00103400 C A BI REGION..............
00103500          INBI=1
00103600          NBB=NBDATA(INBI,1)
00103700          NGP=NBDATA(INBI,3)
00103800          NQSEG=NBB/2
00103900          ISUPS=NBDATA(INBI,2)
00104000          GMOD=0.5*PROPS(1,1)/(1.0+PROPS(1,2))
00104100 C FIND NO OF TRACTION COMPONENTS
00104200          NTBB=NBB
00104300          DO 15 I=1,NQSEG
00104400       15 NTBB=NTBB+INTYP(INBI,I)
00104500          NBB3=2*NTBB
00104600          NBB2=2*NBB
00104700 C CALCULATION FOR INTERIOR PTS........
00104800          READ(5,1020) NOINPT
00104900          WRITE(6,1025) NOINPT
00105000     1025 FORMAT(20H NO OF INTERIOR PTS=,I5)
00105100          IF(NOINPT.EQ.0) GOTO 400
00105200          WRITE(6,1040)
00105300     1040 FORMAT(1X,'INTERIOR DISPLACEMENTS')
00105400          WRITE(6,1029)
00105500     1029 FORMAT(5X,3H ID,5X,2H X,7X,2H Y,6X,3H UX,6X,3H UY /)
00105600          DO 250 IP=1,NOINPT
00105700          XSI=0.0
00105800          READ(5,1030) JSING,XIN,YIN,XSI
```

```
00105900          WRITE(6,1030) JSING,XIN,YIN
00106000 C  SET UP COORDS OF BI REGION......
00106100          J=0
00106200          DO 80 I=1,NBB
00106300          J=J+1
00106400          NUMC=LNODBI(INBI,I)
00106500          XCRDBI(J)=COORD(NUMC,1)
00106600          YCRDBI(J)=COORD(NUMC,2)
00106700       80 CONTINUE
00106800          NBBP=NBB+1
00106900          XCRDBI(NBBP)=XCRDBI(1)
00107000          YCRDBI(NBBP)=YCRDBI(1)
00107100          DO 85 I=1,NQSEG
00107200       85 ISING(I)=0
00107300          IF(JSING.NE.0) ISING(JSING)=1
00107400          CALL INTAB(XIN,YIN,XSI,ISING,DPMAT,TRMAT,INBI)
00107500          IF(JBUG(33).EQ.0) GOTO 95
00107600          WRITE(6,3005)
00107700          DO 91 I=1,2
00107800       91 WRITE(6,2030)(DPMAT(I,J),J=1,NBB2)
00107900          DO 92 I=1,2
00108000       92 WRITE(6,2030)(TRMAT(I,J),J=1,NBB3)
00108100       95 CONTINUE
00108200          DO 230 I=1,2
00108300          UIN(I)=0.0
00108400          TIN(I)=0.0
00108500          DO 225 J=1,NBB2
00108600      225 UIN(I)=UIN(I)-DPMAT(I,J)*BIDIS(J)
00108700          DO 226 J=1,NBB3
00108800      226 UIN(I)=UIN(I)+TRMAT(I,J)*TRACT(J)
00108900      230 CONTINUE
00109000          DO 244 K=1,2
00109100      244 UIN(K)=UIN(K)/GMOD
00109200          WRITE(6,1050) (UIN(L),L=1,2)
00109300      250 CONTINUE
00109400      400 CONTINUE
00109500      300 CONTINUE
00109600          RETURN
00109700     1020 FORMAT(16I5)
00109800     1030 FORMAT(I5,3F10.3)
00109900     3005 FORMAT(6X,'A AND B MATRICES IN BIPTS ')
00110000     2030 FORMAT(1X,8E12.4)
00110100     1050 FORMAT(25X,2E12.4)
00110200          END
00110300          SUBROUTINE GAUSS
00110400          DIMENSION A8(8),AW(12),B8(8),BW(12),C5(5),C6(6),C7(7),C8(8),CT(1(
00110500          DIMENSION CW(12),D5(5),D6(6),D7(7),D8(8),DT(10),DW(12)
00110600          DIMENSION A4(4),B4(4),C4(4),D4(4),GSP(3),GSPW(3)
00110700          COMMON /GSPWT/ ZGSP(3),ZGSPW(3)
00110800          COMMON /MATCON/ RA(5),IA(2),GP(12,12),GPW(12,12),IB(3),RB,IC(5),
00110900         1          ,ID(2),GLP(12,12),GLPW(12,12)
00111000          DATA GSP/0.1127017 , .5, .8872983 /
00111100          DATA GSPW/ .27777778, .44444444, .27777778 /
00111200          DATA A4/ .3834641, .3868753, .1904351, .03922549 /
00111300          DATA A8/.1644166, .2375256, .2268420, .1757541, .1129240, .05787.
00111400         1       , .0209791, .0036864 /
00111500          DATA AW/ .09319269, .14975183, .16655745, .15963356, .13842483,
00111600         1          .11001657, .07996182, .05240695, .03007109, .01424924,
00111700         2          .004899924, .000834029 /
00111800          DATA B4/ .0414485, .2452749, .5561655, .8489824 /
00111900          DATA B8/ .0133202, .0797504, .1978710, .3541540, .5294586,
00112000         1          .7018145, .8493793, .9533264 /
00112100          DATA BW/ .006548722, .03894680, .09815026, .18113858, .28322007,
00112200         1          .39843444, .51995263, .64051092, .75286501, .85024002,
00112300         2          .92674968, .97775613 /
00112400          DATA C4/ .06943184, .33000948, .66999052, .93056816 /
00112500          DATA C5/ .0469101, .2307654, .5000000, .7692347, .9530899 /
00112600          DATA C6/ .0337653, .1693954, .3806905, .6193096, .8306047,
00112700         1          .9662348 /
00112800          DATA C7/ .0254461, .1292345, .2970775, .5000000, .7029226,
00112900         1          .8707656, .9745540 /
```

```
00113000          DATA C8/ .0198550, .1016670, .2372340, .4082825, .5917175,
00113100        1          .7627660, .8983330, .9801450 /
00113200          DATA CT/ .0130468, .0674684, .1602953, .2833023, .4255629,
00113300        1          .5744372, .7166977, .8397048, .9325317, .9869533 /
00113400          DATA CW/ .0092197, .0479414, .1150487, .2063411, .3160843,
00113500        1          .4373833, .5626167, .6839157, .7936590, .8849513,
00113600        2          .9520586, .9907803 /
00113700          DATA D4/ .17392742, .32607258, .32607258, .17392742 /
00113800          DATA D5/ .1184634, .2393143, .2844444, .2393143, .1184634 /
00113900          DATA D6/ .0856622, .1803808, .2339570 , .2339570, .1803808,
00114000        1          .0856622 /
00114100          DATA D7/ .0647425, .1398527, .1909150, .2089796, .1909150,
00114200        1          .1398527, .0647425 /
00114300          DATA D8/ .0506140, .1111905, .1568533, .1813419, .1813419,
00114400        1          .1568533, .1111905, .0506140 /
00114500          DATA DT/ .0333357, .0747257, .1095432, .1346334, .1477621,
00114600        1          .1477621, .1346334, .1095432, .0747257, .0333357 /
00114700          DATA DW/ .0235877, .0534697, .0800392, .1015837, .1167463,
00114800        1          .1245735, .1245735, .1167463, .1015837, .0800392,
00114900        2          .0534697, .0235877 /
00115000          DO 20 I=1,12
00115100          DO 20 J=1,12
00115200          GP  (I,J)=0.0
00115300          GPW (I,J)=0.0
00115400          GLP (I,J)=0.0
00115500       20 GLPW(I,J)=0.0
00115600          GP (1,1)=0.5
00115700          GPW(1,1)=1.0
00115800          GP(2,1)=0.2113249
00115900          GP(2,2)=0.7886751
00116000          GPW(2,1)=0.5
00116100          GPW(2,2)=0.5
00116200          GLP(2,1)=0.1120088
00116300          GLP(2,2)=0.6022769
00116400          GLPW(2,1)=.7185393
00116500          GLPW(2,2)=.2814607
00116600          DO 3 J=1,3
00116700          ZGSP(J)=GSP(J)
00116800          ZGSPW(J)=GSPW(J)
00116900          K=4-J
00117000          GPW(3,J)=GSPW(J)
00117100        3 GP (3,J)=GSP (K)
00117200          DO 4 J=1,4
00117300          GLPW(4,J)=A4(J)
00117400          GLP (4,J)=B4(J)
00117500          GP  (4,J)=C4(J)
00117600        4 GPW (4,J)=D4(J)
00117700          DO 5 J=1,5
00117800          GP (5,J)=C5(J)
00117900        5 GPW(5,J)=D5(J)
00118000          DO 6 J=1,6
00118100          GP (6,J)=C6(J)
00118200        6 GPW(6,J)=D6(J)
00118300          DO 7 J=1,7
00118400          GP (7,J)=C7(J)
00118500        7 GPW(7,J)=D7(J)
00118600          DO 8 J=1,8
00118700          GLPW(8,J)=A8(J)
00118800          GLP (8,J)=B8(J)
00118900          GP  (8,J)=C8(J)
00119000        8 GPW (8,J)=D8(J)
00119100          DO 10 J=1,10
00119200          GP (10,J)=CT(J)
00119300       10 GPW(10,J)=DT(J)
00119400          DO 12 J=1,12
00119500          GLPW(12,J)=AW(J)
00119600          GLP (12,J)=BW(J)
00119700          GP  (12,J)=CW(J)
00119800       12 GPW (12,J)=DW(J)
00119900          RETURN
00120000          END
```

```
TOTAL NO. OF PROBLEMS =     1




    PROBLEM NO.   1              CANTILEVER
000000000000000000000000000000000000000000000000000000000000000000000000000000000000000

NVFIX =   5      NOTRBC=   6      NTYPE =   1      NOEQ =   1


   16     0     8
NODAL POINT COORDINATES
NODE      X          Y
   1     0.000      0.000
   2     0.500      0.000
   3     1.000      0.000
   4     1.500      0.000
   5     2.000      0.000
   6     2.000      0.250
   7     2.000      0.500
   8     2.000      0.750
   9     2.000      1.000
  10     1.500      1.000
  11     1.000      1.000
  12     0.500      1.000
  13     0.000      1.000
  14     0.000      0.750
  15     0.000      0.500
  16     0.000      0.250


RESTRAINED NODES
NODE CODE       FIXED VALUES
   1    11    0.000000    0.000000
  13    11    0.000000    0.000000
  14    11    0.000000    0.300000
  15    11    0.000000    0.300000
  16    11    0.000000    0.300000


MATERIAL PROPERTIES
NUMBER          PROPERTIES
   1          0.230000E 01   0.300000E 00
    NUMBER OF TRACTION B.CS =     6
    TRACTION BOUNDARY CONDITIONS
   1    3    0.0000    0.0000    0.0000    0.0000    0.0000    0.0000
   2    3    0.0000    0.0000    0.0000    0.0000    0.0000    0.0000
   3    3    0.0000   -1.0000    0.0000   -1.0000    0.0000   -1.0000
   4    3    0.0000   -1.0000    0.0000   -1.0000    0.0000   -1.0000
   5    3    0.0000    0.0000    0.0000    0.0000    0.0000    0.0000
   6    3    0.0000    0.0000    0.0000    0.0000    0.0000    0.0000


   NODE    DISPLACEMENTS                    TRACTIONS
    1    0.0000E 00   0.0000E 00    0.0000E 00   0.0000E 00
    2   -0.2043E 01  -0.1866E 01    0.0000E 00   0.0000E 00
    3   -0.3481E 01  -0.5139E 01    0.0000E 00   0.0000E 00
    4   -0.4340E 01  -0.9558E 01    0.0000E 00   0.0000E 00
    5   -0.4681E 01  -0.1467E 02    0.0000E 00   0.0000E 00
    6   -0.2218E 01  -0.1458E 02    0.0000E 00  -0.1000E 01
    7    0.1516E-04  -0.1451E 02    0.0000E 00  -0.1000E 01
    8    0.2218E 01  -0.1458E 02    0.0000E 00  -0.1000E 01
    9    0.4681E 01  -0.1467E 02    0.0000E 00  -0.1000E 01
   10    0.4340E 01  -0.9558E 01    0.0000E 00  -0.1000E 01
   11    0.3481E 01  -0.5139E 01    0.0000E 00   0.0000E 00
   12    0.2043E 01  -0.1866E 01    0.0000E 00   0.0000E 00
```

```
13   0.0000E 00   0.0000E 00      0.0000E 00   0.0000E 00
14   0.0000E 00   0.0000E 00      0.0000E 00   0.0000E 00
15   0.0000E 00   0.0000E 00      0.0000E 00   0.0000E 00
16   0.0000E 00   0.0000E 00     -0.1417E 02   0.4980E 01
17                               -0.4618E 01  -0.3146E-01
18                               -0.9276E-06   0.5047E 00
19                                0.4618E 01  -0.3147E-01
20                                0.1417E 02   0.4980E 01
```

NO OF INTERIOR PTS=   7
INTERIOR DISPLACEMENTS

```
     ID     X        Y        UX        UY

      0    0.250    0.500  0.3735E-05 -0.4144E 00

      0    0.500    0.500  0.5765E-05 -0.1483E 01

      0    0.750    0.500  0.6856E-05 -0.3010E 01

      0    1.000    0.500  0.9185E-05 -0.4894E 01

      0    1.250    0.500  0.8330E-05 -0.7061E 01

      0    1.500    0.500  0.7791E-05 -0.9443E 01

      0    1.750    0.500  0.5176E-05 -0.1197E 02
```

```
00000100          PROGRAM IBEM
00000200 C        INDIRECT BEM : QUADRATIC BOUNDARY SEGMENTS. DISCONTINUOUS
00000300 C        SOURCES AT CORNERS AND LOADING DISCONTINUITIES
00000400          DIMENSION TITLE(12)
00000500          COMMON/DBUG/JBUG(80)
00000600          COMMON/GSPWT/GSP(3),GSPW(3)
00000700          COMMON/CONTRO/NPOIN,NNODE,NDOFN,NTYPE,NGAUS,NPROP,NMATS,NVFIX
00000800         ..NPROB,NTCOMP,NCALCD
00000900          COMMON/LGDATA/COORD(60,2),PROPS(1,5),NOFIX(60),IFPRE(60,2),
00001000         .MATNO(1),PRESC(60,2)
00001100          COMMON/BMATS/NBDATA(1,4),LNODBI(1,60),NBIREG,XCRDBI(61),
00001200         .YCRDBI(61),CNRTOL,INTYP(1,61)
00001300          COMMON/MATCON/RNU,CONS1,CONS2,CONS4,DMAX,NGP,NGGP,GP(12,12)
00001400         .         ,GPW(12,12),ISUPS,NBB,NBBM,CONS3,MAXSZ,NBB2,NBB2P
00001500         .  ,NCOMP,NCLPT,RMAXL,NUNIT,NOEQ,GLP(12,12),GLPW(12,12)
00001600          COMMON/SOLV/ RLHS(122,123),UFIX(120),RHS(122),TFIX(120)
00001700          COMMON /LDISK/ NDISKA,NDISKB
00001800          NDISKA=17
00001900          NDISKB=18
00002000          MAXSZ=122
00002100 C **** INCLUDE FOLLOWING LINES FROM PROGRAM DBEM ****
00002200 C ****   LINE NO 2100 TO 4000 ****
00002300          STOP
00002400          END
00002500          SUBROUTINE INPUT
00002600 C **** INCLUDE COMMON BLOCKS ****
00002700 C **** MATCON CONTRO LGDATA BMATS ****
00002800 C*** READ THE FIRST DATA CARD, AND ECHO IT IMMEDIATELY.
00002900          READ(5,900) NVFIX,NTYPE,NOEQ
00003000    900   FORMAT(16I5)
00003100          WRITE(6,905) NOEQ,NVFIX,NTYPE
00003200    905   FORMAT(//8H NOEQ  =,I4,4X,7H NVFIX=,I4,4X,8H NTYPE =,I4)
00003300          NDIME=2
00003400          NDOFN=2
00003500          NMATS=1
00003600          NUMBI=1
00003700          NUMEL=1
00003800          NPROP=2
00003900          NREG=1
00004000          READ(5,900) (NBDATA(NUMBI,J),J=1,3)
00004100          WRITE(6,900)(NBDATA(NUMBI,J),J=1,3)
00004200          NBB=NBDATA(NUMBI,1)
```

```
00004300           NPOIN=NBB
00004400           NQSEG=NBB/2
00004500           MATNO(1)=1
00004600           DO 10 I=1,NBB
00004700    10     LNODBI(NUMBI,I)=I
00004800           READ(5,900) (INTYP(NUMBI,I),I=1,NBB)
00004900           WRITE(6,900) (INTYP(NUMBI,I),I=1,NBB)
00005000  C*** ZERO ALL THE NODAL COORDINATES, PRIOR TO READING SOME OF THEM.
00005100           DO 20 IPOIN=1,NPOIN
00005200           DO 20 IDIME=1,NDIME
00005300    20     COORD(IPOIN,IDIME)=0.0
00005400  C*** READ SOME NODAL COORDINATES, FINISHING WITH THE LAST NODE OF ALL
00005500           WRITE(6,920)
00005600   920     FORMAT(//25H    NODAL POINT COORDINATES)
00005700           WRITE(6,925)
00005800   925     FORMAT(6H   NODE,7X,1HX,9X,1HY)
00005900    30     READ(5,930) IPOIN,(COORD(IPOIN,IDIME),IDIME=1,NDIME)
00006000   930     FORMAT(I5,5F10.5)
00006100           IF(IPOIN.NE.NPOIN) GO TO 30
00006200  C*** INTERPOLATE COORDINATES OF MID-SIDE NODES
00006300           CALL NODEXY
00006400    40     CONTINUE
00006500           DO 50 IPOIN=1,NPOIN
00006600    50     WRITE(6,935) IPOIN,(COORD(IPOIN,IDIME),IDIME=1,NDIME)
00006700   935     FORMAT(1X,I5,3F10.3)
00006800  C        BOUNDARY   CONDITIONS
00006900           WRITE(6,940)
00007000   940     FORMAT(//20H BOUNDARY CONDITIONS)
00007100           WRITE(6,945)
00007200   945     FORMAT(5H NODE,1X,4HCODE,6X,12H        VALUES)
00007300           DO 60 IVFIX=1,NVFIX
00007400           READ(5,950) NOFIX(IVFIX),(IFPRE(IVFIX,IDOFN),IDOFN=1,NDOFN),
00007500          . (PRESC(IVFIX,IDOFN),IDOFN=1,NDOFN)
00007600    60     WRITE(6,950) NOFIX(IVFIX),(IFPRE(IVFIX,IDOFN),IDOFN=1,NDOFN),
00007700          . (PRESC(IVFIX,IDOFN),IDOFN=1,NDOFN)
00007800   950     FORMAT(1X,I4,3X,2I1,2F10.6)
00007900  C*** READ THE AVAILABLE SELECTION OF ELEMENT PROPERTIES.
00008000           WRITE(6,960)
00008100   960     FORMAT(//21H   MATERIAL PROPERTIES)
00008200           WRITE(6,965)
00008300   965     FORMAT(8H   NUMBER,7X,10HPROPERTIES)
00008400           READ(5,930)   NUMAT,(PROPS(NUMAT,IPROP),IPROP=1,NPROP)
00008500           WRITE(6,930)  NUMAT,(PROPS(NUMAT,IPROP),IPROP=1,NPROP)
00008600           RETURN
00008700           END
00008800           SUBROUTINE NODEXY
00008900  C **** INCLUDE COMMON BLOCKS ****
00009000  C **** LGDATA CONTRO BMATS ****
00009100  C **** INCLUDE FOLLOWING STATEMENTS FROM PROGRAM DBEM ****
00009200  C **** LINES 14300 TO 16000
00009300           RETURN
00009400           END
00009500           SUBROUTINE BELMAT(M,NBBI)
00009600           DIMENSION AMT(2,120),BMT(2,120),ISING(120),SCOL(3)
00009700  C **** INCLUDE COMMON BLOCKS ****
00009800  C **** MATCON CONTRO LGDATA BMATS LDISK DBUG ****
00009900           SHN1(S)= 2.0*(S-0.5)*(S-1.0)
00010000           SHN2(S)=-4.0*S*(S-1.0)
00010100           SHN3(S)= 2.0*S*(S-0.5)
00010200           DSH1(S)=4.*S-3.
00010300           DSH2(S)=-8.*S+4.
00010400           DSH3(S)=4.*S-1.
00010500           NBB=NBDATA(NBBI,1)
00010600           ISUPS=NBDATA(NBBI,2)
00010700           NGP=NBDATA(NBBI,3)
00010800           NBBM=NBB
00010900  C  SET UP MATERIAL CONSTANTS........................
00011000           MT=MATNO(M)
00011100           PI=3.14159
```

```
00011200          IF(NTYPE.EQ.1)  GOTO 1
00011300          RNU=PROPS(MT,2)
00011400          RNU=RNU/(1.0+RNU)
00011500          ES=PROPS(MT,1)
00011600          ES=ES/(1.0-RNU*RNU)
00011700          PROPS(MT,2)=RNU
00011800          PROPS(MT,1)=ES
00011900     1    CONTINUE
00012000          RNU=PROPS(MT,2)
00012100          CONS1=-1.0/(8.0*PI*(1.0-RNU))
00012200          CONS2=3.0-4.0*RNU
00012300          CONS3=-1.0/4.0/PI/(1.0-RNU)
00012400          CONS4=1.0-2.0*RNU
00012500          J=0
00012600          DO 2 I=1,NBB
00012700          J=J+1
00012800          NUMC=LNODBI(NBBI,I)
00012900          XCRDBI(J)=COORD(NUMC,1)
00013000          YCRDBI(J)=COORD(NUMC,2)
00013100     2    CONTINUE
00013200          NBBP=NBB+1
00013300          XCRDBI(NBBP)=XCRDBI(1)
00013400          YCRDBI(NBBP)=YCRDBI(1)
00013500 C FIND MAX ELEMENT LENGTH .........
00013600          RMAXL=0.0
00013700          NBBQ=NBB/2
00013800          DO 3 I=1,NBBQ
00013900     3    INTYP(NBBI,I)=0
00014000          DO 5 I=1,NBBQ
00014100          I1=2*I-1
00014200          I2=2*I+1
00014300          IP=I1+1
00014400          IM=I1-1
00014500          IF(I.EQ.1)IM=NBB
00014600          IF(I.EQ.NBBQ) I2=1
00014700          RL=SQRT((XCRDBI(I2)-XCRDBI(I1))**2 +
00014800        .          (YCRDBI(I2)-YCRDBI(I1))**2)
00014900          IF(RL.GT.RMAXL) RMAXL=RL
00015000          XA=XCRDBI(I1)-XCRDBI(IM)
00015100          YA=YCRDBI(I1)-YCRDBI(IM)
00015200          XB=XCRDBI(IP)-XCRDBI(I1)
00015300          YB=YCRDBI(IP)-YCRDBI(I1)
00015400          DOT=XA*XB+YA*YB
00015500          RA=SQRT(XA*XA+YA*YA)
00015600          RB=SQRT(XB*XB+YB*YB)
00015700          DOT=DOT/RA/RB
00015800          IF(DOT.LT.0.86666) INTYP(NBBI,I)=1
00015900 C        IF ANGLE.LT.30 DEG.  NO CORNERS
00016000     5    CONTINUE
00016100          IF(JBUG(5).EQ.0)  GOTO  77
00016200          WRITE(6,2004)(XCRDBI(I),YCRDBI(I),I=1,NBBP)
00016300     2004 FORMAT(1X,10E12.4)
00016400     77   CONTINUE
00016500 C PERFORM BOUNDARY COLLOCATION
00016600          NCLPT=0
00016700 C        FIND NO. OF COMP. IN  A   AND   B  MATRICES
00016800          NCOMP=NBB
00016900          DO 10 I =1,NBBQ
00017000     10   NCOMP=NCOMP+INTYP(NBBI,I)
00017100          NCOMP2=2*NCOMP
00017200          DO 70 ISEG=1,NBBQ
00017300 C SET UP 3 NODAL COORDS ON ASEG............
00017400          ISA=2*ISEG-1
00017500          ISB=ISA+1
00017600          ISC=ISB+1
00017700          IF(ISEG.EQ.NBBQ) ISC=1
00017800          XA=XCRDBI(ISA)
00017900          YA=YCRDBI(ISA)
00018000          XB=XCRDBI(ISB)
```

```
00018100          YB=YCRDBI(ISB)
00018200          XC=XCRDBI(ISC)
00018300          YC=YCRDBI(ISC)
00018400 C        SET UP COLLOCATION POINTS
00018500          ISEGP=ISEG+1
00018600          IF(ISEG.EQ.NBBQ) ISEGP=1
00018700          ITYP1=INTYP(NBBI,ISEG)
00018800          ITYP2=INTYP(NBBI,ISEGP)
00018900          IF(ITYP1.NE.1) GO TO 15
00019000          NCOL=3
00019100          SCOL(1)=1.0/6.0
00019200          SCOL(2)=0.5
00019300          SCOL(3)=1.0
00019400          IF(ITYP2.EQ.1) SCOL(3)=5.0/6.0
00019500          GO TO 16
00019600       15 NCOL=2
00019700          SCOL(1)=0.5
00019800          SCOL(2)=1.0
00019900          IF(ITYP2.EQ.1) SCOL(2)=5.0/6.0
00020000       16 CONTINUE
00020100 C   NULL SINGULARITY FLAG................
00020200          DO 25 I=1,NBBQ
00020300       25 ISING(I)=0
00020400          ISING(ISEG)=1
00020500          DO 60 IC=1,NCOL
00020600 C   FIND A AND B MATRICES..........
00020700          XSI=SCOL(IC)
00020800          XPT=SHN1(XSI)*XA+SHN2(XSI)*XB+SHN3(XSI)*XC
00020900          YPT=SHN1(XSI)*YA+SHN2(XSI)*YB+SHN3(XSI)*YC
00021000          DXN=DSH1(XSI)*XA+DSH2(XSI)*XB+DSH3(XSI)*XC
00021100          DYN=DSH1(XSI)*YA+DSH2(XSI)*YB+DSH3(XSI)*YC
00021200          RPN=SQRT(DXN*DXN+DYN*DYN)
00021300          XNP=DYN/RPN
00021400          YNP=-DXN/RPN
00021500          IF(JBUG(10).EQ.0)  GOTO 40
00021600          WRITE(6,41) XSI,XPT,YPT,XNP,YNP
00021700       41 FORMAT(6X,'FIELD PTS',5E12.4)
00021800       40 CONTINUE
00021900          CALL INTAB(XPT,YPT,XSI,ISING,AMT,BMT,XNP,YNP)
00022000 C UPDATE TOTAL NO OF COLLOCATIONS
00022100          NCLPT=NCLPT+1
00022200 C   WRITE A AND B MATRICES TO DISK
00022300          NBB2=NCOMP2
00022400          NBB3=NCOMP2
00022500          DO 50 I=1,2
00022600          WRITE(NDISKA)(AMT(I,J),J=1,NBB2)
00022700       50 WRITE(NDISKB)(BMT(I,J),J=1,NBB3)
00022800       60 CONTINUE
00022900          IF(JBUG(15).EQ.0)  GOTO 65
00023000          WRITE(6,1010)
00023100     1010 FORMAT(6X,' A AND B MATRICES ')
00023200          DO 61 I=1,2
00023300       61 WRITE(6,2004)(AMT(I,J),J=1,NBB2)
00023400          DO 62 I=1,2
00023500       62 WRITE(6,2004)(BMT(I,J),J=1,NBB3)
00023600       70 CONTINUE
00023700       65 CONTINUE
00023800          RETURN
00023900          END
00024000          SUBROUTINE INTAB(XPT,YPT,XSI,ISING,AMT,BMT,XNP,YNP)
00024100          DIMENSION  AMT(2,120),BMT(2,120),ISING(120)
00024200          DIMENSION SGP(12),GWPJ(3),ISNGSG(3),AII(3),BII(3)
00024300          DIMENSION SLGP(12),GLWPJ(3),AI(3),BI(3),S4P(3),AAI(3),BBI(3)
00024400          DIMENSION AIII(3),BIII(3)
00024500 C   *** INCLUDE COMMON BLOCKS ****
00024600 C   **** DEBUG GSPWT BMATS MATCON ****
00024700          NBBI=1
00024800          RAT=0.0
00024900          SLNG=0.0
```

```
00025000            SLNGS=0.0
00025100            DO 21 I=1,2
00025200            DO 21 J=1,120
00025300    20 AMT(I,J)=0.0
00025400    21 BMT(I,J)=0.0
00025500 C    **** INCLUDE LINES 30400 TO 32000 FROM DBEM ****
00025600 C    *******************************
00025700 C    **** INCLUDE FOLLOWING LINES FROM DBEM PROGRAM ****
00025800 C    **** LINES 33200 TO 63500 ****
00025900 C COORDS OF VECTOR BETWEEN GUASS AND FIELD PTS
00026000            XPG == XGP +XPT
00026100            YPG =-YGP +YPT
00026200            XLPG=-XLGP+XPT
00026300            YLPG=-YLGP+YPT
00026400 C OUTWARD NORMAL AT FIELD PT
00026500            RPG =SQRT(XPG*XPG +YPG*YPG)
00026600            XNGP=XNP
00026700            YNGP=YNP
00026800            RPG2=1.0/RPG/RPG
00026900            XRP = XPG*XPG*RPG2
00027000            YRP = YPG*YPG*RPG2
00027100            XYRP= XPG*YPG*RPG2
00027200            RNXY= XNGP*XPG + YNGP*YPG
00027300 C INTEGRATION KERNELS
00027400            GLPWJ=GLPW(NGP,IG)*RLJAC*TRANSL
00027500            GLWPJ(1)=SHN1(SL)*GLPWJ
00027600            GLWPJ(2)=SHN2(SL)*GLPWJ
00027700            GLWPJ(3)=SHN3(SL)*GLPWJ
00027800            GPWJ=GPW(NGP,IG)*RJAC*TRANSL
00027900            GWPJ(1)=SHN1(S)*GPWJ
00028000            GWPJ(2)=SHN2(S)*GPWJ
00028100            GWPJ(3)=SHN3(S)*GPWJ
00028200            RTERM = CONS1*CONS2*ALOG(RPG)
00028300            DO 80 K=1,3
00028400            K2=2*K + IS44 + MPOS
00028500            KM=K2-1
00028600            RTERMG=CONS1*CONS2*GLWPJ(K)
00028700            BMT(1,KM)=BMT(1,KM)-GWPJ(K)*CONS1*XRP
00028800            BMT(2,K2)=BMT(2,K2)-GWPJ(K)*CONS1*YRP
00028900            IF(ISING(IS).EQ.1.AND.ISNGSG(ISB).EQ.1) GOTO 71
00029000            IF(ISING(ISM).EQ.1.AND.ISB.EQ.1.AND.ABS(XSI-1.).LT.0.0001) GOTO 71
00029100            IF(ISING(ISP).EQ.1.AND.ISB.EQ.3.AND.ABS(XSI).LT.0.0001) GOTO 71
00029200            GOTO 72
00029300    71   CONTINUE
00029400 C SINGULAR INTEGRATION OF LOG TERMS
00029500            RBI=BI(ISB)*RABC
00029600            IF(ISING(IS).EQ.1) GOTO 73
00029700            IF(ISB.EQ.1) RBI=ABS(BIII(1))*SLNG
00029800            IF(ISB.EQ.3) RBI=ABS(BIII(3))*SLNGS
00029900    73   CONTINUE
00030000            STERM=-RTERMG+CONS1*CONS2*ALOG(RBI)*GWPJ(K)
00030100            BMT(1,KM)=BMT(1,KM)+STERM
00030200            BMT(2,K2)=BMT(2,K2)+STERM
00030300            GOTO 76
00030400    72   CONTINUE
00030500            BMT(1,KM)=BMT(1,KM)+GWPJ(K)*RTERM
00030600            BMT(2,K2)=BMT(2,K2)+GWPJ(K)*RTERM
00030700    76   CONTINUE
00030800            BMT(1,K2)=BMT(1,K2)   -GWPJ(K)*CONS1*XYRP
00030900    80   BMT(2,KM)=BMT(2,KM)
00031000            GWPJ(1)=SHN1(S)*GPWJ
00031100            GWPJ(2)=SHN2(S)*GPWJ
00031200            GWPJ(3)=SHN3(S)*GPWJ
00031300            RNYX= YNGP*XPG -XNGP*YPG
00031400            DO 100 K=1,3
00031500            K2=2*K + IS44 + MPOS
00031600            KM=K2-1
00031700            GTERM=GWPJ(K)*CONS3*RPG2
00031800            AMT(1,KM)=AMT(1,KM)+ GTERM*(CONS4 +2.0*XRP)*RNXY
```

```
00031900          AMT(2,K2)=AMT(2,K2)+ GTERM*(CONS4 +2.0*YRP)*RNXY
00032000          AMT(1,K2)=AMT(1,K2)+ GTERM*(CONS4*RNYX +2.0*XYRP*RNXY)
00032100      100 AMT(2,KM)=AMT (2,KM)+GTERM*(-CONS4*RNYX+2.*XYRP*RNXY)
00032200      130 CONTINUE
00032300      150 CONTINUE
00032400 C        END OF INTEGRATION
00032500      200 CONTINUE
00032600 C  CALCULATION OF DIAGONAL TERMS OF MATRIX AMT
00032700          IF(JSING.EQ.0) GOTO 253
00032800          NCOMP=K2
00032900          SUM11=-0.5
00033000          SUM22=-0.5
00033100          SUM12=0.0
00033200          SUM21=0.0
00033300          MMPOS=0
00033400          DO 205 I=1,JSING
00033500          ITP=INTYP(NREG,I)
00033600          IF(I.NE.1) MMPOS=MMPOS +2*ITP
00033700      205 CONTINUE
00033800          JS4=4*(JSING-1) + MMPOS
00033900          SHP(1)=SHN1(XSI)
00034000          SHP(2)=SHN2(XSI)
00034100          SHP(3)=SHN3(XSI)
00034200          DO 250 I=1,3
00034300          I2=2*I+JS4
00034400          IM=I2-1
00034500          AMT(1,IM)=AMT(1,IM)-SUM11*SHP(I)
00034600          AMT(2,I2)=AMT(2,I2)-SUM22*SHP(I)
00034700          AMT(2,IM)=AMT(2,IM)-SUM21*SHP(I)
00034800          AMT(1,I2)=AMT(1,I2)-SUM12*SHP(I)
00034900      250 CONTINUE
00035000      253 CONTINUE
00035100          IF(JBUG(10).EQ.0) GOTO 300
00035200          WRITE(6,1000)
00035300     1000 FORMAT(6X,'A AND B MATRICES')
00035400          WRITE(6,1010) ((AMT(I,J),J=1,NCOMP),I=1,2)
00035500          WRITE(6,1010) ((BMT(I,J),J=1,NCOMP),I=1,2)
00035600      300 CONTINUE
00035700 C    **** INCLUDE FOLLOWING STATEMENTS FROM PROGRAM DBEM ****
00035800 C    **** LINE NOS 75400 TO 76000 ****
00035900          RETURN
00036000          END
00036100          SUBROUTINE SIME( A,M,N,N1 )
00036200          DIMENSION A(M,123)
00036300 C    **** INCLUDE FOLLOWING LINES FROM PROGRAM DBEM ****
00036400 C    **** LINES 76500 TO 78500 ****
00036500          STOP
00036600          END
00036700          SUBROUTINE  BISOLV
00036800          DIMENSION AMT(2,120),BMT(2,120)
00036900 C    *** INCLUDE COMMON BLOCKS ****
00037000 C    *** DBUG CONTRO LGDATA BMATS MATCON SOLV LDISK ****
00037100          GMOD=0.5*PROPS(1,1)/(1.0+PROPS(1,2))
00037200          NREG=1
00037300          NBB=NBDATA(NREG,1)
00037400          NBBQ=NBB/2
00037500          NBBS=NBB
00037600 C        FIND THE NUMBER OF SOURCES
00037700          DO 10 I =1,NBBQ
00037800       10 NBBS=NBBS+INTYP(NREG,I)
00037900          NBBS2=NBBS*2
00038000          NBB2=NBBS2
00038100          NBB3=NBBS2
00038200          IRANG=1
00038300 C LOOP OVER COLLOCATION PTS
00038400 C AND SET UP MATRIX EQNS
00038500          REWIND NDISKA
00038600          REWIND NDISKB
00038700          IDST=0
```

```
00038800 C LOOP OVER SOURCE NODES ..........
00038900         DO 200 ICP=1,NBBS
00039000         ICP2=2*(ICP-1)
00039100 C   READ A AND B MATRICES
00039200         DO 110 I=1,2
00039300         READ(NDISKA)(AMT(I,J),J=1,NBB2)
00039400     110 READ(NDISKB)(BMT(I,J),J=1,NBB3)
00039500 C SET UP EQNS AND RHSIDES.........
00039600 C LOOP OVER X,Y COMPONENTS    .........
00039700         DO 180 IXY=1,2
00039800         IDST=IDST+1
00039900         RLHS(IDST,NBB3+1)=0.0
00040000         RLHS(IDST,NBB3+2)=0.0
00040100         IF(IFPRE(ICP,IXY).EQ.2) GOTO 150
00040200         IF(IFPRE(ICP,IXY).NE.1) WRITE(6,4000)
00040300    4000 FORMAT(1X,'ERROR NODE BC FLAG HAS INCORRECT VAL')
00040400 C DISP CONDITION .........
00040500         DO 130 J=1,NBB3
00040600     130 RLHS(IDST,J)=BMT(IXY,J)
00040700         NBB3XY=NBB3+IXY
00040800         RLHS(IDST,NBB3XY)=1.0
00040900 C SET UP RHS........
00041000         RHS(IDST)=PRESC(ICP,IXY)*GMOD
00041100         GOTO 180
00041200     150 CONTINUE
00041300 C   TRACTION B . C.............
00041400         DO 170 J=1,NBB3
00041500     170 RLHS(IDST,J)=AMT(IXY,J)
00041600         RHS(IDST)=PRESC(ICP,IXY)
00041700     180 CONTINUE
00041800     200 CONTINUE
00041900 C IMPOSE INTGRAL OF SOURCES =0.0.......
00042000         JP=IDST+2
00042100         DO 202 I=1,2
00042200         IIDST=I+IDST
00042300         DO 202 J=1,JP
00042400     202 RLHS(IIDST,J)=0.0
00042500 C N.B THIS CODE FOR ST LINE SEGMENTS ...........
00042600         JPOS=0
00042700         DO 204 IS=1,NBBQ
00042800         IS1=2*IS-1
00042900         IS2=IS1+1
00043000         IS3=IS2+1
00043100         IF(IS.EQ.NBBQ) IS3=1
00043200         XA=XCRDBI(IS1)
00043300         YA=YCRDBI(IS1)
00043400         XC=XCRDBI(IS3)
00043500         YC=YCRDBI(IS3)
00043600         RL=SQRT((XC-XA)**2+(YC-YA)**2)
00043700         IS6=4*(IS-1)
00043800         IF(IS.NE.1) JPOS=JPOS+2*INTYP(NREG,IS)
00043900         DO 203 I=1,2
00044000         I1=I+JPOS+IS6
00044100         I2=I1+2
00044200         I3=I2+2
00044300         IDSTII=IDST+I
00044400         RLHS(IDSTII,I1)=RLHS(IDSTII,I1)+RL/6.0
00044500         RLHS(IDSTII,I2)=RLHS(IDSTII,I2)+2.0*RL/3.0
00044600     203 RLHS(IDSTII,I3)=RLHS(IDSTII,I3)+RL/6.0
00044700     204 CONTINUE
00044800         RHS(IDST+1)=0.0
00044900         RHS(IDST+2)=0.0
00045000         IDST=IDST+2
00045100         DO 210 I=1,IDST
00045200         RLHS(I,IDST+1)=RHS(I)
00045300     210 CONTINUE
00045400         IP=IDST+1
00045500         NCLPT2=2*NBBS+2
00045600         IF(JBUG(20).EQ.0)  GOTO  215
```

```
00045700          WRITE(6,1030)
00045800     1030 FORMAT(6X,'FINAL SYSTEM OF EQNS',/)
00045900          DO 211 I=1,NCLPT2
00046000      211 WRITE(6,1028) (RLHS(I,J),J=1,IP)
00046100     1028 FORMAT(1X,12E10.3)
00046200      215 CONTINUE
00046300          IF(IDST.NE.NCLPT2) WRITE(6,1000) IDST,NCLPT2
00046400     1000 FORMAT(1X,'***ERROR EQN SET NOT SQUARE***- NO OF COLS
00046500        . =',2I5)
00046600 C SOLVE LINEAR EQNS
00046700 C ROW INTERCHANGE..........
00046800          RMAX=0.0
00046900 C SKIP ROW INTERCHANGES ................
00047000          IF(IRANG.EQ.0) GOTO 251
00047100          DO 240 J=1,IDST
00047200          IROW=J
00047300          Z=0.0
00047400          DO 220 K=J,NCLPT2
00047500          IF(ABS(RLHS(K,J)).LT.Z) GOTO 220
00047600          Z=ABS(RLHS(K,J))
00047700          IROW=K
00047800      220 CONTINUE
00047900          IF(Z.GT.RMAX) RMAX=Z
00048000          DO 230 K=1,IP
00048100          VAR=RLHS(J,K)
00048200          RLHS(J,K)=RLHS(IROW,K)
00048300      230 RLHS(IROW,K)=VAR
00048400      240 CONTINUE
00048500          IF(JBUG(20).EQ.0) GOTO 251
00048600          WRITE(6,1031)
00048700     1031 FORMAT(6X,'REARRANGED MATRIX ')
00048800          DO 250 I=1,NCLPT2
00048900      250 WRITE(6,1028) (RLHS(I,J),J=1,IP)
00049000      251 CONTINUE
00049100          MAXSZ=122
00049200          CALL SIME(RLHS,MAXSZ,IDST,1)
00049300 C  WRITE RESULTS
00049400          WRITE(6,1010)
00049500     1010 FORMAT(1X,'BOUNDARY FICTITIOUS        TRACTIONS')
00049600          INODE=IDST/2
00049700          DO 401 JK=1,INODE
00049800          IJ=(JK-1)*2
00049900          WRITE(6,1020) JK,RLHS(IJ+1,IDST+1),RLHS(IJ+2,IDST+1)
00050000      401 CONTINUE
00050100     1020 FORMAT(I5,2E12.4)
00050200 C FIND BOUNDARY DISPS AND TRACTIONS
00050300          REWIND NDISKB
00050400          REWIND NDISKA
00050500          DO 270 I=1,IDST
00050600          UFIX(I)=0.0
00050700          TFIX(I)=0.0
00050800      270 RHS(I)=RLHS(I,IDST+1)
00050900          DO 300 IS=1,NBBS
00051000          IS2=2*(IS-1)
00051100          DO 280 I=1,2
00051200          READ(NDISKA)(AMT(I,J),J=1,NBB3)
00051300      280 READ(NDISKB)(BMT(I,J),J=1,NBB3)
00051400          DO 291 I=1,2
00051500          IP=IS2+I
00051600          IQ=IDST-2+I
00051700          DO 290 J=1,NBB3
00051800          TFIX(IP)=TFIX(IP)+AMT(I,J)*RHS(J)
00051900      290 UFIX(IP)=UFIX(IP)+BMT(I,J)*RHS(J)
00052000      291 UFIX(IP)=UFIX(IP)+RHS(IQ)
00052100      300 CONTINUE
00052200          DO 311  I=1,NBB3
00052300      311 UFIX(I)=UFIX(I)/GMOD
00052400          WRITE(6,1032)
00052500     1032 FORMAT(1X,' BOUNDARY DISPLACEMENTS')
```

```
00052600          INODE=NBB3/2
00052700          DO 402 JK=1,INODE
00052800          IJ=(JK-1)*2
00052900          WRITE(6,1020) JK,UFIX(IJ+1),UFIX(IJ+2)
00053000    402   CONTINUE
00053100          WRITE(6,1040)
00053200   1040   FORMAT(1X,'    BOUNDARY TRACTIONS')
00053300          DO 403 JK=1,INODE
00053400          IJ=(JK-1)*2
00053500          WRITE(6,1020) JK,TFIX(IJ+1),TFIX(IJ+2)
00053600    403   CONTINUE
00053700          NUNIT=NBB3
00053800          RETURN
00053900          END
00054000          SUBROUTINE BIPTS
00054100          DIMENSION DPMAT(2,120),TRMAT(2,120),UIN(2),TIN(2)
00054200          DIMENSION SPT(3),ISING(40)
00054300   C    **** INCLUDE FOLLOWING COMMON BLOCKS ****
00054400   C      **** DBUG SOLV CONTRO LGDATA BMATS MATCON ****
00054500   C THIS ROUTINE FINDS DISPS AND TRACTIONS IN AND ON
00054600   C A BI REGION.............
00054700          INBI=1
00054800          NBB=NBDATA(INBI,1)
00054900          ISUPS=NBDATA(INBI,2)
00055000          NGP=NBDATA(INBI,3)
00055100          NOSEG=NBB/2
00055200          GMOD=0.5*PROPS(1,1)/(1.0+PROPS(1,2))
00055300   C NUMBER OF FICTITIOUS TRACTION  COMPONENTS
00055400          NBB2=NUNIT
00055500          NBB3=NUNIT
00055600   C CALCULATION FOR INTERIOR PTS........
00055700          READ(5,1020) NOINPT
00055800          WRITE(6,1025) NOINPT
00055900   1025   FORMAT(20H NO OF INTERIOR PTS=,I5)
00056000          IF(NOINPT.EQ.0) GOTO 400
00056100          WRITE(6,1028)
00056200   1028   FORMAT(1X,'INTERIOR DISPLACEMENTS AND STRESSES')
00056300          DO 250 IP=1,NOINPT
00056400          XSI=0.0
00056500          READ(5,1030) JSING,XIN,YIN,XNP,YNP,XSI
00056600          WRITE(6,1029)
00056700   1029   FORMAT(5X,3H ID,5X,2H X,7X,2H Y,6X,3H UX,6X,3H UY
00056800         .,6X,3H TX,6X,3H TY/)
00056900          WRITE(6,1030) JSING,XIN,YIN
00057000   1030   FORMAT(I5,5F10.3)
00057100   C  SET UP COORDS OF BI REGION......
00057200          J=0
00057300          DO 80 I=1,NBB
00057400          J=J+1
00057500          NUMC=LNODBI(INBI,I)
00057600          XCRDBI(J)=COORD(NUMC,1)
00057700          YCRDBI(J)=COORD(NUMC,2)
00057800    80   CONTINUE
00057900          NBBP=NBB+1
00058000          XCRDBI(NBBP)=XCRDBI(1)
00058100          YCRDBI(NBBP)=YCRDBI(1)
00058200          DO 85 I=1,NOSEG
00058300    85   ISING(I)=0
00058400          IF(JSING.NE.0) ISING(JSING)=1
00058500          CALL INTAB(XIN,YIN,XSI,ISING,TRMAT,DPMAT,XNP,YNP)
00058600          IF(JBUG(33).EQ.0) GOTO 95
00058700          WRITE(6,3005)
00058800          DO 91 I=1,2
00058900    91   WRITE(6,2030)(DPMAT(I,J),J=1,NBB2)
00059000          DO 92 I=1,2
00059100    92   WRITE(6,2030)(TRMAT(I,J),J=1,NBB3)
00059200    95   CONTINUE
00059300          DO 230 I=1,2
00059400          UIN(I)=0.0
```

```
00059500          TIN(I)=0.0
00059600          DO 225 J=1,NBB2
00059700     225  UIN(I)=UIN(I)+DPMAT(I,J)*RHS(J)
00059800          DO 226 J=1,NBB3
00059900     226  TIN(I)=TIN(I)+TRMAT(I,J)*RHS(J)
00060000     230  CONTINUE
00060100          UIN(1)=UIN(1)+RHS(NBB2+1)
00060200          UIN(2)=UIN(2)+RHS(NBB2+2)
00060300          DO 244 K=1,2
00060400     244  UIN(K)=UIN(K)/GMOD
00060500          WRITE(6,1050)          (UIN(L),L=1,2),(TIN(L),L=1,2)
00060600     250  CONTINUE
00060700     400  CONTINUE
00060800          RETURN
00060900    1020  FORMAT(16I5)
00061000    3005  FORMAT(6X,'A AND B MATRICES IN BIPTS ')
00061100    2030  FORMAT(1X,8E12.4)
00061200    1050  FORMAT(25X,4E12.4)
00061300          END
00061400  C    **** SUBROUTINE GAUSS IS IDENTICAL TO THAT OF DBEM ****
```

# 15-10 REFERENCES

1. Lachat, J. C. (1975) 'Further developments of the boundary integral technique for elasto-statics', Ph.D. thesis, Southampton University.

2. Lachat, J. C., and Watson, J. O. (1976) 'Effective numerical treatment of boundary integral equations: a formulation for three-dimensional elastostatics', *Int. J. Num. Meth. in Engng*, **10**, 991–1005.

3. Lachat, J. C., and Watson, J. O. (1975) 'Progress in the use of boundary integral equations, illustrated by examples', *Comp. Meth. in Appl. Mech. Engng*, **10**, 273–289.

4. Lachat, J. C., and Watson, J. O. (1975) 'A second generation of boundary integral equation programs for three-dimensional elastic analysis', in T. A. Cruse and F. J. Rizzo (eds), *Proc. ASME Conf. on Boundary Integral Equation Meth.*, AMD-11, ASME, New York.

5. Cruse, T. A. (1974) 'An improved boundary integral equation method for three-dimensional elastic stress analysis', *Int. J. Computers and Structs*, **4**, 741–757.

6. Besuner, P. M., and Snow, D. W. (1975) 'Application of two-dimensional boundary integral equation method to engineering problems', in T. A. Cruse and F. J. Rizzo (eds), *Proc. ASME Conf. on Boundary Integral Equation Meth.*, AMD-11, ASME, New York.

7. Butterfield, R., and Banerjee, P. K. (1971) 'The problem of pile cap–pile group interaction', *Géotechnq.*, **21**(2), 135–142.

8. Banerjee, P. K., and Driscoll, R. M. (1976) 'Three-dimensional analysis of raked pile groups', *Proc. Inst. Civ. Engrs, Res. and Theory*, **61**, 653–671.

9. Banerjee, P. K., and Davies, T. G. (1978) 'The behaviour of axially and laterally loaded single piles embedded in nonhomogeneous soils', *Géotechnq.*, **28**(3), 309–326.

10. Banerjee, P. K., and Davies, T. G. (1979) 'Analysis of some reported case histories of laterally loaded pile groups', *Int. Conf. Num. Meth. in Offshore Piling*, pp. 83–90, Institute of Civil Engineers, London.

11. Watson, J. O. (1979) 'Advanced implementation of boundary element method in two and three-dimensional elasto-statics', in P. K. Banerjee and R. Butterfield (eds), *Developments in Boundary Element Methods*, Chap. III, Applied Science Publishers, London.

12. Mustoe, G. G. W. (1979) 'A combination of the finite element method and boundary solution procedure for continuum problems', Ph.D. thesis, University of Wales, University College, Swansea.

13. Stroud, A. H., and Secrest, D. (1966) *Gaussian Quadrature Formulae*, Prentice-Hall, Englewood Cliffs, N. J.

14. Rizzo, F. J., and Shippy, D. J. (1976) 'An advanced boundary integral equation method for three-dimensional thermo-elasticity', *Int. J. Num. Meth. in Engng*, **11**, 1753.

15. Rizzo, F. J., and Shippy, D. J. (1979) 'Boundary element methods in thermo-elasticity', in P. K. Banerjee and R. Butterfield (eds), *Developments in Boundary Element Methods*, Chap. VII, Applied Science Publishers, London.

16. Tomlin, G. R. (1972) 'Numerical analysis of continuum problems in zoned anisotropic media', Ph.D. thesis, Southampton University.

17. Butterfield, R., and Tomlin, G. R. (1971) 'Integral techniques for solving zoned anisotropic continuum problems', *Proc. Int. Conf. on Variational Meth. in Engng*, Southampton University, pp. 9/31–9/51.

18. Das, P. C. (1978) 'A disc based block elimination technique used for the solution of non-symmetrical fully populated matrix systems encountered in the boundary element method', *Proc. Int. Symp. on Rec. Dev. in Boundary Element Meth.*, Southampton University, pp. 391–404.

19. Davies, T. G. (1979) 'Linear and nonlinear analyses of pile groups', Ph.D. thesis, University of Wales, University College, Cardiff.

20. Cruse, T. A. (1977) 'Mathematical foundations of the boundary integral equation method in solid mechanics', Report AFOSR–TR–77–1002, Pratt and Whitney Aircraft, Connecticut.

# INDICIAL NOTATION, SUMMATION CONVENTION, TRANSFORMATIONS, AND TENSORS

## A-1 INTRODUCTION

The reader unfamiliar with indicial notation, the summation convention, and elementary tensor transformation rules will find that the book progresses from a minimal use of these ideas in the initial chapters to a progressivey more fearsome-looking notation adorned with a multiplicity of suffixes in the later ones. This is virtually unavoidable if we are to handle efficiently symbols with very many components which combine, one with the other, according to very precisely defined rules.

In this appendix we set out the basic features of suffix notation and Einstein's summation convention which, in combination, allows us to deal with arrays of quantities in a manner ideally suited to computer manipulation. In Sec. A-6, we then add some fundamental ideas related to tensor algebra in curvilinear coordinates. This latter topic is rather beyond the range of what we currently require but, since the whole BEM concept rests upon geometrical descriptions of boundaries, internal cells, and functions distributed over them, we feel that the way ahead lies with truly curvilinear coordinate analyses for which tensor calculus is the tool; perhaps some reader may find the simple presentation and the elegance of the symbolism sufficiently attractive to study it further and revolutionize our analyses!

## A-2 INDICIAL NOTATION

The key concept here is that all entities which can be best defined by a number of components are to be labelled by a suffix (or superfix) which indicates the number and form of the components. Thus, for example, coordinate components would be labelled, $x_i$, which, in three dimensions, means the set of components $(x_1, x_2, x_3)$. The *range* of $i = 1, 2, 3$ here, whereas in two dimensions the range would be $i = 1, 2$, representing $(x_1, x_2)$. Similarly, components of a vector **u** would be shown simply as $u_i$, which represents the whole set of them $(u_1, u_2, u_3)$. Note that: (1) suffixes, $i, j, k$, etc., can have any specified range; (2) we now *no longer* use different labels for components of the same set of quantities [such as, $(x, y, z)$ or $(u, v, w)$, etc.].

More complicated quantities might be usefully characterized by multiple suffixes, for example, $\sigma_{ij}$, $T_{ijk}$, $C_{ijkl}$, etc. The first of these $(\sigma_{ij})$ could represent compactly all the nine stress components at a point by permuting all the combinations of the $i, j = 1, 2, 3$ suffixes $(\sigma_{11}, \sigma_{12}, ..., \sigma_{23}, \sigma_{33})$; the known symmetry of $\sigma_{ij} = \sigma_{ji}$ means that only six of the components are independent. [In two dimensions, of course, the *same* symbol $\sigma_{ij}\,(i, j = 1, 2)$ represents just $\sigma_{11}, \sigma_{12}, \sigma_{21}, \sigma_{22}$]. These ideas are already familiar to most applied scientists through matrix algebra but they can be extended indefinitely to provide a convenient way of handling entities such as $T_{ijk}$ (with $3^3$ components in three dimensions) and the elastic compliance $C_{ijkl}$ with $3^4 = 81$ components, particularly when used in conjunction with the following summation convention.

## A-3 THE SUMMATION CONVENTION FOR INDICES

We shall assume here $i, j, k ... = 1, 2, 3$ unless otherwise stated. Consider first the implications of (outer) products of our indexed symbols (for example, $u_i\, v_j$ or $\sigma_{ij}\, n_k$ or $\sigma_{ij}\, \varepsilon_{kl}$); clearly there are nine combinations of $u_i$ and $v_j$ components and we shall have

$$u_i v_j = v_j u_i = w_{ij} \quad \text{(say)} \tag{A-1}$$

in which the ordering of the symbols $(u, v)$ is of *no* importance. Similarly $\sigma_{ij}\, n_k = s_{ijk}$ and $\sigma_{ij}\, \varepsilon_{kl} = E_{ijkl}$, etc. However, we frequently encounter other (inner) products in which some suffixes are repeated, such as, for example, the scalar product of $u$ and $v$ where

$$\phi = \mathbf{u} \cdot \mathbf{v} = u_1 v_1 + u_2 v_2 + u_3 v_3 \quad \text{that is,} \quad \sum_{i=1}^{3} u_i v_i = \phi \tag{A-2}$$

or a strain energy $(U)$ term (with $\varepsilon_{ij}$ strain components) such as

$$U = \tfrac{1}{2}(\sigma_{11}\varepsilon_{11} + \sigma_{12}\varepsilon_{12} + \sigma_{13}\varepsilon_{13} + \sigma_{21}\varepsilon_{21} + \cdots + \sigma_{33}\varepsilon_{33})$$
$$= \sum_{i=1}^{3}\sum_{j=1}^{3} \sigma_{ij}\varepsilon_{ij} \tag{A-3}$$

or, indeed, our usual matrix product for an $n \times n$ system,

$$\sum_{j=1}^{n} A_{ij} x_j = y_i \tag{A-4}$$

If, in all of these examples, we stipulate that *repeated* suffixes imply summation on them over their full range of values we can immediately write $\phi = u_i v_i$, $U = \sigma_{ij} \varepsilon_{ij}$, and $A_{ij} x_j = y_i$ without ambiguity, providing we ensure that no suffix appears more than twice in any expression. The summed indices are known as dummy indices and the symbols used for them are interchangeable, thus:

$$A_{ij} u_j + B_{ik} u_k = A_{il} u_l + B_{il} u_l \tag{A-5}$$

Equations (A-4) and (A-5) are important in that they show how the *free* indices must 'match' on each side of correctly written equations. More elaborate expressions obey the same rules; e.g., the general linear stress strain relationship becomes, say,

$$\sigma_{ij} = C_{ijkl} \varepsilon_{kl} \tag{A-6}$$

and Laplace's equation

$$\frac{\partial^2 \phi}{\partial x_1^2} + \frac{\partial^2 \phi}{\partial x_2^2} + \frac{\partial^2 \phi}{\partial x_3^2} = \frac{\partial^2 \phi}{\partial x_i \partial x_i} = \phi_{,ii} \tag{A-7}$$

where the utmost abbreviation is often achieved by using a comma to indicate partial differentiation, as above, or $\partial \phi / \partial x_i = \phi_{,i}$; $\partial u_i / \partial x_j = u_{i,j}$ etc. Note that whereas, $A_{ij} B_{kl} \equiv B_{kl} A_{ij}$ etc., and the sequence of writing $A, B$ is *not* important, $A_{ij} \neq A_{ji}$ unless $A$ happens to be symmetrical in $i$ and $j$.

Two symbols play a major role in manipulating our indexed quantities:

1. The Kronecker delta (or substitution tensor)

$$\delta_{ij} = \begin{cases} 1 & \text{when } i = j \\ 0 & \text{when } i \neq j \end{cases} \tag{A-8}$$

has, by definition, the property that

$$\left. \begin{array}{l} u_i \delta_{ij} = u_j, \quad u_i v_j \delta_{ik} = u_k v_j, \text{ etc., and also} \\ u_i v_j \delta_{ij} = u_i v_i, \quad \sigma_{ij} \varepsilon_{kl} \delta_{ik} \delta_{lj} = \sigma_{ij} \varepsilon_{ij}, \text{ etc.} \end{array} \right\} \tag{A-9}$$

The latter expressions are '*contracted*' by multiplication with $\delta_{ij}$ (i.e., reduced in rank by two by each multiplication). Scalars ($\phi$) are called rank 0; $u_i$ 'vectors', rank 1; etc.; $C_{ijkl}$ rank 4. Thus the final stress-strain product in (A-9) is reduced from rank 4 to a scalar quantity (energy).

2. The permutation symbol

$$e_{ijk} = \begin{cases} 0 & \text{if } any \text{ two suffixes are equal} \\ 1 & \text{for cyclic suffix order, } 12312... \\ -1 & \text{for anticyclic suffix order, } 321321... \end{cases}$$

is rather less convenient to deal with than $\delta_{ij}$ and arises in evaluating determinants: viz.,

$$\det \| J_{ij} \| = e_{rst} J_{1r} J_{2s} J_{3t} \tag{A-10}$$

or components of vector cross-products, $\mathbf{W} = \mathbf{U} \times \mathbf{V}$ say, as

$$W_i = e_{ijk} U_j V_k \tag{A-11}$$

Whereas $\delta_{ii} = 3$ is an obvious result it is a worthwhile exercise to check that $e_{ijk} e_{jki} = 6$.

In most texts summation over indices is implied unless 'suppressed', for example, by writing $A_{(ii)}$ which then means any of $A_{11}, A_{22}, ...$, not their sum.

## A-4 CARTESIAN TENSORS AND TRANSFORMATION RULES

If we confine ourselves for the moment to orthogonal cartesian coordinate systems and a set of axes $y_i$ which are translated (by $h_i$) and rotated with respect to $x_i$, then, if the direction cosine array of the rotation is $\lambda_{ij}$ (with $\lambda_{12} = \cos y_1 x_2$, etc.), we have the transformation rule

$$y_i = \lambda_{ij}(x_j - h_j) \tag{A-12}$$

or
$$dy_i = \lambda_{ij} dx_j = \frac{\partial y_i}{\partial x_j} dx_j = \frac{\partial x_j}{\partial y_i} dx_j \tag{A-13}$$

since, in our special orthogonal cartesian coordinates, $\lambda_{ij} = \partial y_i/\partial x_j = \partial x_j/\partial y_i$. Note (for comparison with Sec. A-6 and as an example of indicial manipulation) that the length ($ds$) of a line element in $y_i$ can be written as

$$ds^2 = dy_i\, dy_i = \left( \frac{\partial y_i}{\partial x_j} dx_j \right) \left( \frac{\partial x_k}{\partial y_i} dx_k \right)$$

$$= \delta_{jk}\, dx_j\, dx_k = dx_j\, dx_j$$

as expected.

We now define cartesian tensors as entities which transform according to the following rules (dashed symbols refer to the tensor components in $y$; undashed in $x$):

1. Zero rank (scalars):

$$\phi'(y) = \phi(x)$$

2. First rank (true vectors, forces, displacements, gradients of scalars, etc.):

$$V'_i = \lambda_{ij} V_j = \frac{\partial y_i}{\partial x_j} V_j = \frac{\partial x_j}{\partial y_i} V_j \qquad \text{(A-14)}$$

3. Second rank (stress, strain, conductivity, etc.):

$$\sigma'_{ij} = \lambda_{ik} \lambda_{jl} \sigma_{kl} \quad \text{etc.}$$

4. Fourth rank (elastic compliance, etc.):

$$C'_{ijkl} = \lambda_{im} \lambda_{jn} \lambda_{kp} \lambda_{lq} C_{mnpq} \quad \text{etc.}$$

Note that whereas $dx_i$ is first rank, coordinate components $x_i$ are not in general tensor quantities, nor is $\lambda_{ij}$ a second-rank tensor. Thus although we can always write vector components as column matrices and second-rank tensors as matrix arrays the converse is not generally true.

Again, as an exercise, it is useful to demonstrate the invariance of strain energy under transformation:

$$2U = \sigma'_{ij} \varepsilon'_{ij} = \lambda_{ik} \lambda_{jl} \sigma_{kl} \lambda_{im} \lambda_{jn} \varepsilon_{mn} = \delta_{km} \delta_{ln} \sigma_{kl} \varepsilon_{mn}$$

$$= \sigma_{mn} \varepsilon_{mn} = \sigma_{ij} \varepsilon_{ij}$$

Note also that here $\| \lambda_{ij} \| = J = +1$, $[J] = [\lambda]$, and $[J]^{-1} = [\lambda]^{-1} = [\lambda]^T$.

## A-5 USEFUL EXERCISES

This illustrative and very relevant example of indicial manipulation is taken from Ref. 13 of Chapter 11.

Consider once more potential flow, but this time a case in which the anisotropic and inhomogeneous permeability of the system will be $k_{ij} \alpha(x)$ with $k_{ij}$ constant and $\alpha$ a parameter which is assumed to vary continuously in a known manner with $(x)$. The governing equations then are

$$u = v_i n_i = -(k_{ij} \alpha p_{,j}) n_i \qquad \text{(A-15)}$$

and 
$$(k_{ij} \alpha p_{,j})_{,i} = k_{ij}(\alpha p_{,ij} + \alpha_{,i} p_{,j}) = -\psi \qquad \text{(A-16)}$$

We shall also require both the homogeneous free-space Green's function $G(x, \xi)$ and the inhomogeneous one $G^*(x, \xi)$ which satisfy the following equations respectively:

$$k_{ij} G_{,ij} = -\delta(x, \xi) = -\delta \qquad \text{(A-17)}$$

and 
$$(k_{ij} \alpha G^*_{,j})_{,i} = -\delta \qquad \text{(A-18)}$$

If we knew the value of $G^*$ then the standard BEM equation (3-30) would be applicable, simply with $(G^*, F^*)$ replacing $(G, F)$. However, this function is never likely to be known in general and we shall now investigate the consequences of attempting to develop a DBEM solution of (A-16) using the standard integration by parts procedure on the product of (A-16) with $G$, thus

$$-\int_V G\psi \, dV = \int_V G(\alpha k_{ij} p_{,j})_{,i} \, dV = \int_V Gk_{ij}(\alpha p_{,ji} + \alpha_{,i} p_{,j}) \, dV$$

$$= \int_S Gk_{ij}(\alpha p_{,i} + \alpha_{,i} p) n_j \, dS$$

$$- \int_V [(G\alpha k_{ij})_{,j} p_{,i} + (Gk_{ij}\alpha_{,i})_{,j} p] \, dV$$

$$= \int_S [(G\alpha k_{ij}) p_{,i} + (Gk_{ij}\alpha_{,i}) p - (G\alpha k_{ij})_{,i} p] n_j \, dS$$

$$+ \int_V [(G\alpha k_{ij})_{,ij} p - (Gk_{ij}\alpha_{,i})_{,j} p] \, dV$$

$$= \int_S [(G\alpha k_{ij}) p_{,i} - (G_{,i}\alpha k_{ij}) p] n_j \, dS$$

$$+ \int_V p[(G_{,i}\alpha k_{ij} + G\alpha_{,i} k_{ij})_{,j} - (Gk_{ij}\alpha_{,i})_{,j}] \, dV$$

and if
$$F = -k_{ij} G_{,j} n_i$$
then

$$-\int_V G\psi \, dV = \int_S [-uG + \alpha Fp] \, dS + \int_V p[(G_{,ij}\alpha k_{ij}) + (G_{,i}\alpha_{,j} k_{ij})] \, dV$$

Therefore,

$$\alpha p(\xi) = \int_S (\alpha pF - uG) \, dS + \int_V G\psi \, dV + \int_V (p\alpha_{,j} k_{ij} G_{,i}) \, dV \qquad \text{(A-19)}$$

This equation is the extended DBEM statement for an inhomogeneous, anisotropic region, throughout which the directions of the principal axes of permeability are constant.

Establishing this solution without the aid of indicial notation is extremely cumbersome. The result shows that, at the expense of the additional integral on the right-hand side of (A-19), the homogeneous Green's function may be used to solve the anisotropic inhomogeneous problem. How to utilize this result is explained in the above reference, as also is the corollary that the analysis breaks down if the principal axis directions vary from point to point within the region. As a second illustrative example let us consider the derivation of the strain kernel function defined in Chapter 4. The displacement field $u_i(x)$ due to a unit force vector $e_k(\xi)$ is given by

$$u_i(x) = c_1 \left[ c_2 \ln r\, \delta_{ik} - \frac{y_i y_k}{r^2} \right] e_k(\xi) \tag{A-20}$$

where

$$y_i = (x - \xi)_i \quad r^2 = y_i y_i$$

We can obtain $\partial u_i(x)/\partial x_j$ from Eq. (A-20) as

$$\frac{\partial u_i}{\partial x_j} = c_1 \left[ c_2 \delta_{ik} \frac{\partial}{\partial x_j}(\ln r) - \frac{\partial}{\partial x_j} \frac{y_i y_k}{r^2} \right] e_k(\xi)$$

$$= c_1 \left[ c_2 \delta_{ik} \frac{\partial}{\partial r}(\ln r)\frac{\partial r}{\partial x_j} - y_i y_k \frac{\partial}{\partial r}\left(\frac{1}{r^2}\right)\frac{\partial r}{\partial x_j} - \frac{y_i}{r^2}\frac{\partial y_k}{\partial x_j} - \frac{y_k}{r^2}\frac{\partial y_i}{\partial x_j} \right] e_k(\xi) \tag{A-21}$$

Noting that here $\partial y_k/\partial x_j = \delta_{jk}$ and $\partial y_i/\partial x_j = \delta_{ij}$ we can rewrite (A-21) as

$$\frac{\partial u_i}{\partial x_j} = c_1 \left[ c_2 \delta_{ik}\left(\frac{1}{r}\right)\frac{y_j}{r} - y_i y_k \frac{-2}{r^3}\frac{y_j}{r} - \frac{y_i}{r^2}\delta_{jk} - \frac{y_k}{r^2}\delta_{ij} \right] e_k(\xi)$$

$$= \frac{c_1}{r} e_k(\xi)\left( c_2 \delta_{ik}\frac{y_j}{r} + \frac{2 y_i y_j y_k}{r^3} - \frac{y_i}{r^2}\delta_{jk} - \frac{y_k}{r^2}\delta_{ij} \right) \tag{A-22}$$

Interchanging indices $i$ and $j$ we get

$$\frac{\partial u_j}{\partial x_i} = \frac{c_1}{r} e_k(\xi)\left( c_2 \delta_{jk}\frac{y_i}{r} + \frac{2 y_i y_j y_k}{r^3} - \frac{y_j}{r^2}d_{ik} - \frac{y_k}{r^2}\delta_{ij} \right) \tag{A-23}$$

Then using (A-22) and (A-23) we arrive at the strain field as

$$\varepsilon_{ij} = \frac{1}{2}\left(\frac{\partial u_i}{\partial x_j} + \frac{\partial u_j}{\partial x_i}\right)$$

$$= \frac{c_1}{r} e_k(\xi)\left[ 0.5(c_2 - 1)\left(\delta_{ik}\frac{y_j}{r} + \delta_{jk}\frac{y_i}{r}\right) - \delta_{ij}\frac{y_k}{r} + \frac{2 y_i y_j y_k}{r^3} \right] \tag{A-24}$$

## A-6 GENERAL TENSOR TRANSFORMATIONS, CONTRA-VARIANCE, AND CO-VARIANCE

Unfortunately, when we wish to study much more general transformations, for example, from cartesian to curvilinear coordinates, in Chapter 8, only the scalar rule in (A-14) remains absolutely identical. Subtle changes occur in all the others due to the new coordinates being either non-orthogonal or not dimensionally homogeneous, or both. Whereas we negotiated Chapter 8 with merely scalar and geometrical scale changes, any more ambitious transformations will be best dealt with using more general tensor analysis (it was designed for this purpose). We hope that the following very brief introduction to the salient new features may prove to be both readable and useful.

Our general transformation will be proper and admissible (Chapter 8) from $x_i = x_i(\zeta_1, \zeta_2, \zeta_3)$ to $\zeta_i = \zeta_i(x_1, x_2, x_3)$ whence, similarly to (A-13),

$$d\zeta^i = \frac{\partial \zeta_i}{\partial x_j} dx^j \tag{A-25}$$

Note that the indices in $d\zeta^i$ and $dx^j$ are now superfixes (for a reason explained subsequently) and that, as before, $[J] = [\partial \zeta_i / \partial x_j]$. The metric tensor $g_{ij} = (\partial x_k / \partial \zeta_i)(\partial x_k / \partial \zeta_j)$ of the $Z$ space, which plays a major role in what follows, is defined by considering the length of a differential line element $ds$ in $x_i$ and $\zeta_i$:

$$(ds)^2 = dx^i\, dx^j\, \delta_{ij} = dx^i\, dx^i = \left(\frac{\partial x_k}{\partial \zeta_i} d\zeta^i\right)\left(\frac{\partial x_k}{\partial \zeta_j} d\zeta^j\right)$$

$$= \left(\frac{\partial x_k}{\partial \zeta_i}\right)\left(\frac{\partial x_k}{\partial \zeta_j}\right) d\zeta^i\, d\zeta^j = d\zeta^i\, d\zeta^j\, g_{ij} \tag{A-26}$$

and we see that, just as $g_{ij}$ is the 'metric' of $Z$ so is $\delta_{ij}$ the 'metric' of $X$.

It will be found that $[g]$ is diagonal for orthogonal coordinate systems and that cartesian (straight) coordinate lines generate constant components in $g_{ij}$.

Since two-dimensional skew cartesian coordinates arose in Chapter 8, it might prove to be both simple and useful to introduce the extended tensor notation in relation to this system. From the previous remark we would expect the relevant $[g]$ to be $(2 \times 2)$ and fully populated with constant coefficients. We should perhaps add that the one major virtue of tensor analysis is that properly constituted tensor relationships are true in *all* coordinate systems and therefore theorems correctly established in $X$ are true in all $Z$.

Figure A-1$a$ shows a vector **V** in $Z$ with its tensor components $(V^1, V^2)$ together with base vectors $(\mathbf{e}_1, \mathbf{e}_2)$ defining the parallelogram vector components $(\mathbf{e}_1 V^1, \mathbf{e}_2 V^2)$ of $V$. The $V^i$ (superfix) components are called contra-variant tensor components of the vector **V** and $\mathbf{e}_i$ are the related base vectors in $Z$. It is easily shown that $g_{ij} = \mathbf{e}_i \cdot \mathbf{e}_j$ and since, from the geometry of Fig. A-1$a$, the $dx^i \to d\zeta^i$ transformation rule is simply

$$\begin{Bmatrix} dx^1 \\ dx^2 \end{Bmatrix} = \begin{bmatrix} \cos\alpha_1 & \cos\alpha_2 \\ \sin\alpha_1 & \sin\alpha_2 \end{bmatrix} \begin{Bmatrix} d\zeta^1 \\ d\zeta^2 \end{Bmatrix}$$

from (A-26),

$$[g_{ij}] = \begin{bmatrix} 1 & \cos\beta \\ \cos\beta & 1 \end{bmatrix} \qquad \beta = \alpha_2 - \alpha_1 \tag{A-27}$$

If, however, we consider even a simple polar coordinate $(r, \theta)$ transformation $(\zeta_1 = r, \zeta_2 = \theta)$ then

$$\begin{Bmatrix} dx^1 \\ dx^2 \end{Bmatrix} = \begin{bmatrix} \cos\zeta_2 & -\zeta_1 \sin\zeta_2 \\ \sin\zeta_2 & \zeta_1 \cos\zeta_2 \end{bmatrix} \begin{Bmatrix} d\zeta^1 \\ d\zeta^2 \end{Bmatrix} \quad \text{and} \quad [g_{ij}] = \begin{bmatrix} 1 & 0 \\ 0 & (\zeta_1)^2 \end{bmatrix} \tag{A-28}$$

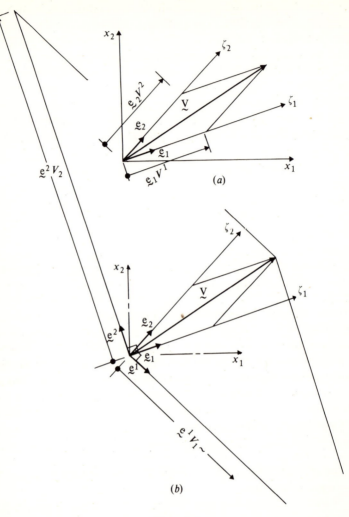

**Figure A-1**

We see now that, whereas $\mathbf{e}_1$ (related to $r$) is still a constant unit vector, $|\mathbf{e}_2| = \zeta_1 = r$ and is therefore no longer unity but varies from point to point. [Note that, since $(r, \theta)$ are orthogonal $g_{ij} = 0$ $(i \neq j)$ and, furthermore, $|\mathbf{e}_2|$ has length units to rectify the intrinsic dimensional inhomogeneity of $(r, \theta)$ in terms like $(d\zeta^2 \mathbf{e}_2)$.]

Consider now Fig. A-1$b$ which shows the same vector $\mathbf{V}$ in $Z$ with a further set of axes perpendicular to $(\zeta_1, \zeta_2)$, defined by base vectors $(\mathbf{e}^1, \mathbf{e}^2)$; in the general case $\mathbf{e}^1$ is orthogonal to the $(\mathbf{e}_2, \mathbf{e}_3)$ plane, etc. The parallelogram components of $\mathbf{V}$ referred to these axes [which are in fact the *resolved* components of $\mathbf{V}$ in relation to $(\zeta_1, \zeta_2)$; Fig. A-1*b*] are designated by $(e^1 V_1, e^2 V_2)$ where the quantities $(V_1, V_2)$ are called the *co-variant* tensor components of **V**. Clearly in orthogonal cartesian coordinate systems $V^i \equiv V_i$ and the contra-variant–co-variant distinction does not arise, whereas in general coordinates it is quite indispensable.

In parallel with Eq. (A-25) we can define co-variant elements $d\zeta_i$ along the new axes with

$$d\zeta_i = \frac{\partial x_j}{\partial \zeta_i} dx^j \quad (dx^i \equiv dx_i)$$

and, similarly,

$$(ds)^2 = dx^i\, dx^j = \left(\frac{\partial \zeta_i}{\partial x_k}\right)\left(\frac{\partial \zeta_j}{\partial x_k}\right) d\zeta_i\, d\zeta_j = g^{ij}\, d\zeta_i\, d\zeta_j \qquad \text{(A-29)}$$

The entity $g^{ij}$, which is also a second-rank tensor, is called the *associated metric tensor* of $Z$ and it is easily shown that $g_{ik}\, g^{kj} = \delta_{ij}$ (that is, $[g_{ij}] = [g^{ij}]^{-1}$).

Reverting to our skew cartesian example, we have

$$[g^{ij}] = \frac{1}{\sin^2 \beta}\begin{bmatrix} 1 & -\cos \beta \\ -\cos \beta & 1 \end{bmatrix}$$

Therefore $|e^1| = |e^2| = \operatorname{cosec} \beta$ are no longer unit vectors. Similarly, in a plane polar system although the $d\zeta^i$ and $d\zeta_i$ axes coincide, $|e^1| = 1, |e^2| = 1/\zeta_1$, and the contra-variant and co-variant components are different ($|e_1| = 1, |e_2| = \zeta_1$).

One consequence of non-unit base vectors is that a distinction has to be made between the *physical* components of a tensor (which are necessarily dimensionally homogeneous) and the *tensor* components (which may *not* be). Consider, say, the vector $(V^2 \cdot e_2)$; if $|e_2| \neq 1$ then the physical magnitude of the vector will not be $V^2$, but, since $|e_2| = \sqrt{g_{22}}$, it will be $(V^2 \sqrt{g_{22}})$. This rule is easily generalized such that if the *physical* components of **V** are $U^i$ and $U_i$ then

$$U^i = V^i \sqrt{g_{(ii)}} \quad \text{and} \quad U_i = V_i \sqrt{g^{(ii)}} \quad i \text{ not summed} \qquad \text{(A-30)}$$

For example, if $|\mathbf{V}| = 10$ at $(5, 0)$ in polar coordinates, then $\sqrt{g_{22}} = 5$ and the contra-variant component $V^2 = 2$, whereas obviously $U^2 = 10$.

In order to accommodate contra-variant and co-variant components, the summation convention is minimally modified such that summation is now implied across lower to upper (or upper to lower) repeated indices. Thus we find that

$$\mathbf{V} = V^i \mathbf{e}_i = V_i \mathbf{e}^i$$

as it must, and

$$|\mathbf{V}|^2 = V^i \mathbf{e}_i \cdot V^j \mathbf{e}_j = V^i V^j(\mathbf{e}_i \cdot \mathbf{e}_j) = V^i V^j g_{ij}$$

or

$$= V_i \mathbf{e}^i \cdot V_j \mathbf{e}^j = V_i V_j(\mathbf{e}^i \cdot \mathbf{e}^j) = V_i V_j g^{ij}$$

$$\left. \right\} = V^i V_i \qquad \text{(A-31)}$$

or

$$= V^i \mathbf{e}_i \cdot V_j \mathbf{e}^j = V^i V_j(\mathbf{e}_i \cdot \mathbf{e}^j) = V^i V_j g_i^j$$

Equations (A-31) establish three further important points:

1. *Mixed* tensors can arise (for example, $g_i^j$, $T_k^{ij}$, $E_{kl}^{ij}$, etc.).

2. $g_i^j = \dfrac{\partial \zeta_j}{\partial x_k} \dfrac{\partial x_k}{\partial \zeta_i} = \begin{cases} 0 & \text{when } i \neq j \\ 1 & \text{when } i = j \end{cases} \equiv \delta_{ij}, \quad \text{always,}$

   whence

$$|\mathbf{V}|^2 = V^i V_j \delta_{ij} = V^i V_i \tag{A-32}$$

   which is only identical with $V_i V_i$ in the cartesian space $X$, whereas generally $|\mathbf{V}|^2 = V^1 V_1 + V^2 V_2 + V^3 V_3$, an expression easily shown from Fig. A-1b to be simply the cosine rule.

3. The $g^{ij}$, $g_{ij}$ tensors have an index raising, or lowering, property such that $V_i = V^j g_{ij}$, $V^i = V_j g^{ij}$ and similarly for second-rank and higher tensors $\sigma^{ij} = \sigma_{kl} g^{ik} g^{jl}$, etc.

Just as we have to distinguish between physical and tensor components of vectors (first rank) so must we distinguish them for second- and higher-rank tensors. At this point one might fairly ask what (apart from the general coordinate invariance of tensor equations mentioned previously) is the practical utility of it all. The answer is that the completely general, tensor transformation rules listed below apply to all admissible transformations of *tensor* (*not* physical) components of tensors (such as displacement, stress, strain, elastic compliance, scalar gradient, etc.).

Scalars transform identically as in (A-14) but

1. First rank: $\qquad (V')^i = \dfrac{\partial \zeta_i}{\partial x_j} V^j \quad (V')_i = \dfrac{\partial x_j}{\partial \zeta_i} V_j$

$$\left. \vphantom{\begin{array}{c} a \\ b \\ c \end{array}} \right\} \tag{A-33}$$

2. Second rank: $\qquad (\sigma')^{ij} = \dfrac{\partial \zeta_i}{\partial x_k} \dfrac{\partial \zeta_j}{\partial x_l} \sigma^{kl} \quad (\sigma')_{ij} = \dfrac{\partial x_k}{\partial \zeta_i} \dfrac{\partial x_l}{\partial \zeta_j} \sigma_{kl}$

with an obvious extension to both mixed and higher ranks. For example, our ordinary differential element $d\zeta^i = (\partial \zeta_i / \partial x_j) dx^j$ transforms as a contra-variant vector, whereas a scalar gradient transforms co-variantly

$$(\partial \phi / \partial \zeta_i) = (\partial x_j / \partial \zeta_i)(\partial \phi / \partial x_j);$$

also, our conventional strain energy expression $U = \frac{1}{2}\sigma^{ij} \varepsilon_{ij}$ implies that the stress and strain tensors shall be of opposite variance. Note that all the above operations are purely 'mechanical' once the coordinate transformation (and hence the metrics) have been decided. Anyone who is now inspired to study this further will find lucid accounts in Y. C. Fung, *Foundations of Solid Mechanics*, Prentice-Hall (1965), and I. S. Sokolnikoff, *Tensor Analysis Theory and Applications to Geometry and Mechanics of Continua*, Wiley (1964).

# ON THE INTEGRAL IDENTITIES

## B-1 THE GENERAL FORM OF GAUSS' THEOREM

Consider the function $F$ to be a typical scalar component of a general tensor field $F_{jkl...}$ of any rank, all components of which are defined, continuous and differentiable as required, within the region $V$ and on its surface $S$. Then, from Fig. B-1, considering a volume element with its axis along $x_1$, we have

$$\int_V \frac{\partial F}{\partial x_1} dV = \int_S (F^* - F^{**}) dx_2 \, dx_3 \tag{B-1}$$

where, throughout, single-starred quantities refer to their values at the 'right-hand' end of the volume element and double-starred ones to those at the 'left-hand' end. Thus, for outward unit normals $\mathbf{n}$ and surface area elements $ds$, we have

$$dx_2 \, dx_3 = n_1^* \, ds^* = -n_1^{**} \, ds^{**}$$

and, therefore,

$$\int_V \frac{\partial F}{\partial x_1} dV = \int_S (F^* n_1^* \, ds^* + F^{**} n_1^{**} \, ds^{**}) = \int_S F n_1 \, dS$$

whence

$$\int_V \frac{\partial F}{\partial x_i} dV = \int_S F n_i \, dS$$

and similarly for all other components of $F_{jkl...}$ to give, finally,

$$\int_V \frac{\partial F_{jkl...}}{\partial x_i} \, dV = \int_S F_{jkl...} \, n_i \, dS \qquad \text{(B-2)}$$

which is the general form of Gauss' theorem.[1]

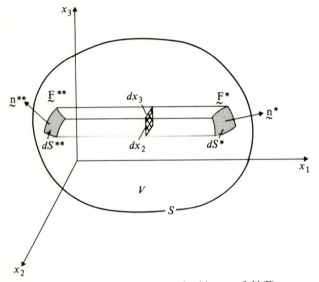

**Figure B-1** Elemental volume core parallel to the $x_1$, axis with vector field (**F**) components on surface area elements ($dS$).

Many other classical integral theorems and statements of conservation laws follow directly from Eq. (B-2) which is, of course, applicable in any number of dimensions.[1, 2]

1. For a scalar function $F$:

$$\int_V F_{,i} \, dV = \int_S F n_i \, dS \qquad \int_V \text{grad } F \, dV = \int_S F \mathbf{n} \, dS \qquad \text{(B-3)}$$

2. For a vector function $F$:
   (a) The divergence theorem,

$$\int_V F_{i,i} \, dV = \int_S F_i n_i \, dS \qquad \int_V \text{div } \mathbf{F} \, dV = \int_S \mathbf{F} \cdot \mathbf{n} \, dS \qquad \text{(B-4)}$$

if, say, $F_i = v_i = -kp_{,i}$ and $u = v_i n_i$ and $F_{i,i} = -kp_{,ii} = \psi$, which are the potential flow equations of Chapter 3, then

$$-k \int_V p_{,ii} \, dV = \int_V \psi \, dV = \int_S v_i n_i \, dS = \int_S u \, dS$$

that is,

$$\int_V \psi \, dV = \int_S u \, dS$$

which expresses the conservation equation in potential flow.

(b) Stokes' theorem,

$$\int_V \varepsilon_{ijk} F_{k,j} \, dV = \int_S \varepsilon_{ijk} F_k n_j \, dS \qquad \int_V \text{curl } \mathbf{F} \, dV = \int_S \mathbf{n} \times \mathbf{F} \, dS \qquad \text{(B-5)}$$

3. For second-rank tensor components $F_{ij}$ :

$$\int_V F_{ij,j} \, dV = \int_S F_{ij} n_j \, dS$$

if, say, $F_{ij} = \sigma_{ij}$ (i.e., stress components) and $\sigma_{ij,j} + \psi_i = 0$ and $t_i = \sigma_{ij} n_j$ (the stress equilibrium equations), then

$$\int_V \psi_i \, dV + \int_S t_i \, dS = 0 \qquad \text{(B-6)}$$

which can be considered as either a stress-flux conservation equation or a statement of overall equilibrium.

## B-2 GREEN'S IDENTITIES

These follow directly from (B-4). If a vector field $\mathbf{F}$ is related to two scalar fields $(\phi, \chi)$, say, via

$$\mathbf{F} = \phi \, \text{grad } \chi \quad \text{or} \quad F_i = \phi \chi_{,i}$$

then

$$F_{i,i} = \phi_{,i} \chi_{,i} + \phi \chi_{,ii} \qquad F_i n_i = \phi \chi_{,i} n_i$$

Substitution into (B-4) yields Green's first identity:

$$\int_V (\phi \chi_{,ii} + \phi_{,i} \chi_{,i}) \, dV = \int_S (\phi \chi_{,i} n_i) \, dS$$

or

$$\int_V (\phi \nabla^2 \chi + \text{grad } \phi \cdot \text{grad } \chi) \, dV = \int_S \left( \phi \frac{\partial \chi}{\partial n} \right) dS \qquad \text{(B-7)}$$

Interchanging $\phi$ and $\psi$ and subtracting the result from (B-7) eliminates the grad $\phi$ . grad $\psi$) terms to produce Green's second identity:

$$\int_V (\phi \chi_{,ii} - \chi \phi_{,ii}) \, dV = \int_S (\phi \chi_{,i} - \chi \phi_{,i}) n_i \, dS$$

or

$$\int_V (\phi \nabla^2 \chi - \chi \nabla^2 \phi) \, dV = \int_S \left( \phi \frac{\partial \chi}{\partial n} - \chi \frac{\partial \phi}{\partial n} \right) dS \qquad \text{(B-8)}$$

## B-3 IDENTITIES FOR DIRECT BOUNDARY ELEMENT METHODS

This latter expression can be used to generate the basic identity for Laplace's equation which leads directly to the standard DBEM statement.

Thus, if we adopt, in dimensionless variables, the potential $p = \phi$ (that is, $p_{,ii} = -\psi$) and the flux $u = v_i n_i = -p_{,i} n_i = -\partial p / \partial n$ together with $\chi = G(x, \xi)$, the free-space Green function introduced in Chapter 3, in which case $\partial \chi / \partial n = \chi_{,i} n_i = -F(x, \xi)$, and substitute these values into (B-8) we obtain the DBEM statement [Eq. (3-30)]: viz.,

$$p(\xi) - \int_V G \psi \, dV = \int_S (pF - Gu) \, dS \quad (\xi \in V) \tag{B-9}$$

In fact, if we substitute for $G$ and $F$ the three-dimensional source kernel values and put $\psi = 0$, this equation reduces immediately to Green's third identity (1828)! Alternatively, the idea of expressing $F_i$ as a product of two functions, used to develop Green's identities, can be used with products of higher-rank quantities.[3] For example, writing $F_j = u_i \sigma_{ij}^*$ and $F_j^* = u_i^* \sigma_{ij}$ where $(u_i, \sigma_{ij})$ and $(u_i^*, \sigma_{ij}^*)$ are two corresponding, differentiable displacement and stress fields in an elastic body then, using (B-4) again exactly as for the development of (B-7), we obtain

$$\int_V (u_{i,j} \sigma_{ij}^* + u_i \sigma_{ij,j}^*) \, dV = \int_S (u_i \sigma_{ij}^* n_j) \, dS = \int_S (u_i t_i^*) \, dS$$

$$\tag{B-10}$$

and 
$$\int_V (u_{i,j}^* \sigma_{ij} + u_i^* \sigma_{ij,j}) \, dV = \int_S (u_i^* \sigma_{ij} n_j) \, dS = \int_S (u_i^* t_i) \, dS$$

If we subtract these two equations the first terms on the left-hand side cancel since

$$u_{i,j} \sigma_{ij}^* = \tfrac{1}{2} \, C_{ijkl}(u_{i,j} u_{k,l}^* + u_{i,j} u_{l,k}^*)$$

and 
$$u_{i,j}^* \sigma_{ij} = \tfrac{1}{2} \, C_{ijkl}(u_{i,j}^* u_{k,l} + u_{i,j}^* u_{l,k})$$

are identical, because of the symmetries of $C_{ijkl}$ in the various suffixes. Therefore we deduce that

$$\int_V (u_i \sigma_{ij,j}^* - u_i^* \sigma_{ij,j}) \, dV = \int_S (u_i t_i^* - u_i^* t_i) \, dS$$

For equilibrium, $\sigma_{ij,j} + \psi_i = 0$, etc. Therefore this equation is the Betti–Maxwell identity

$$\int_V (u_i^* \psi_i - u_i \psi_i^*) \, dV = \int_S (u_i t_i^* - u_i^* t_i) \, dS \tag{B-11}$$

Once more the DBEM statement for elasticity follows by specifying the starred state to be that of our free-space (unit solution) Green function [Chapter 4, Eqs (4-35) and (4-37)].

We see therefore that, for all the differential operators which we have been concerned with, there are two common factors in the analyses leading to the identities from which the DBEM stem:

1. The symmetry of the various terms in the identities [see Eqs (B-7), (B-9), and (B-10)]
2. The divergence (Gauss) theorem and its assumed validity in $(V+S)$.

## B-4 INTEGRATION OF DIFFERENTIAL OPERATORS

The first point above is related specifically to the particular (self-adjoint) type of differential operator which has arisen in all of our equations. In the earlier chapters we developed the DBEM equations using integration by parts, which is a more general procedure than invoking the symmetrical products for $F_i$ leading to Eqs (B-6) and (B-9). Thus if our general differential equation were, say,

$$L(u) = \psi \tag{B-12}$$

and we were to investigate the integration by parts of the product of (B-12) with some other, adequately differentiable and continuous, function $(v)$ over $V$ we could, step by step, transpose the differential operations from $u$ to $v$ whilst simultaneously generating integrations of functions of $u$ and $v$ over $S$. [See, for example, Eqs (2-20), (2-28) to (2-32) and (2-36) to (3-29).] The general form of equation obtained will always be

$$\int_V vL(u)\,dV = \int_V uL^*(v)\,dV + \int_S [M^*(v)N(u) - M(u)N^*(v)]\,dS$$

where the operator $L^*$ is called the adjoint of $L$ and $(M, N, M^*, N^*)$ are also differential operators which arise from the integration by parts procedure.

In all the cases which we have considered $L = L^*$, $M = M^*$, $N = N^*$ and therefore the operator is self-adjoint. For example, in potential flow, the integration by parts procedure yields

$$\int_V (Gp_{,ii})\,dV = \int_V (pG_{,ii})\,dV + \int_S [(Gp_{,i}n_i) - (pG_{,i}n_i)]\,dS$$

$$= \int_V (pG_{,ii})\,dV + \int_S \left[\left(G\frac{\partial p}{\partial n}\right) - \left(p\frac{\partial G}{\partial n}\right)\right]\,dS \tag{B-13}$$

[the three-dimensional form of Eq. (3-28) and (3-30)] and we see that the Laplacian operator $(\nabla^2)$ is self-adjoint and $M^* = M$, $N^* = N$. Furthermore, the desired result has been obtained by using the free-space Green function $(G)$ as the multiplying function $(v)$.

As a second example consider the simple, one-dimensional, beam problem [Eq. (2-30)] for which the equivalent equation is (writing $D^m w = d^m w/dx^m$)

$$\int_L (GD^4 w)\, dL = \int_L (wD^4 G)\, dL + [(GD^3 w - DGD^2 w) - (wD^3 G - DwD^2 G)]_0^L$$

(B-14)

Biharmonic operators are therefore self-adjoint also.

A further point of interest is that in such cases the sets $M(u)$ prescribed are called the essential boundary conditions and $N(u)$ the natural boundary conditions. Although either type can be specified on $S$ the essential boundary conditions must at least be enforced at one point on $S$ if the solution is to be unique.[4] Thus, from (B-13), in potential flow the potentials $(p)$ are the essential and the fluxes $[-k(\partial p/\partial n)]$ the natural boundary value components. For the biharmonic operator (B-14) when the four boundary operators have been arranged in the symmetrical form shown, we see that the displacement and displacement gradient (slope) comprise the essential boundary conditions with moments and shears the natural ones. Clearly, in elasticity problems these two sets of quantities will be the boundary displacements and tractions respectively.

The above operators are not only self-adjoint but also positive definite, which is defined as the property that

$$\int_V uL(u)\, dV \geqslant 0$$

(B-15)

for all $u$ and only equal to zero when $u$ is identically zero. Problems governed by operators possessing these properties can certainly be solved by the methods outlined in this book, whereas this is not yet proven for all other types of differential equations.

## B-5 REFERENCES

1. Fung, Y. C. (1965) *Foundations of Solid Mechanics*, Prentice-Hall, Englewood Cliffs, N. J.
2. Milne-Thompson, L. M. (1960) *Theoretical Hydrodynamics*, 4th ed., Macmillan, London.
3. Watson, J. (1979) 'Advanced implementation of the boundary element method for two and three-dimensional elasto-statics', in P. K. Banerjee and R. Butterfield (eds), *Developments in Boundary Element Methods*, Chap. III, Applied Science Publishers, London.
4. Tottenham, H. (1979) 'The boundary element method for plates and shells', in P. K. Banerjee and R. Butterfield (eds), *Developments in Boundary Element Methods*, Chap. VIII, Applied Science Publishers, London.

# GAUSSIAN QUADRATURE

## C-1 INTRODUCTION

The purpose of this appendix is simply to present a short summary of nodal positions and their weights for gaussian quadrature formulae which will be useful in evaluating various integrals over elements and cells in BEM. The fundamental improvement which Gauss–Legendre methods of numerical integration achieve over the commoner methods (Simpson's trapezoidal rules, etc.) is that equivalently precise results can be obtained whilst using essentially only half the number of ordinates. This is a consequence of adopting as parameters in the formulae not only the weights applied to each ordinate but also the location (nodes) of the ordinates with the range of integration (Fig. C-1).

Therefore, if we wish to approximate the following integral (over a one-dimensional domain) by the summation shown, as we modify the number of ordinates used not only will their weights ($A_i$) change but also the nodal coordinates ($x_i$) will have specific optimal values:

$$I_1 = \int_{-1}^{1} f(x)\,dx \simeq \sum A_i\, f(x_i) \tag{C-1}$$

As an example consider three nodes ($i = 3$); they will be located at points ($x_i$) $(0, \pm\sqrt{15}/5)$ within the range of integration with weights ($A_i$) (8/9, 5/9, 5/9); see Fig. C-1. In equations like (C-1) summation over the range ($i$) will be implied.

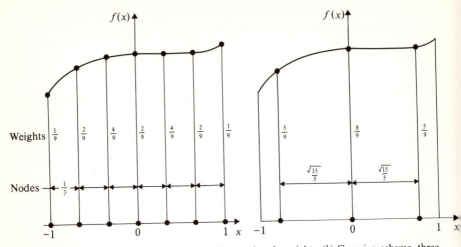

**Figure C-1** (a) Equally spaced ordinates with Simpson's rule weights. (b) Gaussian scheme, three nodes, exact for fifth-order polynomial.

Such gaussian formulae will integrate polynomial expressions exactly up to the order $(2i - 1)$ (that is, $i = 3$ is exact for a quintic polynomial) and the errors are therefore of order $d^{2i} f/dx^{2i}$.

Discussion of the underlying mathematics will be found in Lanczos[1] and Stroud and Secrest.[2] All the tables in this appendix have been taken from either Ref. 2 or Ref. 3.

## C-2 BASIC NUMERICAL INTEGRATION FORMULAE

Repeated application of Eq. (C-1) allows us to use the basic weights and nodes given in Table C-1 for both two- and three-dimensional integration via, for rectangles,

$$I_2 = \int_{-1}^{1} \int_{-1}^{1} f(x_1, x_2)\, dx_1\, dx_2 \simeq \sum_i \sum_j A_i A_j f(x_i, x_j) \qquad \text{(C-2)}$$

and, for hexahedra,

$$I_3 = \int_{-1}^{1} \int_{-1}^{1} \int_{-1}^{1} f(x_1, x_2, x_3)\, dx_1\, dx_2\, dx_3 \simeq \sum_i \sum_j \sum_k A_i A_j A_k f(x_i, x_j, x_k) \qquad \text{(C-3)}$$

with the summation pattern following precisely that explained in Appendix A.

Tables C-2 and C-3 are taken from Ref. 3 (and also Ref. 4) and are a development of Radau's[5] application of gaussian quadrature to triangular and tetrahedral cells. By combining a triangular pattern with linear ones it is clearly possible to derive integration schemes for triangular-prismatic elements much as related parametric representations were combined in Chapter 7.

Finally, Table C-4[2] provides weights and nodal coordinates for a form of (C-1) particularly useful in planar BEM problems: viz.,

$$I_4 = \int_0^1 \ln\left(\frac{1}{x}\right) f(x)\, dx \simeq \sum A_i\, f(x_i) \tag{C-4}$$

Since the fundamental solutions in two-dimensional problems all contain logarithmic terms, (C-4) is useful for integrating products involving them when the load and field points fall within the same element. Note that the range of integration is (0, 1) and therefore the boundary element has to be divided into two subelements in order to use (C-4) as explained in Chapter 15.

## C-3 TABLES OF WEIGHTS AND NODAL COORDINATES

**Table C-1** $\displaystyle\int_{-1}^{1} f(x)\, dx \simeq A_i\, f(x_i)$

| $x_i$ | $A_i$ |
|---|---|
| $i = 2$ | |
| 0.5773502691 8962576450 9148780502 | (1) 0.1000000000 0000000000 0000000000 |
| $i = 3$ | |
| 0.7745966692 4148337703 5853079956 | 0.5555555555 5555555555 5555555556 |
| 0.0000000000 0000000000 0000000000 | 0.8888888888 8888888888 8888888889 |
| $i = 4$ | |
| 0.8611363115 9405257522 3946488893 | 0.3478548451 3745385737 3063949222 |
| 0.3399810435 8485626480 2665759103 | 0.6521451548 6254614262 6936050778 |
| $i = 5$ | |
| 0.9061798459 3866399279 7626878299 | 0.2369268850 5618908751 4264040720 |
| 0.5384693101 0568309103 6314420700 | 0.4786286704 9936646804 1291514836 |
| 0.0000000000 0000000000 0000000000 | 0.5688888888 8888888888 8888888889 |
| $i = 6$ | |
| 0.9324695142 0315202781 2301554494 | 0.1713244923 7917034504 0296142173 |
| 0.6612093864 6626451366 1399595020 | 0.3607615730 4813860756 9833513838 |
| 0.2386191860 8319690863 0501721681 | 0.4679139345 7269104738 9870343990 |
| $i = 7$ | |
| 0.9491079123 4275852452 6189684048 | 0.1294849661 6886969327 0611432679 |
| 0.7415311855 9939443986 3864773281 | 0.2797053914 8927666790 1467771424 |
| 0.4058451513 7739716690 6606412077 | 0.3818300505 0511894495 0369775489 |
| 0.0000000000 0000000000 0000000000 | 0.4179591836 7346938775 5102040816 |

$i = 8$

0.9602898564 9753623168 3560868569

0.7966664774 1362673959 1553936476

0.5255324099 1632898581 7739049189

0.1834346424 9564980493 9476142360

0.1012285362 9037625915 2531354310

0.2223810344 5337447054 4355994426

0.3137066458 7788728733 7962201987

0.3626837833 7836198296 5150449277

$i = 9$

0.9681602395 0762608983 5576202904

0.8360311073 2663579429 9429788070

0.6133714327 0059039730 8702039341

0.3242534234 0380892903 8538014643

0.0000000000 0000000000 000000000

$(-1)$ 0.8127438836 1574411971 8921581105

0.1806481606 9485740405 8472031243

0.2606106964 0293546231 8742869419

0.3123470770 4000284006 8630406584

0.3302393550 0125976316 4525069287

$i = 10$

0.9739065285 1717172007 7964012084

0.8650633666 8898451073 2096688423

0.6794095682 9902440623 4327365115

0.4333953941 2924719079 9265943166

0.1488743389 8163121088 4826001130

$(-1)$ 0.6667134430 8688137593 5688098933

0.1494513491 5058059314 5776339658

0.2190863625 1598204399 5534934228

0.2692667193 0999635509 1226921569

0.2955242247 1475287017 3892994651

$i = 11$

0.9782286581 4605699280 3938001123

0.8870625997 6809529907 5157769304

0.7301520055 7404932409 3416252031

0.5190961292 0681181592 5725669459

0.2695431559 5234497233 1531985401

0.0000000000 0000000000 0000000000

$(-1)$ 0.5566856711 6173666482 7537204425

0.1255803694 6490462463 4694299224

0.1862902109 2773425142 6097641432

0.2331937645 9199047991 8523704843

0.2628045445 1024666218 0688869891

0.2729250867 7790063071 4483528336

## Table C-2 Numerical integration formulae for triangles

| Order | Figure | Error | Points | Triangular coordinates | Weights $2W_k$ |
|-------|--------|-------|--------|------------------------|----------------|
| Linear | | $R = 0(h^2)$ | $a$ | $\frac{1}{3},\frac{1}{3},\frac{1}{3}$ | $1$ |
| Quadratic | | $R = 0(h^3)$ | $a$ <br> $b$ <br> $c$ | $\frac{1}{2},\frac{1}{2},0$ <br> $0,\frac{1}{2},\frac{1}{2}$ <br> $\frac{1}{2},0,\frac{1}{2}$ | $\frac{1}{3}$ <br> $\frac{1}{3}$ <br> $\frac{1}{3}$ |
| Cubic | | $R = 0(h^4)$ | $a$ <br> $b$ <br> $c$ <br> $d$ | $\frac{1}{3},\frac{1}{3},\frac{1}{3}$ <br> $\frac{11}{15},\frac{2}{15},\frac{2}{15}$ <br> $\frac{2}{15},\frac{11}{15},\frac{2}{15}$ <br> $\frac{2}{15},\frac{2}{15},\frac{11}{15}$ | $-\frac{27}{48}$ <br> $\frac{25}{48}$ |

This formula not recommended due to negative weight and round-off error

| Order | Figure | Error | Points | Triangular coordinates | Weights $2W_k$ |
|-------|--------|-------|--------|------------------------|----------------|
| Cubic | | $R = 0(h^4)$ | $a$ <br> $b$ <br> $c$ <br> $d$ <br> $e$ <br> $f$ <br> $g$ | $\frac{1}{3},\frac{1}{3},\frac{1}{3}$ <br> $\frac{1}{2},\frac{1}{2},0$ <br> $0,\frac{1}{2},\frac{1}{2}$ <br> $\frac{1}{2},0,\frac{1}{2}$ <br> $1,0,0$ <br> $0,1,0$ <br> $0,0,1$ | $\frac{27}{60}$ <br> $\frac{8}{60}$ <br> $\frac{3}{60}$ |
| Quintic | | $R = 0(h^6)$ | $a$ <br> $b$ <br> $c$ <br> $d$ <br> $e$ <br> $f$ <br> $g$ | $\frac{1}{3},\frac{1}{3},\frac{1}{3}$ <br> $\alpha_1,\beta_1,\beta_1$ <br> $\beta_1,\alpha_1,\beta_1$ <br> $\beta_1,\beta_1,\alpha_1$ <br> $\alpha_2,\beta_2,\beta_2$ <br> $\beta_2,\alpha_2,\beta_2$ <br> $\beta_2,\beta_2,\alpha_2$ | $0.225$ <br> $0.13239415$ <br> $0.12593918$ |

with
$\alpha_1 = 0.05971587$
$\beta_1 = 0.47014206$
$\alpha_2 = 0.79742699$
$\beta_2 = 0.10128651$

## Table C-3 Numerical integration formulae for tetrahedra

| No. | Order | Figure | Error | Points | Tetrahedral coordinates | Weights |
|---|---|---|---|---|---|---|
| 1 | Linear | | $R = 0(h^2)$ | $a$ | $\frac{1}{4}, \frac{1}{4}, \frac{1}{4}, \frac{1}{4}$ | $1$ |
| 2 | Quadratic | | $R = 0(h^3)$ | $a$ $b$ $c$ $d$ | $\alpha, \beta, \beta, \beta$ <br> $\beta, \alpha, \beta, \beta$ <br> $\beta, \beta, \alpha, \beta$ <br> $\beta, \beta, \beta, \alpha$ <br> $\alpha = 0.58541020$ <br> $\beta = 0.13819660$ | $\frac{1}{4}$ <br> $\frac{1}{4}$ <br> $\frac{1}{4}$ <br> $\frac{1}{4}$ |
| 3 | Cubic | | $R = 0(h^4)$ | $a$ $b$ $c$ $d$ $e$ | $\frac{1}{4}, \frac{1}{4}, \frac{1}{4}, \frac{1}{4}$ <br> $\frac{1}{3}, \frac{1}{6}, \frac{1}{6}, \frac{1}{6}$ <br> $\frac{1}{6}, \frac{1}{3}, \frac{1}{6}, \frac{1}{6}$ <br> $\frac{1}{6}, \frac{1}{6}, \frac{1}{3}, \frac{1}{6}$ <br> $\frac{1}{6}, \frac{1}{6}, \frac{1}{6}, \frac{1}{3}$ | $\frac{4}{5}$ <br> $\frac{9}{20}$ <br> $\frac{9}{20}$ <br> $\frac{9}{20}$ <br> $\frac{9}{20}$ |

**Table C-4** $\displaystyle\int_0^1 \ln(1/x)\, f(x)\, dx \simeq A_i\, f(x_i)$

| $x_i$ | $A_i$ |
|---|---|
| $i = 2$ | |
| 0.1120088061 6697618295 7205488948 | 0.7185393190 3038444066 5510200891 |
| 0.6022769081 1873810275 7080225338 | 0.2814606809 6961555933 4489799109 |
| $i = 3$ | |
| (−1) 0.6389079308 7325404996 1166031363 | 0.5134045522 3236332512 9300497567 |
| 0.3689970637 1561876554 6197645857 | 0.3919800412 0148755480 6287180966 |
| 0.7668803039 3894145542 3682659817 | (−1) 0.9461540656 6149120064 4123214672 |
| $i = 4$ | |
| (−1) 0.4144848019 9383220803 3213101564 | 0.3834640681 4513512485 0046522343 |
| 0.2452749143 2060225193 9675759523 | 0.3868753177 7476262733 6008234554 |
| 0.5561654535 6027583718 0184354376 | 0.1904351269 5014241536 1360014547 |
| 0.8489823945 3298517464 7849188085 | (−1) 0.3922548712 9959832452 5852285552 |
| $i = 5$ | |
| (−1) 0.2913447215 1972053303 7267621154 | 0.2978934717 8289445727 2257877888 |
| 0.1739772133 2089762870 1139710829 | 0.3497762265 1322418037 5071870307 |
| 0.4117025202 8490204317 4931924646 | 0.2344882900 4405241888 6906857943 |
| 0.6773141745 8282038070 1802667998 | (−1) 0.9893045951 6633146976 1807114404 |
| 0.8947713610 3100828363 8886204455 | (−1) 0.1891155214 3195796489 5826824218 |
| $i = 6$ | |
| (−1) 0.2163400584 4116948995 6958558537 | 0.2387636625 7854756972 2268303330 |
| 0.1295833911 5495079613 1158505009 | 0.3082865732 7394679296 9383109211 |
| 0.3140204499 1476550879 8248188420 | 0.2453174265 6321038598 4932540188 |
| 0.5386572173 5180214454 8941893993 | 0.1420087565 6647668542 1345576030 |
| 0.7569153373 7740285216 4544156139 | (−1) 0.5545462232 4886290015 1353549662 |
| 0.9226688513 7212023733 3873231507 | (−1) 0.1016895869 2932275886 9351162755 |
| $i = 7$ | |
| (−1) 0.1671935540 8258515941 6673609320 | 0.1961693894 2524820752 5427377585 |
| 0.1001856779 1567512158 6885031757 | 0.2703026442 4727298214 5271719533 |
| 0.2462942462 0793059904 6668547239 | 0.2396818730 0769094830 8072785252 |
| 0.4334634932 5703310583 2882482601 | 0.1657757748 1043290656 0869687736 |
| 0.6323509880 4776608846 1805812245 | (−1) 0.8894322713 7657964435 7238403458 |
| 0.8111186267 4010557652 6226796782 | (−1) 0.3319430435 6571067025 4494111034 |
| 0.9408481667 4334772176 0134113379 | (−2) 0.5932787015 1259239991 8517844468 |
| $i = 8$ | |
| (−1) 0.1332024416 0892465012 2526725243 | 0.1644166047 2800288683 1472568326 |
| (−1) 0.7975042901 3894938409 8277291424 | 0.2375256100 2330602050 1348561960 |
| 0.1978710293 2618805379 4476159516 | 0.2268419844 3191912636 8780402936 |
| 0.3541539943 5190941967 1463603538 | 0.1757540790 0607024498 8056212006 |
| 0.5294585752 3491727770 6149699996 | 0.1129240302 4675905185 5000442086 |
| 0.7018145299 3909996383 7152670310 | (−1) 0.5787221071 7782072398 5279672940 |
| 0.8493793204 4110667604 8309202301 | (−1) 0.2097907374 2132978043 4615241150 |
| 0.9533264500 5635978876 7379678514 | (−2) 0.3686407104 0276190133 5232127647 |

## C-4 REFERENCES

1. Lanczos, C. (1961) *Linear Differential Operators*, Van Nostrand, New York and London.
2. Stroud, A. H., and Secrest, D. (1966) *Gaussian Quadrature Formulas*, Prentice-Hall, Englewood Cliffs, N. J.
3. Hammer, P. C., Marlowe, O. P., and Stroud, A. H. (1956). 'Numerical integration over simplexes and cones', *Math. Tables Aids Comp.*, **10**, 130–137.
4. Zienkiewicz, O. C. (1971) *The Finite Element Method in Engineering Science*, 2nd ed., McGraw-Hill, London.
5. Radau (1880). *J. de Math.*, **3**, 283.

# AUTHOR INDEX

# SUBJECT INDEX